(E9 602

Role of Cement Science In Sustainable Development

Proceedings of the International Symposium dedicated to
Professor Fred Glasser, University of Aberdeen, Scotland
held on 3-4 September 2003 at the University of Dundee, Scotland, UK

Edited by

Ravindra K. Dhir
Director, Concrete Technology Unit
University of Dundee

Moray D. Newlands
CPD / Consultancy Manager, Concrete Technology Unit
University of Dundee

and

Laszlo J. Csetenyi
Research / Teaching Fellow, Concrete Technology Unit
University of Dundee

Thomas Telford

Published by Thomas Telford Publishing, Thomas Telford Ltd, 1 Heron Quay, London E14 4JD.
www.thomastelford.com

Distributors for Thomas Telford books are
USA: ASCE Press, 1801 Alexander Bell Drive, Reston, VA 20191-4400, USA
Japan: Maruzen Co. Ltd, Book Department, 3–10 Nihonbashi 2-chome, Chuo-ku, Tokyo 103
Australia: DA Books and Journals, 648 Whitehorse Road, Mitcham 3132, Victoria

First published 2003

The full list of titles from the 2003 International Symposium 'Celebrating concrete' and available from Thomas Telford is as follows

- Role of cement science in sustainable development. ISBN: 0 7277 3246 3
- Role of concrete in sustainable development. ISBN: 0 7277 3247 1
- Role of concrete bridges in sustainable development. ISBN: 0 7277 3248 X

A catalogue record for this book is available from the British Library

ISBN: 0 7277 3246 3

Printed and bound in Great Britain by MPG Books, Bodmin, Cornwall

PREFACE

Portland cement concrete is arguably the most important construction material used today. However, to maintain its unrivalled place within the construction industry, the driving force of concrete, namely cement, must embrace all environmental issues, whilst still achieving desired long-term performance. The cement industry is responding rapidly to the need for reduced CO_2 emissions by increasing the production of blended Portland cements, using supplementary cementitious materials that are principally derived from industrial by-products, such as blast-furnace slags and coal-combustion fly ashes. This is vital to ensure that the use of natural resources is minimised and value added sustainable cementitious products are derived from existing waste materials.

The Concrete Technology Unit at the University of Dundee has organised this unique series of three International Symposia to promote the contributions that concrete has made to our lives. Under the umbrella title *of Celebrating Concrete: People and Practice,* the Symposium entitled *Role of Cement Science in Sustainable Development* is dedicated to Professor Fred Glasser, University of Aberdeen, UK, who has made an outstanding contribution in the area of cement science.

The event was organised in collaboration with three of the world's most recognised institutions: the Institution of Concrete Technology, the American Concrete Institute and the Czech Society of Structural and Civil Engineers. The proceedings contain fifty-four papers presented in six themes: (i) Modern Cement Manufacture, (ii) Hydration of Portland Cement, (iii) Supplementary Cementing Materials, Activators and Inhibitors, (iv) Performance of Portland Cement Concrete in the Environment, (v) Special and Non-Portland Cements and (vi) Exploiting Science, comprehensively summarising the state-of-the-art in cement science and sustainability

The Opening Paper was given by Professor Fred Glasser, University of Aberdeen, UK and six Keynote Papers were given by Dr Martin Deussner, Heidelberg Cement, Germany, Professor Guiseppi Frigione, University della Calabria, Italy, Professor Stan Wild, University of Glamorgan, UK, Professor Jacques Marchand, Laval University, Canada, Dr Ellis Gartner, Lafarge, France and Dr Willem van Loo, CEMBUREAU, Belgium. The Closing Paper was given by Professor Della M Roy, Pennsylvania State University, USA

The support of the Collaborating Institutions, Sponsoring Organisations and International Professional Associations was a major contribution to the success of the Symposia. The work of the event was an immense undertaking and all of those involved are gratefully acknowledged, in particular, the members of the Organising Committee for managing the event from start to finish; members of the Scientific and Technical Committee for advising on the selection and reviewing of papers; the Authors and the Chairmen of Technical Sessions for their invaluable contributions to the proceedings and the exhibiting organisations who took part in the Trade Fair.

All of the proceedings have been prepared directly from the camera-ready manuscripts submitted by the authors and editing has been restricted to minor changes where it was considered absolutely necessary.

Dundee
September 2003

Ravindra K Dhir
Moray D Newlands
Laszlo J Csetenyi

ORGANISING COMMITTEE

Concrete Technology Unit

Professor R K Dhir OBE (Chairman)

Dr M D Newlands (Secretary)

Professor P C Hewlett
British Board of Agrément

Professor T A Harrison
Quarry Products Association

Dr M R Jones

Dr M J McCarthy

Dr T D Dyer

Dr K A Paine

Dr L J Csetenyi

Dr J E Halliday

Dr L Zheng

Dr S Caliskan

Ms K Giannakou

Mr M C Tang

Mrs E Csetenyi

Ms P I Hynes (Symposia Assistant)

Mr S R Scott (Unit Assistant)

SCIENTIFIC AND TECHNICAL COMMITTEE

SCIENTIFIC AND TECHNICAL COMMITTEE

Professor Roger J Kettle,
Aston University, UK

Professor Johann Kollegger,
Vienna University of Technology, Austria

Mr Paul Livesey,
Castle Cement Limited, UK

Dr Donald E Macphee,
University of Aberdeen, UK

Mr Douglas G McBeth,
WSP Group, UK

Mr Rod McClelland,
Alfred McAlpine plc, UK

Professor Byung Hwan Oh,
Seoul National University, South Korea

Professor Shunsuke Otani,
Chiba University, Japan

Dr Donald Pearson-Kirk,
Parsons Brinckerhoff Ltd, UK

Professor Jean Pera,
INSA Lyon, France

Dr William F Price,
Lafarge Cement, UK

Professor Musa Resheidat,
Jordan University of Science & Technology, Jordan

Dr Vlastimil Sruma,
Czech Concrete Society, Czech Republic

Dr Tappur N Subba Rao,
Construma Group, India

Dr Geoff Taplin,
Maunsell Australia Pty Ltd, Australia

Professor Taketo Uomoto,
University of Tokyo, Japan

Dr Olafur Wallevik,
Icelandic Building Research Institute, Iceland

Dr Roger P West,
Trinity College Dublin, Ireland

COLLABORATING INSTITUTIONS

Institute of Concrete Technology, UK
American Concrete Institute
Czech Society of Structural and Civil Engineers

SPONSORING ORGANISATIONS WITH EXHIBITION

Angus and Dundee Convention Bureau
British Board of Agrément
British Cement Association
Castle Cement Limited
Degussa Construction Chemicals, Italy
Dundee City Council

SUPPORTING INSTITUTIONS

Brazilian Concrete Society
Concrete Institution of Australia
Czech Concrete Society
Indian Concrete Institute
Irish Concrete Society
Japan Concrete Institute
Netherlands Concrete Society
New Zealand Concrete Society

CONTENTS

THEME 3 SUPPLEMENTARY CEMENTING MATERIALS, ACTIVATORS AND INHIBITORS

OPENING PAPER

FUTURE DIRECTIONS IN THE CEMENT INDUSTRY

F P Glasser

University of Aberdeen

United Kingdom

ABSTRACT. Concrete is the most important structural material of the built environment. Nevertheless it faces challenges, including need to reduce its high environmental impact and specific CO_2 emission. While the industry continues to take steps to reduce impacts and CO_2 emission, only modest improvement is forecast over the next one or two decades. Perhaps the greatest environmental benefit could be obtained by improving the durability of concrete constructions. Durability is an abstract term but its underlying attributes are well known. Design and workmanship contribute; however, the weak link is testing and test methods used to assess durability. Existing standards are heavily based on empiricism and often fail to reflect an understanding of the underlying physical and chemical mechanisms. Selected examples of sulfate attack are given to illustrate deficiencies in existing standards. It is suggested that the way forward will be modelling, linked to field studies and revised test methods appropriate to the service environment, thus enabling concrete performance to be optimised.

Keywords: Durability, Standards, Test methods, Sulfate attack.

F P Glasser is a Professor in the School of Engineering and Physical Science at the University of Aberdeen. He is the author or co-author of more than 400 peer-reviewed publications in the fields of materials science, solid-state science and cements. He has received numerous distinctions including fellowship of the Royal Society of Edinburgh. Married with three children and nine grandchildren (and still counting) he enjoys life and numerous hobbies.

INTRODUCTION

Portland cements comprise an important - arguably the most important - constructional material used by mankind; world production is on the order of 10^9 tonnes/year. Considering that cement is usually mixed with aggregate materials to form concrete, world production of concrete must exceed 10^9 tonnes/year.

Although Portland cement made today is superficially similar to that made in 1903, i.e., a century ago, important changes have been made to its formulation and processing. In the course of the past century:

- Lime saturation factors have increased, giving a clinker which contains a higher ratio of tricalcium silicate to dicalcium silicate.
- Clinker is more finely ground, giving on average much higher specific surface areas for modern products.
- Owing to changes in mineralogy and fineness, a set retarder - normally gypsum - has to be added to cement.
- Standards which define the composition of "Portland cement" have been changed so that Portland cement clinker can be interground with other potentially reactive materials, e.g., fly ash, slag or limestone. The maximum limits of substitution vary depending on national standards but often allow up to 10% replacement.
- Designers of concrete systems have become much more willing to use a wide range of chemical additives to optimise specific aspects of concrete performance: water-reducing, air entraining, anti-bleed and other admixtures are widely used, either singly or in combination.

Manufacturing methods for clinker have also changed with the result that greater capital investment is required per tonne of clinker. Wet process kilns have been or are being replaced by dry process kilns and suspension preheaters installed to minimise energy requirements. Excess kiln ventilation has reduced and sensors installed to monitor performance, often under computer control, and improved burner designs developed which limit NO_x emissions. This list of process changes is only partial but suffices to demonstrate the industrial commitment to product improvement as well as to reduction in pollution and effluents.

Nevertheless the industry comes under increasing pressure to improve its environmental performance. Worldwide, cement production and utilisation, including indirect energy, for example transportation, produces 6 - 10% of global, man-made CO_2 emissions. This relatively high range of values arises from the need to form calcium silicates cheaply: the cheapest and lowest cost raw material for cement manufacture is calcium carbonate, $CaCO_3$. Elimination of CO_2 from $CaCO_3$ during pyroprocessing, coupled with fossil fuel combustion and other energy inputs (quarrying, grinding, transport as well as energy associated with aggregate extraction, processing and delivery) cumulatively result in substantial specific CO_2 emissions. It may be argued that these emissions are modest relative to competitive structural materials, e.g., steel, but this misses the point: if climate change models are to be believed, CO_2 emissions must be drastically reduced. Moreover, because of the success of the cement industry, its cumulative emissions are large in a global context and become a priority target for action.

The production of cement can be viewed in thermodynamic terms: Portland cement is an energy-rich material. As with any thermodynamic process, energy balances can be constructed. These show that the best modern plants are relatively efficient: only modest scope exists for betterment. However, much plant in operation today is not state-of-the-art and, as old plant wears, economic factors will conspire with legislation and taxation policy to ensure replacement with improved plant. As a result, a slow overall incremental drift to greater efficiency, resulting in less emission, will be achieved. However the gains thus achieved will be modest, typically on the order of a few percent, projected over the next decade.

The energy and emission impacts of cement production are not the only factors which need to be taken into account. Socio-economic factors are also very relevant. Much construction is done on the basis of a competitive tender process: clients will seek maximum value for money and the actual construction will be let according to negotiated or sealed bids. This system, operated without checks and balances, has resulted in some remarkably non-durable construction. Much has been written about durability (or non-durability) which I do not intend to review in detail. But I note that improvements in durability probably represent the greatest remaining area of benefit: by increasing durability and hence performance lifetime, energy inputs associated with construction are diluted. 'Durability' is an abstract word: it can only be defined operationally in terms of specific attributes. Nevertheless, *improvements to durability comprise the most significant contribution which could be made towards reducing the impact of cement production and use.* A conflict does arise in terms of time scales: the favourable impact of enhanced lifetimes for concrete constructions is attained slowly, over decades, while legislation generally has amongst its objectives achieving betterment over much shorter time scales.

The writer views the industry holistically: cement production is tied to construction and this is, in turn, tied to investment. Alternatives exist to concrete in many applications, so infrastructure investment does not - as a complacent industry often assumes - automatically result in more concrete being made. Moreover, the entire construction sector is being overtaken by a plethora of policies intended to lower energy inputs, reduce consumption, reduce maintenance and running costs, as well as fostering reuse and recycling. The issues, as well as the means whereby change can be implemented, are too complex to deal with here: I concentrate on the enabling science and technology. Although I cannot foretell the future in detail, certain things are apparent: we cannot simply continue as in the past; change is required. Moreover, simple continuation of past responses may be approaching natural limits. New ways and new approaches are needed and I discuss selected examples.

DURABILITY

Present Concepts Relating To Durability

"Durability" and its associated aspects are an important field of current research as evidenced by the number of conferences and publications taking this as their theme. It has been argued that, since socio-economic needs change, we do not need durable construction. Except in a few instances, this proposition is demonstrably untrue. The social and economic costs and disruptions associated with infrastructure renewal (tunnels, bridges, roads, dams, sewers, etc.) are enormous. At a more local scale, many better-built early-to-mid 20th century constructions continue in service, although often with change of use. This tendency, to renew and refurbish rather than replace, has spread even to 'young' countries such as the U.S., Australia, etc. Thus durability and longevity of service are important issues for the industry. During past decades, empirical research on durability has greatly advanced the formulation and production of durable concretes. Thus we have specifications for cement content and type, preferred water:cement ratios, cure conditions and duration, etc. While these and other procedures appear to have been introduced on a piecemeal basis they all conform to a common pattern. In stage 1, test data on the impact of changing a parameter, e.g., w/c ratio, is researched. In stage 2, empirical testing of product durability is undertaken, often by permanent, not-for-profit organisations such as trade associations or government departments. In stage 3, the parametric charge, having indicated promise, is used to guide changes to specification. In stage 4, in conjunction with permissive changes to codes of practice, the changes are applied to real structures. The pace of change is such that the cycle of concept, trials and implementation may require 10 - 20 years (or longer) to achieve and implement all four stages. Moreover little transfer of knowledge occurs across national boundaries: research done in one country tends to be duplicated elsewhere before it is deemed acceptable.

In the 21st century, processes for technological advancement must change. Firstly, a key element - the not-for-profit and government laboratories capable of maintaining and monitoring long-term trials - have either disappeared or become focussed on shorter-term objectives. Secondly the pace and tempo of change has speeded; environmental pressures usually demand rapid responses. Thirdly, empiricism has limitations; as pressure grew to improve concrete properties and achieve more economic construction the initial stages of improvement were fairly obvious. Historic examples include specification of lower water:cement ratios, better compaction and minimum cement composition. But subsequent improvements become progressively more difficult to verify: how can these be proven to be effective? Finally, users and stakeholders increasingly want quantitative service life predictions and, in some instances, guarantees of service lifetimes. But if empiricism is insufficient to achieve change and assure stakeholders, how can progress be made in achieving more durable construction? How can specification, including codes of practise and test methods, adequately reflect these pressures?

To progress we need to return to basics. Some aspects of these basics e.g., design and detailing, are beyond the scope of this presentation; I confine myself to the cement matrix. Figure 1 shows conceptually a holistic approach to the problem. On the one hand we have the concrete itself: its formulation, degree of maturation and cure history. Its chemistry, physics, mineralogy and structure may to varying extents require additional characterisation: "structure" in this content includes a range of length scales, from coarse (joints, cracks, voids, larger pores) to small (microcracks interphase contacts, microtexture, etc. Furthermore an increasing body of evidence suggests that in some instances we may require structural information at the nanoscale level to assess performance: many parameters which ultimately control engineering properties range across length scales and extend into the nanoscale range.

However, the proportionality factors are not readily isolated for separate study. Empirical test methods, typical of those in use today, do not adequately address the problem; indeed in many instances they show little sign of having considered all relevant factors.

In the quest to determine these proportionalities, an additional concept has to be introduced: the definition of the "system" being measured. Both factors very much affect the nature of testing and specification of standard test methods; a specific example, of sulfate attack, will be developed subsequently.

Proportionality factors and interaction coefficients are most readily explored in a closed system (top row, Figure 1). By a 'closed' system is meant one which is closed with respect to transport of matter in or out of the concrete. Relatively few examples of totally closed systems occur in practice; Table 1 gives selected examples. Nevertheless, closed systems afford a valuable starting point for analysis: degradation processes in closed systems enables us to conserve mass; mineralogical reconstitution may occur, but with the restriction that bulk chemistry is unaffected.

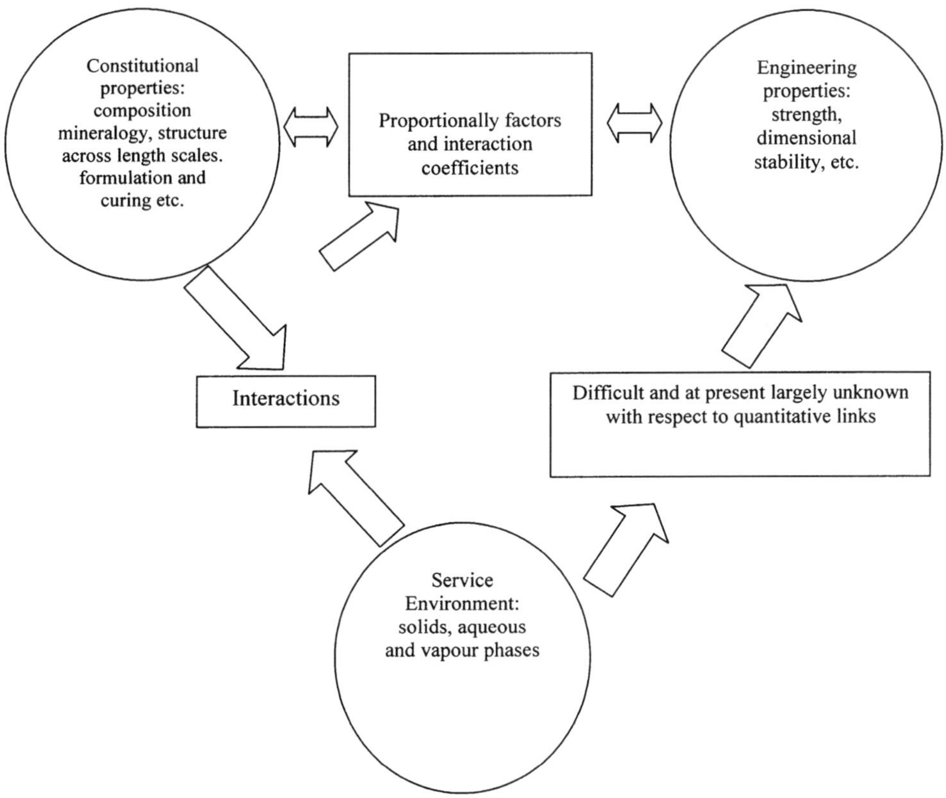

Figure 1 Modelling the performance of concrete

However, many commonly observed types of degradation occur in open or partially open systems. Familiar examples are sulfate attack, sea water attack and carbonation. In the latter, CO_2 from the atmosphere, or dissolved in groundwater, or both, convert the cement phases ultimately to calcium carbonate, hydrated amorphous silica and alumina hydrate, either gel or gibbsite. Table 1 distinguishes the relative degree of openness of selected systems. This openness affects the choice of a test method. From the Table it can be seen that a minimum of three types of test may be needed to characterise carbonation: the fourth relates to the adjunct role of carbonation in connection with other processes and is discussed subsequently. In the first case, of atmospheric carbonation, chemical mass, other than of CO_2 and water, is conserved. Thus the system, while technically open, is very restricted in its openness. Note however that transport of both water and CO_2 needs to be considered. As an example, consider the carbonation of $Ca(OH)_2$:

$$Ca(OH)_2 + CO_2 \rightarrow CaO_3 + H_2O$$

Water is liberated and, depending on humidity and the tortuosity of diffusion paths, is partially retained and partially lost.

Table 1 Carbonation of cement concretes

MECHANISM OF CARBONISATION	DEGREE OF OPENNESS AND COMMENT
Uptake of atmospheric CO_2 in moist conditions	Only mass gain of CO_2 occurs, although oxygen uptake may also be relevant to state or condition of embedded steel. Water is lost.
Atmospheric carbonation with intermittent wetting	As above but leaching, particularly of alkali, may affect chemical balances in leachate as a function of time and as alkali is exhausted.
Carbonation in groundwater/flowing waters containing CO_2.	CO_2 is only occasionally the major contributor to degradation. In the absence of other components, the process is readily modelled.
Carbonation occurs but other modes of attack are dominant.	Few restrictions on the system: open to transport of many species. Not yet a function which can be modelled.

At present, the progress of carbonation in concrete is monitored by using a chemical test. Phenolphthalein reagent is applied to a fresh surface and the depth of carbonation inferred from colour change; the reagent gives a pink-purple colour at high pH but remains colourless at lower pH. Owing to the mechanism of colour change, zones which retain even a little $Ca(OH)_2$ or other source of alkalinity remain pink-purple, thus the method underestimates the extent to which carbonation has penetrated. This is shown by comparing the depth of carbonation determined by indicator with that determined by thermogravimetric analysis for $Ca(OH)_2$ and $CaCO_3$ [1]. The discrepancies are perhaps not too serious for individual determinations, but when data are used to calibrate models of the time dependence of carbonation, and results are extrapolated into the future, serious error can result. To follow this process and eventually obtain data suitable for modelling, we are in the process of constructing equipment to image by NMR both carbon (as carbonate) and water in samples undergoing dynamic carbonation: new experimental techniques are required to assist a fundamental understanding of the rate-limiting kinetics.

The second carbonation process (Table 1) might be experienced by concrete which, in addition to carbonation, is directly impinged by liquid water; examples might be exterior cladding panels, roofing tiles, etc. In this latter case, of roof tiles, we have explored the connection between cement alkali content and the appearance of $CaCO_3$ efflorescence [2]. The third process depends greatly on water chemistry; some waters are particularly aggressive. They are reviewed and modelled in [3] and carbonation in cracks is modelled in [4].

Carbonation by groundwater is often more complicated: not only can cement substances dissolve, but depending on water composition, the principal reaction product, $CaCO_3$, may also be significantly soluble. Sea water attack involves not only dissolution reactions but also uptakes of initially soluble sea water components - magnesium, sulfate, chloride and carbonate. Key reactions are described subsequently.

Because open systems, with potential for mass loss as well as gain, are intrinsically more complicated than closed systems, they are placed separately in Figure 1. However, restrictions placed on the degree of openness are commonly encountered; utilising these restrictions is an important adjunct to selection and design of appropriate test methods.

Table 2 considers further examples. Many structures experiencing AAR also gain water from an external source: dams are an example. In an isolated system the rate of AAR may reduce with time if the system is depleted in water, whereas if it remains open with respect to water transport, reaction may continue. Thus access to water, specimen geometry and permeability are important test parameters.

Table 2 Classification of concrete deterioration

CLOSED SYSTEM BEHAVIOUR	
Brief description	Nature of reaction
Alkali-aggregate (AAR) reaction	Hydroxide ions in cement and cement pore fluid attack siliceous aggregates with expansion as water is imbibed.
Delayed ettringite formation (DEF)	Subsequent to warm cure, ettringite reforms in cement matrix which expands.
OPEN SYSTEM BEHAVIOUR	
Sea water attack	Chloride, sulfate carbonate and magnesium diffuse inward. Some leaching occurs. Complex mineralogical zoning occurs.
Carbonation	Atmospheric and dissolved CO_2 carbonate cement phases. Water is lost

The examples given relate to standards and standard test methods relevant to durability. Our present standards and standard test methods have grown up over a number of decades: they represent a consensus view of what should be done and how it should be done. Much attention has been devoted to standardisation of tests, including collection and recording of data and its interpretation. As a result, the body of standard test methods appears superficially to be "scientific". However, a closer examination reveals that very often little scientific justification can be given for the test in terms of underlying mechanisms: this is particularly so when the test purports to accelerate the underlying mechanism(s) of degradation. As a result, our scientific knowledge of degradation, degradation mechanisms, kinetics and interactions, together with the potential to integrate and extrapolate these data by computer has expanded while at the same time, test methods and guidance have remained virtually static and largely unmindful of relevant advances made elsewhere. Increasingly, the inadequacy of existing test methods and lack of fundamental physiochemical input will handicap the development of more durable concretes by obscuring the fundamental principles which underlie optimisation.

Presently available standards and test methods are seen by the writer as having reached a dead end: at best the philosophies presently employed do not permit either quantitative evaluations or extension to conditions other than those specified in the test and at worst they prevent or handicap the more widespread development of a genuine predictive capability.

Thus, in summary, cement deterioration may occur isochemically although examples of purely isochemical change are relatively uncommon. Although open system behaviour is more complex, in practise relatively few systems are totally open: chemical change may be restricted to relatively few components. Closed systems are susceptible to a complete physiochemical description and changes can be predicted using well-established computational iterations. Good progress is being made in modelling systems which are open to transport, particularly where mass transport is confined to a few components.

The Way Forward

Over the past two decades the writer has analised problems associated with the stability and performance of cement and concrete in the context of nuclear waste isolation. Potential waste repository environments include almost all those encountered in civil engineering. Results of both experiments and calculations disclose that *the impact of the environment on cement is such that changes to the cement are driven by the approach to equilibrium between cement, aggregate (if any) and the geochemistry of the local environment.*

Recognition of this postulate is an important step in the way forward. It provides the rationale for modelling cement performance on a thermodynamic basis. Thermodynamics tells us what the end products of reaction will be. Of course thermodynamics in its purest form needs to be guided by experience; in isolation, it will not necessarily predict the *route* by which reaction occurs. Thus a thermodynamic approach to durability has to be coupled with empirical observation and controlled laboratory experiments in order adequately to assess and include within the scope of calculation the -

- Degree of openness of the system.
- Relative diffusibility of reacting components.
- Extent to which local equilibria, leading to formation of zoned structures, dominate reaction patterns.
- Nature and severity of local density changes which may lead to potentially expansive reactions. Within hardened concrete.

Three principal factors handicap our predictive capability (i) deficiencies in the database for cement substances, (ii) relatively poor knowledge of reaction kinetics and coupled kinetic – phase equilibria models and (iii) inability quantitatively to define relationships between chemical/mineralogical parameters and associated engineering properties e.g., strength and permeability.

Once these data are available at least in part, because some data deficiencies can be bridged, models can be developed which permit quantitative calculations. Perhaps the most progress in developing such models has occurred amongst metallurgists. Metallurgists have long relied on thermochemical and thermodynamic databases for alloy development. Recent reviews of progress describe the real advances which have been made and continue to be made [5]. The writer, in conjunction with access to metallurgical databases, was able to reconcile the phase composition of a clinker with its chemical composition and show how the cooling history of a clinker affects its phase composition [6]. The model shows how non-equilibrium cooling enhances alite formation and whý the Bogue calculation systematically underestimates clinker alite contents.

Both metallurgists and geochemists have much experience of the treatment of open or partially open systems. Well-established computer-based iterative routines are available; some commercial, others free. The algorithms iterate to minimise the free energy of the system although, by assignment, it is also possible to designate metastable phenomena for study. Mass transport with reaction is coupled to equilibrium calculation enabling composition gradients and the impact of gradients on local mineralogy to be calculated as well as development either of porosity or overfilling, the latter with potential for expansion. One-dimensional transport models are usually sufficient; two and three dimensional models have been developed but require progressively more computing power.

In summary, tests and test methods have to be reproducible from laboratory to laboratory and generally perform well with respect to this criterion to achieve reproducibility. Tests tend to be prescriptive: test conditions are specified in great detail supposedly to assist reproducibility. However failure to analyse in detail the underlying mechanisms often means that relatively unimportant factors are specified in great detail while at the same time, some important factors remain uncontrolled: examples will be given.

However the writer concludes that the focus on defining test conditions, while important, misses the goal of testing: to give a realistic measure of cement and concrete durability when exposed to potentially aggressive environments. The extent to which tests succeed in achieving this objective is debatable. The literature abounds in observations concerning the value of standard tests and test methods but it is not generally possible to reach conclusions regarding the fitness for purpose of tests other than to record that:

- Substantial uncertainty attaches to the value of existing standard tests and test methods.
- Conditions specified for the test are not usually defined with respect to a physiochemical foundation: they are arbitrary.
- Tests give little or no guidance on extrapolation between performance in specified conditions and "real" conditions, in which non-standard concentrations and temperatures may obtain.
- Test results, including parameters which must be measured, are often incomplete, or are not suitable for incorporation into predictive exercises, or some combination thereof.
- Although some tests purport to achieve acceleration, the acceleration factor cannot be unequivocally determined, nor is it possible to determine if the underlying mechanisms of deterioration are uniformly accelerated.

Standards are defined by consultation and it is therefore inappropriate for the writer to define new standards. However as a constructive contribution to the ongoing dialogue, I take sulfate attack as an example and show what factors must enter into the design of a test or tests, and how predictive computer-based models contribute to enhancing concrete performance.

Sulfate Attack

The problem of devising suitable tests and test methods for "sulfate attack" lies initially in defining what is meant by sulfate attack. While this might at first sight seem to be a trivial problem, it is not; the EN Standards Committee are at the time of writing unable to agree a definition. But in keeping with the return to basics, I analyse several examples. Table 3 serves as an illustrative guide to the problems.

Table 3 Classification of sulfate attack mechanisms: classification and mechanisms

CLOSED SYSTEM: MASS TRANSPORT ACROSS THE CEMENT INTERFACE WITH ITS ENVIRONMENT IS EITHER ZERO OR NEGLIGIBLE	OPEN SYSTEM: MASS TRANSPORT OCCURS ACROSS THE BOUNDARY BETWEEN CEMENT AND ITS SERVICE ENVIRONMENT
Thermally cycled cement matrices: ettringite partially dissolves at elevated temperature but reprecipitates at or near ambient. The mechanism of reformation may promote physical expansion.	Attack by sulfate ground waters, sea water, etc. Sulfate contents increase. Depending on environment, leaching may also occur.
	Acid attack: $H_2S + O_2$ forming H_2SO_4. Mediated by bacteria with dissolution and gypsum precipitation occurring.

The first example, of so-called DEF, is typically encountered when fresh, incompletely hydrated cement-based product is cured at elevated temperature, normally <100°C, and subsequently experiences service in ambient conditions, say 0 - 30°C. After some time, the concrete is observed to expand. It is apparent that the initial stages of formation and cure, together with the associated thermal excursion, comprise essentially a closed system; matter is neither lost nor gained. As such, the relevant parameters are water:cement ratio, cement sulfate content, maximum temperature excursion and its relationship to degree of hydration. Note however that the thermal excursion characteristically occurs early in the curing history, perhaps within the first few hours, so that the reactive fraction of the system changes in the course of the thermal excursion. This is of course one motivation for elevated temperature cure; to increase the rates of hardening and strength gain by increasing the rate at which cement reacts. It is noteworthy that large masses of concrete may achieve similar thermal cycling as a consequence of heat release, albeit not necessarily on a planned basis. But the added complications resulting from extensive reaction in the course of the thermal excursion means that we also need to monitor the % hydration achieved at various points in the thermal cycle; unreacted clinker cannot participate in mineralogical transformations amongst hydrates, or affect solid-aqueous phase balances except insofar as it continues to hydrate. Moreover, the nature of the hydration products are temperature dependent.

On account of the closed system behaviour characteristic of thermally cycled concrete, this type of degradation is sometimes termed "internal sulfate attack" because total sulfate remains unchanged throughout. I accept this as a valid concept but a better description is 'closed system sulfate reaction'. Although we know the origin of the problem [7] much data are still missing concerning the equilibrium and kinetic aspects of DEF. For example, although the warm-cured commercial product contains much pore space in which ettringite might crystallise, upon return to ambient temperature, ettringite often nucleates within the solid parts of the matrix rather than in pores; we require a better understanding of the kinetics and mechanisms of ettringite nucleation to determine its impact on dimensional stability.

At present the DEF problem has been approached by observing large numbers of cements, subject to cycling to various maximum temperatures, and extending the observations over months or years with measurement of the expansion, if any. In this way "safe" conditions are established. But this empirical approach must of necessity be very conservative. It inhibits precasters from making the most effective use of warm cure and imposes limits on designers of massive concrete structures with concomitant economic penalty. Thus the purely empirical foundation for the avoidance of DEF prevents achieving the most economical construction. While the writer accepts that it is better safe than sorry, it would be even better to have scientific guidance to avoid DEF.

One conclusion about DEF is possible: while ettringite is invariably present in expanding concretes, ettringite is not necessarily the cause, or the sole cause, of expansion. For example, calcium sulfoaluminate cement-gypsum mixes continue to form copious ettringite in the post-hardening period yet are not normally expansive [8]. In heat-cured Portland cement, the cure normally induces a relatively coarse pore structure relative to the same cement cured isothermally at normal (10 – 40°C) temperatures. Despite the abundance of coarse porosity, examination of concretes which are believed to have undergone DEF reveal pores which are empty or only partially filled with ettringite. As ettringite forms, other reactants including solid cement substances must be consumed. Most cement paste components have densities in the range 2.2 – 2.4 g/cm^3, i.e., much greater than ettringite, *ca* 1.77 g/cm^3. These density relationships are frequently cited as proof that too much sulfate is expansive, but the argument is incomplete. On account of the high water content of ettringite. *ca* 50 wt% H_2O, water is also required to form ettringite. Only complete calculation of the relevant mass and density balances can determine if potential for expansion exists. Qualitative consideration of these mass balances shows that the potential for expansion will always be increases if water necessary to form ettringite comes from an external source i.e., if the system is open with respect to water transport into the concrete. The probability that water can come from an external source depends on service environment, but also relates to the original w/c ratio and the possibility that drying may have occurred during the thermal excursion, with reduction in the effective w/c ratio. Experimental studies do not always control these factors. But the need for a rigorous examination of the phenomena remains: tests purporting to reproduce DEF are not generally realistic.

External "sulfate attack" is of course more complicated; not only does sulfate migrate into the concrete, but the sulfate is accompanied by cations. Moreover, migration of sulfate salts suggests that water may also be mobile. I illustrate the importance of the chemical nature of the charge-balancing cation and its essential nature in determining the course of reaction by selecting three extreme examples, of H_2SO_4 attack and of sulfate groundwaters by either Na_2SO_4 or $MgSO_4$. Table 4 contrasts closed and open system behaviour. Consider first H_2SO_4 attack. Sewers often carry sulfide-containing sludges and liquids. However, sewers are of necessity vented for safety and hygienic reasons and the oxygen thus introduced reacts:

$$H_2S + 2O_2 = H_2SO_4$$

The reaction is catalysed by bacteria which colonise the concrete. As attack progresses, bacterial colonisation spreads inwards with the result that much H_2SO_4 is generated in interstices and the recesses of partially closed pores. The acid thus generated is neutralised locally by cement substance and a balance occurs between dissolution, mainly of acid soluble substances, and precipitation of gypsum (and possibly other sulfate-containing phases) with the result that porosity increases. One reaction product, gypsum, is often found precipitated in or near zones undergoing attack but which are partially protected from dissolution. The process is difficult to simulate in tests because much sulfuric acid is generated internally, within the concrete: simply adding sulfuric acid to the test solution does not necessarily reproduce the concentration gradients occurring during the actual attack process. Moreover, while conventional wisdom might suggest the use of sulfate-resistant cement in such applications, field evidence suggests that aluminous cements actually perform well. This is probably because alumina hydrate inhibits growth and reproduction of the bacterial species responsible for acid generation.

As far as I am aware, the degradation of concrete materials has not been simulated except by immersion in dilute sulphuric acid. This type of test, as I have shown, is somewhat unrealistic with respect to its ability to predict. For example, it fails to predict the enhanced resistance of aluminous cements, apparent in the field. Moreover, problems arise concerning the choice of acid strength in simulants. Dilute acid is inconvenient to use: it is quickly neutralised and requires frequent renewal or replenishment. Moderately concentrated acid accelerates reaction but also changes the mineralogical nature of the products: with increasing ionic strength gypsum is replaced by anhydrite, and the contribution of dissolution processes is artificially reduced by insolubilisation of $CaSO_4$. Thus the balance between precipitation and dissolution, as well as the mineralogical nature of the solid reaction products and the resulting volume changes, are difficult to reproduce with the result that an apparently simple form of chemical attack proves unexpectedly difficult to simulate.

Table 4 Cations associated with "sulfate attack"

PROCESS	NATURE OF CATION	CONSEQUENCES
Delayed Ettringite Formation (DEF)	Na, K, Ca: essentially fixed by cement composition	Sulfate solubility is temperature dependent. Mineralogy, and hence the distribution of sulfate, is also temperature dependent but is reversible, at least in part.
Sulfuric acid attack	H^+(or H_3O^+)	Biogenic sulfuric acid dissolves cement substance and in part converts Ca to gypsum. Leaching occurs.
Sulfate-rich groundwater	Na, K, Mg, Ca: external sources.	Alteration and dissolution occur. Mg for Ca exchange occurs. (Sulfates (ettringite, gypsum) are deposited.

I now consider the case of sulfate attack in open or partially open systems, such that sulfate is charge balanced by Na, K, Ca and Mg, either singly or in combination. In all cases known to the writer, standard tests which are designed to measure reaction by immersion in an aqueous solution emphasise mass gain by the solid without at the same time attempting to measure mass loss. The situation is shown in Figure 2 for two environments: Na_2SO_4 and $MgSO_4$. The geometry is schematic. While the setups appear superficially similar, they are not. Firstly, no attempt is generally made in tests to match the ionic strength of the solutions to that of the ionic strength of the pore fluid within the cement. Natural groundwaters tend to have low $MgSO_4$ concentrations (sea water is *ca* 0.05M with respect to $MgSO_4$) and thus tests often specify relatively weak $MgSO_4$ solution whereas typically more concentrated solutions are specified for Na_2SO_4. However an initially rapid movement of ions across the interface is driven by differences in ionic potential and these forces are likely to be more important in concentrated solutions e.g., Na_2SO_4, whose ionic strength typically exceeds that of cement pore fluid, sometimes unrealistically so. Secondly, the pH of the solutions differ; initially Na_2SO_4 is characteristically alkaline while $MgSO_4$ is characteristically neutral to acidic. Since the rate of transfer of CO_2 across an interface, other factors being equal, depends on pH, the Na_2SO_4 solution will uptake CO_2 more rapidly and attain a higher equilibrium CO_2 solubility than the $MgSO_4$ solution. This in turn, creates differences in CO_2 availability in the vicinity of the cement-solution interface, with the result that carbonation will in general be more significant in Na_2SO_4 than in $MgSO_4$. This contrasts with sea water which has a stable carbonate content. There is need either to restrict access of CO_2 or else make allowance for the chemical differences between simulants and real environments.

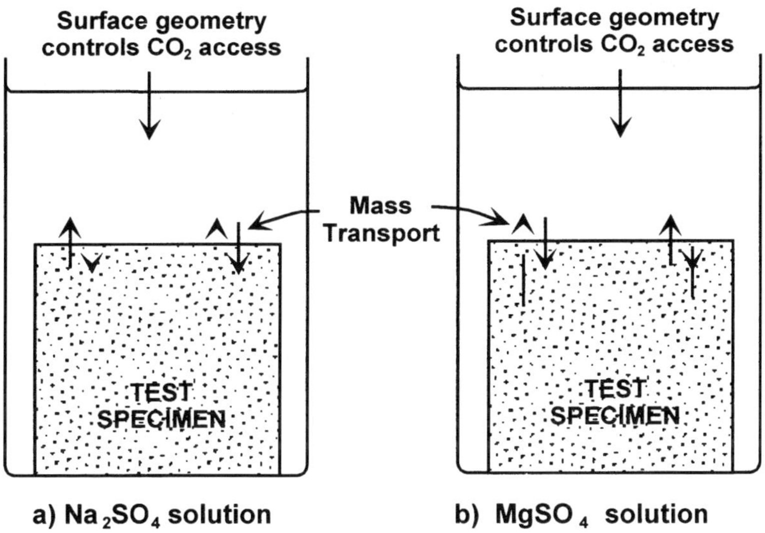

Figure 2 Immersion testing for sulfate resistance. Despite the similarities, CO_2 dissolves more rapidly in Na_2SO_4 than $MgSO_4$. The test describes other differences.

However as reaction progresses, further differences arise. Sodium is of course present in the cement and, in any event, will interact only weakly with cement phases, e.g., C-S-H, although rather more strongly with blended cements containing a low Ca:Si ratio C-S-H (9,10). Depending on aqueous concentrations, reactivity and the relative masses of solution and cement, sulfate tends to be removed from solution faster than sodium. In order for this process to continue, the cement must furnish OH^- ions to charge balance sodium, i.e., the solution pH increases as reaction progresses. This in turn increases the absolute solubility and transport of CO_2 across the air-solution interface such that its surface area relative to volume ratio as well as limits imposed on CO_2 ingress (sealed, partly sealed, unsealed) modify the course of reaction by spontaneously forming a pH-buffered solution. Moreover, while the solid gains sulfate from the aqueous phase, leaching also occurs: the solubility and stability of ettringite are, however, dependent on solution concentration. This is important because if ettringite is stable and being precipitated, it tends to block pores and impede diffusion. In the initial stages of reaction, calcium solubility may be relatively high but as the aqueous phase pH increases, Ca solubility decreases while that of sulfate increases. Hence conditions may not be steady state. In partial recognition of the fact that complex changes occur to the solution, many test methods specify renewal of the leachant. But intervals between renewals often seem to be chosen on an arbitrary basis, without consideration of the changing mass balances.

Compare this with attack by "real" Na_2SO_4 - containing groundwater. Firstly, the groundwater composition may fluctuate seasonally or even daily, depending on rainfall, but is not renewed totally in the same sense as occurs in laboratory conditions. Moreover, access of CO_2 occurs directly from the aqueous phase; transport across an air interface is less important or unimportant. The amount of dissolved CO_2 fluctuates and is difficult to determine accurately without perturbing the system but may have ranges quite different from those achieved by equilibrating laboratory test solutions with the atmosphere. The flow direction is also important. In many real examples, e.g., floor slabs at grade, sulfate-containing solutions enter mainly at one face and, as water is transpired, the solution precipitates: the importance of dissolution is accordingly diminished except for local transport occurring within the thickness of the slab. Since almost any process occurring in cement concrete is expansive if the balance of deposition exceeds that resulting from dissolution, transportation processes coupled with loss of water are much more likely to be disruptive than would be predicted by immersion testing. Thus it may not be enough to simulate the chemical chemistry of the service environment; appropriate mechanistic considerations need also to be incorporated in test design.

Consider now differences between Na_2SO_4 and $MgSO_4$; whereas sodium (Na) interacts only weakly with cement paste and sulfate more strongly, magnesium (Mg) interacts strongly. The principle products of reaction between soluble Mg and cement paste, namely brucite ($Mg(OH)_2$), a hydrotalcite-like Mg-aluminate hydrate and a magnesia-silica-water gel, are all very insoluble, whereas the sulfate reaction product, gypsum (or at elevated temperatures, hemihydrate or anhydrite, $CaSO_4 \cdot 0.5H_2O$ and $CaSO_4$ respectively) are relatively soluble, with the result that Mg is removed from the test solution more rapidly than sulfate. Solution charge imbalance is prevented by dissolution of calcium, furnished from cement whereas with sodium sulfate, charge balance is maintained by dissolution of OH^- and atmospheric CO_2 to form carbonate. The aqueous phase balances are much affected by geometry: size and surface area of cement, as well as by relative masses of test solution and cement and the frequency of renewal (if any) of test solution. In many published descriptions of cement - $MgSO_4$ interactions it is probable that during much of the test period the aqueous phase became strongly depleted in Mg with respect to its nominal composition. One cannot therefore emphasise too much the importance of monitoring the composition of the aqueous phase. For example, contrast the artificial situation sometimes achieved in the laboratory, as described above, with sea water: sea water comprises an infinite reservoir of constant Na, Mg, SO_4 and Cl concentrations.

I have not addressed the temperature factor. Most tests of concrete durability are either performed in the temperature range 18°C – 25°C or at elevated temperature, to accelerate reaction. The thaumasite form of sulfate attack has been recognised to be characteristic of low temperature service environments, less than 18°C [11]. Thus where thaumasite attack is a possible concern, low temperature (< 15°C) testing is required. The occurrence of thaumasite illustrates the importance of coupling test conditions to physiochemical data on phase formation and stability.

SUMMARY

It is apparent that existing tests and standards are inadequate tools with which to enhance the performance of cement concrete in a wide range of service environments. Existing performance-related tests rest on an empirical foundation and often fail to control important variables or accord with relevant physico-chemical considerations.

At the same time a new methodology is emerging for prediction of durability. This utilises chemical-mineralogical-kinetic models which are essentially solved using computer-basted iterations. Not all the necessary data are as yet available with which to implement such models, but rapid progress means that almost the entire body of existing standards for specification and testing of the durability of concrete are obsolete.

The time has arrived to support, extend, complete and verify predictive modelling. Only in this way will a fully flexible and comprehensive tool be devised with which to tailor concrete specifications for optimum performance and achieve meaningful service life predictions.

REFERENCES

1. RAHMAN, A. H., GLASSER, F. P., Comparative Studies of the Carbonation of Hydrated Cements, Advances in Cement Research, 2 No 6, 45 – 54, 1989.

2. DOW, C., GLASSER, F. P., Calcium Carbonate Efflorescence on Portland Cement and Building Materials, Cement and concrete Research, 33, No 1, 145 – 154, 2003.

3. COWIE, J., GLASSER. F. P., The Reaction Between Cement and Natural Waters Containing Dissolved Carbon Dioxide, Advances in Cement Research, 4, 119 – 134, 1992.

4. BRODERSEN, K., CRACK 2 – Modelling Calcium Carbonate Deposition from Bicarbonate Solution in Cracks in Concrete, Risø National Laboratory (Denmark) Risø-R-1143(EN), March 2003.

5. SAUNDERS, N., MIODOWNIK, A. P., CALPHAD- Calculation of Phase Diagrams: A comprehensive Guide, Pergamon Materials Series, Volume 1, 19998, ISBN O-08-0421296.

6. BARRY, T. I., GLASSER, F. P., Calculation of Portland Cement Clinkering Reactions, Advances in Cement Research, 12, No. 1, 19 – 28, 2000.

7. SKALNY, J., MARCHAND, J., ODLER, I., Sulfate Attack on Concrete, Modern Concrete Technology Series, Volume 10, E. and F. Spon, (London), 2002. ISBN 0-419-24550-2.

8. GLASSER, F. P., Zhang, L., ZHOU, Q., Reactions of Aluminate Cement with Calcium Sulfate, Prod. 2nd Intl. Symposium on Calcium Aluminates (Ed., Mangahabi, R., and Glasser, F. P.), 551 – 564. Institute of Materials (London), 2001. ISBN 1-86125-142-4.

9. HONG, S.-Y., GLASSER, F. P., Alkali Binding in Cement Paste. I: The C-S-H Phase, Cement and Concrete Research, 29, 1893-19003, 1999.

10. HONG, S.-Y., GLASSER, F. P., Alkali Sorption by C-S-H and C-A-S-H: Part II, Cement and Concrete Research, 32, 1101 – 1111, 2002.

THEME ONE:

MODERN CEMENT MANUFACTURE

Keynote Paper

MODERN CEMENT MANUFACTURE AND THE CEMENT INDUSTRY'S SUSTAINABILITY INITIATIVE

M Deussner

Heidelberg Cement

Germany

ABSTRACT. The Sustainability Initiative of the cement industry is a joint action of 10 internationally active cement producers, who have formed the Cement Sustainability Initiative under the chairmanship of the World Business Council for Sustainable Development (WBCSD). The Initiative had a study carried out by the Batelle Memorial Institute, covering all aspects of sustainability: economics, ecology and social aspects, in order to develop workable and durable strategies for the industry. The Initiative drew up an Action Plan, describing precise measures to be realized within the next several years by the companies engaged. The spectrum of the agenda items embraces all fields of modern cement manufacture: from climate protection and emission reduction, use of primary and alternative raw materials and fuels, workers' health and safety, to internal business processes. HeidelbergCement already introduced quite a number of projects, which shows that there is no contradiction between sustainability and economic cement production: improved energy efficiency by reconstruction of kiln lines, use of alternative fuels including biomass, production of cement types with reduced clinker content by the use of slag, and introduction of innovative processes. These are examples of a successful way to be continued in the future, which is further emphasised by HeidelbergCement's unhesitating commitment to CO_2 reduction.

Keywords: Cement, Sustainability Initiative, Manufacture, Economics, Ecology

Martin Deussner is Regional Coordinator, Western Europe with Heidelberg Cement. After holding various positions in both industry and academia including Managing Director of RMC West German cement plants, he joined Heidelberg in 2000.

INTRODUCTION

Sustainable Development is understood to mean a "development that meets the needs of the present without compromising the ability of future generations to meet their own needs" [1].

The term „sustainability" increasingly determines the discussion on the local and national development, especially the development of the industry. As the President of German Industry and Trade explained at a conference in 2002, sustainability is not only environmental politics, and environmental politics is not only regulation, but sustainability also contains a third leg – trade and industry, besides the environmental and social matters [2].

However, sustainability is also of global interest. Ten years after the first world summit conference in Rio de Janeiro, more than 180 states came together in Johannesburg in August 2002, in order to draw the interim balance and to discuss further action. On this global level, the development is among others promoted by the World Business Council for Sustainable Development (WBCSD). The WBCSD is an association of more than 160 internationally active companies, which have a common interest in sustainable development of different industries.

THE CEMENT INDUSTRY'S SUSTAINABILITY INITIATIVE

Under the chairmanship of the WBCSD ten major cement companies got together in the Cement Sustainability Initiative [3] to meet these challenges.

The purpose of the Initiative is to:

- explore what sustainable development means for these ten companies and the cement industry
- identify and facilitate actions that companies can take as a group and individually to accelerate the move toward sustainable development
- provide a framework through which other cement companies can participate
- provide a framework for engaging external stakeholders

The Initiative had a study carried out by the Batelle Memorial Institute, covering all aspects of sustainability: economics, ecology and social aspects, in order to develop workable and durable strategies for the industry (Figure 1).

The Initiative drew up an Agenda For Action, describing precise measures to be realized within the next several years by the companies engaged [3]. The reason for this is that sustainability is understood as a challenge for a prosperous future business, among other things by

- making more efficient use of natural resources and energy.
- meeting the expectations of stakeholders through greater transparency of operations and effective engagement with society, and
- building new market opportunities through process innovations, which achieve improved efficiency and long term cost savings.

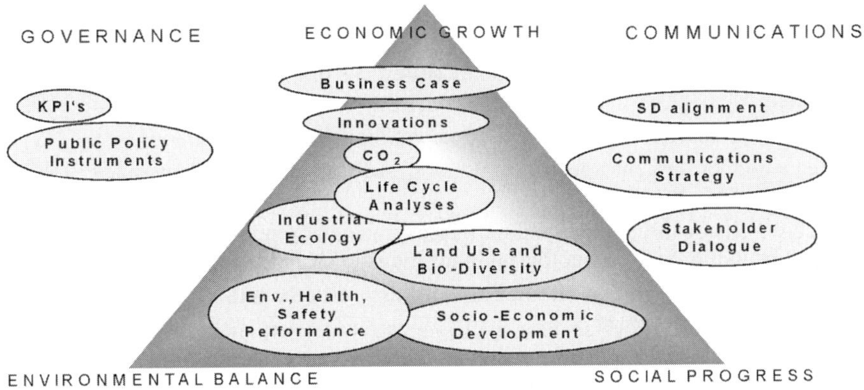

Figure 1 Sub-Studies and the Sustainable Development Triangle

The spectrum of the Agenda For Action embraces all fields of modern cement manufacture:

- climate protection
- (primary and alternative) fuels and raw materials
- employee health and safety
- emission reduction
- local impacts
- internal business processes

Although dealing commonly is central to the initiative, the companies involved have an individual responsibility to implement the action items. Guidelines are established in the common project, which are applied as individual action items. The initiative is therefore not simply an abstract declaration of intent, but concrete and verifiable action items with clear responsibilities.

HEIDELBERG CEMENT'S BOARD COMMITMENT: SUSTAINABLE VALUE CREATION – OUR COMMITMENT

Environmental protection has a tradition within HeidelbergCement [4]. Natural resources are the foundation of our company. We will only be able to secure our basic needs and those of future generations if we manage to retain and look after these resources.

It is our objective to be sustainable – to reach balance between economy, ecology and social responsibility. Thinking long-term is another significant criterion for sustainability – and characteristic of our corporate strategy. Through our products, our partnerships across the Group and our social commitment we have already made the first steps in the right direction. In many cases our activities exceed the legal requirements.

As a global company, we strive towards reaching solutions, which improve the local situations in our different strategic business units and conform to our company's philosophy.

Our path towards becoming a sustainable company is based on the following principles:

1. protection of resources through an increased use of both alternative raw materials and fuels
2. climate protection through reducing carbon dioxide (CO_2) emissions
3. reduction of cement specific emissions, such as dust, nitrogen oxides (NO_x) and sulphur dioxide (SO_2)
4. eco-efficiency through continuous improvements of production processes and product quality and through technological innovations in order to increase energy efficiency
5. step-by-step integration of the sustainability criteria into all decision-making and management processes
6. dialogue and co-operation with our partners (employees, customers, neighbours, investors, interest groups, politicians and governmental organisations) in order to reach sustainable corporate objectives

We have developed an ambitious environmental programme that we aim to implement by 2006. To reach this goal we need the commitment of each employee, regardless of the department, the location and the level. We want to be judged according to our success in reaching these goals.

MODERN CEMENT MANUFACTURE

HeidelbergCement already introduced quite a number of projects, which show that there is no contradiction between sustainability and economic cement production. Recent projects highlight the following:

- improved energy efficiency by reconstruction of kiln lines,
- increased use of alternative fuels including biomass,
- production of cement types with reduced clinker content by the use of slag, and
- the introduction of innovative processes.

Cost Efficient Kiln Lines

Optimisation of the efficiency and productivity of our cement plants contributes both to beneficial cost structures and improved environmental conditions. Beyond an intensified continuous improvement process to tap the full potential of existing installations, a substantial progress is achievable by replacing long wet and long dry kilns by modern precalciner kilns. In one go installing high capacity units might rectify the production scheme of a group of plants. In connection with HeidelbergCemernt's kiln project in Union Bridge / USA (5,500 tpd), Lixhe / Belgium (4,200 tpd), Gorazde / Poland (6,000 tpd) and Padeswood / UK (2,650 tpd) the operation of 13 long wet and long dry kilns will be stopped.

Kiln Conversion at Lixhe Plant

As an example, the kiln conversion in Lixhe [5] exemplifies the coincidence of ecological and economic goals:

- shut down the long wet kiln of Lixhe
- increase the capacity of the preheater kiln from 3,400 to 4,200 tpd
- maximize the use of suitable waste fuels
- ensure and improve environmental protection
- reduce operational and fixed costs of the plant

This involved major modifications in the whole plant:

- complete revision of the raw material grinding section
- installation of a modern calciner capable of burning a variety of secondary fuels
- aiming for low pressure drop of the preheater, the kiln inlet or smoke chamber, all cyclones and all gas ducts have been replaced. The new system comprises of a single calciner and a two string, four stages preheater
- in order to maintain a high preheater waste gas temperature of approx. 420°C for the raw materials rotary dryer a particular design feature is the "split raw meal feeding", by which a part of the raw meal can be fed directly to cyclone 3.
- in addition, the clinker cooler was fundamentally modernized for improved recuperation to the kiln and supply of hot exit air to the dryer.
- furthermore, two bag filters were installed to dedust kiln exhaust and cooler exit gases. The choice for bag filters was made because of its insensitivity to CO peaks or to different process conditions.

Secondary Fuels at Lixhe Plant

The modernized kiln system in Lixhe was designed for a maximum use of secondary fuels. This growing use of secondary fuels is endowing the ILC SLC calciner with more and more significance. This combination of an *in-line calciner* and a *separate line calciner* is outlined by Figure 2, which show the construction principle and its arrangement in the Lixhe plant.

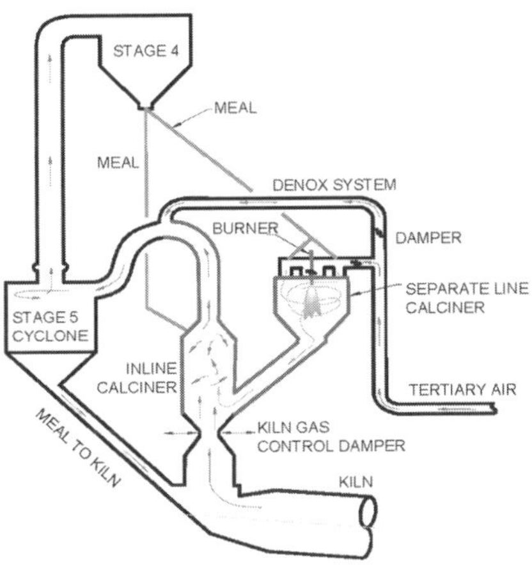

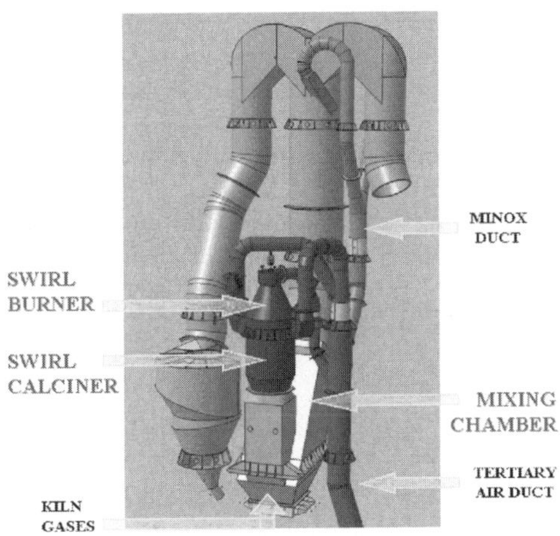

Figure 2 Principle Diagram of the "Hot Spot" Calciner and Arrangement at Lixhe Plant

The "hot spot" in the SLC, which acts as a pre-burning chamber, is working at temperatures between 1.200 and 1.400°C by burning the fuels in pure tertiary air in the absence of raw meal. This solves the problem of igniting secondary fuels with poor ignitability. Subsequent, raw meal is scattered in the swirl calciner to control the temperature and protect the refractory. The burn out of the (secondary) fuels takes place in the in-line calciner, where gas velocities are very low. Therefore, large particles of secondary fuels remain in this calciner part for some minutes and can burn out completely.

The CO rich SLC gases (in case of air deficiency operation) are mixed with NO_x rich kiln gas in the mixing chamber. Intensive mixing and sufficient retention time in the in-line calciner effect the desired NO_x reduction. The NO_x emission of the calciner is controlled by a combination of fuel- and air-staging. In the afterburning zone of the ILC, an adapted oxygen-rich tertiary air portion assures the final burnout of residual CO.

In particular, a typical fuel mix consists of

Main Burner	Coal	6 tph	Kiln inlet	Tyres	2tph
	Liquid waste	2 tph	Precalciner	Coal	1,2 tph
	Animal meal	8 tph		Resofuel	8 tph

In order to be able to burn secondary fuels with high chlorine content, the installation of a bypass is essential. In Lixhe it was designed to withdraw 5% of the kiln exit gas. The kiln gas is quenched with cold air to approx. 320°C, pre-dedusted in a cyclone and finally dedusted in the clinker cooler filter. This design offers two simultaneous advantages: reduced investment needs, and easy handling of the mixture of bypass dust and clinker dust for further use.

Summarizing, it can be stated that the Lixhe conversion was a big success. All technical, ecological and economic goals were reached or even exceeded. The project was professionally managed and finalized below planned budget. Within a short time, nominal capacity was reached, and today the kiln is generally operated between 4,400 and 4,600 tpd. All technical guarantees were met. The substitution factor amounts to 65% secondary fuels of the total fuel package leading to zero- or negative fuel cost. All emissions of the plant have been substantially reduced, thus meeting environmental demands of sustainability.

Use of Biomass and Alternative Fuels

The manufacture of cement consumes substantial amounts of heat or thermal energy and approximately 12 t of typical coal per 100 t Ordinary Portland Cement. Fuel usage and the associated CO_2 from combustion can be reduced in at least two ways. One is to use extenders, such as slag or fly ash, to reduce the amount of cement used per cubic metre of concrete; the other is to use waste derived fuels that would otherwise be incinerated for no energy benefit, or sent to landfill.

Utilization of alternative fuels in cement plants is much more advantageous than other methods of utilization or disposal:

- saving of natural resources
- saving of valuable primary fuels
- non-combustible components are used as raw material

- reduction of total emissions, since additional waste incineration is dispensable
- avoidance of high investments in waste incinerator plants helps to lower waste fees
- less space is required for landfill
- the high quality of cement is maintained

This report can only highlight some examples of actual developments. For a comprehensive overview to this topic please refer to [6].

Sewage Sludge

In Central Europe approx. 1 to 2 litres of moist sewage sludge accumulate per head per day. The quality of the sewage sludge significantly improved over the past years, compelled by stricter legal requirements In particular, this is true for heavy metals, dioxins/furans and absorbable organic halogen compounds. The reduction of mercury to values of 1.5 mg/kg dry substance made sewage sludge usable for cement production in an environmentally compatible and harmless manner.

Due to its calorific value, ash composition and content, dry sewage sludge is suitable for use as fuel and raw material in the clinker burning process. Dried sewage sludge has typically 10 % residual moisture, thus showing a calorific value of 11 MJ/kg. Figure 3 shows the composition of sewage sludge ash in a ternary system. The ash, which accounts for around 40 % of the dry material, is similar to hard coal ash and replaces sand and clay as raw material. However, the relatively high phosphate content is limiting the substitution to 5 to 20 % of other fuels.

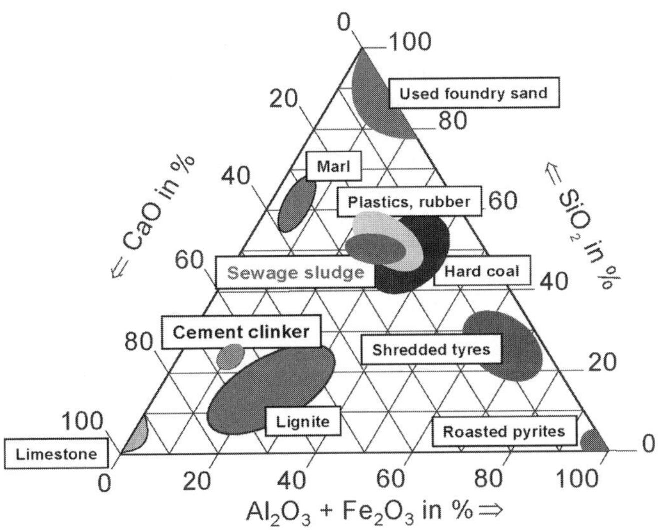

Figure 3 Composition of AFRs and AFR ashes

Sewage Sludge at Maastricht Works of ENCI

HeidelbergCement's ENCI Maastricht cement plant has gathered extensive experience by the use of around 35,000 tpy dried sewage sludge [7]. The calorific value amounts to 14 MJ/kg. It is used in a long dry kiln (178 m) with a two-stage preheater, producing 975,000 t clinker per year. Dry sewage sludge pellets are purchased to be pulverized at plant site. Studies of its combustion characteristics, carried out by ENCI and the International Flame Research Institute (IFRF / Ijmuiden), revealed that a fineness of max. 25 % < 0.09 mm is appropriate. Finally, a Claudius Peters ring-ball mill was selected, which shows reliable operation and good results. It should be particularly stressed, that ENCI uses up to 10 different fuels simultaneously, mainly secondary ones. Accurate feeding to the kiln is realized by the combination of a multi-channel burner with a consecutive dosing concept of some of the different fuels to one transport line. In particular, these are ground sludge, lignite, fine coke, bed coke, coal dust and anode dust.

To ensure reliable kiln operation, no increased heavy metals emissions, as well as high clinker quality, the chlorine, mercury and phosphate input have to be limited. By means of a small ESP-dust purge, the chlorine content in the hot meal is maintained to an acceptable level. Emission measurements prove the strict compliance with regulatory limits. Due to the good results, ENCI is going to further increase the amount of sewage sludge in its fuel package.

Animal Meal

Due to well-known reasons the use of animal or bone meal has recently gained importance. Depending on its fat content animal meal shows calorific values up to 21 MJ/kg (Table 1). This high value and the chemical composition of the ash make animal meal ideal for the use in cement kilns. However, on closer examination the chlorine and phosphate contents are the limiting factors for the quantities to be burned. Low phosphate contents in clinker increase the strength development in mortar and concrete. Exceeding approx. 1% phosphate the hydration and setting reactions will increasingly be retarded. It should be observed that bone meal contains nearly twice as much phosphate as animal meal.

Table 1: Composition of animal meal and bone meal (dry basis)

Sulphur	20 to 40 %	Ash	20 to 40 %
Phosphate	0.4 to 1.0 %	Nitrogen	to 8 %
Chlorine	0.6 to 1.6 %	Loss on ignition	to 18 %
		Calorific value	18 to 21 MJ/kg

Animal meal is usually stored in silos, which are pneumatically filled. Transport problems can arise if the fat contained in the animal meal liquefies at higher temperatures and causes clogging. To avoid this, the transport air has to be cooled. Development of odour can be avoided by exhausting the silo filter air to the kiln hood or clinker cooler. Animal meal normally replaces up to 25 % of the clinker burning fuel consumption.

Six plants of HeidelbergCement in Western Europe are going to burn more than 100.000 t animal meals this year. It should be explicitly noted, that this application is a substantial contribution to solve community tasks in an ecologically and economically sound manner.

INNOVATIVE PROCESSES

HOTDISC at Kjøpsvik Works of Norcem

Kjøpsvik in Norway operates a 1800 tpd kiln equipped with an ILC precalciner, a 5 stages preheater and a 30 % Bypass for alkali reduction. It has been normal for decades to burn used tyres in cement kilns. So, Kjøpsvik was also feeding whole car tyres to the kiln inlet chamber, which of course limited the maximum primary fuel substitution rate.

In order to further increase the alternative fuel rate, but to avoid a costly used-tyre shredding device needed to prepare the tyres for burning in the existing ILC calciner, an innovative system has been installed. The new external heat generator for lumpy fuels (up to 1200 mm) is an incineration chamber for use with alternative fuel „as it is" and is operated in combination with the calciner.

Figure 4 shows the scheme of the so-called HOTDISC pre-burning chamber from F.L. Smidth, installed upstream to the calciner. The HOTDISC is a cylindrical burning chamber of 6 m in diameter and 2.5 m burning space height, which is suitable not only for tyres, but also for wood, bags of urban waste, etc..

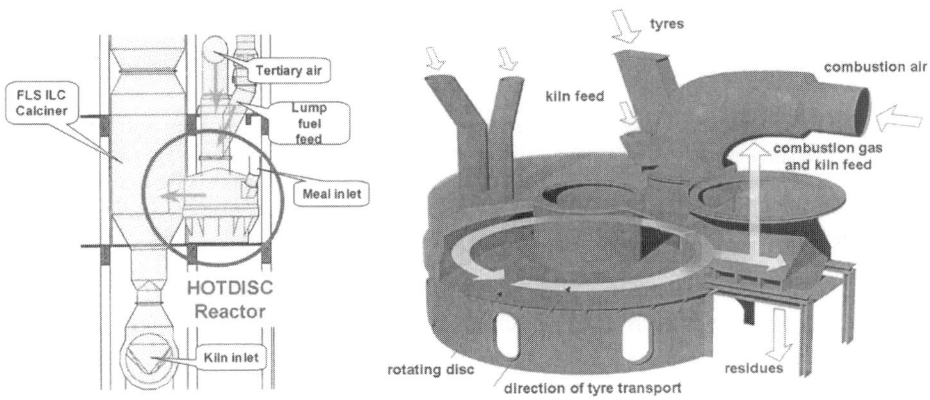

Figure 4 HOTDISC, Kjøpsvik

A rotary disc in the lower part of the reactor transports the whole tyres, which are incinerated by means of the hot tertiary air. The disc speed is controlled with up to three rotations per hour, resulting in a minimum retention time of 20 minutes for the tyres. This time is sufficient for complete burn out. To use the heat produced during this process, part of the meal from the next to last cyclone stage is fed together with the tyres. Thus, concurrently the temperature in the HOTDISC is controlled below 1200 °C. The HOTDISC is flanged to the existing calciner into which the solid pyrolysis residues fall.

This prototype installation is capable of burning up to 4 t/h used tyres (or other lumpy fuels), amounting to approx. 70 % of the calciner fuel energy or 40 % of the total fuel consumption.

PROFUEL Project at Ketton Works of Castle Cement

This project processes waste paper and plastic into a form that can be used to replace pulverised coal as a fuel in cement manufacture. The processed fuel, hence PROFUEL, is then delivered in precisely metered amounts to two cement kilns.

Raw material is taken from a great number of different waste streams, for most of which recycling is either impracticable or uneconomic. Raw materials include for instance, off-cuts from paper nappy manufacture, paper, plastic, cellophane, shredded contraband cigarettes from Customs and Excise, plastic car bumpers, shoe off-cuts, carpet, cloth and associated fibres, sticky labels, old railway tickets etc. etc.

The material has to meet a chemical specification so that neither the quality of cement nor emissions to the environment are put at risk. Extensive testing has shown that emissions from the cement-making thermal process improve when replacing part of the pulverised coal fuel with PROFUEL. There is a significant reduction in NOx and negligible change to other emissions. Chloride, which comes mainly from the plastics, is accepted up to 0.5% in PROFUEL. The chlorine reacts with spare potassium in the cement making process to form the relatively harmless compound potassium chloride.

Raw material streams are identified and tested for calorific value, chlorine and relevant metals. Acceptable streams are delivered into the warehouse where the recipe is controlled by selecting suitable proportions of the available feedstocks.

The material is reduced in size in three stages, de-baling, coarse shredding and fine shredding. The finished PROFUEL is typically less than 25 mm square. Because so much of the source material is fundamentally in sheet or filament form this size is sufficiently small for good combustion. The PROFUEL is tested to ensure that it meets the specification that has been agreed with the Environment Agency and that forms part of the site authorisation.

The plant is equipped with fire detection and fire suppression equipment. Most of the plant is enclosed. Open hoppers, such as the discharge from the de-baler, are provided with mist spray dust suppression. The remainder of the plant is provided with relatively conventional reverse jet bag filters. Metal separation before fine shredding is achieved using an eddy-current separator.

The plant is now performing at its design rating and the use of the material as a fuel is increasing steadily. It is confidently expected that the planned usage of 1,000 t/week will be achieved.

CO2 Reduction by Lowering the Clinker Factor in Cement

The use of cement intergrinding materials, like slag, fly ash and limestone offers a substantial potential for the reduction of CO_2 emissions. Due to its latent-hydraulic characteristics and consequently the reactivity, granulated blast furnace slag (GBFS) can almost substitute Portland cement clinker in a proportion of nearly 1:1 under certain outline conditions. This substitution contributes to the saving of natural resources, fuel and electrical energy as well as to the reduction of emissions.

Energy Consumption

Cement production is an energy-intensive process. Around 75 % of the total primary energy consumption for the production of Portland cement is necessary for cement clinker burning, only 25 % of primary energy for electrical energy generation is necessary for grinding cement, raw material and fuels. The substitution of clinker by GBFS substantially reduces the fuel energy consumption of the burning process (Figure. 5). There is practically no reduction of the electrical energy consumption, since cements containing GBFS usually need a higher fineness, thus cement grinding having higher power consumption.

Several limited effects on the energy consumption when using GBFS were not taken into consideration. The transport of the GBFS to the cement plant, higher wear rates of the mills or additional logistics (silo capacities, transports) can be mentioned here as some examples. Nevertheless, this summary shows the possibility of a considerable decrease of the energy consumption by using GBFS.

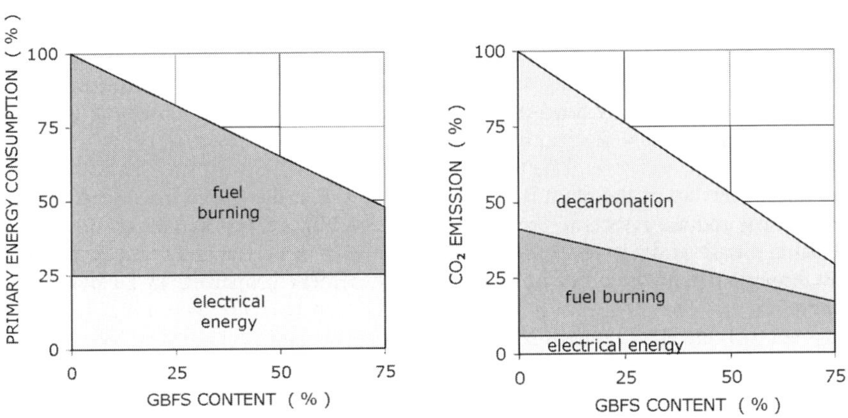

Figure 5 Relative reduction of primary energy consumption (left) and of CO_2 emissions (right) dependent on the amount of GBFS

Reduction of CO_2 Emission

During the manufacturing of ordinary Portland cement (OPC), the essential CO_2 emissions result from raw materials decarbonization (55 to 60 %) and fuel burning. The contribution of electric energy consumption to CO_2 emissions is comparably low with less than 10 % (Figure 5). So, the main CO_2 emission reduction effect by using GBFS is achieved in the cement plants, while the emissions from electrical energy generation remain constant.

Saving of Natural Resources

Partly substituted Portland cement clinker requires less raw materials for its production, which obviously contributes to the saving of natural resources. For further detailed information on this topic please refer to a comprehensive report regarding product application, recycling, disposal and the development of GBFS production [8].

Slag Grinding by Roller Presses with V-Separator

An innovative alternative to ball mills and vertical roller mills is the combination of roller press and V-separator. HeidelbergCement introduced this technology to grinding plants in Gorazde (Poland), Büyükcekmece (Turkey), Gent (Belgium) and recently in Ijmuiden (Netherlands). Whereas the grinding systems at Poland and Turkey are producing mainly OPC's and composite cements with up to 30 % trass and limestone, Gent and Ijmuiden plants are exclusively producing slag cements.

Figure 6 shows the principal flow sheet of the new cement grinding plant at Gent. It consists of a KHD roller press, grinding a mixture of wet granulated slag and dry separator rejects in closed circuit with the V-separator and a dynamic separator. The dynamic separator allows the grinding circuit to produce products with high finesses, so it is applicable not only for finish grinding slag meal, but also cements. The production rate for pre-ground slag varies between 125 t/h (normal operation 3000 Blaine) and 50 t/h for a fineness of max. 5000 Blaine. Subsequently, for cement finish grinding the slag meal is fed together with clinker to an existing and a new Polysius ball mill in closed circuits.

The static V-separator makes a convincing impression with its simple construction and the way it can be used simultaneously to separate, to cool or dry, and to deglomerate (Figure 6). Compared to conventional systems it promises quite lower wear and higher energy savings at lower investment costs.

During initial start up of the plant it became apparent, that the geometrical feeding situation of the wet slag and the rejects to the roller press feed bin, as well as the moisture control in the grinding circuit is decisive for the performance. Besides rated and reliable production, it is most important that there are no differences in cement properties to be noticed by our customers.

Our latest installation at the Ijmuiden plant was even equipped with a so-called VSK separator, in which a dynamic cage wheel takes over the task for final separation. This feature made the dynamic separator dispensable. Results will be reported at the conference.

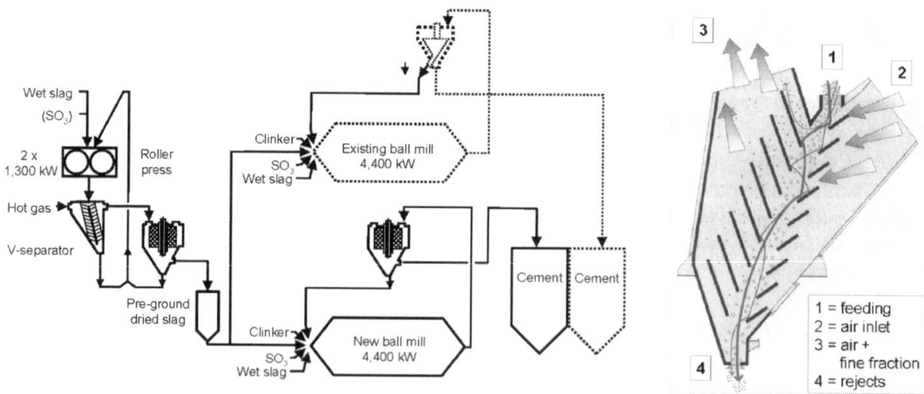

Figure 6 Roller press and V-Separator at Gent (left), and static V-separator (right)

Commitment of HeidelbergCement to Carbon Dioxide Reduction

For HeidelbergCement, the carbon dioxide emission reduction represents a major element of sustainable development. By reducing our consumption of natural resources as well as by optimising our production methods and products, we would like to make a contribution to reaching the targets of the Kyoto Protocol.

Related to the base year 1990, we set ourselves the following reduction target as a voluntary commitment:

15 % reduction of the specific net emissions
(kg CO_2 / kg cement) group wide by 2010.

Net emissions include the indirect saving potential through the utilization of alternative fuels. Gross CO_2 emissions contain the total amount of CO_2 emitted from our cement production activities, excluding indirect savings such as the use of alternative fuels. The above target corresponds to a 10 % reduction of our specific gross CO_2 emissions (kg CO_2 / kg cement) group wide by 2010.

This reduction target refers to fully or proportionately consolidated cement plant, including upcoming acquisitions up to 2010. For the latter, emissions of the acquired plants from 1990 on are included. Divestments will be treated the same way as acquisitions. The CO_2 emissions are determined based on the CO_2 Protocol of the World Business Council for Sustainable Development, i.e. the CO_2 emission from the decarbonization of the raw material is included.

Already today we have had considerable success in CO_2 emission reduction. This success can be put down mainly to the following three measures:

- Utilization of alternative fuels

  In 1990, the share of alternative fuels in the Group was only 4 %, of the total fuel energy consumption of the clinker burning process. We increased this in recent years; in 2002 the degree of substitution was more than 11 %. Based on generally accepted scientific studies, the utilization of alternative fuels is considered to be CO_2 neutral, as it avoids the release of green house gases elsewhere. Furthermore this energy recovery does not cause any waste materials.

- Reduction of the specific fuel energy consumption

  From 1990 to 2002, HeidelbergCement could decrease the specific fuel energy consumption from 4.100 kJ/kg clinker to 3.600 kJ/kg clinker. This was achieved through the modernization of production plants, as well as the shutdown of old kilns and an improved heat recovery for the drying process of raw materials.

- Reduction of the clinker / cement factor

  From 1990 to 2002 we reduced the clinker / cement ratio from 0,86 to 0,83. The clinker was mainly substituted by blast furnace slag and fly ash, i.e. by waste or by-products arising in other industries.

In order to reach our reduction target, we will continue and intensify the above measures. In 2010, alternative fuels will supply 30 % of our energy consumption. Our average fuel energy consumption will then be under 3,500 kJ/kg clinker. For cement grinding, at least 20% of CO_2 neutral intergrinding material, such as blast furnace slag or fly ash, will be used. Furthermore, we will utilize more CaO containing raw materials as a substitution for natural limestone. We will also intensify our efforts in research and development in order to push the development of sustainable and high quality cement further.

REFERENCES

1. Our common future. World Commission on Environment and Development (Ed.), Oxford Univ. Pr., Oxford (1987), ISBN 0-19-282080-X.

2. ROGOWSKI, M.: Nachhaltigkeit: Modewort oder Leitbild für Politik und Wirtschaft? Speech on the occasion of the BMU/BDI-conference „Nachhaltiges Wirtschaften als unternehmerische Herausforderung“ Auswärtiges Amt, 15. Mai 2002, http://www.bdi-online.de, Rev. 2002-05-29

3. The cement sustainability initiative – our agenda for action; WBCSD, July 2002; ISBN 2-240940-42-6

4. Heidelberg Cement – Group Environmental Report 2002

5. BOULANGÉ, B.: HeidelbergCement's Lixhe Plant, CBR, Belgium; World Cement 9 / 2002

6. SCHEUER, A.: Utilization of alternative fuels and raw materials (AFRs) in the cement industry; Cement International 1 / 2003

7. TAKX, A., PLUIJMEN, T.H.H.: Klärschlamm als Sekundärbrennstoff und –Rohstoff; Proceedings of the 5. International VDZ Congress 2002; Verein Deutscher Zementwerke, Düsseldorf

8. HÄRDTL, R.: Utilization of granulated blast furnace slag – A contribution to sustainable development in construction industry; fib-Symposium on Concrete and Environment; May 2001; Berlin; Germany

ANHYDRITE IN CLINKER

W Kurdowski
Research Centre of Building Materials
Poland

ABSTRACT. There are contradicting opinions about the behaviour of concrete made of cement produced of clinker containing anhydrite. Some authors claim that this concrete will expand with or even without heat treatment after long water curing. On the other hand, many researches have shown that anhydrite in clinker is quickly dissolved and reacts with aluminates during early hydration and can not be a source of expansion. In this paper, results of the examination of cements prepared of clinkers rich in sulphates are presented. Some clinkers contained anhydrite as a separate phase, and in others anhydrite formed numerous inclusions in belite. These inclusions were the product of decomposition of sulphate spurrite. Heat treated mortar made of cement from clinker containing anhydrite inclusions in belite presented expansion after water curing during 400 days. However cement made of clinker containing anhydrite as a separate phase did not expand under the same conditions. Also mortar of cement made of clinker of similar mineralogical composition, but low in sulphate and with interground gypsum, did expand only when the SO_3 content was equal to 4.0%. The results of these experiments allow the final conclusion that clinker containing anhydrite forming inclusions in belite gives mortars which expand after heat treatment during long water curing.

Keywords: Ettringite, Anhydrite, Belite, Sulphate, Tricalcium aluminate, Brownmillerite

W Kurdowski is a Professor at the Research Centre of Building Materials in the University of Mining and Metallurgy, Poland

INTRODUCTION

Many papers have been devoted to the problem of delayed ettringite formation (DEF). Day [1] and Thomas [2] presented very interesting reviews of this literature. One of the hypotheses of DEF and concrete expansion concerns the presence of anhydrite in clinker or, generally speaking, the delayed release of sulphates from clinker compounds [3-5]. On the other hand, Michaud and Suderman [6] found that "dead burned" anhydrite showed slower solubility. However, most of it dissolved after 2 days and all of it after 4 days. Michaud and Suderman [7] prepared also laboratory clinker with an extremely high level of sulphate and found that 2.1% SO_3 was present as calcium langbeinite, 0.7% was in silicate and 0.65% as anhydrite. Most of this anhydrite was dissolved quickly within a few hours and nearly all of it was gone after a few days. Herfort et al [8] found that the suphate in silicate phases correlates well with the total SO_3 and $[SO_3]/[Na_2O_e]$ ratio and was relatively very high, amounting to 1.45% SO_3. This high SO_3 content in silicates increased also the Al_2O_3 content and the ratio $[Al_2O_3]/[SO_3]$ was greater than 1. Herfort et al [8] expressed an opinion that thay are released at the same rate thus favouring the formation of monosulphate rather than ettringite. The dissolution of anhydrite from clinker was rather quick and a substantial part of it was dissolved after 10 min of hydration.

This paper presents the results of examination of five clinkers, three of them rich in sulphate, prepared in laboratory scale rotary kiln.

MATERIALS

The chemical and mineralogical composition of prepared clinkers is presented in Table 1.

Table 1 Chemical and mineralogical composition of clinkers

COMPOUND	CLINKER 1	2	3	4	5
	Percent, % by mass				
SiO_2	22.43	23.57	22.29	23.29	21.75
Al_2O_3	3.74	3.81	3.62	3.81	4.92
Fe_2O_3	4.28	1.62	4.02	4.01	2.79
CaO	63.28	66.26	62.56	63.04	65.23
MgO	0.70	0.58	1.05	0.70	2.00
SO_3	3.62	2.48	5.34	3.81	1.23
K_2O	0.35	0.34	0.39	0.41	1.23
Na_2O	0.08	0.08	0.34	0.35	0.16
Na_2O_e	0.32	0.30	0.34	0.35	0.97
$[SO_3]/[Na_2O_e]$	9.05	6.46	15.7	10.9	0.98
C_3S	42	51	40	37	55
C_2S	33	29	34	39	21
C_3A	3	7	3	3	8
C_4AF	13	5	12	12	9
$CaSO_4$*	4.5	2.6	6.7	4.0	0

**Calculated in the order: Na_2O as Na_2SO_4; K_2O as $Ca_2K_2[SO_4]_3$; remaining SO_3 as $CaSO_4$. In clinker 5 K_2SO_4; remaining 0.02% K_2O in solid solution in clinker phases.*

The first three clinkers are belitic and 5 is alitic-belitic. Clinkers 1-4 are rich in sulphate, but poor in alkalis. These clinkers are certainly not characteristic of industrially produced ones but they were synthesised with high amounts of sulphate in order to study the behaviour of the anhydrite content in cement paste after heat treatment.

X-ray examination during burning showed that sulphate spurrite was formed which gave the characteristic lines at 3.19 and 3.04 2θ. Clinker 3 was reburned three times for 20 minutes at 1450°C in order to provide the condition in which inclusions form a separate anhydrite phase.

In SEM examination of polished section of clinkers, it was observed that clinkers 1 and 4 contained numerous anhydrite inclusions in belite, but clinker 3 contained a separate anhydrite phase (Figures 1 and 2). The repeated heating of this clinker changed the situation and the majority of inclusions formed separate anhydrite phase was formed. Clinker 2 contained only calcium langbeinite and clinker 5 only potassium sulphate, but no anhydrite was found. From these clinkers seven cements were prepared: clinkers 1 to 4 were ground without gypsum and to clinker 5 gypsum was added during grinding in such a quantity to obtain 2%, 4% and 5% of SO_3 in cement. These samples were marked 5-2, 5-4 and 5-6 respectively. All cements had specific surface equal 400 m^2/kg according Blaine.

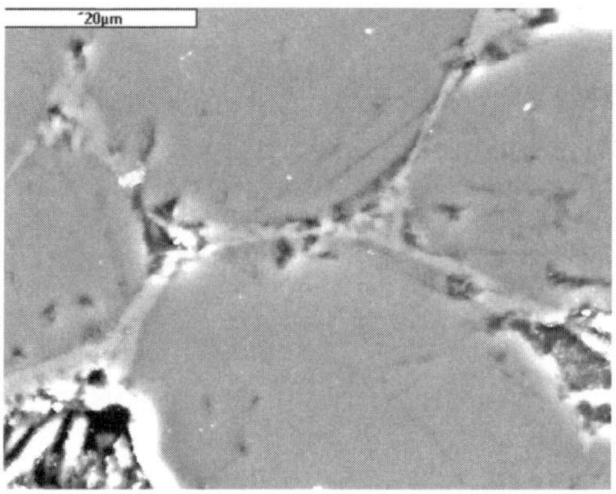

Figure 1 Inclusions of anhydrite (dark spots) in belite in Clinker 4

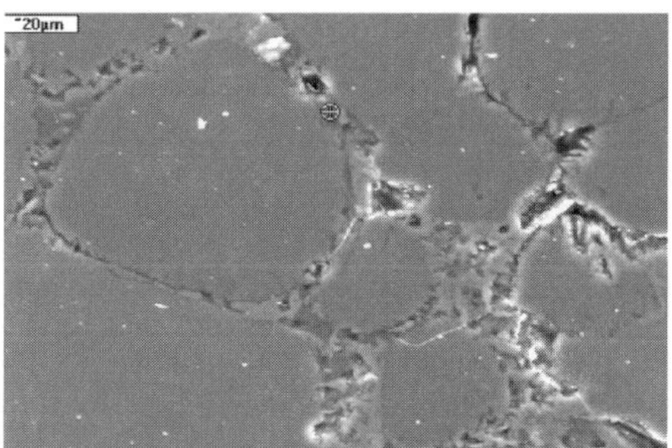

Figure 2 Anhydrite as main interstitial phase in Clinker 3

Mortars were prepared using these cements in accordance with European Standard EN-196-1. The mortars were cast in form of standard bars of the dimension 4x4x16 cm. These mortars bars were cured for 2 h in humid conditions (100% RH) and then subjected to heat treatment at 90°C. The heat treatment program is shown in Figure 3. After heat treatment, all samples were cured in water at 20°C.

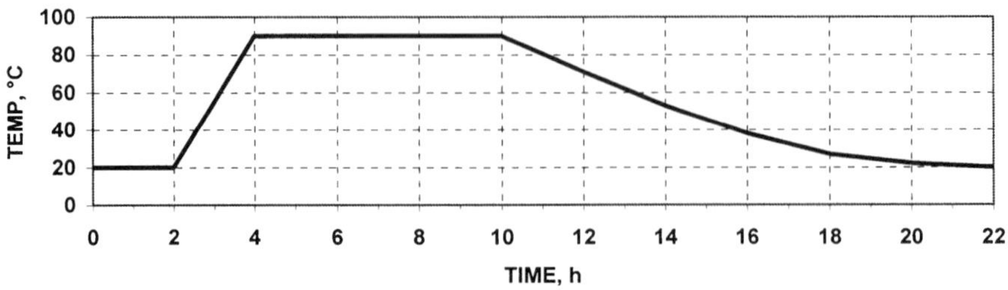

Figure 3 Programme of heat treatment

METHODS

The change of linear dimension of mortar bars were measured according to the Graf-Kaufmann method. SEM was applied for the examination of polished sections of the mortars. Finally XRD was used for mortar examination. The samples for this examination were prepared by grinding in an agate mortar and sieving the sand fraction larger than 40 μm.

RESULTS AND DISCUSSION

Figure 4 shows the expansion of two mortars: 1 and 5-4. Mortar 3 was cured for a relatively short time (50 days), especially when we compare this with mortar 1 which showed expansion after about 400 days of water curing. So the situation will be clear after a longer water curing.

SEM observations showed that in expanded mortars massive bands of ettringite are formed around the sand grains (Figure 5). However, also numerous conglomerates of big ettringite crystals were found in the cement gel (Figure 6). One may also see some microcracks inside the C-S-H dark gel, very similar to ones described by Yang [9]. It is evident that very small ettringite crystals must have been formed in the C-S-H matrix during water curing.

Mortar number 5 behaved classically according to the conclusion taken by Kelham [10]. It is a pessimium at 4% of SO_3 and mortars of cements containing higher and lower content of gypsum do not expand. Also the alkali content is well above the threshold value i.e. above 0.7%. The expansion of mortar 1 presents a different behaviour: the tricalcium aluminate content is very low as well as the alkali content.

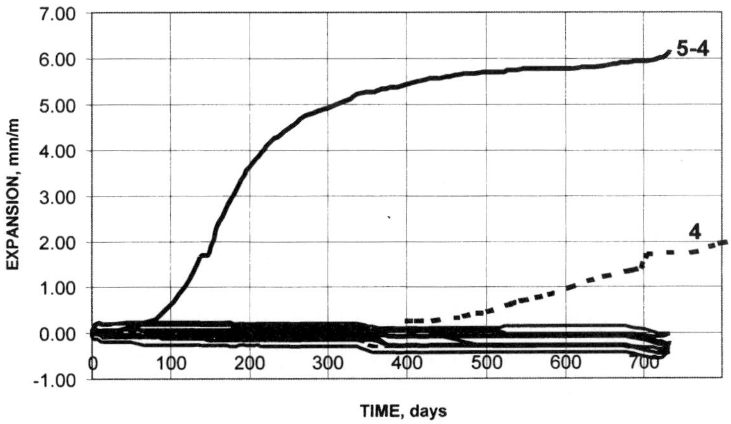

Figure 4 Expansion of the mortars

Kelham claims that sulphate resisting cements do not expand after heat treatment. One can conclude that only a late release of sulphate can be the reasons for this expansion. The inclusion of anhydrite in belite grains will hydrate together with this phase thus very slowly and the sulphate ions will dissolve slowly giving the condition for DEF. The small alkalis content does not play any role in this mechanism of DEF.

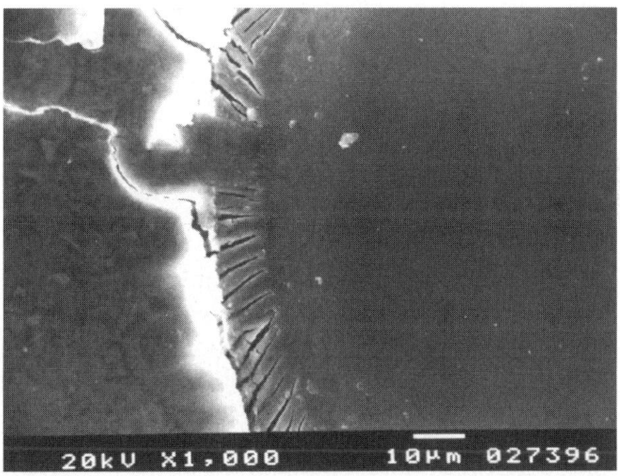

Figure 5 Bands of ettringite around sand grains in mortar 5-4

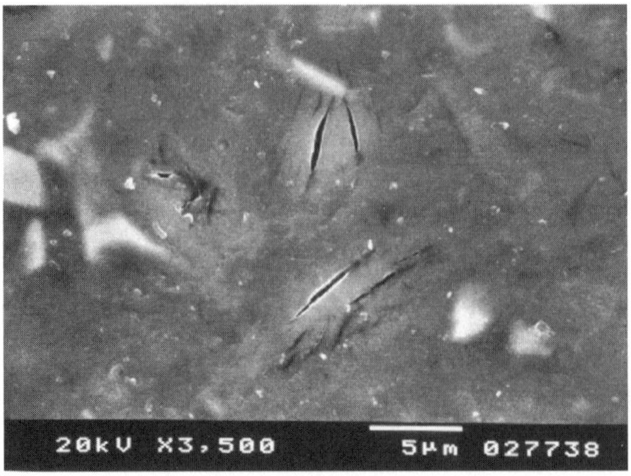

Figure 6 Ettringite crystals in cement gel

The expansion of mortar 1 takes place independently of a very high SO_3/Na_2O_e molar ratio which according to Glasser's calculations, should give the stability of ettringite during heat treatment. From the examination of this mortar further conclusion can be drawn. The expansion of mortar 1 shows that the tricalcium aluminate content can not play the decisive role in expansion. Cement 1 contained only 3% of C_3A and relatively high content of brownmillerite, namely 14%. It appears that brownmillerite can also participate in DEF [13].

In clinker 2 no anhydrite inclusions in belite were found, presenting no expansion till 810 days of water curing. This clinker contained only anhydrite and calcium langbeinite as separate phases (Figure 7). Numerous inclusions in belite waere found in clinker 3, but the time of water curing of mortar 3 after heat treatment is too short. The same is true for mortar 4, but clinker 4 contained also, as clinker 2, calcium langbeinite and anhydrite as separate phases.

Figure 7 Calcium langbeinite in clinker 2

CONCLUSIONS

From the experimental results the following conclusions can be drawn:

- Anhydrite in clinker is not the cause of mortar expansion after heat treatment,
- Anhydrite forming inclusions in belite causes the mortar to expand after heat treatment
- The expansion of mortar caused by anhydrite inclusions in belite is independent of low alkalis and tricalcium aluminate content in clinker. The source of ettringite can be brownmillerite.

REFERENCES

1. DAY R.L., Research and Development Bulletin RD108T, Portland Cement Association, Skokie, IL, 1992

2. THOMAS M. in S. Mindess, J. Skalny (Eds.), Delayed Ettringite Formation in Concrete: Recent Developments and Future Directions, University of Toronto, 1998, vol. VI, American Ceramic Society, Westerville, OH, 2001.

3. MIELENZ R.C., MARUSIN S., HIME W.G., ZUGOVIC Z.T., Concr. Int. 17 (1995) 62-68 (December).

4. HIME W.G., in R.K. Dhir, T.D. Dyer (Eds.), Concrete for Environment Enhancement and Protection, E&FN Spon, London, 1996, pp. 387-395.

5. MARUSIN S.L., HIME W.G., Supplementary Papers of the 4th CANMET/ACI Int. Conf. Durability of Concrete, Sydney, Australia, 1997, pp. 1-12.

6. MICHAUD V., SUDERMAN R., 10th ICCC, Goeteborg, vol. II, 1997 (paper 2ii011).

7. MICHAUD V., SUDERMAN R., Paper presented at the ASTM Symposium on Internal Sulfate Attack on Cementitious Systems: Implications for Standards Development, San Diego, December 1997 (after M. Thomas (2)).

8. HERFORT D., SOERENSEN J., COULTHARD E., Paper presented at ACI Spring Convention, Seattle, April 1997 (after M. Thomas (2)).

9. YANG R., LAWRENCE C. D., SHARP J.H., Delayed ettringite formation in 4-year old cement pastes, Cem. Concr. Res. 26 (1996) 1649-1659.

10. KELHAM S., Cem. Concr. Compos. 18 (1996) 171-179.

11. KELHAM S., Conf. Cement and Concrete Science, 2-3 September 1999, Keele University, Book of abstracts.

12. GLASSER F.P., DAMIDOT D., ATKINS M., Adv. Cem. Res. 7 (1995) 57-68.

13. KURDOWSKI W., Cem. Concr. Res. 32 (2002) 401-407.

MICROSTRUCTURE OF CEMENT CLINKER: ITS RELATION WITH PROCESS PARAMETERS AND EFFECT ON HYDRAULIC BEHAVIOUR

H S Patel

Y Z Pathak

Gujarat Ambuja Cements Ltd

India

ABSTRACT. A thorough microscopic investigation was carried out at our plant in order to understand the correlation of clinker strength development vis-à-vis the raw material characteristics as well as the process conditions. During the period of study, the clinker quality was maintained at constant C_3S (or alite) content. The kiln system corresponds to a state of the art technology. The preheater consists of 6 stages of low pressure drop cyclone. The kiln is lined with Mg Spinel bricks to withstand the short and intense flame in the burning zone. The cooler is a high efficiency grate cooler. The kiln system operates at high volumetric and thermal loads. Good correlation was also found between the Burning Zone Temperature and the Clinker Litre weight. Very good microstructure development was correlated with the optimum development of compressive strength. The process parameters conducive to the good development of microstructure are identified. Microscopic investigation of clinker has been identified as a tool to analyze the clinker quality and subsequent strength development pattern and further improvements in the clinker or cement quality can be achieved.

Keywords: Clinker quality, Litre weight, Microstructure, C_3S, C_2S, Burning time, Heating rate, Calcination, Coarse quartz.

Sri H S Patel is the Joint President, and heads the Central Technical Cell of Gujarat Ambuja Cements. He has an illustrious career in different fields of plant operation and possesses varied experience in cement technology.

Dr Y Z Pathak is the Head of Research and Development Center based at the plant site in Gujarat. His research interests include areas of alternate fuel, fly ash utilization and plant optimization.

INTRODUCTION

A thorough microscopy investigation vis-à-vis raw materials and process conditions was carried out at one of our plant site in Gujarat, India. This exercise was carried out especially with regard to the strength development pattern or hydration behavior of the clinker produced at our plant site. The idea behind this exercise was to identify and freeze down the relevant parameters when the kiln was running in optimum condition producing optimum quality of clinkers.

The basic principles of strength development may be envisaged as following :

- ✓ Among the four major chemical / mineralogical compounds in the cement clinker C_3S (or alite) & C_2S (or belite) are mainly responsible for the strength development.
- ✓ C_3S is responsible for early strength and C_2S for the late strength. Consequently, smaller size of C_3S is desirable for the early hydration i.e. early strength whereas bigger C_2S for slower hydration.
- ✓ If the size distribution pattern of C_3S & C_2S are disturbed or reversed i.e. bigger C_3S and smaller C_2S then the proper development of strength is likely to be disturbed e.g. reduction in early strength.

The alite (or C_3S) size is a composite function of degree of calcination, heating rate as well as the length of burning zone besides other. Whereas C_2S size is dependent on the retention time i.e. kiln speed and the length of the effective burning zone [1]. Heating rate may be defined as the rate of increase in temperature of the material from around 900°C to 1350°C (or the maximum burning zone temperature). For the optimum clinker microstructure development the schematic temperature time relationship is illustrated below [2].

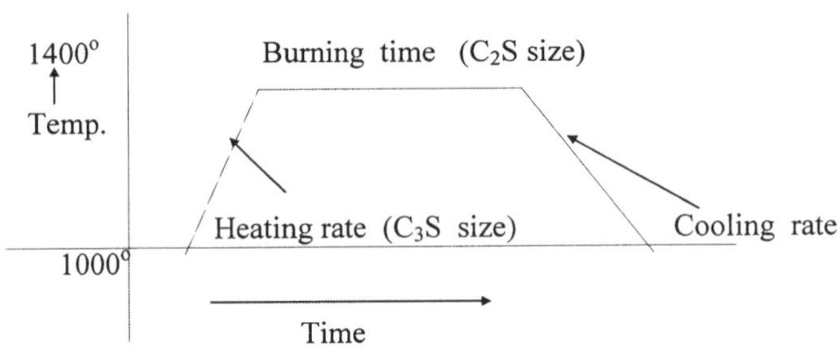

Figure 1 Schematic temperature-time relation of clinker burning.

RAW MATERIAL PROPORTIONING AND COARSE QUARTZ

Raw material proportioning as derived from the stock pile showed some minor variation during this period of study e.g.

- Marl varied from 6.2 % to 8.8 %,
- Sandstone varied from 1.0% to 1.5%, while
- Laterite varied from 0.4% to 0.6%
- limestone mostly hovered around 82 - 83% (including high & low grade).

The detail composition of the set of kiln feed is given in the Table 1.

Table 1 Raw meal proportioning

SAMPLE	DATE	KILN FEED COMPOSITION				
		Limestone	Marl	Solaj clay	Laterite	Sandstone
# 1	011001	83.14	6.94	7.88	0.62	1.42
# 2	280901	82.40	7.20	8.50	0.60	1.30
# 3	270901	81.50	8.80	7.80	0.60	1.30
# 4	260901	81.75	8.55	7.80	0.62	1.28
# 5	250901	82.28	7.23	8.76	0.58	1.15
# 6	240901	83.28	6.20	9.00	0.50	1.02
# 7	230901	83.28	6.20	9.00	0.50	1.02
# 8	220901	82.36	7.26	8.80	0.46	1.12
# 9	210901	82.20	7.30	8.85	0.40	1.25
# 10	200901	83.02	7.20	8.00	0.40	1.38
# 11	190901	82.95	7.00	8.20	0.35	1.50

The coarse quartz (i.e. quartz grains > 45 micron) content as derived for some of the corresponding Kilnfeed samples is as given below in the Table 2:

Table 2 Details of % Quartz (<45 micron) and % Sandstone in Kilnfeed

SAMPLE DATE	> 45MICRON QUARTZ	SANDSTONE % IN KILNFEED
23 Sep 01	1.31 %	1.02
24 Sep 01	1.49 %	1.02
25 Sep 01	1.43 %	1.15
26 Sep 01	1.47 %	1.28
27 Sep 01	1.25 %	1.30
28 Sep 01	1.25 %	1.30
01 Oct 01	1.72 %	1.42

PROCESS PARAMETER OBSERVATIONS

The salient observations on process parameters made during the study are as following:

- feed rate varied between 260 to 290 tph
- burning zone temperature hovered around 1300+25 °C.
- the kiln revolutions per minute (RPM) was more or less constant at 4.5
- Subsequent to all these the resultant litre weight was also more or less constant around 1300+20 gm/lit. As the above figure shows that there is a positive correlation between the burning zone temperature & clinker litre weight (but for exception like 21st & 19th Sep). The degree of calcination was more than 90% in all the days and hovered mostly around 95%.

MICROSCOPY ANALYSIS

Microscopy was carried out following the standard statistical procedure of sampling and analysis [3]. The summary of mineralogical distribution in the spot clinkers studied during the period of study is given in Table 3.

Microscopy Observations

a) Alite and Belite includes both pure and impure C_3S and C_2S respectively.

b) Alite size and Belite size (in microns) indicates the modal size i.e. the range in which most of the crystals occur.

c) OMS indicates overall microstructure of the clinker studied.

d) The primary modal size of Alite is around 30 - 40 micron. Presence of occasional big alite crystals (greater than 50 micron size) may be due to comparatively high degree of calcination (i.e is greater than 90%).

e) The modal size of Belite primarily is around 20 - 30 micron.

f) Occasional presence of ring belites around pores indicates presence of coarse silica in the raw meal.

g) Presence of raggedy or dendritic structures could rarely be observed in C_2S grains. These dendritic structures were restricted in the C_2S grains occurring in clusters or groups. This may happen due to moderately slow cooling rate through alpha prime belite (C_2S) temperature range.

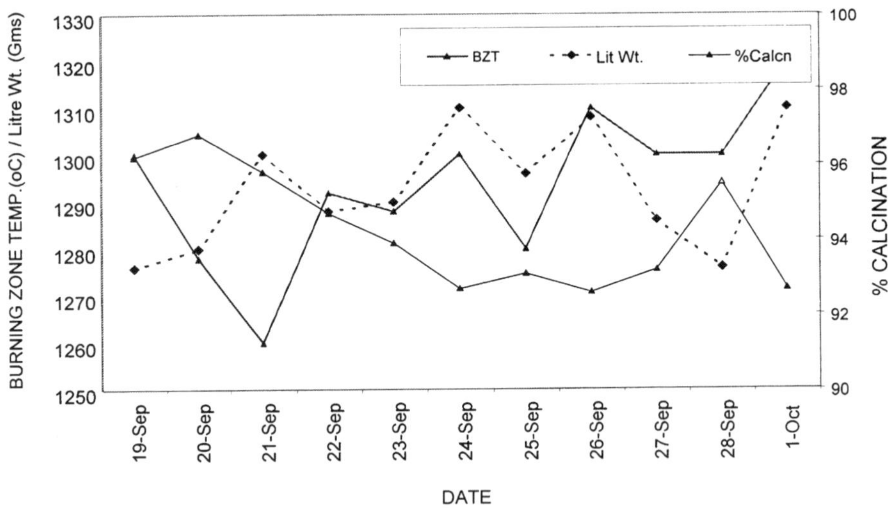

Figure 2 Relationship between different process parameters

Table 3 Summary on the plant clinker microscopy analysis

DATE	C_3S (Alite)	C_2S (Belite)	$C_3A + C_4AF$	Free CaO	Alite Size, um	Belite Size, um	Overall Microstructure
011001	56.4	26.4	15.9	1.3	30 - 40	25 - 35	Good -Very Good
280901	56.9	25.4	16.1	1.6	40 - 50	15 - 25	Good -Very Good
270901	53.7	28.9	15.5	1.9	40 - 50	15 - 25	Good - Moderate
260901	58.2	24.8	16.0	1.0	30 - 40	35 - 45	Very good
250901	55.4	26.4	16.2	2.0	35 - 45	40 - 50	Good
240901	56.6	25.5	15.9	2.0	30 - 40	35 - 45	Good -Very Good
230901	57.7	24.2	16.3	1.8	35 - 45	30 - 40	Good -Very Good
220901	58.0	24.5	15.7	1.8	30 - 40	25 - 35	Very good
210901	54.9	27.4	15.5	2.2	35 - 45	20 - 30	Good
200901	56.8	25.4	15.9	1.9	25 - 35	20 - 30	Good -Very Good
190901	58.1	24.0	16.0	1.9	30 - 40	30 - 40	Very good

STRENGTH DEVELOPMENT PATTERN

OPC were prepared for all the eleven samples in the laboratory ball mill and subsequently tested for their physical strength properties. The Figure 3 depicts the strength development pattern for these lab ground clinker.

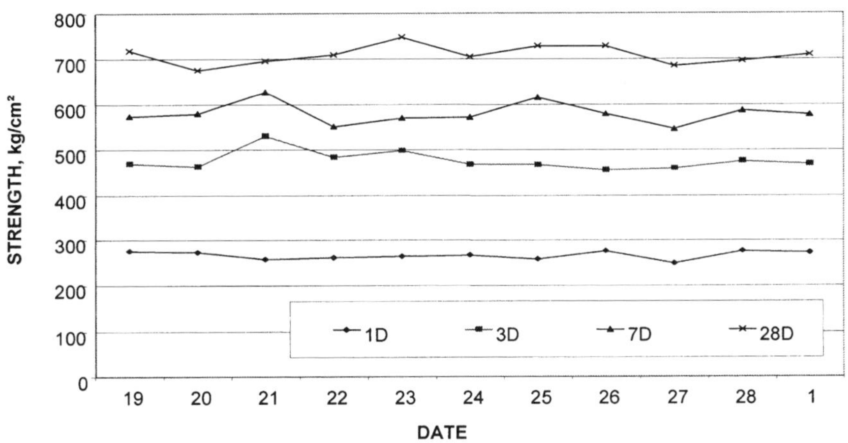

Figure 3 Strength development in the Gajambuja plant clinkers studied.

CONCLUSIONS

- The microscopy analysis of clinker suggests that the clinker structure and subsequent quality is very good, Figure 4. In fact, the only room left for improvement is in lowering the alite size. This may be achieved by decreasing the degree of calcination.
- Lowering in the size of alite may help in further gaining the early strength [4].
- Lowering of coal in the pre-calciner disturbs the kiln condition and subsequent clinker quality and as such it is be better to continue in the existing condition of the kiln.
- Considering the microstructure and strength development pattern it seems that process parameters of either the days of 23rd, 24th, 25th or 26th September may be adhered to for getting optimum microstructure. The relevant process parameters are as given in Table 4.

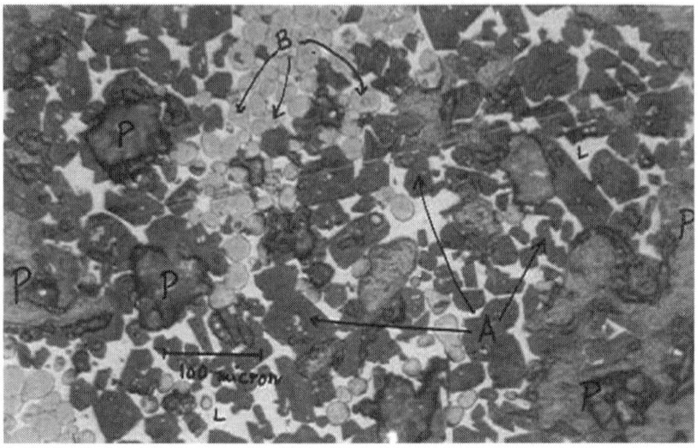

Figure 4 Very good development of microstructure in the clinkers studied. In the photomicrograph A = alite or C3S, B = C2S and L = C3A + C4AF. Note the shape and distribution of alite and belite.

Table 4 Process parameters

DATE	FEED RATE (tph)	BURNING ZONE TEMP.	KILN RPM	INLET O_2 (%)	INLET CO (%)	LIT. WT.
230901	274	1288	4.5	2.60	0.00	1290
240901	260	1300	4.0	2.00	0.00	1310
250901	260	1280	4.4	1.80	0.00	1296
260901	275	1310	4.5	2.00	0.00	1308

REFERENCES

1. SANYAL, M.R., SINGH, U.P. AND CHOUHAN, R.H., Strength development pattern in Gajambuja clinker, Unpublished R&D Report, Gujarat Ambuja Cement, November-2001.

2. ONO, Y., Microscopical estimation of burning condition and quality of clinker, 7th International Congress on the Chemistry of Cement, Paris, Vol-2, Theme I, 1980c, pp 206-211.

3. CAMPBELL, D.H., Microscopical examination and interpretation of portland cement and clinker, 2nd Edition, pp 19 – 27.

4. CESAR ARBELAEZ, Problems in cement strength decrease and its solution using microscopy, Proceedings of the International Cement Microscopy Association, 1988, pp 178-201.

NEW CEMENTS FOR SUSTAINABILITY

S Sarkar
Sarkar and Associates
J-C Roumain
Holcim (US) Inc
United States of America

ABSTRACT. This paper provides a brief historical perspective on the development of cement, a look at common problems and solutions, and a statement of the current status of cement production and applications. The authors further give their assessment of the need for performance-based cements and offer a look forward to characteristics that the cements of the future will need. As we move into the new millennium and portland cement reaches its 178 anniversary, global cement consumption stretches over the 1.5-billion-ton mark. Although cement strength has increased significantly since the Aspdin era, new challenges for the cement industry loom as we enter the new millennium. One of these is to abide by the Kyoto Agreement to reduce atmospheric CO_2 emissions. Binary-, ternary-, and quaternary-component blended cements appear to offer viable solutions for achieving this environmental target. There is an imminent need to make provisions for performance-oriented blended cements to deal with the key issue of the low-clinker-factor cement.

Keywords: Binary, Ternary blended cement, Performance-based, Environment, Emission, Low clinker factor.

Shondeep L Sarkar, PhD, PE, is a principal of Sarkar & Associates, Inc., a Houston-based construction materials engineering, consulting, forensic services and research and development company. A specialist in the durability aspect of concrete, he is active in research on high-performance cement and concrete.

Jean-Claude Roumain is the Product Manager of Holcim (US) Inc. He holds a bachelor's degree in civil engineering from the University of Denver and a management degree from the University of Michigan. He is a member of a number of professional organizations, including ACI, PCA, ACPA, NRMCA, ACBM, AASHTO.

INTRODUCTION

Throughout history, builders have tried to find or develop materials for binding stones or bricks together to make building more flexible and practical. Evidence of binder use can be found as early as the Natufian age [1]. The structures, typically circular in shape, had walls made of unhewn fieldstones. Mud, one of the oldest binders, is still used in different parts of the world. It is mixed with chopped straw and other vegetable fibers to enhance the binding capacity of building blocks and units. Various ancient cultures made strides in making cementing materials. The Egyptians used gypsum mortar produced by burning impure gypsum [2] to build the Pyramid of Cheops (~3000 B.C. and similar structures. Both the Greeks and Romans produced hydraulic lime by calcining limestone containing clayey impurities. Apparently, this pyroprocessed lime had hydraulic characteristics. Lime-plastered floors were widespread in Levant and Turkey, while gypsum-plastered floors were more common in Egypt [3]. Moreover, they discovered how to produce a cementing material that yielded higher strength than ordinary lime mortar and was resistant to water. The process consisted in finely grinding specific volcanic deposits, which were then mixed with lime and sand. Ronen et al. [4] determined that the strength of some of these materials was as high as 45 MPa (6,500 psi). Despite these advances, science didn't try to determine the mechanisms at play until the 18th century.

Work on developing cementing materials fairly flourished starting in the 18th century. Finally, on October 21, 1824, Joseph Aspdin, a builder from Leeds, England, patented what he referred to as portland cement [5]. It was named portland cement not because it was first made in Portland, Oregon, USA, or in Portland Hill, England, but because it resembled portland stone in color. It is reported that repeated destruction of the Eddystone lighthouse made of timber, in Plymouth in England inspired Aspdin to develop this improved construction material. Thereafter, the use of hydraulic cement spread rapidly throughout Europe and North America, paralleled by advances in manufacturing the equipment for producing portland cement. Early development of the rotary kiln started about 1877 in England, while Fredrick Ransome is usually credited with the first successful rotary kiln, which he patented in 1885 [5]. While Ransome's kiln stood out as a major breakthrough in the cement industry of the time, many years passed before a successfully operating rotary kiln was actually put into production. A few years after Ransome's discovery, the pioneering work of some American engineers brought the concept of the rotary kiln out of its infancy. The first cost-effective rotary kiln in America, developed by Hurry and Seaman of Atlas Cement Company, went into production in 1895.

Once production of portland cement began gearing up, more attention was paid to testing and characterizing the raw component materials as well as the cement itself. This process led to standardization of basic cement tests by around 1900. Since then, many tests have been revised and several new ones added to cement standards across the world. Today's cement industry faces growing environmental challenges as illustrated by the fact that producing one ton of cement clinker generates one ton of atmospheric CO_2. Furthermore, cement consumes 3 billion tons of natural resources per year. The concrete industry also happens to be the largest consumer of natural resources (rocks and sand) and is one of the largest consumers of water. This industry consumes natural aggregates at the rate of 8 billion tons per year. Cement and concrete together consume 600 to 700 billion gallons of water per year.As we move forward into the 21st century, we will have to deal with challenges that include reducing CO_2 emissions from cement production as per the Kyoto Agreement and producing cement with a low clinker factor and fewer durability problems. The latter means developing well-formulated cements that will be less susceptible to different types of internal and external chemical attack. It is not unreasonable to state that the future of hydraulic cement lies in the development of a value-added product for the customer that is environmentally friendly. Irrespective of the Kyoto protocol, we have a societal responsibility to fulfil: introducing earth-friendly materials that require less energy to produce. In doing so, we do a better job of balancing two of society's important needs: building an infrastructure capable of supporting our desired standard of living while protecting the environment and efficiently using our natural resources.

THE CREATIVE REVOLUTION

The focus in developing the cement for the new millennium should be on quality with the prime goal of infrastructure durability; cost-effectiveness should not be equated to "least expensive." Since Aspdin's day, cement compressive strength has increased at least eight-fold, yet improvements in concrete's mechanical properties have failed to keep pace in terms of durability and service life. Compositional factors associated with portland cement account for much of this discrepancy. The gap between current and future infrastructure needs reflects neither a technological shortfall nor an engineering windfall. Neither does it demand that environmental responsibility be sacrificed for economic growth. The commitment should be to building the best (rather than the least expensive) highways, bridges, and other cement-based structures. We need only look to the large number of ancient monoliths from the Neolithic, Roman, Byzantine, and Ottoman periods, some of which are still in excellent condition, to find inspiration for building durable structures. Their very existence should make us wonder about the materials that went into these magnificent pieces of architecture as well as the construction practices. Since the designer, the contractor, and the concrete supplier all look to the "performance" requirement, why isn't there focus on *the performance of cement*, which is the most important ingredient in concrete? Cement manufacturers face a dual dilemma: on the one hand, projections indicate that the demand for cement will outstrip our supply capacity (see Figure 1); on the other, the restriction on CO_2 emissions under the Kyoto Agreement is coming into effect in the very near future. Since environmental concerns, for the most part, are making it increasingly difficult to get authorizations for expanding production capacity, producing more cement with less clinker is undeniably an attractive solution.

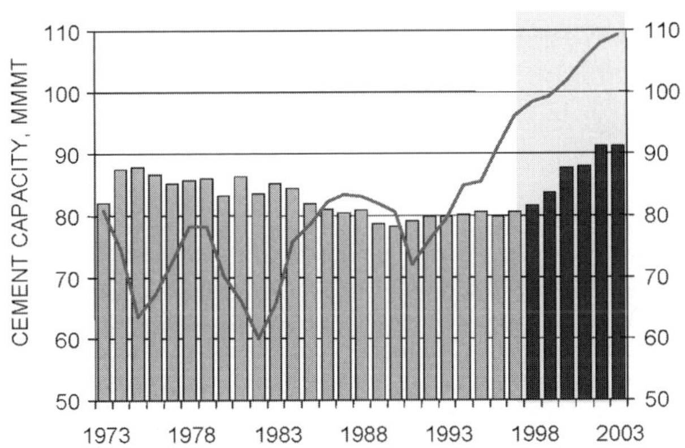

Figure1 Demand for cement in the US exceeding production within the early part of this century

In1994, Gebhart [6] conducted a survey of portland cements marketed in North America. His main objectives were to provide modern data on cement characteristics and to evaluate the need for developing standards. Gebhart concluded that modern cements tend to have higher sulfate content, fineness, and strength, particularly for Types I and II, than those that produced 50 years ago. While Aspdin's clinker and a modern clinker may look very similar (Figures 2 and 3) microstructurally, on average, today's cement has a higher lime saturation factor (LSF) and alite (C_3S) content, more uniform composition, and higher alkali content, often occurring as alkali sulfate minerals. In turn, these factors have impacted on cement properties, yielding a much faster rate of strength development, with higher early strength and less variation in cement properties. Yet today's concrete fails to delivers the high degree of durability seen in some of historical concrete structures.

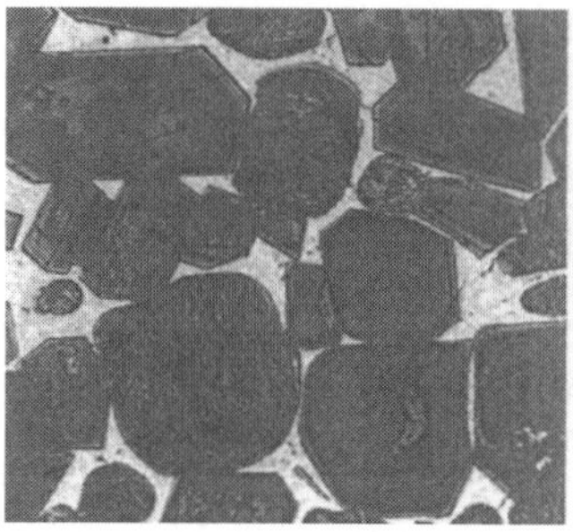

Figure 2 An optical micrograph of unhydrated cement particles from Aspdin's clinker. (By courtesy of Dr. Don Campbell).

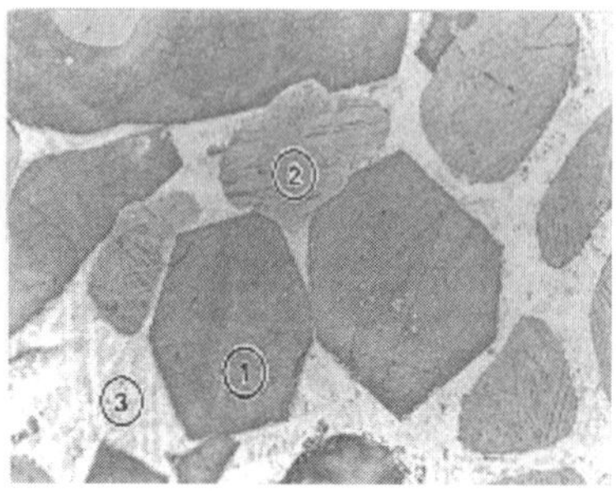

Figure 3 An optical micrograph of today's clinker showing mostly euhedral alite (1); round, lamellar belite (2); and interstitial matrix (3) comprising C_3A and C_4AF.

Driven by markets, the cement industry has historically used a number of solutions to satisfy market needs:

Market Need	Solution
Rapid construction	High early strength cement
Sulfate attack resistance	Low C_3A cement
	>8% C_3A for moderate sulfate environment
	>5% C_3A for high sulfate environment
ASR resistance	Low-alkali cement

Low-Alkali Cement

Producing low-alkali cement places a heavy burden on the cement manufacturer. It requires selective quarrying, thus depleting natural resources. Producing low-alkali cement consumes more energy, leading to a tendency for creating a reducing condition in the kiln that produces inferior clinker. This also increases production cost, because the clinker is harder to grind. Higher amounts of kiln dust are generated, yet it cannot be reinsufflated into the kiln because of its high alkali content. Each tonne of portland cement requires 1.5 to 1.9 tonnes of raw materials, whereas up to 2 to 2.5 tonnes of materials are used to produce low-alkali cement in some manufacturing facilities. The message is clear: we need to upgrade our knowledge about the relationship between cement clinker composition/properties and concrete durability.

Changes in Manufacturing Processes

Well-ventilated wet-process kilns are giving way to the dry process, with suspension preheaters, heat regenerators, and a minimum of ventilation. While this has reaped very significant fuel savings, the penalty is paid in more alkalis, fluorine, chlorine, and other volatiles from the raw materials or fuel turning up in the clinker [7]. Environmental considerations have also led a number of cement plants to to minimize plant dust generation by reinsufflating kiln dust, which is a partly-calcined material usually rich in alkali sulfates. While alkali sulfates are classified as minor clinker constituents, they play an important role in concrete durability. High alkali content in portland cement has been associated with alkali silica reaction (ASR). When used in combination with reactive aggregates, high-alkali cements can cause ASR in concrete. The alkalis combine with sulfates to form a number of alkali sulfate minerals (Figure 4), such as arcanite (K_2SO_4), thenardite (Na_2SO_4), calcium sulfate, apthitalite (a double sulfate salt of sodium and potassium), and calcium langbeinite (calcium-potassium sulfate). Since all these compounds are highly soluble, the sulfate component can be a major source of internal sulfate attack of concrete, also known as delayed ettringite formation (DEF). Figure 5 provides an example of DEF in concrete.While there may not be many reported occurrences of DEF in comparison to other types of deterioration, such as corrosion of reinforcing steel, the absolute value of damage caused by DEF is phenomenal, since it mostly occurs in expensive precast concrete elements.

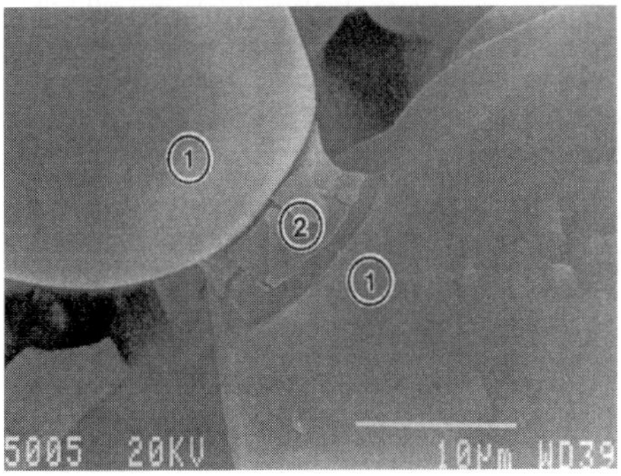

Figure 4 A scanning electron micrograph illustrating profuse distribution of alkali sulfate in the interstitial matrix of a clinker

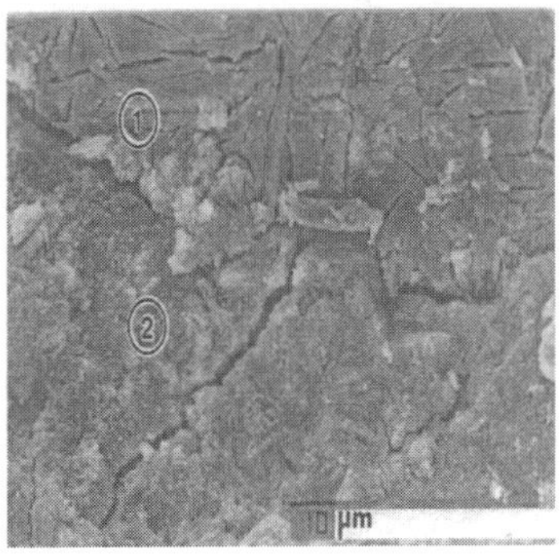

Figure 5 An example of DEF in concrete. The air voids and wide microcracks in the paste and the interstitial zone are full of ettringite crystals

Energy Considerations

Cement is the most important ingredient of concrete, the building material that literally is the foundation of a nation's infrastructure. Manufacturing cement, however, requires significant amount of energy, both to power the plant and fuel the kilns. Aggregate expenditures for fuel and electricity represent some 30% to 35% of all operating costs. Kiln fuel accounts for about 83% of a plant's total energy requirement. The remaining 17% is electrical power used to operate grinding mills and other electrical equipment. The primary purpose of the energy used is to convert the raw materials into cement clinker, which, in turn, is interground with a proportionate amount of gypsum to make *cement*. Producing cement also depletes a large amount of natural resources, primarily limestone, clay, and other materials that provide the required chemical components of CaO, SiO_2, Al_2O_3, and Fe_2O_3.

Fortunately, the cement industry is highly competitive, which drives companies to increase their efforts to reduce total operating costs, with energy being first on the list. Alternate fuels and the substitution of raw materials have been at the forefront of cutting energy costs. Both options have put the cement industry in an excellent position to help in managing America's waste management problems. In both cases, using alternative materials reduces the cost of energy and raw materials as well as conserves natural resources, while effectively disposing of waste in an environmentally sound manner.

While the issue of using alternative fuels still meets with some public resistance and significant regulatory mazes, raw material substitution is now available to the entire industry. Industry can use a variety of materials in the waste stream to make cement that still yields the same high quality required by ASTM specifications. National and international growth in using substitute raw materials in cement makes concrete a more environmentally-friendly building material in meeting the needs of the world's future construction requirements.

GLOBAL CEMENT AND CONCRETE MARKETS

According to Cembureau's annual report for 1994-1995 [8], world cement production in 1994 reached 1,390 mmt, representing an increase of about 50% over 10 years. The annual growth rate, then more than 4%, remained fairly steady up to 1997, before slumping to -2% in 1998, coinciding with the Asian economic crisis. Table 1 illustrates the annual global cement consumption rate. By 2005, consumption is expected to top 1,800 mmt.

Table 1 Projection of Worldwide Cement Consumption (mmt)

CONTINENT	1994	2000	2005
Europe	313	393	432
Asia	680	853	1,000
Middle East	65	79	82
Africa	63	71	77
North America	90	118	142
South and Central America	92	118	142
Others	7	9	10
Total	1,310	1,625	1,835

Over the last 25 years, the cement industry has invested heavily in upgrading its manufacturing facilities. Nevertheless, US production capacity from 1975 till the very end of the 20th century (Figure 1) remained nearly unchanged. Cement consumption now exceeds capacity by over 20 mmt. When all the annual additional capacities expected over the next five years are taken into account, the US will still have a 15-mmt capacity shortfall. Portland cement consumption and ready-mix concrete (RMC) used in different construction sectors in the US are given in Table 2 [9]. According to this table, RMC consumes 85% of the cement that is used in construction. In 2000, over 441 million cubic yards of RMC was produced in the US, of which single family units alone consumed as much as 100 million cubic yards. The largest consumption, however, has been highway and street construction. Western Europe reported steady but modest consumption, with total consumption rising from 313 mmt in 1994 to 393 mmt in 2000. Predictions point to 432 mmt by 2005, for an average growth of 2.5%.

Table 2 Forecast of Sector-wise Portland Cement and Ready-mix Concrete Consumption in the US [9] (Portland Cement (mmt x1000))

SECTOR \ YEAR	1999	2000	2001	2002	2003
Residential buildings	22,310	22,610	20,560	20,145	20,770
Nonresidential buildings	21,520	22,100	22,040	22,570	24,125
Public construction	48,990	56,045	56,900	57,340	57,820
Total	98,820	100,760	99,520	100,060	102,720
RMC (millyd^3)					
Residential buildings	123.6	125.3	114.0	111.6	115.1
Nonresidential buildings	95.3	97.8	97.6	99.9	106.8
Public construction	190.4	218.2	221.5	223.4	224.9
Total	409.2	441.4	433.1	435.0	446.8

As a conglomerate industry, the US cement industry annually generates $4.2B in sales and over 17,000 jobs. The figures for the concrete industry are $20B in sales and over 275,000 jobs. The concrete construction industry accounts for $100B annually in sales and more than 2 million jobs in the cast-in-place and precast concrete construction, repair, retrofit and renovation, reinforcement, falsework, formwork, placement, finishing, assembly, and equipment sectors.

HYDRAULIC CEMENTS OF THE NEW MILLENNIUM

ASR-Resistant Cement

The burden to suppress ASR has often been left to the cement industry. Table 3 indicates that ASR affects from 5 to 25% of concrete in the US. One of the first deliverables we must develop is a cement that can inhibit ASR.

Table 3 Estimated Yearly Production of ASR-affected Concrete (Holcim US, Inc.)

TYPE OF CONSTRUCTION	CEMENT (t)	PERCENTAGE OF ASR-AFFECTED CONCRETE	ASR-AFFECTED CONCRETE (yd^3)
Private Construction	43.8	-	9.9 million
Residential	21.2	5%	3.2 million
Commercial	19.3	10%	5.8 million
Others	3.1	10%	0.9 million
Public	47.0	-	19.6 million
Buildings	10.2	5%	1.5 million
Streets/Highways	29.6	15%	11.1 million
Water/Wastewater	8.4	25%	6.3 million
Utilities	1.0	10%	0.3 million
Others	2.7	5%	0.4 million

Various research findings indicate that additions of lithium compounds can delay or mitigate ASR in hardened concrete [10]. In a study conducted to determine if intergrinding lithium compound with portland cement to produce ASR-resistant cement, cements containing lithium carbonate and lithium hydroxide monohydrate were produced and tested to determine the optimum sulfate content. Trial production runs on a scale-model, closed-circuit mill indicated slightly higher long-term compressive strength with lithium hydroxide, while less ASR gel was evident in concrete produced with lithiated cement.

Precautions Against DEF

Delayed ettringite formation (DEF), which has drawn significant attention recently, means internal sulfate attack [11]. In addition, it implies that ettringite has been formed at a temperature that either prevented the initial formation of ettringite or destroyed the ettringite that was originally present. DEF is normally associated with both precast and cast-in-place concrete. Laboratory studies on mortars and concretes have shown that DEF associated with expansion or cracking only occurs under certain conditions. The material must have been at a temperature of more than 60°C (140°F) and the cement must have a SO_3/Al_2O_3 ratio by mass of at least 0.5 to 0.6 [12]. Very many cements, including almost all modern ASTM Type III cements, satisfy these conditions. After the heat treatment, the concrete must be stored either in water or in air that is saturated or very nearly so. These are *necessary*, not *sufficient* conditions. The rate or ultimate extent of expansion, or both, depend on several factors, including cement composition. High-alkali cements yield greater expansion, which can be attenuated and possibly eliminated if a sufficient fraction of the cement is replaced by blending agents such as slag or fly ash [13]. One proposal to avoid DEF is to produce low-alkali cement that is also low in sulfate (<3.6%) and MgO (1.6 $\pm$ 0.2%). Reducing the cement's specific surface area can also be beneficial [13].

Filler Cement

The co-grinding of portland cement and sand has long been practiced [14]. The fillers complete the finer portion of the particle-size distribution (PSD) curve, which improves concrete workability, shrinkage, density, and strength. This can be enhanced if the filler is calcareous. Skepticism about cement containing filler stems from oversimplified reasoning. For instance, it has been posited that when cement is mixed at a constant W/C, the ratio of water-to-clinker is higher for a cement containing a filler than for a standard cement containing only clinker and gypsum. Consequently, porosity increases, while strength decreases. This reasoning, however, does not take into account that, for a given strength class, the clinker fraction of the cement is more finely ground in the cement with filler than in a standard cement. Moreover, since the coarse particles in the cement do not hydrate at an early age, they also act as a filler. The cement's hydraulic potential, therefore, is actually much higher [14]. Masonry cement, which contains a relatively high percentage of added limestone powder, accounts for only a small portion of the global market. Even Type III cement, which takes a smaller market share than general-purpose Type I cement, is produced in much larger volumes than other specialty cements.

Within the last decade, several countries have modified their standards to permit limestone additions in different amounts, up to 35% replacement. In modern milling systems equipped with high efficiency separators, the steep particle-size grading can result in increased bleeding and delayed initial setting. The addition of a small amount of limestone powder broadens the particle-size distribution, thereby offsetting these disadvantages.

Limestone filler can partially substitute for gypsum in the cement. Hypothetically, adding 2.25% limestone to 1.75% gypsum yields a total of 4.0% gypsum, which should still be acceptable under portland cement specifications. The effectiveness of this level of substitution, however, strongly depends on the content and polymorphic form of tricalcium aluminate (C_3A). The gypsum/limestone mixture in the cement should provide for adequate regulation of normal setting. Optimum gypsum content increases with cement fineness, since an increase in specific surface area raises the rate of reaction.

Limestone can react with C_3A to form monocarboaluminate, which is a stable compound. The advantage of limestone filler over silica is that limestone added to a cement with high C_3A content will effectively yield more carboaluminate in the hydrated paste instead of sulfoaluminates such as ettringite and monosulfate. Since this implies a theoretical reduction in internal sulfate attack or DEF, this type of cement must be investigated further as a viable alternative to existing Type III cement.

Though finer cement tends to generate more heat, fine particles have a positive effect on the interfacial transition zone (ITZ) by improving the bond between the aggregate and the paste. To date, improving the bond has been achieved through the combined use of mineral and chemical admixtures, such as silica fume, slag, fly ash, and superplasticizer, yielding considerably heightened resistance to chemical attack. Using a reactive filler to replace or even dilute the finest fraction of portland cement can positively affect the long-term performance of concrete. Restriction of the initial reaction between C_3A and gypsum can reduce heat evolution, but not necessarily at the expense of delayed setting. A number of European and Japanese cement manufacturers [15] are already producing filler cement by substituting filler in the cement clinker in order to improve the cement's hydraulic potential and save energy. Manufacturing filler cement, however, is harder than it appears: merely adding a fixed amount of limestone fines to the cement will rarely achieve the targeted purpose.

Low Clinker Factor Cement

The most outstanding achievement in this category has been the development of blended cement. In fact, blended cement has now come to be recognized as a basic material that meets the requirements of modern society [16]. In addition to offering performance that ordinary portland cement simply does not have, it makes good use of by-products from the steel, electric-power, and silicon industries. Moreover, producing this type of cement saves considerable resources and energy in the cement plant.

In 1984, Duntran [17], then chairman of the European Committee for Standardization of Cements, predicted that, by 2000, the use of blending ingredients would increase substantially in many regions of the world, including North America. Duntran felt that eventually "pure portland cements will come to be regarded as special cements reserved for application where exceptional performance is required ..." Unfortunately, that goal is still a long way off. The volume of blended cement manufactured in the United States remains small compared to the volume of portland cement, although the trend toward using blended cement is increasing. A renewed interest in blended cements that had come about following the oil embargo in 1973 appears to have tapered off. In Europe, however, blended cements have been in use for decades, and, in some countries, represent nearly 50% of the total cement production. Blended cements are used as an alternative to low C_3A cement for construction in sulfate environments and have been found to represent a valid defense against expansion in concrete made with reactive aggregates [18].

The most widely quoted incentive for using blended cements is financial and energy savings of the final concrete mixture. Since manufacturing portland cement is energy-intensive, the use of blended cements stands out as an important energy-saving means [19]. Some of the distinct benefits to cement manufacturers are lower production energy, fewer gaseous emissions, and less solid waste. Manufacturing blended cement requires less raw material, yet plant production capacity can be boosted with low capital outlay.

As structures become more complex in design and larger in size every year, the need for high-performance, multi-functional types of cement will grow [20]. Conventional binary-component blended cement containing slag, fly ash, silica fume, or natural pozzolans no longer meets the requirements of performance and function. As a result, multi-component blended cement, which has started to attract attention, will probably be the focus of intense research in the early part of the new millennium. Figure 6 shows the dense microstructure of a concrete containing three cementitious components. A process for manufacturing ternary and even quaternary blended cement will probably soon be established; the use of such systems will develop and spread in the 21st century.

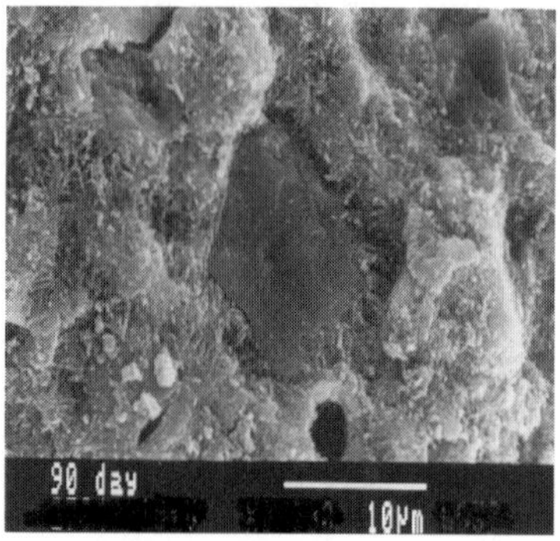

Figure 6 The dense microstructure of a concrete in which the cementitious components are slag, portland cement, and silica fume (not visible because of its ultrafine) size.

The construction industry foresees using cost-effective but not necessarily inexpensive rapid construction based on state-of-the art technologies for large-scale infrastructure projects, such as ultra high-rise buildings, deep subterranean structures, very large bridges, marine airports, and marine cities. If social requirements for comfortable living space, blended with natural environments and relatively easy maintenance of these structures are to achieved, then a breakthrough in cement is essential [20].

Although large volumes of concrete have been placed under water, a typical example is the state-of-the art anti-washout underwater concrete for very large cross-channel bridges. This type of concrete requires strict properties such as a slump flow of 50 cm or more retained for 8 hours or longer, a setting time of 30 hours or less, an adiabatic temperature rise of 30°C or lower, and a compressive strength of at least 280 kg/cm^2 at 91 days [21]. Although these properties can be achieved by modifying clinker composition, blended cement is more practical when compared to the high cost for manufacturing a new type of clinker, the long development period, and the savings in resources and energy when an appropriate blended cement is used. Conventional binary blended cement using slag, silica fume, or fly ash, however, cannot meet these performance criteria. These properties cannot be produced without resorting to a ternary blended cement.

Ternary systems generally have higher dispersibility, fewer coarse voids, and a more uniform and thinner water film around cementitious particles, thus resulting in much less segregation, and permitting concrete self-consolidation. The composition of C-S-H in the hardened body of a ternary blended cement depends upon the composition of the cementitious components (Figure 6). The setting characteristics and strength development depend on the mixing ratio, the character of the three components, and their fineness. Although the early strength decreases as the content of blending components increases, the mortar strength at 28 days is equal to or higher than that of plain portland cement mortar.

Since fly ash acts as a diluent, the adiabatic temperature rise is lower than in a concrete prepared with slag cement; the temperature decreases as the fly-ash content increases [22]. Although the permeability of an adequately cured hardened ternary blended cement is lower than that of ordinary portland cement, its rate of carbonation is higher [23]. Uchikawa and Okamura [24] have shown that alkali-ion diffusion and movement are affected by the pore structure in the hardened paste. Capillary pores do more than gel pores in reducing the diffusion coefficient. The diffusion coefficient of alkali ions through hardened paste of portland cement is highest, followed by that of slag cement and fly-ash cement paste. The diffusion coefficient of chloride ions is lower in a ternary blended cement than in a binary blended cement [24].

VISIONS OF THE FUTURE

Today's cement industry faces growing environmental challenges. The industry is dealing with the issue of reducing NO_x and SO_x emissions. Additionally, as per the Kyoto Agreement, cement companies have to respond to increasing restrictions and limits on CO_2 emissions. Low-clinker factor cement offers a sensible solution, since it essentially means producing blended cements that utilize non-CO_2 emitting components. In addition, this will increase the total supply of hydraulic cement needed to meet the growing demand of the construction industry during the century. From the cement producer's standpoint, production of blended cements should increase plant capacity with low capital investment. Today, the trend is to produce ternary and quaternary blended cements rather than replacing portland cement with only one blending component because the more complex systems are proving to be more efficient. Benefits have been realized by replacing a portion of the cement with limestone powder, while limestone-filled cements have been introduced in several countries. Researchers have demonstrated that controlled particle-size distribution helps enhance the properties of blended cements. Moreover, cement plants are in an advantageous position and better equipped than ready-mix concrete plants to monitor the quality of raw materials and the final product.

Cements need to be formulated to control ASR, give better sulfate resistance so as to replace current Type II and Type V cements, or reduce the heat of hydration for hot-weather concreting. Restrictions must not be placed on the composition of the cements or its constituents; the new standards must only set performance requirements.

CONCLUDING REMARKS

This paper presents a comprehensive review of historical developments of cements and their evolution leading to present-day trends towards performance-based hydraulic cements. It is also an informative position paper on the eco-efficiency of current and future cement systems. The requirements for technical innovation in cement science are also discussed. The chronology of the discussions in this paper provides a cogent insight into the standards requirements for new innovation.

REFERENCES

1. BAR-YOSEF, O. in The Architecture of Ancient Israel from the Prehistoric to the Persian Period, ed. E. Katzentain, The Society for Survey of Israel and Its Antiquities, Jerusalem, 1987, p. 29.

2. SNELL, L.M. AND SNELL, B.G., The Early Roots of Cement, Concrete International, February 2000, pp. 83-85.

3. BONEN, D., TASHDEMIR, M.A. AND SARKAR, S.L., The Evolution of Cementitious Materials Through History, Materials Research Society Symposium Proceedings, Vol. 370, 1995, pp. 150-156.

4. RONEN, A., BENTUR, A. AND SOROKA, I., Paleoorient, Vol. 17, pp. 149-150, 1991.

5. PERAY, K.E., The Rotary Cement Kiln, 2nd Edition, Chemical Publishing Co., Inc., New York 1987, pp. 3-4.

6. GEBHART, R.F., Survey of North American Portland Cements: 1994, Cement, Concrete, and Aggregates, Vol. 17, 1995, pp. 145-189.

7. GLASSER, F.P., Advances in the Chemistry of Portland Cement, Proceedings, 18th Annual Convention of The Institute of Concrete Technology, Coventry, 1990, pp 1-10

8. CEMBUREAU ANNUAL REPORT, 1994-1995, European Cement Association, World Cement, August 1995, pp. 20-21.

9. SUPRENANT, B. AND MALISCH, W.R., Concrete Market Y2K, Concrete Construction, October 2000, pp 30-42.

10. PORTLAND CEMENT ASSOCIATION, Development of a Cement to Inhibit Alkali-Silica Reactivity, PCA Bulletin RD112, PCA, Skokie, IL, 1996, 53 p.

11. SARKAR, S.L., Delayed Ettringite Formation - In a Nutshell, Concrete Repair Bulletin, July/August 1998, pp. 3-7.

12. HEINZ, D. AND LUDWIG, U., Mechanism of Subsequent Ettringite Formation in Mortars and Concretes after Heat Treatment, Proceedings of the 8th International Congress on Chemistry of Cements, Vol. 5, Rio de Janeiro, 1986, pp. 189-194.

13. TAYLOR, H.F.W., Delayed Ettringite Formation, Advances in Cement and Concrete, ed. S.L.Sarkar and M.W. Grutzeck, ASCE, New York, 1994, pp. 122-131.

14. COCHET, G. AND SORRENTINO, F., Limestone Filled Cements: Properties and Uses, Mineral Admixtures in Cement and Concrete, ed., S.L. Sarkar, ABI Press, New Delhi, 1993, pp. 266-295.

15. DETWILER, R.J. AND TENNIS, P.D., The Use of Limestone in Portland Cement: A State-of-the-Art Review, PCA RP118U, 1996, PCA, Skokie, 38 p.

16. UCHIKAWA, H. AND OKAMURA, T., Binary and Ternary Components Blended Cement, Mineral Admixtures in Cement and Concrete, ed. S.L. Sarkar, ABI Press, N. Delhi, 1993, pp. 1-83.

17. DUNTRAN, P., Present Situation of Cement Standardization in Europe, Blended Cements, ed. G. Frohnsdorff, ASTM ST 897, 1986, pp. 144-154.

18. FRIGIONE, G., Manufacture and Characteristics of Portland Blast-Furnace Slag Cements, Blended Cements, ed. G. Frohnsdorff, ASTM STP 897, 1986, pp. 15-28.

19. Federal Energy Administration, Energy Conservation Potential in the Cement Industry, Conservation Paper No. 26, NT15 PB 245 159/AS, Washington, D.C., 1975.

20. UCHIKAWA, H., The Way to Creation of High-performance, Multi-Functional Cement, Cement and Concrete (in Japanese), No. 535, 1991, pp. 112-118.

21. OKADA, R., FUJIKI, T., UCHIDA, N., OMORI, Y. AND SONE, T., Characteristics of Anti-washout Underwater Concrete Using Low Heat Cement, Cement and Concrete (in Japanese), No. 45, 1991, pp. 442-447.

22. SUMIOYOSHI, H., TOBINAI, K. AND HARUTA, Y., Properties of Cement Containing Both Fly Ash and Slag, Cement and Concrete (in Japanese), No. 406, 1980, pp. 8-13.

23. VON DER WEGEN, G., BIJEN, J. AND VON SELST, R., Powder Coal Fly Ash as a Filler in Portland Blast Fume Slag Cement Concrete, Intron Report, 1988, pp. 88-105.

24. UCHIKAWA, H., Effect of Blast Furnace Slag and Fly Ash Bonding on Diffusion of Alkali Ions through Hardened Cement Paste, Cement and Concrete (in Japanese), No. 460, 1985, pp. 20-27.

CEMENT KILN DUST (CKD): CHARACTERISATION AND UTILISATION IN CEMENT AND CONCRETE

M S Konsta-Gdoutos

Democritus University of Thrace, Greece

S P Shah

Northwestern University

United States of America

ABSTRACT. Cement kiln dust (CKD) is a by-product of the cement manufacturing process. Despite the fact that usually its composition is similar to partially calcined raw feed, the variability in raw materials, fuel, type of process, dust collection systems and product specifications influence both the physical and chemical characteristics resulting in a highly variable material. As a result, a thorough characterization of the material has not been completed and commercial use of CKD in industry and products has been limited. The relatively high alkaline content of CKD is the predominant factor preventing its recycling in cement manufacturing. One effective way to utilize CKD is to use it in the production of blended cements. However research has been sporadic regarding the applicability of CKD as such an activator in concrete. Past research of blended cements containing CKD as a partial replacement of cement and ground granulated blast-furnace slag and/or fly ash has produced results that are variable and contradictory. The chemical composition of the CKD used most likely had a direct impact on the results obtained. The objective of this paper is to provide a general overview of the research that has taken place into the CKD production, composition, and characterization and the keys to utilization of CKD in the cement and concrete industry.

Keywords: Cement kiln dust, Alkalis, Strength, Heat of hydration, X-ray diffraction, Durability.

Dr M S Konsta-Gdoutos, is an Assistant Professor in the Department of Civil Engineering at Democritus University of Thrace, Greece.

Professor S P Shah, is a Walter P. Murphy Professor of Civil Engineering at Northwestern University and Director of the Center for Advanced Cement Based Materials.

INTRODUCTION

Cement kiln dust is a by-product of cement production. The chemical composition of kiln dust depends on numerous factors, such as the composition of the raw materials, the kiln process incorporated for cement manufacturing, and the fuel used to heat the kiln, while the particle size distribution is determined by the collection method employed. It is estimated that 30 million metric tons of cement kiln dust are generated worldwide per year [1]. The cement industry in North America generates about 4.1 million metric tons of cement kiln dust every year, approximately 14% of the world total. Of these 3.35 million tons is placed in landfills and only 0.75 million metric tons enter commerce as by-products [2]. Due to its high alkali content, kiln dust is often used an alkali source, such as a raw material for extraction of potassium oxide (K_2O) and its soluble salts. Cement kiln dust also contains a large amount of lime, thus making it a substitute for fertilizer and lime used in waste-water streams. Studies have shown that cement kiln dust can be effectively used as hydraulic filler for coal mine shafts. Kiln dust has been used as a partial replacement of soda in glass production. European countries have used cement kiln dust to improve the properties of the slurries used in gas and oil wells [2]. Recently, the production of active β-C_2S, $C_{12}A_7$, and C_2AS cements using different proportions of a low-CaO fly ash and a lime-rich cement kiln dust as raw materials was proposed [3].

In general, the amount of cement kiln dust used has not significantly impacted the amount available for use through production. Costs due to the production and elimination of kiln dust add up quickly. They are incurred on many levels, such as the loss of raw materials, the cost of grinding the raw materials that end up as cement kiln dust, the loss of heat and energy from the kiln for the calcination of the dust, management of the storage of the kiln dust, and the transportation and disposal of the kiln dust. The financial loss due to these factors has been put anywhere from $5 to $20 per ton of cement kiln dust generated (ASTM D 5050-90) [4] to 15% of the total cost of production [2]. Due to these high costs, future storage space problems, and environmental considerations, the reduction and use of kiln dust has become an important effort.

COMPOSITION AND CHARACTERISTICS OF CEMENT KILN DUST

Chemical Composition

Cement kiln dust is defined as "composed of variable mixtures of calcined and uncalcined feed materials, fine cement-clinker formed during the high temperature processing, fuel combustion by-products, and condensed alkali compounds," by ASTM D 5050 – 90. Thus, as described above, the chemical composition of kiln dust is compiled from unreacted phases from the raw material, partially calcined raw feed, condensed volatiles, fine cement clinker particles and product phases like free lime and poorly crystalline belite. Secondary products formed on storage or treatment like hemihydrate may be present. The variability in chemical composition is heavily influenced by the variability in raw feed, fuel type, kiln process type, and product specifications for the cement produced. Table 1 presents an excessive listing of the chemical compositions of the kiln dusts reported in the literature. The chemical composition of ordinary Portland cement was included also for comparative reasons. From this table, it can be observed that generally all cement kiln dusts contain alkalis (Na_2O, K_2O) and sulfates (SO_3^-) in much higher percentages than in Portland cement. The major portion of the alkali sulfates found in the cement kiln dust is formed in the kiln's hot zone. If fuel containing sulfur is used, SO_3 will be present in the combustion gases, and both sodium and potassium will react at these temperatures to form sulfate salts. Given equimolar quantities of Na_2O and K_2O in the presence of SO_3, three times as much K_2SO_4 is formed as Na_2SO_4. Consequently, dusts collected from gas or oil-fired kilns contain higher proportions of soluble K_2O compared to coal-fired ones [5].

Some components such as free lime and chlorides, which normally appear in minimal quantities in OPC, show a very large variation. Free lime concentration is typically highest in the coarser particles captured closest to the kiln. In stockpiled CKD that has been exposed to the environment for long periods there is very little or any free lime. Chlorides are introduced in the kiln system with the fuel where fuel is coal, and with the raw materials, particularly when the clays are of a marine origin. Chlorides readily volatilize in the burning zone and condense in the heat exchanger to combine with alkalis and sulfates. When a kiln system has an effective by-pass, a deliberate addition of chlorides is used (either in the form of calcium chloride or by firing the kiln with a proportion of a chlorinated solvent) in order to reduce the amount of alkalis present in the clinker.

In practical terms, it is almost impossible to categorize different cement kiln dusts according to the type of kiln systems however, it has been reported that long kilns generally produce CKD with relatively higher percent of loss on ignition (LOI). Dusts originated in higher temperature kilns such as the ones collected from alkali by-pass with precalciner kilns present a relatively low LOI percentage and a relatively high free lime content [6]. Finally, in addition to the bulk primary constituents, cement kiln dusts often contain trace metals such as antimony, cadmium, lead, mercury, selenium, silver, zinc, arsenic and strontium, at levels within the range commonly found in native soils (USEPA 1993).

Mineralogical Characteristics

Since kiln dust is the by-product resulting from the calcination of high calcium or dolomite limestone, the predominant phase is expected to be partially calcined or uncalcined calcite ($CaCO_3$). X-Ray diffraction techniques have been used to identify different phases. Daugherty determined the compositions of 27 different kiln dusts [7]. Although the quantitative composition varied widely, the majority of the kiln dusts had calcite ($CaCO_3$) as a major component, in a percentage varying from 20 to almost 70%. Also present in significant amounts were K_2SO_4 and $CaSO_4$, while lime (CaO) and quartz (SiO_2) appeared in minor quantities. Other components such as Na_2SO_4 and Al_2O_3 were identified, but always as minor constituents. Recently Konsta-Gdoutos performed x-ray diffraction analysis in 4 different cement kiln dusts [8]. A list of the crystalline compounds found in the 4 different cement kiln dusts is given in Table 2. CKD's from wet process plants (CKD(E) and (P)) are characterized by the presence of substantial quantities of calcite. Modern plants equipped with precalciners and alkali by-pass (CKD(X)) would present a relatively high free lime content. Among the other phases found in kiln dust are arcanite (K_2SO_4), sylvite (KCl) and apthitalite ($K_3Na(SO4)_2$). Other researchers reported similar results [9].

Physical Characteristics

In the majority of studies, physical properties such as fineness, particle shape, size and distribution and specific gravity, are rarely reported. In general, kiln dust is a fine powder, which is off-white or light brown in appearance. The specific gravity of kiln dust has been reported to be in the range of 2.6 to 2.9, less than that of cement (3.15), however the cement specific gravity was often used interchangeably for kiln dust in the research reviewed.

Table 1 Composition of various Cement Kiln Dusts

Kiln Dust Type	SiO2	Al2O3	Fe2O3	CaO	MgO	SO3	Na2O	K2O	Total Alkalis	Free Lime	LOI
Kiln Dust (UK)[1]	15.30	2.10	1.70	43.00	0.90	6.20	0.70	6.00	NA	ND	ND
Kiln Dust (USA)[1]	11.10	5.50	2.90	44.00	2.50	5.60	0.90	6.00	NA	ND	21.50
Kiln Dust (Malaysia)[1]	12.20	5.80	2.30	42.70	1.30	6.50	0.80	4.30	NA	6.10	22.10
Kiln Dust A[2]	13.30	3.80	1.60	36.90	1.30	8.00	0.44	9.10	6.41	NA	26.40
Kiln Dust B[2]	14.00	4.50	1.60	41.60	2.70	5.20	2.06	4.20	4.84	NA	25.50
Kiln Dust C[2]	15.30	2.70	3.40	42.10	1.00	2.80	1.01	6.00	4.98	NA	27.10
Kiln Dust D[2]	16.10	5.40	2.20	39.70	2.30	0.30	0.17	1.50	1.16	NA	7.10
Kiln Dust E[2]	14.70	3.90	1.40	48.00	0.50	4.40	3.61	3.60	5.99	NA	21.10
Kiln Dust F[2]	10.90	3.90	1.50	37.00	1.50	8.70	0.36	14.30	9.80	NA	21.40
Kiln Dust G[2]	7.30	2.60	1.10	23.40	0.80	24.10	4.65	2.50	18.92	NA	13.90
Kiln Dust H[2]	13.90	3.90	1.70	42.90	1.40	8.50	1.82	4.60	4.83	NA	20.10
Kiln Dust I[2]	6.30	2.20	0.70	19.50	0.80	10.30	2.76	33.10	24.60	NA	55.40
Kiln Dust J[2]	17.10	6.30	2.20	49.10	1.10	3.30	2.43	3.90	5.01	NA	17.10
Kiln Dust K[2]	15.40	5.30	1.90	44.60	1.20	3.10	1.72	5.10	5.09	NA	24.50
Kiln Dust L[2]	15.70	5.20	2.30	40.30	2.10	6.30	1.54	8.40	7.09	NA	19.40
Kiln Dust M[2]	17.00	5.10	2.20	40.30	1.20	7.40	3.64	18.10	15.50	NA	22.00
Kiln Dust N[2]	11.00	4.10	2.00	40.70	0.50	9.30	1.48	9.10	7.45	NA	20.20
Kiln Dust O[2]	16.30	4.40	2.20	47.00	0.60	3.70	3.92	3.80	6.40	NA	21.10
Kiln Dust P[2]	14.20	4.90	26.00	42.20	0.90	7.20	1.54	7.50	6.49	NA	20.00
Kiln Dust Q[2]	14.50	5.00	2.00	52.60	0.90	2.70	1.21	1.70	2.34	NA	21.10
Kiln Dust R[2]	16.70	4.40	2.00	47.00	1.80	1.80	0.10	2.10	1.49	NA	24.70
Kiln Dust S[3]	14.70	3.31	2.04	49.30	1.03	3.07	0.23	2.60	1.94	10.43	23.43
Kiln Dust H[3]	15.50	3.84	2.13	51.70	1.59	11.10	0.22	2.50	1.86	21.72	10.50
Kiln Dust C[3]	14.00	2.85	1.51	41.70	2.73	5.55	0.27	3.60	2.64	4.31	27.81
Kiln Dust I[3]	16.50	3.71	1.79	44.50	1.43	5.67	0.31	3.50	2.61	7.46	21.33
Kiln Dust L[3]	15.40	2.65	1.51	41.90	1.35	0.25	0.23	4.90	3.45	3.00	34.06
Kiln Dust[4]	15.78	3.95	1.09	45.71	0.98	2.32	NA	NA	NA	NA	27.83
Kiln Dust 1[5]	22.90	16.04	2.40	32.70	1.00	3.24	NA	NA	NA	NA	21.60
Kiln Dust 2[5]	23.22	15.46	3.60	32.76	1.00	1.83	NA	NA	NA	NA	22.20
Kiln Dust 3[5]	12.54	5.12	1.20	42.56	1.20	0.55	NA	NA	NA	NA	36.46
Kiln Dust[6]	16.16	4.71	2.83	53.85	2.18	4.79	NA	NA	3.80	14.12	9.10
Kiln Dust G[7]	15.02	3.85	1.88	41.01	1.47	6.27	0.74	2.57	NA	0.85	25.78
Kiln Dust H[7]	9.64	3.39	1.10	44.91	1.29	6.74	0.27	2.40	NA	0.52	30.24
Kiln Dust S[7]	15.23	3.07	2.00	61.28	2.13	8.67	0.34	2.51	NA	27.18	4.48
Kiln Dust 1[8]	20.60	5.30	3.10	54.60	2.60	7.10	0.50	3.40	2.70	NA	14.10
Kiln Dust 2[8]	14.00	3.20	1.40	32.80	2.60	9.60	0.48	7.05	5.11	NA	14.90
Kiln Dust[9]	16.16	4.71	2.83	53.85	2.18	4.79	6.69	NA	NA	NA	9.10
Kiln Dust A[10]	10.12	4.07	2.89	43.95	0.95	0.27	0.18	0.82	NA	NA	NA
Kiln Dust B[10]	13.68	4.36	2.30	42.59	1.23	0.10	0.28	0.79	NA	NA	NA
Kiln Dust A[11]	34.30	3.50	2.00	34.30	0.80	11.40	1.20	8.20	NA	NA	NA
Kiln Dust B[11]	12.20	3.20	1.80	34.80	0.90	10.60	1.60	7.50	NA	NA	NA
Kiln Dust[12]	11.95	1.12	2.45	9.75	1.86	6.35	3.87	2.66	NA	NA	17.92
Portland Cement	20.35	5.24	3.58	64.29	1.13	2.56	0.11	0.60	0.50	NA	1.10

NA = Not Available
ND = Not Detected

[1] Ravindrarajan, "Usage of Cement Kiln Dust in Concrete"
[2] Daugherty and Funnell, "The Incorporation of Low Levels of By-Products in Portland Cement and the Effects on Cement Quality
[3] Bhatty, "Use of Cement-Kiln Dust in Blended Cements"
[4] Ramakrishnan, "Evaluation of Kiln Dust in Concrete"
[5] Ghosh, Padmanathan, and Mohan, "Study on Effects of Kiln Dust Addition to Portland and Blended Cements"
[6] Abo-El-Enein, Hekal, Gabr, and El-Barbary, "Blended Cements Containing Cement Kiln Dust"
[7] Bhatty, "Alternative Uses of Cement Kiln Dust"
[8] Ozyildirim and Lane, "Effects of Blends of Cement Kiln Dust and Fly Ash on Properties of Concrete"
[9] Amin, E-Ebied, and El-Didamony, "Activation of Granulated Slag with Calcined Cement Kiln Dust"
[10] Batis, Katsiamboulas, Meletiou, and Chaniotakis, "Durability of Reinforcde Concrete Made with Composite Cement Containing Kiln Dust"
[11] Dyer, Halliday, and Dhir, "An Invstigation of the Hydration Chemistry of Ternary Blends Containing Cement Kiln Dust"
[12] Shoaib, Balaha, and Abdel-Rahman, "Influence of Cement Kiln Dust Substitution on the Mechanical Properties of Concrete"
Note: Kiln dust contains other compounds not listed here, therefore, composition may not total 100%

Table 2 Mineralogical phases in Cement kiln Dust

	CKD(E)	CKD(A)	CKD(P)	CKD(X)
Lime	+	++	+	++++
Calcite	+++	++	++++	++
Dolomite	---	---	++	+
Quartz	+	+	+	+
Anhydrite	+	++++	+	+
Arcanite (K_2SO_4)	+	++	---	---
Sylvite (KCl)	+	+	++	++
Apthitalite $K_3Na(SO4)_2$	+	++	+	+++

The particle size distributions reported in the literature confirm that kiln dust on average has a smaller than cement particle size. CKD particle sizes generally vary by kiln process type and the collection methods and range from 0-5 μm to greater than 50 μm. Blaine fineness ranging from about 4,600 to 14,000 cm^2/g has been reported. A particle size distribution of kiln dusts representing three typical kiln systems (long-wet kiln (Dust G), long-dry kiln (Dust H), alkali by-pass operation (Dust S)) was compiled by Todres [6] and appears in Figure 1.

EFFECTS ON PROPERTIES OF PASTES, MORTARS AND CONCRETE

Workability

In general, kiln dust has a larger surface area than cement due to its finer particle size distribution. The volume of solids also increases as a result of the fineness of the kiln dust, thus increasing demand for water to achieve a constant consistency. Ravindrarajah [10] examined the consistency of cement paste containing cement kiln dust as a partial replacement of cement. To produce a mix of standard consistency, the water to cementitious material ratio was increased steadily as the amount of cement kiln dust increased in proportion to cement. He concluded that more water is required for absorption by the larger surface area and for the lubrication of the particles to achieve better flow.

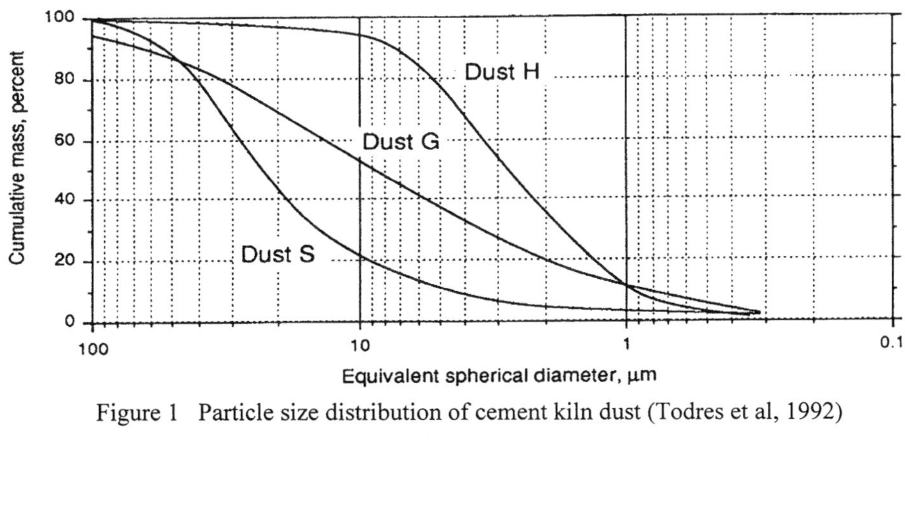

Figure 1 Particle size distribution of cement kiln dust (Todres et al, 1992)

Bhatty [9] determined the workability of mixes with five different types of cement kiln dust, prepared by adding up to 20% kiln dust to cement. One of his most important results was that workability of cement paste decreased with addition of cement kiln dust and other means were required to improve the workability. His results were consistent for all four CKD's he used, with the exception of the CKD containing 3.47% Cl^-. As the amount of high chloride dust increased in the blends, workability also increased. Addition of fly ash to cement-kiln dust blends generally resulted in better workability than cement and kiln dust blends, however contrary to this, addition of slag generally decreased workability compared to both cement and cement-kiln dust blends.

Time of Set

The degree to which the time of setting is affected is dependent on the proportion of the blend used, the water to cementitious materials ratio and the initial temperature. Alkalis in the mix play an important role as well, since the presence of alkalis in a mix was found to accelerate the hydration process, change the rate of nucleation and crystallization of hydration products, and consequently modify the setting process [11]. However, despite the fact that cement kiln dust contains high amounts of alkali salts, retardation of the setting time in blends containing cement kiln dust has been observed by the majority of researchers. Ravindrarajah [10] conducted tests on cement pastes, which had replacement of cement by kiln dust in 25% increments. Initial set times were longer with each incremental increase in kiln dust percentage, but the rate of increase slowed after 50% of cement had been replaced. This was the opposite of final set, which showed the rate of increase rising as the amount of cement kiln dust increased. Bhatty [9] determined the time of initial setting of pastes containing up to 30% cement kiln dust. Time of initial set generally decreased for blends when compared to cement. When Class F fly ash was added to the blends, time of initial set generally decreased as compared to cement and blends without fly ash.

Time of set of blended cement paste containing CKD was evaluated by Ramakrishnan [12], according to ASTM C 191. He observed that initial set for the pastes, which contained 5% cement kiln dust by weight of cement, took 22 minutes longer on average than normal cement paste. Final set occurred 40 minutes after final set of normal cement paste. Time of set was also measured by Abo-El-Enein et al [13] in cement pastes using a local Egyptian cement (ordinary Portland cement and 20% fine sand) and CKD. The weight ratios used were from 10% to 30% kiln dust by increments of five percent. They found that both initial and final setting times were decreased with the blended cements.

Heat of Hydration

There are very few data available on heat of hydration of blends containing kiln dust and the very recent and limited test data show contradictory results. Dhir et al. [14] reported heat of hydration results using blends with a constant Portland cement content of 65% and various proportions of CKD and ground granulated blast furnace slag using isothermal conduction calorimetry. Giving results of only the total heat evolved at 24, 48, and 72 hours, they concluded that blends containing both CKD and slag generated more heat than blends containing only one of these components, which is a sign of enhanced activity. In a subsequent paper by the same authors [1], the effect of the same cement kiln dusts in blends containing Portland cement and pulverized fuel ash was examined. They concluded that an increased degree of pozzolanic reactions with PFA was observed and it was attributed to the presence of CKD. The values of the maximum rate of the heat development indicated that both types of CKD's used accelerated the hydration process. However, contrary to the rate of heat evolution values, the data also implied that cement kiln dust had a retarding effect on hydration.

Recently Wang, Konsta-Gdoutos and Shah [15] measured the rate of hydration of OPC-CKD blends under adiabatic conditions. Results are shown in Figure 2. Compared to pure Portland cement system, a higher accelerating peak appears for 15% cement replacement, which implies that alkalis coming from CKD may facilitate dissolution of the silicates and accelerate formation of C-S-H. The rate of heat evolution also depends on the quantity of the cement. As the amount of cement is reduced by using 25% CKD as cement replacement, the rate of heat evolution is reduced, probably because of the low hydraulic property of CKD. Blends made with 50% OPC and 50% CKD, exhibit a much lower rate of heat development than OPC.

MECHANICAL AND DURABILITY PROPERTIES OF HARDENED PASTES, MORTARS AND CONCRETE

Compressive Strength of Pastes and Mortars

A great deal of work on compression testing of cement pastes containing cement kiln dust was presented by Bhatty in a series of reports [9, 16-18]. Using in the beginning 5 different cement kiln dusts in mixes with 10, 15 and 20% kiln dust, he demonstrated that particular ratios of alkalis, chlorides and sulfates are important for better performance of cement blends. Higher amounts of calcium carbonates in the dust appeared detrimental to strength development, but higher free lime contents appeared to be beneficial. In general, Bhatty concluded that compressive strengths of pastes and mortars decrease with the addition of CKD, but do not decrease as much with the use of either slag or fly ash. Ramakrishnan [12], consistently found that mortar cubes, containing 5% per weight cement kiln dust, had lower strengths than the normal cement mortar cubes. Similar results were reported by Dyer et al. [1], who had characterized ternary blends containing two kiln dusts, PFA and OPC. The same authors, in a more recent paper [14] have tested ordinary Portland cement, CKD, and slag blends with a constant Portland cement content of 65%. Higher strengths were consistently observed in blends containing cement kiln dust and slag. The optimum performance was obtained with cement kiln dust to slag ratios of 1/3. Blends containing only Portland cement and CKD possessed the lowest strengths, demonstrating the absence of any cementitious components in the kiln dust used. Similar results were reported from other investigators using different types of cement such as sand Portland cement blend [13] and phosphorous furnace slag cements.

The limited available test results on compressive strength of concrete containing CKD are in accordance with results reported from experiments with pastes and mortars. Tensile strength generally followed the same pattern as compressive strength. Ravindrarajah [10] was the first one to point out that the magnitude of compressive strength reduction increased as the amount of CKD increased. Ozyildirim and Lane [19] reported higher strength for his control concrete, made only with Portland cement, at all ages tested. The addition of 5% to 25% cement kiln dust in concrete with fly ash did not result in benefits in strength development. Finally, most recently, Shoaib [20] indicated that for concrete containing Portland cement, the ultimate compressive strength decreases with increasing dust percent. Slightly higher strengths were observed for concrete made with blast furnace slag cement (65% slag) and sulfate resistant cement (80-85% slag). The trend indicates that generally with increasing the amount of replacement of CKD, compressive strength decreases, possibly due to the loss of hydraulic properties of the cement kiln dust and the presence of alkalis (Figure 3).

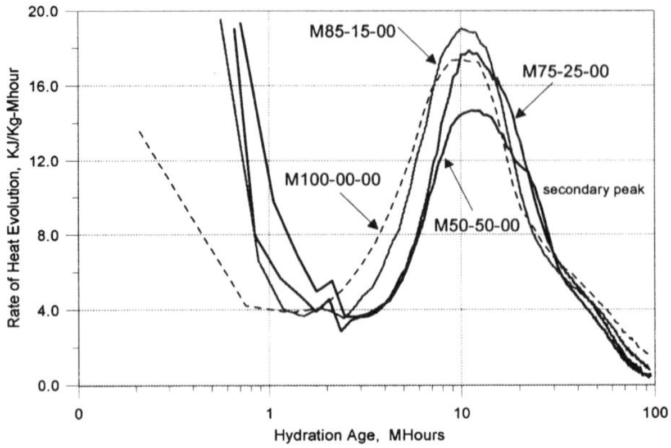

Figure 2 Effect of CKD on rate of heat development of OPC-CKD blends [15]

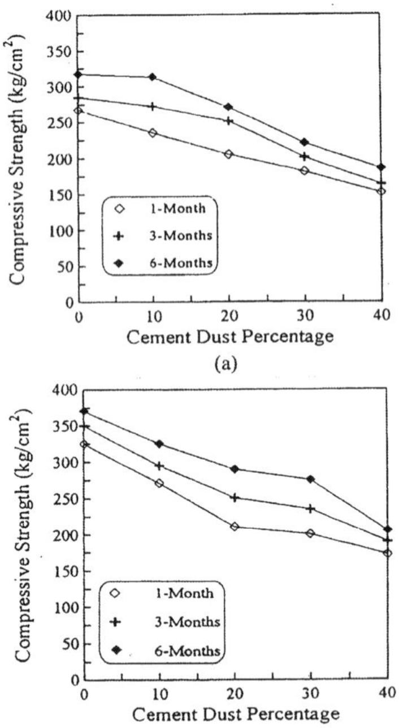

Figure 3 (a) Rate of compressive strength development for blends with OPC and cement kiln dust, (b) Effect of kiln dust percentage on the compressive strength of blends with OPC and cement kiln dust [20]

Durability Testing

The concern about the long term durability of concrete has been addressed by almost all investigators. Blended cements made from OPC, kiln dust and fly ash or slag or both were studied for alkali-silica reaction mostly using ASTM C 227[1]. In general, alkali mortar-bar expansions of binary cement-kiln dust and ternary blends containing cement kiln dust and slag or class F fly ash were below the ASTM limit of 0.1% at six months [17-19]. Conflicting results were obtained and trends in the results were reversed when class C fly ash or slag was used instead of class F. Addition of class C fly ash increased the expansion significantly compared to cement alone and to ternary blends containing class F fly ash.

Partial replacement of Portland cement with cement kiln dust is found to improve sulfate resistance of concrete. Mortar prisms were tested for sulfate expansion by Bhatty [18], in accordance with ASTM C 1012. The binary blend of 10% cement kiln dust and ordinary Portland cement resulted in expansions that were significantly lower than for the plain cement control mix at nine months and one year. The ternary blends containing both classes of fly ash up to 30% showed lower expansions than the binary system. Ternary blends containing slag showed also lower expansions than the binary blend, but only when the amount of slag was increased to 30% and 40%. Additional testing performed by Ozyildirim and Lane [19], also using ASTM C 1012, showed that kiln dust with Type I cement and 15% fly ash class F had high sulfate resistance as opposed to the control batch (Portland cement alone), which had only moderate sulfate resistance. Conversely, when a second test was completed according to ASTM C 342 (modified) at a cycling temperature of 54°C, large expansions were encountered in the control mix and mixes containing 20% of cement kiln dust. When the temperature was raised to 80°C, mixes containing twenty percent kiln dust experienced expansions of 0.2%, twice the set limit of 0.1%. Since the type of conditions in this test simulated steam cured precast concrete used in transportation systems, their results suggested that the curing temperature of concrete containing CKD should be controlled to avoid exceeding 54°C.

Corrosion tests have been completed on concrete specimens containing cement kiln dust in a percentage of 6% per wt, with three different water to cement ratios, 0.50, 0.65 and 0.75. Testing specimens exposed to a solution of 3.5% NaCl, Batis, et al., [21], reported that the use of cement kiln dust in concrete decreased the permeability of the concrete, reduced the porosity and that the mass loss in the reinforcing bars was approximately the same as for the reference specimens. However, corrosion rates were accelerated using another CKD with a higher Cl^- content in all water to cement ratios, as expected.

Little information is available on the chloride permeability of concrete containing kiln dust. Ozyildirim and Lane [19] tested the chloride permeability of concrete containing blends of up to 25% fly ash and 20% cement kiln dust, prepared in the laboratory, and two samples of commercially available pre-blended mixes of fly ash and cement kiln dust (FA/CKD) at the fixed ratios of 70/30 and 45/55. All tests were done in accordance with AASHTO T 277. Results indicated that coulomb values at 28 days were in the high range for the control (Portland cement), fly ash concretes (fly ash up to 25%), and 10% cement kiln dust-Portland cement blend, indicating low resistance to the penetration of chlorides. At 1 year, control mixes attained moderate coulomb values, but all fly ash concretes displayed low values

[1] ASTM C 227: Standard Test Method for Potential Alkali Reactivity of Cement-Aggregate Combinations" also known as "Mortar-Bar Method"

indicating high resistance to the penetration of chlorides. Concretes made with the commercial blends exhibited moderate to low permeability values at 28 days, which declined to very low values at later ages indicating very high resistance to chloride penetration. Similar results were obtained with concrete containing cement with high alkali levels (1.25%) indicating that the early coulomb values can be reduced by increasing the alkali content of concrete by incorporating CKD.

Finally, results of the limited studies conducted to measure the resistance of concrete with kiln dust to freezing and thawing generally indicated that concrete containing 5% to 6% cement kiln dust tested in comparison with Type I cement exhibited either the same or lower resistance [12]. Conversely, Batis et al. [21] reported different results when they performed freeze-thaw testing according to ASTM C 666, on specimens containing 6% cement kiln dust of two different types. The mixes containing cement kiln dust with high chloride and sulfate content lost significantly more mass than the control mix indicating very low resistance to freezing and thawing. Nevertheless, they reported behavior similar to the reference concrete for another type of kiln dust.

CONCLUSIONS

CKD is a composite material consisting of highly variable crystalline phases. This makes CKD a complex material to classify and characterize in specific terms. Depending on its reactivity and although results vary widely, it was found that the addition of CKD as a cement replacement has a limit value. With increasing dust value, it adversely affects the physical and mechanical properties of cement pastes. The general trend is that mixes with cement kiln dust in quantities of 15% or less, as a cement replacement, exhibit a very promising behavior, since results indicate that CKD at these quantities act as a filler and has the effect of accelerating hydration. In addition, most work with kiln dusts has shown that kiln dust can be used as a perfect activator in pozzolanic blended cements. However, because aspects of the chemical composition of the CKD may induce reactions such as alkali silica reaction, specifications are required to define both the physical and chemical characteristics needed to ensure acceptable field performance.

ACKNOWLEDGMENTS

The authors gratefully acknowledge the support of the Center for Advanced Cement Based Materials (ACBM), Northwestern University, in carrying out this study.

REFERENCES

1. DYER T.D., HALLIDAY J.E., and DHIR, R.K., An Investigation of the hydration chemistry of Ternary blends Containing Cement Kiln Dust, Journal of Materials Science, Volume 34, Number 20, (1999) pp. 4975-4983.

2. BHATTY, J. I., Alternative Uses of Cement Kiln Dust, RP 327 Portland Cement Association, Skokie, Illinois, (1995).

3. XU, A. and SARKAR, S. L., Active β-C_2S Cement from Fly Ash and Kiln Dust, Fly Ash, Silica Fume, Slag, and Natural Pozzolans in Concrete, Proceedings of the Fifth International Conference, American Concrete Institute, SP 153-12, (1995), pp. 213-227.

4. ASTM D 5050-90, Standard Guide for Commercial Use of Lime Kiln Dusts and Portland Cement Kiln Dusts, Annual Book of ASTM Standards, Philadelphia, Pennsylvania, USA. (1990).

5. DAUGHERTY, K. E., FAUST, R. J., ZGAMBO, T. P., and GRIFFIN, J. G., Potassium Sulfate in Kiln Dust, Alkalis in Concrete, Research and Practice, Proceedings of the Sixth International Conference, Technical University of Denmark, (1983), pp. 85-91.

6. TODRES, H. A., MISHULOVICH, A., and AHMAD, J., Cement Kiln Management: Permeability, Research and Development Bulletin, RD103T, PCA, Skokie, Illinois, (1992).

7. DAUGHERTY, K. E. and FUNNELL, J. E., The Incorporation of Low Levels of By-Products in Portland Cement and the Effects on Cement Quality, Cement, Concrete, and Aggregates, Vol. 5, No. 1, (1983), pp. 14-20.

8. KONSTA-GDOUTOS, M.S., and SHAH, P.S., Hydration and Properties of Novel Blended Cements based on Cement Kiln Dust and Blast Furnace Slag, Cement and Concrete Research, 2003, (accepted for publication).

9. BHATTY, M.S.Y., Use of Cement-Kiln Dust in Blended Cements, World Cement, Vol. 15, No. 4, (1984), pp. 126-128, 131-134.

10. RAVINDRARAJAH, R. S., Usage of Cement Kiln Dust in Concrete, The International Journal of Cement Composites and Lightweight Concrete, Vol. 4, No. 2, (1982), pp. 95-102.

11. JAWED I. and SKALNY J., Alkalies in Cement: A Review, Cement and Concrete Research, Vol. 8, (1978), pp.37-52.

12. RAMAKRISHNAN, V., Evaluation of Kiln Dust in Concrete. Fly Ash, Silica Fume, Slag, and Natural Pozzolans in Concrete, ACI SP-91, American Concrete Institute, Detroit, (1986), pp. 821-839.

13. ABO-EL-ENEIN, S.A., HEKAL, E.E., GABR, N.A., and EL-BARBARY, M.I., Blended Cements Containing Cement Kiln Dust, Silicates Industriels, Volume 9-10, (1994), pp. 265-269.

14. DHIR, R.K., DYER,T.D., and HALLIDAY, J.E., Activation and acceleration of Portland Cement/GGBS blends using Cement Kiln Dust, Modern Concrete Materials: Binders, Additions and Admixtures, Proceedings of the International Conference, Dundee, Scotland, UK, Ed. Dhir and Dyer, 1999, pp.361-370.

15. WANG, K., KONSTA-GDOUTOS, M.S., and SHAH, S.P., Hydration, Rheology and Strength of OPC-CKD-Slag binders, ACI Materials Journal, Vol. 99, No 2, 2002, pp.173-179.

16. BHATTY, M.S.Y., Kiln Dust Cement Blends Evaluated, Rock Products, October 1985, (1985), pp. 47-52, 65.

17. BHATTY, M.S.Y., Use of Cement Kiln Dust in Blended Cements – Alkali-Aggregate Reaction Expansion World Cement, Vol. 16, No. 10, (1985), pp. 386-391.

18. BHATTY, M.S.Y., Properties of Blended Cements Made with Portland Cement, Cement Kiln Dust, Fly Ash, and Slag, Proceedings of the Eighth International Congress on the Chemistry of Cement, Communications Theme 3, Vol. IV, Brazil, (1986), pp. 118-127.

19. OZYILDIRIM, C. and LANE, D. S., Effects of Blends of Cement Kiln Dust and Fly Ash on Properties of Concrete, VTRC96-R1, Virginia Transportation Research Council (1995).

20. SHOAIB, M. M., BALAHA, M. M., and ABDEL-RAHMAN, A. G., Influence of Cement Kiln Dust Substitution on the Mechanical Properties of Concrete, Cement and Concrete Research, Volume 30, (2000), pp. 371-377.

21. BATIS, G., KATSIAMBOULAS, A., MELETIOU, C. A., AND CHANIOTAKIS, E., Durability of Reinforced Concrete Made with Composite Cement Containing Kiln Dust, Concrete for Environment Enhancement and Protection, Proceedings of the International Conference, Dundee, Scotland, UK, R.K. Dhir and T.D. Dyer eds. (1996), pp. 67-72.

THEME TWO:

HYDRATION OF PORTLAND CEMENT

Keynote Paper

SLAG CEMENT: AN ECONOMIC DURABILITY

F Parissi

Cementir

G Frigione

University of Calabria

Italy

ABSTRACT. The use of alternative raw materials and fuels in the production of cement is of growing importance. However, it highlights the problem of the presence of hazardous compounds as heavy metals and chloride. Slag cement - a blended cement containing at least 65 ÷ 70% of granulated blast-furnace slag - has two advantages: low energy consumption in its manufacture and high durability compared to Portland cement. Slag cement concrete durability is due to high impermeability. The dramatic decrease in the ability of water to penetrate concrete made with slag cement is the result of the successful development of a highly refined pore structure of the cement paste. Based on this, the leaching of heavy metals from concrete can also be dramatically reduced if slag cement is used.

Keywords: Alternative fuel, Ashes, Chloride, Durability, Heavy metals, Portland cement, Slag cement, Waste.

F Parissi is the General Technical Manager of Cementir - Rome, (Italy). Cementir has more than fifty years experience in the production of Portland, slag and pozzolanic cements. The company currently has four factories in Italy and two in Turkey.

G Frigione is Professor of Material Science at the University of Calabria (Italy). His 40 years' research extends over a wide range including the chemistry of Portland and blended cements. He has written numerous publications comprising chapters on blended cements, their use and long term durability.

INTRODUCTION

Waste management policies in the EU countries are enhancing the utilisation of alternative materials in the cement production process due to environmental safety and economic profitability. Alternative materials may be used as fuels and as raw materials.

The range of alternative fuels is very large and, among others, the following ones can be mentioned: used tyres and rubbers, used oils, paints, organic industrial solvents from paint and coatings, ink and printing, cosmetics, toys, paint thinners. Waste plastics and wood are another fuel sources. The list of candidate materials for use in a waste fuels program continues to expand. RDF (Refuse derived fuel) is amongst the utilised ones in cement kilns. The refuse consists of several components depending upon the life style of the population.

The valorisation of wastes presents a problem of increased importance due to the presence of heavy metals in the cement and therefore there are concerns about their leaching from concrete.

To use alternative materials in the cement production, the following has to be avoided:

- Any release of hazardous element from concrete to prevent water and soil contamination;
- Corrosion of steel reinforcement due to elevated chloride content,
- Shattering of concrete by aggressive agents.

The aim of this paper is to present how hazardous components like heavy metals and chloride are blocked in concrete, mainly when produced utilising slag cement.

WASTE IN CEMENT KILNS

Many investigations [1-4] indicate that most hazardous elements are incorporated in the clinker phases during the burning process, particularly in silicate and aluminate networks [5-7]. Distributions of Cr and Ni are of particular interest.

Cr is preferentially found in the silicate phase, especially in belite. Because of the oxidising condition during the clinker burning process, it is possible that at least some of the Cr is oxidised to hexavalent Cr [7-9], while according to Feng Xiuji and Yan Peiyu [10] under oxidant conditions, Cr is present in C_2S as Cr^{4+} and Cr^{5+} co-ordinated in Si distorted tetrahedra. During hydration, Cr^{4+} and Cr^{5+} disproportionate to mixtures of Cr^{3+} and Cr^{6+}.

According to Stephan et al [7] a new compound $MgNiO_2$ was detected in Portland clinkers. Possibly the $MgNiO_2$ is dissolved in MgO and forms a nonstoichiometric compound. The remaining Ni is in the matrix.

Andrade et al [11] have synthesised experimental clinkers with the addition of vanadium, zinc and lead to a standard raw meal, from which a control clinker was obtained for comparison. The three metals were chosen as they are present in the petcoke (vanadium) and in industrial wastes (zinc and lead) commonly burned in cement kilns. It was observed that vanadium has shown a preferential partition towards C_2S. Zinc appears in C_3S and in C_2S although zinc appears in higher amounts in periclase. Lead concentrates in minute spherules and also into C_3S in small amounts. The Authors [11] report "Leaching experiments performed on cements made from these vanadium, zinc and lead added clinkers revealed a low mobility of the metals incorporated in the clinker phases".

WASTE IN HYDRATED CEMENT

During the hydration process, the hazardous components remain blocked in the unhydrated portion or in the newly hydrated compounds.

The source of the chemical immobilisation potential arises from:

- Sorption on high surface area cement phases , e.g. CSH;
- Structural incorporation into one or more of the cement hydrated phases;
- Precipitation, by reaction with one or more of the cement constituents.

The crystalline cement hydrated phases have structures that permit some crystallochemical substitution. Since the level of toxics, which need to be incorporated, is frequently low, solid solution could be an important mechanism for fixation. Ettringite, the AFt phase, has a range of structural sites favourable for incorporation of both anions and cations [12]. Precipitation is, however, the single most effective chemical mechanism of fixation. Elementary chemical considerations suggest that a wide range of metal ions will precipitate at the high pH characteristic for cement systems [13].

Both sulphate containing phases, monosulfate and ettringite are known to allow ionic substitution including Fe^{3+} and Cr^{3+} for the Al^{3+} and SeO_4^{2-}, CrO_4^{2-}, AsO_4^{3-}, VO_4^{3-}. for SO_4^{2-} [14-18]. The widespread use of cement and concrete and the ability of monosulfate to structurally incorporate toxic ions make it important with regard to the mobility of pollutants in the environment. Substitution of hexavalent chromium in monosulfate is of particular interest as chromium is a widespread pollutant that is acutely toxic. A number of reports suggest a variety of bivalent cations (Sr^{2+}, Ba^{2+}, Cd^{2+}, Co^{2+}, Ni^{2+} Zn^{2+} and Pb^{2+}) that replace the M^{2+} site in the ettringite crystal structure [13-15]. According to Gougar et al [19] a summary of these substitutions is provided in Table 1.

Table 1 Waste ion substitutions in Ettringite.

$(Ca^{2+})^{VIII}$ SITE	$(Al^{3+})^{VI}$ SITE	(SO_4^{2-}) SITE
Sr^{2+}	Cr^{3+}	$B(OH)_4$
Ba^{2+}	Ni^{3+}	Cl^-
Pb^{2+}	Co^{3+}	CrO_4^{2-}
Cd^{2+}	Ti^{3+}	AsO_4^{3-}
Co^{2+}		SeO_4^{2-}
Ni^{2+}		VO_4^{3-}
Zn^{2+}		MoO_4^{2-}

Alkali cations are thought to substitute in interlayer sites of C-S-H [20]. C-S-H gel isolates ions and salts of the waste species through a variety of mechanisms, including sorption, admixing of other phases, and substitution at interlayer sites. Table 2 summarises [19] the ion incorporation by C-S-H. The waste species are listed in elemental form but may actually be incorporated into the C-S-H structure as a monatomic ion, a polyatomic oxy-anion, or a salt.

The porosity and pore structure of cement pastes exert major control on the diffusion of potentially deleterious substances, or the release rate of potentially harmful species. The pore structure and microstructure constitute major rate-controlling factors for potentially harmful transport processes. However, the pore size distribution is probably the most important factor of material transport and is not a direct function of the porosity [21].

Table 2 Species immobilised by C-S-H.

Li, Na, K	Mg, Ni, Co, Hg, Zn, Cd	Cr, Pb

Serclérat et al [22] made a clinker from an industrial raw meal that had been enriched in chromium, zinc and lead. They made extraction tests on mortar samples crushed to 100 μm to assess the fixation of the metals in the matrix itself by purely chemical mechanisms, after elimination of the entire transport phenomenon. They reached the following conclusions:

- Zinc is bound in the solid and is nearly insoluble in the chemical conditions applied; therefore its release is very low. The result has been extended to leachants in the pH range 7-13;
- Lead is nearly unreleased: the cement matrix insolubilizes this metal, provided the leaching pH is under 12.5. It must be pointed out that in real conditions of use of cement materials, the contact water never reaches such elevated pH values.
- The trivalent chromium is bound in the cement matrix.
- The chromate ions are partly retained in the matrix owing to their fixation in the ettringite.

According to Roethel and Breslin [23], applications on the field have shown that particulate residues of combustion can be combined with cement and stabilised to form a solid block. The stabilisation process has been shown to effectively prevent the release of metals from stabilised residues of combustion block, and may provide an alternative method for the disposal or reuse of the by-products of combustion. The blocks mix consisted of 85% combined ashes from waste incinerator and 15% Portland (ASTM Type II) cement. After one year the blocks - destined for reef construction - were submerged in seawater. Periodically, analysis was performed to detect any release of metals from the blocks. Results show that even after prolonged seawater exposure these metals of environmental concern were effectively retained within the cementitious matrix. The reef structures were rapidly colonised by a diverse community of marine organisms and provided suitable habitat for a wide variety of fishes and crustaceans.

Dyer et al [24] used the incinerator bottom ash (IBA) in the production of masonry units. The blocks were cast with IBA as aggregate and different cement contents (140, 170 and 200 kg). Samples from the concrete masonry units were subjected to the CEN protocol-leaching test. This test is designed to investigate the leaching behaviour of concrete in contact with drinking water. The results (Table 3) show that the Pb, Zn, Cu and Cd leachate service concentrations were well below the EU maximum.

Table 3 Service concentrations derived from leach tests (CEN protocol)

MIX		LEACHATE SERVICE CONCENTRATION				
Cement kg	MWIBA %		Pb ppb	Zn ppb	Cu ppb	Cd Ppb
		EU limit max conc.	50	5000	3000	5
200	100		3	8.5	13.2	0
170	100		3.7	10.9	11.7	0
140	100		3.1	20	4.4	0

More recently, ashes from thermal treatment plants of sludge from municipal wastewater have been introduced as cement or sand replacement in concrete. Tay [25] used ashes (10%) from the incineration of municipal sludges as filler in concrete. Monzò et al [26] used 15% sewage sludge ash as cement replacement in mortars. The possibility of reusing pelletised industrial waste as coarse aggregate in concrete has been reported in the literature [27]. The sludge contained 1.38% wet wt of Zn, 1.41% wet wt of Cu and 0.62% wet wt of Ni: however, the leaching of metals from the specimens was similar to the control mix, which used granite as coarse aggregate.

Halliday and Dhir [28] used incinerator bottom ash as a replacement for lightweight natural aggregates in the production of lightweight block. The percentage of ash used was 20% and 50%. The service concentrations, derived using CEN leachate test, showed that all species, that are detrimental to the environment, were found in quantities below the limits set by the European Union Drinking Directive [29].

SLAG CEMENT ADVANTAGES

Slag cement - this term is used for blended cements containing at least 65 ÷ 70% of granulated blast-furnace slag - has many advantages, beyond that of reducing raw materials, energy consumption and emission of CO_2. The C_3A phase in Portland cement is known to react with chloride ions to produce the AFm phase $C_3A.CaCl_2.10H_2O$ or Friedel's salt. There is generally a good correlation between C_3A content and chloride binding capacity [30]. Dhir et al [31] reported that "GGBS pastes have a higher chloride binding capacity compared with the Portland cement". The high aluminium oxide fractions in the slag, which are able to bind chloride in the form of the not readily soluble Friedel's salt, are assumed to be responsible for the increased chemisorptive and chemical binding capacity [32]. This is the reason why the slag cement has the capacity of reducing the free chloride, potentially responsible for corrosion of steel reinforcement.

Moreover, the diffusion phenomena of metal ions in hardened cement pastes are important to the immobilisation of heavy metals within cementitious materials. The diffusion rate of ions is affected by the microstructure of hardened pastes, particularly by the pore size distribution and porosity. The diffusion coefficient of ions through porous materials has a close relation to the volume of very small pores - with a radius of less than 2 nm - and does not to the total volume. Slag cement pastes have a high gel pore volume and significantly lower capillary pore fraction as compared with Portland cement pastes [33]. It results in lower water permeability [34]. So it may be expected to be a good material for solidifying chloride and heavy metal ions.

Moreover, the pozzolanic reactions between clinker portland and slag increase the quantity of CSH gel produced and, compared to Portland cement, increases the ability of concrete to immobilise hazardous anions and cations firmly. According to Gheorghe et al [35] stabilisation of Zn seems to be favourable influenced by increasing the granulated blast furnace slag in blended cements.

Kindness et al [36] showed that the ability of Portland cement matrices to immobilise Cr-containing wastes depends on the oxidation state of the chromium. Cr(III) is most effectively incorporated into the solid phases, leaving low levels of Cr in the pore fluid of cements. Cr(VI) is less readily incorporated into cement hydrate phases, hence pore fluid from Cr(VI)-spiked cements contain relatively high levels of Cr. The reducing conditions in the hydrated slag cement pastes (presence of S ions) have the ability to reduce Cr(VI) to Cr(III), which subsequently becomes incorporated into the cement hydrates [37, 38]. Consequently, the concentration of Cr in the pore fluid of slag blended cements, irrespective of its original form, is very low.

Since bottom ashes have elevates alkali content, the use of slag cement will reduce the potential of ASR occurring by:

1. Consuming alkalis in the hydration process, making them unavailable for the alkali silica reaction.
2. Reducing pore size and mobility of the alkali.

Still, the Al metal in the ashes oxidises in the alkaline cement environment, producing H_2, which has an expansive effect [39]. Since the setting time of slag cement is very long, the possible expansion happens when the mass is still plastic, without any cracks.

The presence of $Ca(OH)_2$ in the cement matrix is of major importance for the leaching. After dissolution, the removed $Ca(OH)_2$ generates extra porosity facilitating transport of species by diffusion through the leached shell and resulting in an increasing progress of the dissolution front in the matrix. In the slag cement, blast furnace slag reacts with the $Ca(OH)_2$ released during Portland clinker hydration and forms CSH. According to [40, 41] cements containing more than 50-60% of slag have a pozzolanic behaviour and satisfy the pozzolanicity test (EN 196-Part 5) index of the practical absence of $Ca(OH)_2$ in the hydrated paste.

The action of aggressive agents is of special importance due to theirs effects on the integrity of the cement solid phases. The degraded material originated by the water action may facilitate the transport of toxic components from the material to the environment. Therefore, there is an interest to inglobe hazardous waste incineration in cements that have high durability. Slag cement concrete is known for its high resistance to several aggressive agents [42].

Sulphate and seawater resistance: The use of slag cement protects concrete from sulphate and seawater attack in many ways:

1. Slag cement contains a reduced amount of C_3A,
2. Slag cement reduces concrete permeability, making it harder for sulphates and chlorides to penetrate into concrete.
3. Pozzolanic reaction increases the total amount of calcium-silicate hydrate gel and reduces the amount of $Ca(OH)_2$ in the system.
4. Other authors [43] believe that kinetics and degree of crystallisation of ettringite in slag cements are different from those in Portland cement.

Acid Rain resistance: Due to the previously mentioned of $Ca(OH)_2$ content and high impermeability, slag cement concretes present high resistance to dilute acids (pH range 4.0 – 5.5) [44, 45].

Thermal cracking: The heat of hydration of slag cement is lower than that of Portland cement. The reduction of temperature rise and the lower elastic modulus at early ages, prevent the development of cracks. Microcracks are the primary reason for concrete permeability.

Alkali-aggregate expansion: Smolczyk [45] indicates that 2.0% of equivalent Na_2O is a safe limit if slag content in the cement is higher than 65%. Frigione et al [46] have confirmed that cements with 70% slag show excellent resistance to alkali-aggregate reaction in spite of high-alkali contents, and the alkali content in the clinker does not influence the degree of expansion.

CONCLUSIONS

Alternative raw materials and fuels are increasingly present in the cement production process arising the problem of presence of hazardous element. The production of concrete made with slag cement reduces the risks of leaching due to higher durability and to blocking effect against hazardous element. It is also to be mentioned the reduction of energy required to produce slag cement compared to Portland cement.

REFERENCES

1. STEPHAN, D., MALEKI, H., KNOFEL, D., EBER, B., HARDTL, R., Influence of Cr, Ni and Zn on the properties of pure clinker phases. Part I C_3S. Cement Concrete Research, 1999, Vol. 29, pp 545-552.

2. SPRUNG, S., RECHENBERG, W., Schwermetallgehalte im klinker und im zement. Zement. Kalk Gips, 1994, Vol. 47, pp 258-263.

3. TREZZA, M.A., SCIAN, A.N., Burning wastes as an industrial resource. Their effect on Portland cement clinker. Cement Concrete Research, 2000, Vol. 30, pp 137-144.

4. KIRCHNER, G., Behaviour of heavy metals during cement clinker burning. Zement. Kalk Gips, 1986, (10), pp 555-557.

5. KIRCHNER, G., Reaction of cadmium in the clinker burning process. Zement. Kalk Gips, 1985, (9) pp 535-539.

6. MURAT, M., SORRENTINO, F., Effect of large additions of Cd, Pb, Cr, Zn, to cement raw meal on the composition and the properties of the clinker and the cement. Cement Concrete Research, 1996, Vol. 26, pp 377-385.

7. STEPHAN, D., MALLMANN, R., KNOFEL, D., HARDTL, R., High intakes of Cr, Ni and Zn in clinker. Part I. Influence on burning process and formation of phases. Cement Concrete Research, 1999, Vol. 29, pp 1949-1957.

8. FRIAS, M., SÀNCHES DE ROJAS, M.I., Total and soluble chromium nickel and cobalt content in the main materials used in the manufacturing of Spanish commercial cements. Cement Concrete Research, 2002,.Vol. 32, pp 435-440.

9. STEPHAN, D., MALEKI, H., KNOFEL, D., EBER, B., HARDTL, R., Influence of Cr, Ni and Zn on the properties of pure clinker phases. Part II C_3A and C_4AF. Cement Concrete Research, 1999, Vol. 29, pp 651-657.

10. FENG XIUJI, and YAN PEIYU, Effect of the states of chromium ion on the colour characteristics of doped βC_2S. Advances in Cement Research, 1990, Vol. 10, pp 85-88.

11. ANDRADE, F:R:D:, MARINGOLO, V., KIHARA, Y., Incorporation of V, Zn and Pb into the crystalline phases of Portland clinker. Cement Concrete Research, 2003, Vol. 33, pp 63-71.

12. MILLER, F.MG., KLEMM, W.A., Mechanistic considerations for the efficiency of Portland cement in waste stabilization. Proceedings of the 10th International Congress on the Chemistry of Cement, Göteborg, 1997, Vol 4, p 4iv041.

13. GLASSER, F.P., Application of cements to the treatment and conditioning of toxic wastes. 9th International Congress on the Chemistry of Cement, New Delhi, Publisher: National Council for Cement and Building Materials, New Delhi, 1992, Vol VI, pp 114-118.

14. BONEN, D., SARKAR, S.L., The present state-of-the-art of immobilization of hazardous heavy metals in cement-based materials. Advances Cement & Concrete Proceedings. Engineering Foundation Conference, 1994, pp 481-498; American Society of Civil Engineers, New York.

15. BENSTED, J., VARMA, S.P., Ettringite and its derivatives. II. Chromate substitution. Silicates Industriels, 1973, Vol. 37, pp 315-318.

16. KUMARATHASAN, P., MCCARTHY, G.J., HASSETT, D.J., PFLUGHOEFT-HASSET, D.F., Oxyanion substituted ettringites: synthesis and characterization and their potential role in immobilization of As, B, Cr, Se, and V. Fly Ash and Coal Conversion By-Products: Characterization, Utilization and Disposal. VI. Materials. Research Society Symposium Proceedings.1990, pp 83-104.

17. POELLMAN, H., AUER, St., KUZEL, H.J., WENDA, R., Solid solution of ettringites: Part II: Incorporation of $B(OH)_4^-$ and CrO_4^{2-} in $3CaO.Al_2O_3.3CaSO_4.32H_2O$. Cement Concrete Research, 1993, Vol. 23, pp 422-430.

18. REARDON, E.J., An ion interaction model for the determination of chemical equilibria in cement/water systems. Cement Concrete Research, 1990, Vol. 20, pp 175-192.

19. GOUGAR, M.L.D., SCHEETZ, B.E., ROY, D.M., Ettringite and C-S.H- Portland cement phases for waste ion immobilization: a review. Waste Management, 1996, Vol. 16, pp 295-503.

20. TAYLOR, H.F.W., Nanostructure of C-S-H: Current status. Advances Cement Basic Material, 1993, Vol. 1, pp 38-46.

21. ROY, D.M., Cementitious materials in nuclear waste management. 9th International Congress on the Chemistry of Cement, New Delhi, Publisher: National Council for Cement and Building Materials, New Delhi, 1992, Vol VI, pp 88-113.

22. SERCLÉRAT, I., MOSZKOWICZ, P., POLLET, B., Retention mechanisms in mortars of the trace metals contained in Portland cement clinkers. Waste Management, 2000, Vol. 20, pp 259-264.

23. ROETHEL, F.J., BRESLIN, V.T., MSW Combustor Ash Utilization in Marine and Terrestrial Systems. . In: Use incineration ash, Editors RK Dhir, TD Dyer and KA Plaine. Publisher: Thomas Telford, London, 2000, pp 331-339.

24. DYER, T.D., DHIR, R.K., COLLIER, T.C., Precast Concrete Products Produced Using Incinerator Bottom Ash. . In: Use incineration ash, Eds RK Dhir, TD Dyer and KA Plaine. Publisher: Thomas Telford, London, 2000, pp 341-353.

25. TAY, J.H., Sludge Ash as a Filler for Portland Cement Concrete. Journal of Environmental Engineering ASCE, 1987, Vol. 113, pp 345-351.

26. MONZÒ, J., PAYÀ, J., BORRACHERO, M.V., CÒRCOLES, A., Use of Sewage Sludge Ash(SSA)-Cement Admixtures in Mortars. Cement Concrete Research, 1996, Vol. 26, pp 1389-1398.

27. TAY, J.H., HONG, S.H., SHOW, K.Y., Reuse of Industrial Sludge as Pelletized Aggregate for Concrete. Journal of Environmental Engineering ASCE, 2000, Vol. 126, pp 279-287.

28. HALLIDAY, J.E., DHIR, R.K., Full Scale Trials Using Incinerator Bottom Ash in Cement Based Products. In Suitable Concrete Construction, R.K. Dhir, T.D. Dyer, J.E. Holliday Edts, Publisher: Thomas Telford, London, 2002, pp 429-438.

29. EUROPEAN UNION, Quality of water intended for uman consumption. EU Council Directive (98/83/EC), 1998.

30. RASHEEDUZZAFAR, AL-SAADOUN, S.S., AL-GAHTANI, A.S., DAKHIL, F.H., Effect of Tricalcium Aluminate Content of Cement on Corrosion of Reinforcing Steel in Concrete. Cement Concrete Research, 1990, Vol. 20, pp 723-738.

31. DHIR, R.K., EL-MOHR, M.A.K., DYER, T.D., Chloride Binding in GGBS Concrete. Cement Concrete Research, 1996, Vol. 26, pp 1767-1773.

32. LUO, R., CAI, Y., WANG, C., HUANG, X., Study of Chloride Binding and Diffusion in GGBS Concrete. Cement Concrete Research, 2003, Vol. 33, pp 1-7.

33. UCHIKAWA, H., Effect of Blending Components on Hydration and Structure Formation. 8th International Congress on Chemistry of Cement, Rio de Janeiro, 1986, Vol. 1, pp 249-280.

34. ROY, D.M., Hydration, Microstructure and Chloride Diffusion of Slag Cement Pastes and Mortar. In: Fly Ash, Silica Fume, Slag and Natural Pozzolans in Concrete. Ed. V.M. Malhotra. ACI SP-114, American Concrete Institute, Detroit, 1983, Vol 2, pp 1265-1281.

35. GHEORGHE, M., VOINITCHI, D., TEODORESCU, R., The Hardening and Leaching Properties of the Cementitious Matrix and of the Concrete with Electroplating Waste Admixtures. In: Suitable Concrete Construction, R.K. Dhir, T.D. Dyer, J.E. Holliday Editors, Publisher: Thomas Telford, London, 2002, pp 373-382.

36. KINDNESS, A., MACIAS, A., GLASSER, F.P., Immobilization of Chromium in Cement Matrices. Waste Management, 1994, Vol. 14, pp 3-11.

37. DEJA, J., Immobilization of Cr^{6+}, Cd^{2+}, Zn^{2+} and Pb^{2+} in alkali-activated slag binders. Cement Concrete Research, 2002, Vol. 32, pp 1971-1979.

38. ALLAN, M.L., KUKACKA, L.E., Blast furnace slag-modified grouts for in situ stabilization of chromium-contamined soil. Waste Management, 1995, Vol. 15, pp 193-202.

39. JOHNSON, C.A., A geochemical perspective on the factors that control long-term emission from municipal solid waste incinerator residues. In: Use incineration ash, Editors RK Dhir, TD Dyer and KA Plaine. Publisher: Thomas Telford, London, 2000, pp 77-86.

40. PARISSI, F., BARONE, P., Dimostrazioni sperimentali per vie diverse della pozzolanicità delle loppe d'altoforno. Industria Italiana del Cemento, 1964, Vol. XXXIV, pp 887-892.

41. FRIGIONE, G., Pozzolanicity test on pozzolan and blast-furnace slag cements: the influence of gypsum content. Il Cemento, 1995, Vol. 92, pp 71-75.

42. FRIGIONE, G., Manufacture and Characteristics of Portland Blast-Furnace Slag Cements. In: Blended Cements, ASTM STP 897,. G. Forhnsdorff Edt., ASTM, Philadelphia, 1986, pp 15-28.

43. REGOURD, M., HORNAIN, H., MORTUREUX, B., (1977), Resistance a l'Eau de Mer des Ciments au Laitier. Silicates Industriels, 1977, Vol. 42, pp 19-27.

44. GUTT, W.G., HARRISON, W.H., Chemical Resistance of Concrete. Concrete, 1977, Vol. 11 (5), pp 35-37.

45. SERSALE, R, FRIGIONE, G., BONAVITA, L., Acid Depositions and Concretes Attack: Main Influences. Cement and Concrete Research, 1998, Vol. 28, pp 19-24.

46. SMOLCZYK, H.G., Structure et Charactérisation des Laitiers. 7° Congrès International de la Chimie des Cìments, Paris, 1980, Vol 1, III-1/3.

47. FRIGIONE, G., MAROTTA, R., Portland blast- furnace slag cements and resistance to the alkali-aggregate reaction. World Cement Technology, 1981, Vol. 12, March, pp 73-78.

NON-CONTACTING RESISTIVITY MEASUREMENT FOR HYDRATION OF CEMENT-BASED MATERIALS

X Wei

Z Li

Hong Kong University of Science and Technology

China

ABSTRACT. The change of electrical resistivity can be used to monitor the hydration in cement-based materials. A non-contacting electrical resistivity measurement device was invented to detect the hydration of cement-based materials. This measurement overcomes the polarization effect and contacting problem between specimen and electrode in conventional method. The principle, equation derivation, experimental process about the non-contacting method is also given in this paper. An application of electrical resistivity used to interpret the effect of calcium sulfate on the hydration of cement.

Keywords: Non-contacting, Resistivity, Hemihydrate, Gypsum, Hydration

Xiaosheng Wei is a PhD student in the Department of Civil Engineering at the Hong Kong University of Science and Technology. His research interests include cement chemistry and concrete modelling.

Zongjin Li is a associate professor in the Department of Civil Engineering at the Hong Kong University of Science and Technology. His research interests include constitutive relationship, cement chemistry, HPC and non-destructive.

INTRODUCTION

The resistivity has used to study and analyse the hydration of cement-based materials since the 1930s [1-5]. The electrical methods for cement-based material can be basically divided into three stages based on their property.

DC Method

In 1827, Georg Ohm defined an empirical relationship between the current flowing through a wire and the voltage potential required to drive that current.

$$V=IR \quad (1)$$

In principle, it is relatively simple to measure the resistance of a sample. Connect a battery to a wire of known voltage and then measure the current flowing through the sample. The voltage divided by the current yields the resistance of the sample. The geometrical-independent quantity that is used is called resistivity and is usually indicated by the Greek symbol (ρ).

In the case of a sample, resistivity (ρ) is defined as the resistance in the sample, times the cross-sectional area of the sample, divided by the length of the sample. The units associated with resistivity are thus, ohm - m (ohm - meters).

Resistivity (() is a fundamental parameter of the material making up the sample that describes how easily the sample can transmit an electrical current. High values of resistivity imply that the material making up the sample is very resistant to the flow of electricity. Low values of resistivity imply that the material making up the sample transmits electrical current very easily.

The DC approach is to place two metal contacts (called electrodes) on the end of casting cement-based material sample, apply direct current voltage V and measure the current I. The relationship between (and R is seen by current flowing in a sample of material of cross section S and length L (Figure 1) .The sample resistance, R, is V/I and the resistivity can calculated based on the equation (2).

$$\rho = V/I*(S/L) \quad (2)$$

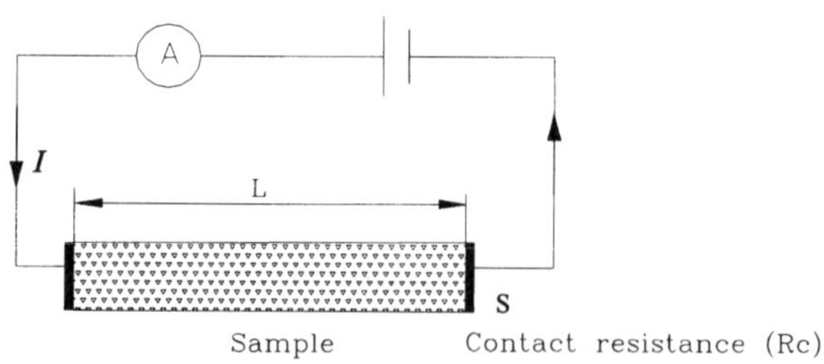

Figure 1 The contacting method

The basic problem with this approach is that there is an electrical resistance associated with the contact between the metal electrode and the sample. The current is carried by the electrons in the metal and by ions in the pore solution of the sample and charge is carried across the interface by an electrochemical reaction, which appears a contact resistance (Rc). At the same time, the contact resistance is frequency dependent. For high conductivity sample such as young concrete, the contact resistance is bigger than the sample resistance. It is difficult to measure them independently. And electrode polarization will occur when applied a DC. A metallic electrode, like a copper or steel rod, in contact with an electrolyte other than a saturated solution of one of its own salts, like cement liquid, will generate a measurable contact potential. These contact potentials can be larger than the natural potential that trying to record. These potentials can be a significant fraction of the total potential measured. For those reasons the two-electrode method is not suitable for most DC and low frequency AC.

AC Method

To overcome the DC electrodes problem, using AC voltage as the probing signal is more reasonable. Recent years, there are many papers concerned this subject [6, 7]. The AC analogy to Ohm's law is:

$$I=V/Z \tag{3}$$

where Z is the impedance of the circuit, V and I are the rms or effective values of the voltage and current. Associated with the impedance Z is a phase angle φ, so that even though Z is the also the ratio of the voltage and current peaks, these peaks do not occur at the same time. The phase angle is necessary to characterize the circuit and allow the calculation of the average power used by the circuit.

Obviously, the contact resistance (Rc) between electrodes and specimen plays an important role in determining the accuracy of the measurement. Loose contact will lead to inaccurate (Figure 2). To improve the contact, some one insert fresh cement paste or colloidal graphite between the electrodes and concrete [8] [9], and others use external force to fasten the electrodes and concrete specimen from two ends.

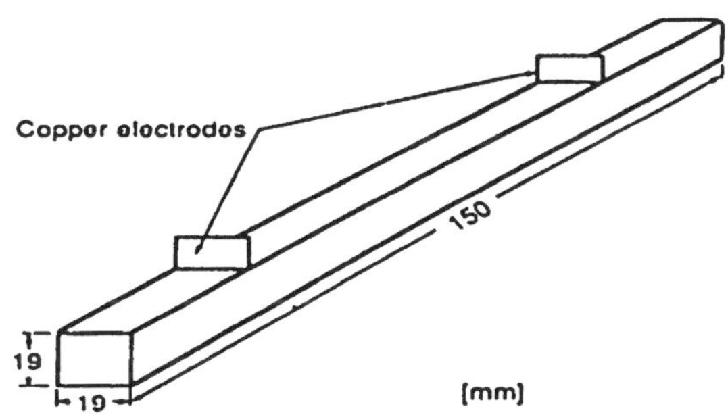

Figure 2 The typical method using by early researchers

These measures are only effective at the beginning of experiments. As the hydration of cement continues, the shrinkage will occur, which can lead to some cracks appeared in the electrode-concrete boundary. If this phenomenon occurred, the experiment cannot be continued. In addition, the cement paste is high alkaline (pH: 12.5-13.5), the electrodes can be gradually corroded within several days. Due to the poor connection problem, the data obtained by this kind of method are inaccurate. Besides, fresh cement paste and colloidal graphite can diffuse toward inside the measured specimen; this situation affects the measured results.

Non-contacting method

To overcome the contacting problem, a non-contacting electrical resistivity measurement device was invented recently by Zongjin Li and Wenlai Li[10]. It successfully solves the contact problem and accurately monitor the electrical resistivity ρ(t) of cement-based material.

The schematic of resistance measurement for novel method is shown in Figure 3. The novel method has no electrodes. The cement-based specimen is a ring with a trapezoid or rectangular section. The specimen ring acts as a secondary of transformer. Transformers are used to increase or decrease AC voltages and currents in circuits. The operation of transformers is based on the principal of mutual inductance. A transformer usually consists of two coils of wire wound on the same core. The primary coil is the input coil of the transformer and the secondary coil is the output coil. Mutual induction causes voltage to be induced in the secondary coil.

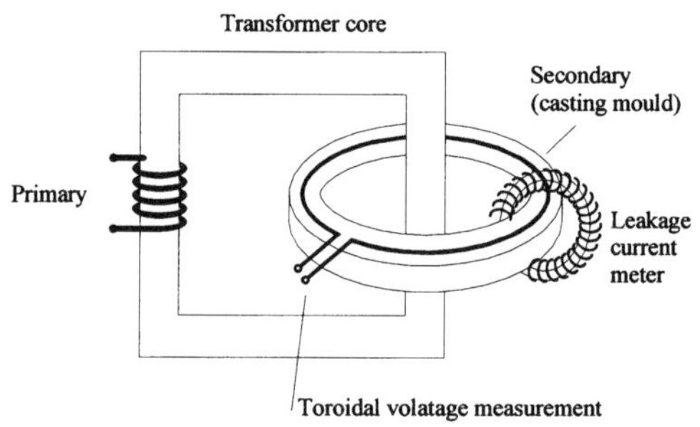

Figure 3 Schematic of non contact resistivity measurement

When an AC voltage is applied on the primary of transformer, a toroidal voltage (V) will be inducted in the secondary, i.e. in specimen ring. Subsequently, a toroidal current (I) will be occurred inside the specimen ring. Supposing the impedance of specimen ring is Z;

$$Z=V/I \tag{4}$$

For such a specimen ring, its inductance can be negligible for 1KHz frequency used. If a capacitance component caused by polarization effect exists. This principle can be represented by an equivalent circuit as showed in Figure 4.

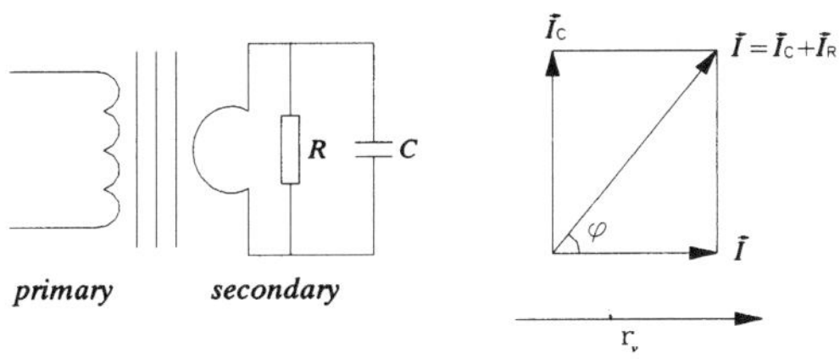

Figure 4 Equivalent circuit

In Figure 4, R is resistance, C is capacitance

$$V = R\, I_R\ ,\quad \vec{I} = \vec{I}_c + \vec{I}_R$$

The impedance of specimen is the parallel connection of R and C.

$$Z = R[(\omega C R)^2+1]^{-1/2} = R[(2\pi f C R)^2+1]^{-1/2}$$

Where:

Z impedance of specimen ring
R the resistance of specimen ring
C the capacitance of specimen ring
f probing frequency
According to the phase difference between toroidal voltage V and toroidal current I, the R can be obtained.

$$R = V/(I \cos\varphi) = V/I \qquad (5)$$

Where V: toroidal voltage
I: current passed through specimen ring
φ: the phase difference between the V and I

There is no electrode, so the I_c=0 and φ=0. R=V/I, based on this formula, the following electrical resistivity can be obtained based on the equation (5).

DERIVATION OF RESISTIVITY CALCULATION FORMULA

The resistance can be derived from (5). Since the resistance of specimen depends on the geometric parameters. So, resistance is not significant for material researches, the resistivity is a necessary parameter for comparing each other between different specimens. Due to the geometric form of a ring, the resistivity cannot be directly calculated from R. In this circumstance, a formula for resistivity calculation was derived from integral equations.

As showed in Figure 5, the cross section of specimen is trapezoid. In fact, the cross section of specimen ring can be either rectangular or trapezoid. However, a trapezoid section is recommended due to the easy of demolding sample. The procedures of formula derivation for resistivity calculation are as follows.

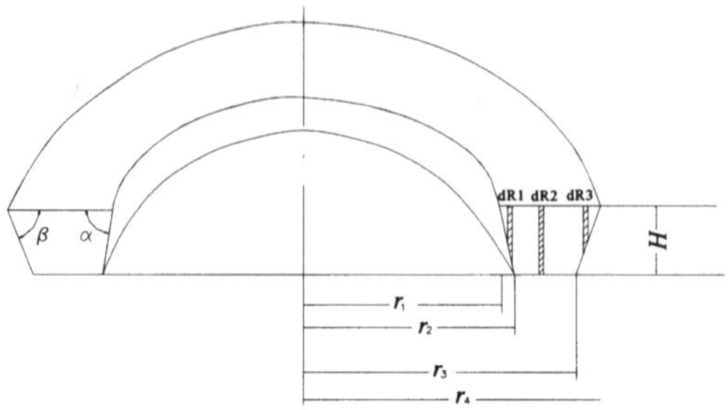

Figure 5 Section of concrete specimen

For a conductor, such as a metal wire, the resistance R can be represented by

$$R=\rho L/S$$

ρ: electrical resistivity of conductor
L: length of conductor
S: cross section of conductor

For a conducting ring with trapezoid section (Figure 5), the trapezoid section can be divided to three area-elements that will be integrated.

dR1, dR2, dR3 are the resistance-elements;
r1, r2, r3, r4 are the radius of sample that are shown in the Figure 5;
dS1, dS2, dS3 are area elements that are shown in Figure 5;
dS=hdr.

The circumference for different radius,

$$dR1=\rho L/dS1=\rho 2\pi r/(h(r)dr)=2\pi r/((r-r1)tg\alpha dr),$$
$$dR2=\rho L/dS2=\rho 2\pi r/(h(r)dr)=2\pi r/(Hdr),$$
$$dR3=\rho L/dS3=\rho 2\pi r/(h(r)dr)=2\pi r/((r4-r)tg\beta dr),$$

Supposing the whole resistance is R, which equals all resistance-elements connecting in parallel.

$$1/R=1/R1+1/R2+1/R3,$$

$$= \int_{r1}^{r2} d(1/R1) + \int_{r2}^{r3} d(1/R2) + \int_{r3}^{r4} d(1/R3)$$

$$= \int_{r1}^{r2} \frac{(r-r1)tg\alpha}{2\pi\rho r} dr + \int_{r2}^{r3} \frac{H}{2\pi\rho r} dr + \int_{r3}^{r4} \frac{(r4-r)tg\beta}{2\pi\rho r} dr$$

$$= \frac{tg\alpha}{2\pi\rho}[(r2-r1) - r1\ln\frac{r2}{r1}] + H\ln\frac{r3}{r2} + \frac{tg\beta}{2\pi\rho}[r4\ln\frac{r4}{r3} - (r4-r3)]$$

$$= -\frac{Hr1}{2\pi\rho(r2-r1)}\ln\frac{r2}{r1} + \frac{H}{2\pi\rho}\ln\frac{r3}{r2} + \frac{Hr4}{2\pi\rho(r4-r3)}\ln\frac{r4}{r3}$$

The resistivity can be obtained

$$\rho = \frac{HR}{2\pi}[-\frac{r1}{r2-r1}\ln\frac{r2}{r1} + \ln\frac{r3}{r2} + \frac{r4}{r4-r3}\ln\frac{r4}{r3}]$$
$$= \frac{H}{2\pi}[-\frac{r1}{r2-r1}\ln\frac{r2}{r1} + \ln\frac{r3}{r2} + \frac{r4}{r4-r3}\ln\frac{r4}{r3}]\frac{V}{I} \qquad (6)$$

The ring with trapezoid section can be made of cement-based materials. Because the resistivities of cement-based material are larger comparing with those of metals (metal conductor $\rho 10^{-8} \sim 10^{-6} \Omega$m). In addition, the electrical current is produced by ion movement and the frequency used in the experiments is lower (1-20kHz). The current density in the specimen section is proportional to 1/r. These initial conditions can satisfy the requirements of the equations used in derivation process. When the applied voltage V is known, and the electrical current I is detected by the leakage current meter, the resistivity ρ will be calculated based on equation (6).

MEASURING SYSTEM

The measuring system of novel method is showed in Figure 6, which consists of generator, amplifier, oscilloscope, transformer core, current sensor, data acquisition interface and computer.

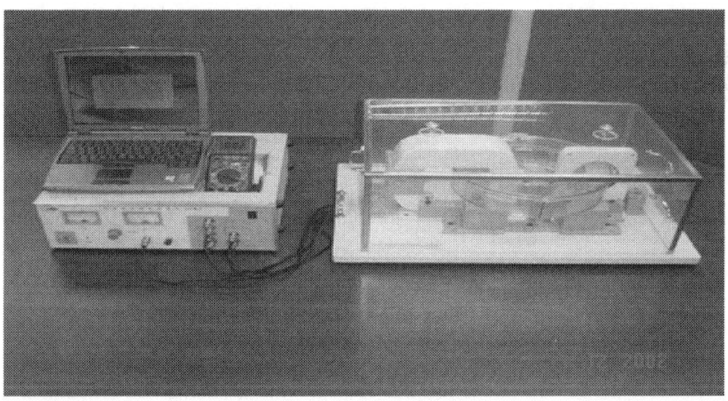

Figure 6 No-contacting resistivity measurement system

The generator produces a sine wave with 1KHz frequency. Amplifier magnifies the sine wave to match with transformer primary. Data acquisition interface samplings the data needed and transfers them to computer. The sampling period is 1 minute or 5 minutes.

Oscilloscope is used to observe the phase between toroidal voltage and current. This system can operate continuously. In order to understand this system clearly, it is necessary to give the relevant parameters of this system:

- The signal applied on primary of transformer is a sine wave of 1KHz frequency.
- The turn number of transformer primary is 36. Secondary toroidal voltage is 1.2v. The measuring range of leakage current clamp are 0-100mA, resolution is 0.1mA.
- Before measuring resistances of cement-based specimens, a circuit of parallel connection by a resistor and a capacitor has been used to inspect accuracy of the measuring system. The values measured by this method, such as resistance, capacitance, are perfectly coincident with nominal values. The phase relations between V and I are correct. When using a pure resistance to simulate the cement specimen. The phase measured is zero. Those results proved that the system is believable.

Generally, the resistivity of cement-based materials is large, when applying a voltage on specimen measured, the current passing through specimen is very weak. A high voltage cannot be applied on it for prevention of electrochemical reaction caused.

The current passed through specimen is measured by a leakage current clamp with high sensitivity. This measuring system is very delicate. In this situation, the electromagnetic shielding has to be needed. Concurrently, it is necessary to prevent vibration of test platform. Besides the problem in contact between specimen and electrodes, two points related to resistivity measurement have to be emphasized: One is polarization effect; other is gas release effect. In the conventional measurement, if using DC probing, an electrical field formed by polarization effect appears in the specimen-electrode boundary, concurrently a significant gas layer will be generated at electrodes. If using AC probing, the polarization effect manifested in a different form, the polarization is in the form of introducing a capacitor in series or in parallel with resistance. Using a small AC voltage of low frequency as the probing signal, which can minimized those affects (electrical field of polarization and gas layer). The two questions are caused by electrode. For novel method, there is no electrode. Ions inside specimen can not get or lose electrons, thus ions can not become atoms or molecules, namely can not yield a gas layer. The specimen measured is a ring, in which the polarized electric-moments are connected in head-end each other, those polarized moments can not formed a strong polarized field. Therefore this method can almost eliminated those affects mentioned above.

Regarding the relation between resistivity measured and probing frequency was done by previous author [11]. The resistivity measured for cement-based materials are independent of probing frequency in the 1K-10MHz range.

HYDRATION OF CEMENT PASTE WITH CALCIUM SULFATE

The non-contacting electrical resistivity can be used to interpret the hydration of cement paste[12], the electrical resistivity $\rho(t)$ drops at the initial stage of hydration, reaches a minimum point, nearly levels and then gradually increases with time. The hydration process can be divided into dissolving period, competition period, setting period and hardening period based on the minimum critical point $M(m,\rho(m))$, critical point $L(l,\rho(l))$ and maximum inflection point $I(i, \rho(i))$. The mineral admixture and chemical influence on the hydration of cement paste were tested, too[13][14]. The following experiment is about the effect of hemihydrate and gypsum on cement hydration.

Materials

In this experiment, 3 different cement samples from the same Portland cements are prepared.

1. Cement [C-0] is Portland cement from Hong Kong; the BET surface area is 1.226m^2/g. The Differential Scanning Calorimetry (DSC) confirmed that it mainly contains gypsum (calcium sulfate dihydrate $CaSO_4.2H_2O$).
2. Cement [C-1] is the machine grinding cement from Cement [C-0]. The procedure is by weighting around 80 g in batches of Cement [C-0] and put into a mechanical ball mill for grinding of 5 minutes each. Because of mechanical collusion during grinding, the ball is so hot that results in gypsum ($CaSO_4.2H_2O$) dehydrating to form the calcium sulfate hemihydrate ($CaSO_4.0.5H_2O$). The phase change during machine grinding is tested by DSC. The BET surface area is 1.855 m^2/g.
3. Cement [C-2] is the hand grinding cement from Cement [C-0]. The procedure is by weighting about 5 g in batches of Cement [C-0] and put into a mortar by hand grinding for 1.5 to 2 minutes each. It mainly contains gypsum as cement [C-0]. The BET surface area is 1.864 m^2/g.

Figures 7 shows the resistivity development of the various cement samples with w/c ratio of 0.4 over the time length of 1440 minutes. While figures 8 focus on the resistivity development of the various cement samples with w/c ratio of 0.4 in the first 250 minutes. Table 1 is a summary showing the global minimum resistivity, resistivity at time 1440 minutes and the time required to reach the global minimum resistivity. And Table 2 shows the initial and final setting time and 1 day compression strength of the 3 different cement samples.

Comparing Resistivity Development for [C-1] with [C-0] & [C-2] Under Same w/c Ratio

Focus on the 1440 minutes resistivity development, [C-0] and [C-2] in Figure 7 shows a dorp at first, convex upward resistivity development, reaching a point of inflextion and then followed by a concave downward trend [12]. With and exception of the cement sample [C-1], it shows a local maximum peak at the beginning of the test, it then disappeared and keep following the convex upward trend, pass through the point of inflextion and then goes into the concave downward region. And we could observe that [C-1] has a lower 1440 mins resistivity compare with [C-0] and [C-2]. We believe that this is due to the effect of hemihydrate and we will further explain this in the latter section.

From the data shown in Table 1, it shows a clear pattern on the minimum resistivity and the 1440 mins resistivity among the 3 samples of cement. By studying the minimum resistivity, we could observe that [C-1] have a higher value than [C-0] & [C-2] at the earlier time. Moreover, from Table 1, we could observe that the time of reaching the minimum point for [C-1] is 2-3 times earlier than [C-0] and [C-2] and this shows a unreasonable stiffening effect and that contributed to rehydration of hemihydrate back to gypsum.

Since solubility of the hemihydrate is 2-3 times more than that of gypsum, they are supersaturated with respect to gypsum in solution. As a result, when they rehydrate back to gypsum, the gypsum being formed is precipitated to some extent from solution, giving the observed stiffening effect.

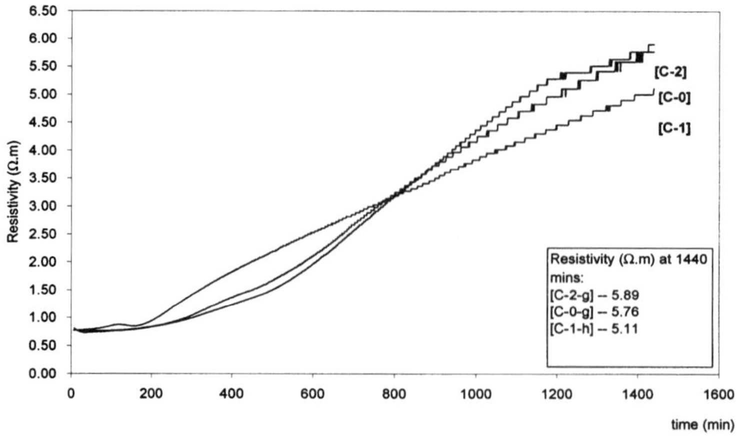

Figure 7 1440-minutes resistivity development for w/c ratio = 0.40 by mass

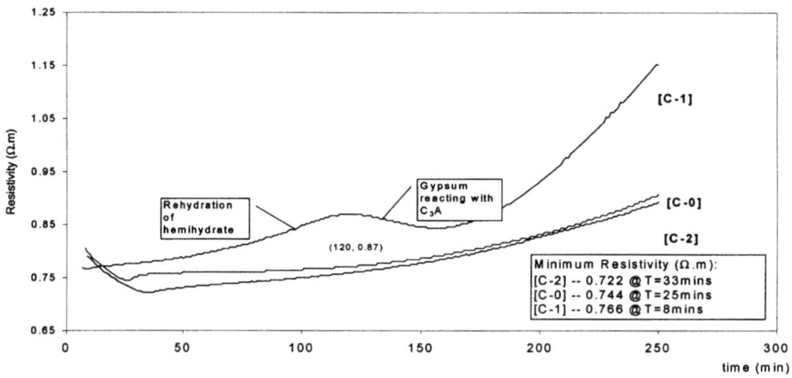

Figure 8 250-minutes resistivity development for w/c ratio = 0.40 by mass

Table 1 Influence of cement sample on minimum resistivity and resistivity at 1440 minutes with different water-cement ratio

CEMENT SAMPLE	MINIMUM RESISTIVITY (Ωm)	TIME TO REACH MINIMUM RESISTIVITY (mins)	RESISTIVITY AT 1440 MINS (Ω.m)
[C-0]	0.744	25	5.76
[C-1]	0.766	8	5.11
[C-2]	0.722	33	5.89

Table 2 Initial setting time, final setting time and 1 day compression strength

CEMENT SAMPLE	INITIAL SETTING TIME (mins)	FINAL SETTING TIME (mins)	COMPRESSION STRENGTH (MPa)
[C-0]	238	318	19.15
[C-1]	147	200	12.91
[C-2]	224	259	25.33

From the Figure 7, the difference on the rate of change of the hydration rate throughout the experimental timeline can be observed. The rate of change of the hydration rate is defined as the slope of the curve at a particular time. By observing Figure 7, we can see that they all show a similar pattern which is [C-1] always has a steeper slope in the convex upward regime than [C-0] & [C-2], while it shows a smaller slope in the concave downward regime than the other 2, result in a lower 1440 minutes resistivity. The lower in the 1440 mins resistivity of [C-1] compared with the other 2 hints a larger macroscopic porosity and a lower in pore tortuosity [12].

Comparing fineness effect the resistivity development for [C-2] with [C-0]

Compare with [C-0], [C-2] have a finer cement content and from the result in Table 1, the minimum resistivity of [C-2] always lower than that in [C-0] and from Figure 8, the curve show a similar rate of dissolution and the time in reaching the minimum resistivity. For [C-2], it needs the longest time to reach the minimum resistivity than that of [C-0]. From these 3 characteristics, we could conclude that they are both undergo the dissolution process and [C-2] has a higher ionic concentrations during this early period due to more particle surface to participate dissolve. As [C-2] has a lower global minimum resistivity compare with [C-0], it means that the conductivity of the matrix solution is higher for [C-2] thus result in a reduction in resistivity. Moreover, the dissolving period for [C-2] is longer than that of [C-0] also proved that the ionic concentrations of [C-2] is higher which need more time to form precipitator to compete with it in order to increase the resistivity. After that, the reaction rate of [C-2] will increase and the slope is larger than that of [C-0].

Setting Time and compressive Strength

Setting Time is the time duration that the cement paste has gained a particular sufficient strength throughout its hydration period. With the variation in the cement fineness and composition especially calcium sulfate state, the setting time will vary, so this is a good measurement for quantifying the early strength achievement. Table 2 illustrates the affect of cement fineness, calcium sulfate state on the initial set, final set and 1 day compression strength of the cement paste. The result shows that [C-1] always have both the earliest initial set and the final setting time compared with [C-0] and [C-2]. Furthermore, [C-2] also has a shorter setting time than [C-0] although the time of these 2 brands are somewhat close to each other. Here we could see that the gypsum state (Dihydrate vs. Hemihydrate) and the fineness have an effect in reducing the setting time of the cement paste which means attaining sufficient strength in a shorter time. Compressive strength is one of the mechanical properties of the cement; from the resistivity measurement we could predict the compressive strength capacity of the cement cube. Table 2 shows the 24-hours compressive strength of the cement cubes made from: [C-0], [C-1] & [C-2] with dimension 40mm x 40mm x 40mm. The results show similar patterns with the 1440 mins resistivity results (Table 1) which is [C-2] cement cubes have the highest compressive strength while [C-1] have the lowest value among the 3 cement samples and this is reflected in the result of the resistivity measurement and the compressive strength test. By comparing the results obtained from the compressive test and the resistivity measurement, it shows a good match in the trend between the 2 tests and they are positively correlated. Thus the resistivity measurement could give an accurate prediction in the trend of strength development of the cement cube.

CONCLUSIONS

The main conclusions from the paper are:

1. Non-contacting resistivity measurement gives an accurate method due to no electrodes being used
2. A formula for resistivity calculation has been derived from an integral equation.
3. Rehydration of calcium sulfate hemihydrate back to gypsum and gypsum reacting with C_3A is reflected in a peak of the resistivity measurement.
4. Resistivity measurement technique is a good tool for measuring (i) the false set of cement and (ii) hydration of cement of different fineness.

REFERENCES

1. DORSCH, K.E., The hardening and corrosion of cement-IV, Cement and Cement Manufacture, 6(4), April, 131-42 (1933)
2. MCCARTER, W.J., The a.c. impedance response of concrete during early hydration, Journal of Material Science, 31 (1996) 6258-92
3. KHALAF, F.M., Wilson, J.G., Electrical Properties of Freshly Mixed Concrete; Journal of Materials in Civil Engineering,11(3)(1999) 242-248
4. ABO, S.A., EI-ENEIN et al, Electrical Conductivity of Concrete Containing Silica Fume; Cement and Concrete Research, 25(8) (1995)1615-1620
5. LEVITA, G. et al, Electrical properties of fluidified Portland cement mixes in the early stage of hydration, Cement and Concrete Research, 30(2000)923-930
6. HUGHS, B.P., SOLEIT, A.K.O., BRIERLEY, R.W., New Technique for Termining the Electrical Resistivity of Concrete, Magazine of Concrete Research, 37(133)(1985) 243-248
7. MCCARTER, W.J., FORDE, M.C., H.W.Whittington, Resistivity Characteristics of Concrete; Proc. Instn Civ. Engrs., Part 2 ,71(1981)107-117
8. WHITTINGTON, H.W.,.MCCARTER, W. J., FORDE, M.C., The conduction of electricity through concrete, Magazine of Concrete research, 33(114)(1981) 48-60
9. BANTHEA, N., DJERIDANE, S., PIGEON, M., Electrical Resistivity of carbon and Steel Micro-Fiber Reinforced Cements; Cement and Concrete Research, 22(1992)804-814
10. LI, Z., LI ,W., No-Contacting Method for Resistivity measurement of Concrete Specimen. US and China Patent in process, 2001
11. WILSON, J.G., WHITTINGTON, H.W., Variations in the Electrical Properties of Concrete with Change in Frequency; JEE Proceedings, 137(5)(1990)246-254
12. LI, Z., WEI, X., LI, W., Interpretation of hydration process of Portland cement using resistivity measurement, Accepted by ACI Material Journal, 2002
13. LI, Z., WEI, X., The electrical resistivity of cement pastes incorporated with mineral admixture, submit to Cement and Concrete Research, 2
14. LI, Z., WEI , X.,The electrical resistivity of cement paste incorporated with retarder, accepted by Journal of Wuhan University of Techology, Materials Science Edition, 2002.

THE HYDRATION PROCESS AND THE MICROSTRUCTURE DEVELOPMENT OF PORTLAND LIMESTONE CEMENTS

P Türker

A Yeşilkaya

A Yeğinobali

TÇMB

Turkey

ABSTRACT. In recent years the utilization of limestone powder as an addition in cement production has increased . The European as well as Turkish standards on "general cements", list "Portland limestone cements" in Group CEM II together with other cements containing mineral additions up to 35 % by mass of cement. Also the cements with other types of additions or the cements in the other groups are allowed to have limestone addition up to 5 % by mass, as a minor constituent. Considered as an inert mineral addition until recently, the increased usage of limestone in cement has also increased the research work on the subject. The findings of these studies indicated that in some instances limestone could actually participate in hydration reactions and increase the rate of hydration. Also it could react with C_3A and form insoluble mono carbo-aluminate which could partially replace ettringite and monosulfate, the usual sulfoaluminates. On the other hand, like other mineral additions, limestone powder replacing cement can cause a reduction in hydration and strength development rates. Appearently the net effect of limestone powder may depend on the type of clinker and the amount of addition. In this study cement pastes were prepared from two types of clinkers (high and low C_3A) ground with gypsum and limestone. Each mixture series will contained 0-35 % limestone powder addition. Periodic heat of hydration measurements were accompanied with studies on paste structure using scanning electron microscope. The test results and observations will be used to determine the effects of limestone addition over the hydration process of cement as influenced by the type of clinker and the amount of cement replacement.

Keywords: Cement performance, Hydration, Microstructure, Portland limestone cements

Pelin Türker is a Senior Microscopist at TÇMB,Ankara,Turkey

Ayşen Yeşilkaya is a Chemical Analysis Specialist TÇMB, Ankara,Turkey

Professor Asım Yeğinobali is a Director of R&D Institute,TÇMB, Ankara,Turkey

INTRODUCTION

Nowadays, blended cements have been produced to an increasing extent in the cement industry. In addition to conventional interground additives, the use of finely ground additives, such as limestone is becoming increasingly important. Since limestone is avaliable at all cement plants and often has no additional transportation costs and also caused enviromental problems, it has been suggested for the usage as a filler in cement. Therefore, to understand the effects of limestone addition on hydration and properties of cement is of main importance.

The primary role of limestone in a cement is that of a filler. Some studies in the literature [1, 2] have been pointed out that limestone is designated inert or quasi-inert, during cement hydration it shows little or no reaction with cement constituents. Accordingly, the filler effect of limestone in cement paste is expected primarily through the reduction of paste porosity. On the other hand, many researchers have been suggested that limestone can participate in the hydration reactions rather than being an inert filler. According to Grandet and Olivier [3], the chemical reaction takes place between calcite and Portland cement paste, leading to carboaluminates formed by the reaction between hydrated calcium aluminate and carbonate ions. Another paper [4] related that calcareous filler has important binding properties which develop during hydration of a mixture of calcite and C_3A by the formation of a hydrated calcium carboaluminate. Confirming this finding, Longuet [5] has reported that according to the definition in the standards, it is necessary to distinguish between inert and active fillers. He has concluded that the effect of limestone filler in cement properties was not only physical but also chemical.

Some authors have tried to explain the influence of finely ground limestone on the physicomechanical properties of filled cement [6]. According to the paper [7] related with the effect of limestone on heat evolution of cement, it was concluded that the addition to 2-10 wt.% limestone to low heat Portland cements enhanced the heat of hydration up to 90 days. Also, in another work referred to in this paper, it was found that the increase in the amount of limestone filler in Portland cements progressively decreases the rate of hydration heat.

In the light of these pieces of information, the present study focused on the effects of limestone filler as also influenced by the type of clinker and the amounts of cement replacement from microstructure and cement performance points of view. Since the effect of limestone powder on the hydration heat of cement is not fully understood, this study would also provide information on the hydration heat of Portland limestone cements.

MATERIALS AND METHODS

In the study, granular limestones were mixed with two types of Portland cement clinker (high and low C_3A) of industrial origin, in 5, 15, 25 and 35 % replacement amounts (by weight). Portland limestone cements were produced by intergrinding clinker and limestone in a ball mill for constant time. Blaine fineness of Portland limestone cements varied in the range between 316-432 m^2/kg. Also all the cements contained 4 wt.% gypsum as a set retarder. The properties of the materials used are given in Table 1(a-b).

Table 1 (a) Chemical and mineralogical composition of the materials used in the study

	LİMESTONE (LF)	CLİNKER 1 (HİGH C_3A)	CLİNKER 2 (LOW C_3A)	GYPSUM
Minerals	Calcite	C_3S - % 59.7 C_2S - % 15.0 C_3A - % 8.6 C_4AF-% 11.8	C_3S - % 68.6 C_2S - % 8.1 C_3A - % 2.2 C_4AF-% 16.4	Gypsum
MgO %	0.09	1.80	1.45	0.24
Al_2O_3%	0.13	5.71	4.28	0.00
SiO_2%	0.57	20.94	20.86	0.00
CaO%	54.61	64.55	64.75	33.68
Fe_2O_3%	0.23	3.87	5.40	0.00
SO_3 %	0.00	0.93	1.14	45.51
K_2O%	0.00	0.72	0.74	0.02
Na_2O%	0.00	0.09	0.11	0.04
LOI%	42.90	0.77	0.67	19.51

Table 1 (b) Blaine fineness of Portland limestone cements, m^2/kg

C1 CEMENT	C1-5LF	C1-15LF	C1-25LF	C1-35LF	C2 CEMENT	C2-5LF	C2-15LF	C2-25LF	C2-35LF
316	341	371	400	420	344	350	390	413	432

(C1 Cement: Cement with high C_3A ; LF: Limestone Filler ; C2 Cement:Cement with low C_3A)

For the microstructural investigations cement pastes with 0.5 water-cement ratio were prepared by the same mixing and moulding method. After curing periods of 7 and 28 days, the hardened cement paste samples were treated with acetone, dried and coated by a thin layer of gold in 10^{-3} torr vacuum. SEM investigations were carried out by a LEO Variable Pressure Scanning Electron Microscope, equipped with an OXFORD Link Isis Energy Dispersive Analyser system by which compounds in the hydrated pastes can be identified.

Heats of hydration of cements at 7 and 28 day, were determined in accordance with ASTM C186 which is a standard based on the measurement of the temperature rise of cement solution as a function of time.

RESULTS AND DISCUSSION

Hydration studies on the samples of hydrated cement paste which differed in the content of the limestone filler, were carried out at 7-day and 28-day.

SEM Observations of Pastes at 7 Days

At 7-day of hydration, it was observed that in cement pastes with 5 and 15% limestone additive, regardless of clinker type, limestone particles effectively fill the voids and reduce the paste porosity when compared with Portland cement pastes (Figures1,2). In 5 % limestone added pastes with C1, it was seen that clinker grains were surrounded by acicular features of Type I C-S-H and limestone particles were covered with the hydration products (Figure 3).This densification around the calcite particles can be attributed to nuclei effect of limestone filler, as confirmed by theliterature [1]. Small sized layered morphology of calcium hydroxides (CH) were widely dispersed in the paste, generally locating very close to calcite particles (Figure 4). The observation of a large amount of small sized CH crystals can be related to accelerated hydration of C_3S and C_2S, by means of limestone. Needle shaped ettringite crystals appeared in porous areas in the paste (Figure3).

Also, around the calcite particles, carboaluminate crystals were found, as detected by EDS analysis(Figures5 a-b). 5 % limestone added pastes with C2 show microstructure a similar to pastes with C1, except that the carboaluminate crystals are not observed. As illustrated in Figure 6, the calcite particles were coated with C-S-H phase and covered with the reaction products. Also layered CH formations occur at the grain boundaries of calcite, revealing partial dissolution of calcite (Figure7). In 15% limestone containing pastes, microstructure were found to be somewhat densified by improved filling of voids by means of calcite particles (Figure 8). In all cement pastes, the increase in limestone addition to 25 and 35 % generally brings about relatively high porosity, as seen in Figure 9. This can be attributed to the dilution of clinker, accompanying with the decrease in hydration products from main clinker constituents.

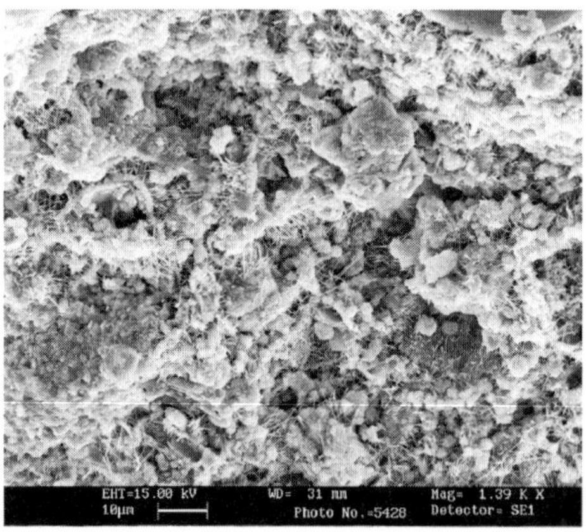

Figure 1 Cement paste, at 7-day

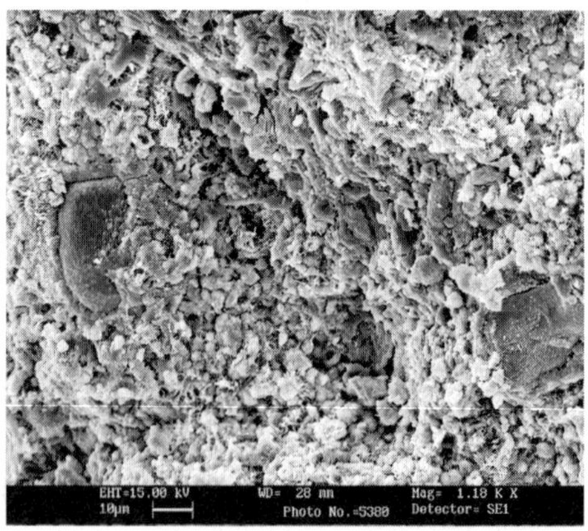

Figure 2 Cement paste with 5% limestone, at 7-day

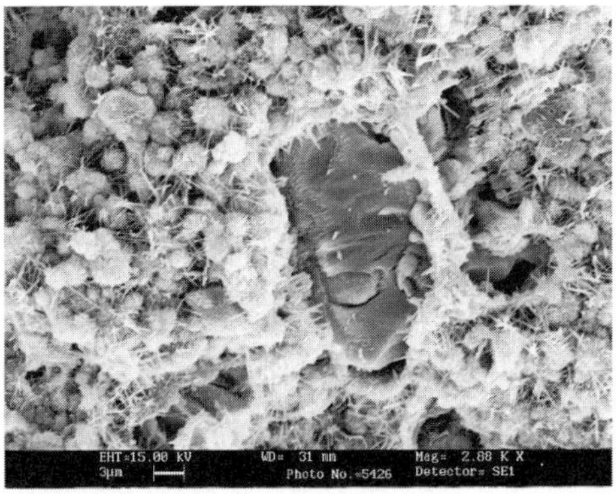

Figure 3 Cement(C1) paste with 5% LF, at 7-day

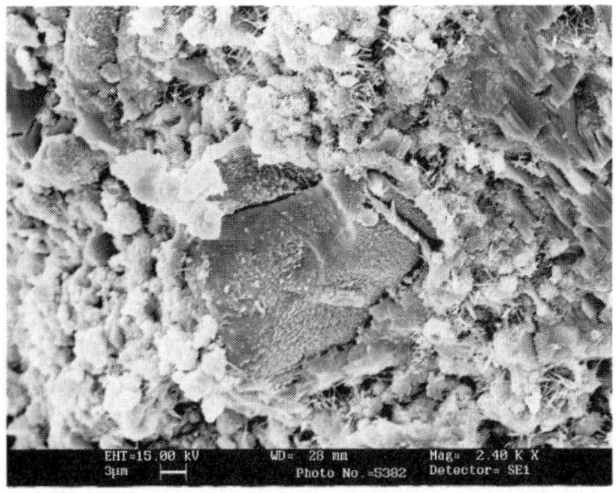

Figure 4 Cement(C1) paste with 5% LF, at 7-day

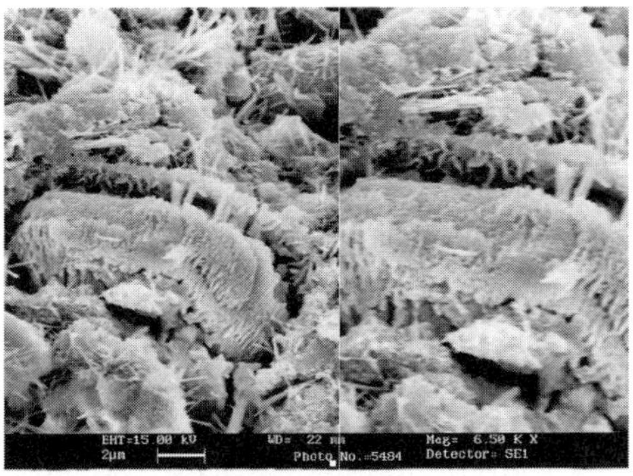

Figure 5(a) Cement(C1) paste with 5% LF, at 7-day

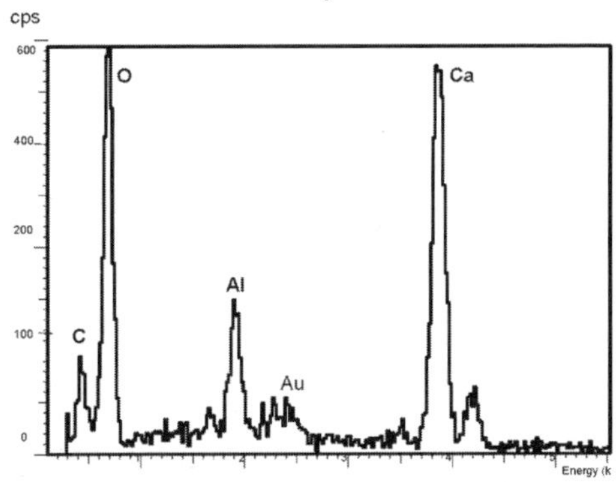

Figure 5(b) EDS spectrum related with Figure 5(a).

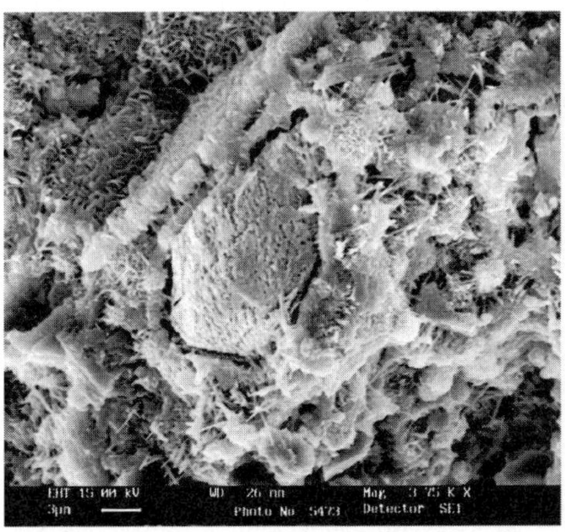

Figure 6 Cement (C2) paste with 5% LF, at 7-day

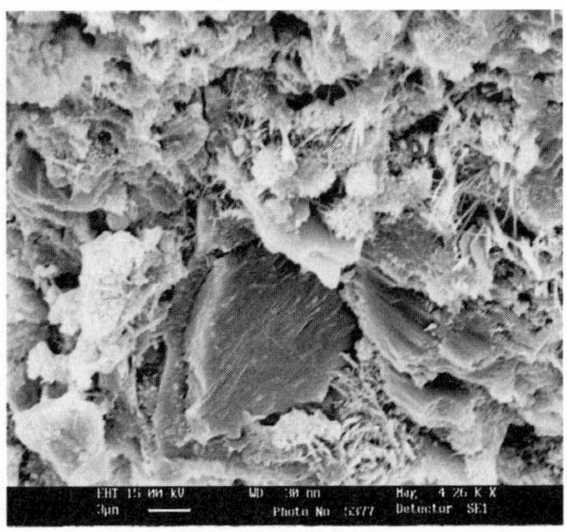

Figure 7 Cement (C2) paste with 5% LF, at 7-day

Figure 8 Cement (C1) paste with 15% LF, at 7-day

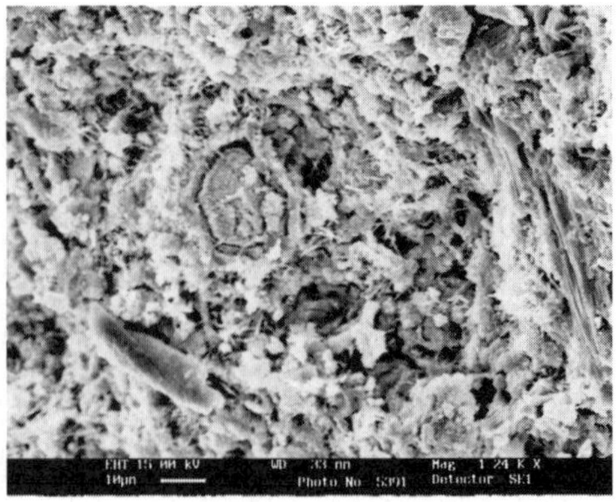

Figure 9 Cement (C2) paste with 35% LF, at 7-day

Figure 10 Cement (C1) paste, at 7-day

Figure 11 Cement (C1) paste, at 7-day

Figure 12 Cement (C2) paste, at 7-day

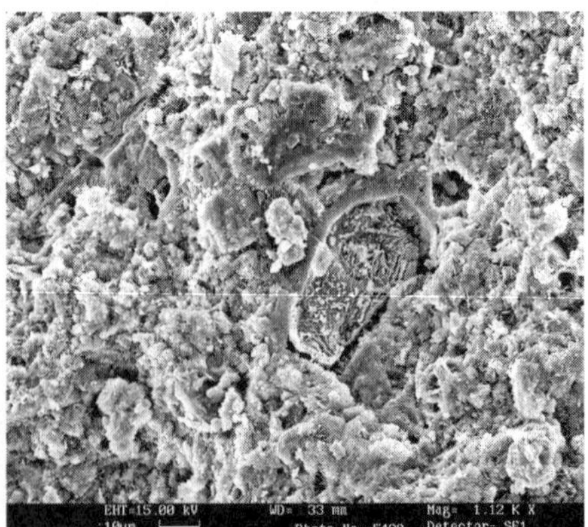

Figure 13 Cement (C1) paste with 5%LF, at 28-day

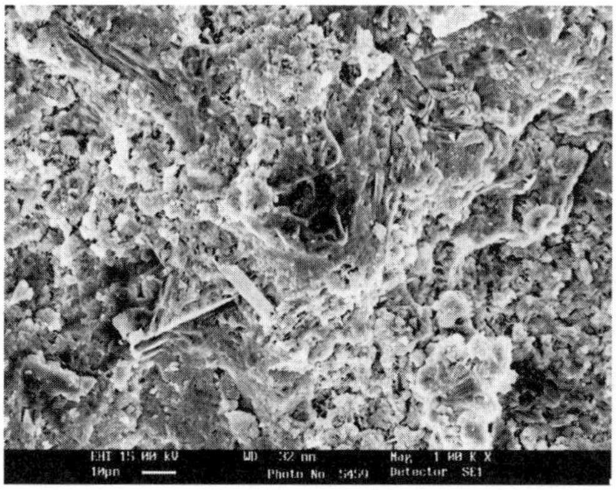

Figure 14 Cement (C2) paste with 5%LF at 28-day

Figure 15 Cement (C1) paste with 25%LF at 28-day

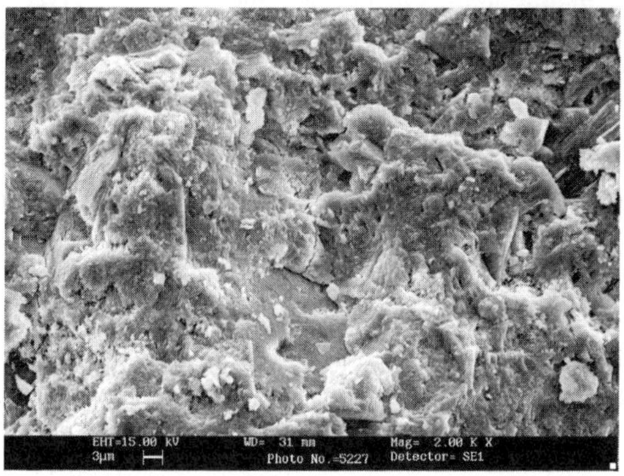

Figure 16 Cement (C1) paste, at 28-day

As to hydration of Portland cement (C1, high C_3A clinker), its microstructure is mainly composed of radiating fibers of C-S-H resembling in pattern to Type I C-S-H, hexagonal crystals with large size of CH, needle shaped ettringites and unhydrated constituents (Figures10, 11).The structure of Portland cement paste has relatively higher porosity than that of limestone added pastes (5 and 15 %) at the same age. This can be due to the fact that as its known , low degree of hydration occurs in the cement paste at early ages. Accordingly, it causes porous areas in the paste (Figure1). When the pastes of cement (C2, low C_3A clinker), it was observed that clinker grains surrounded by radiating fibers of calcium silicate hydrates (Figure12). Calcium hydroxide crystals and small amount of prismatic ettringite crystals spread into the paste.

SEM Observations of Pastes at 28 Days

Portland cement paste (C1, high C_3A clinker) with 5 % limestone replacement shows that a nearly amorphous and compact gel structure is predominant in the paste and the amount of hydration products near the calcite particles somewhat increased (Figure13). The overall porosity in the paste remarkably decreases owing to the advanced state of hydration of Portland cement at 28-day. In addition to this, calcite particles contribute to the decrease in the paste porosity. As to hydration of 5 % limestone added pastes of cement (C2, high C_3A clinker), the microstructure was found to be more densified by means of compact gel morphology (Figure14).

Due to the dense structure, it was difficult to distinguish cement products from calcite particles. In high level limestone added pastes (25-35 % limestone), independent of cement type, the microstructure exhibits relatively high porosity when compared with low level limestone added pastes,as seen in Figure15. In cement pastes, it was observed that the major structure of pastes was small irregular equnant particles of C-S-H phase, resembling in pattern to Type III CSH which fills in all the spaces between the grains and gives certain stability to the structure (Figure16).

Results of Heats of Hydration of Cements

The results of heats of hydration of cements, at 7 and 28-day are given in Table 2. As seen in this table, hydration heat values of limestone added cements, except for 5 % limestone addition, were found to decrease with increasing proportion of limestone. The increase in heat of hydration with the addition of 5 % limestone to cement can be due to accelarated hydration of cement constituents by means of fine calcite particles. Comparing the heat of hydration values of pastes of C1 cement (high C_3A –Clinker) and C2 cement (low C_3A – Clinker) it was detected that both cements showed a nearly similar trend with increasing proportion of limestone.

Table 2 The results of heats of hydration of cements ,at 7 and 28-day
Heats of hydration are in kilocalorie per kilogramme

	HEAT OF HYDRATİON	
	7-day	28-day
C1 Cement	74.0	78.1
C1-05LF	78.1	81.1
C1-15LF	77.0	80.0
C1-25LF	74.0	77.0
C1-35LF	64.9	71.2
C2 Cement	67.0	86.0
C2-05LF	73.0	90.8
C2-15LF	65.2	81.9
C2-25LF	59.8	67.0
C2-35LF	57.7	63.0

CONCLUSION

The following conclusions can be drawn from the study:

1. Some differences in the microstructure of cement pastes and the limestone added cement pastes, as a function of hydration time, were observed. At early ages, microstructure of low level limestone added cement pastes, regardless of the cement type, show lower porosity with respect to that of cement pastes. This can be attributed to the filler effect of limestone. But, after 28 days of hydration, cement paste has a compact structure due to the advanced state of hydration, corresponding to the lower pore sizes for calcite particles to fill effectively.

2. Reaction products were found to be tending to accumulate around the calcite particles and also these products have lower porosity than the products located in other areas of the paste. The reason for this can be due to nucleation effect of calcite particles.

3. Comparing low C_3A contaning cement pastes and cement pastes with high C_3A, it was observed that high C_3A containing pastes have only asmall amount of carboaluminates formed, specially at early ages. Nevertheless, in pastes with low C_3A, these formations were not detected.

4. In more than 15 wt.% limestone added pastes, microstructure exhibited high porosity. Accordingly, it can be stated that the filler effect of limestone was more pronounced for low level additions.

5. Heats of hydration of limestone added pastes were decreasing with increasing proportion of limestone. However, especially in 5% limestone added pastes, the heat hydration of cements increased considerably, indicating accelerated hydration by means of calcite particles.

REFERENCES

1. KRISTULOVIC, P., KAMENIC, N., POPOVIC K., A new approach in evaluation of filler effect in cement, Cement and Concrete Research, Vol. 24, No .4, 1994, pp 721-727

2. SCHMIDT, M., Cement with Interground Additives - Capabilities and Environmental Relief , Zement Kalk Gips, No. 8, 1992, pp 64-69

3. GRANDED, J., OLIVIER, J.P, Etude de la formation du monocarboaluminate de calcium hydrate au Contact D'un granulat calcerie dans une pate ciment Portland, Cement and Concrete Research, Vol. 10, 1980, pp 759-770

4. JAMBOR J., Influence of $3CaO.Al_2O_3.CaCO_3.nH_2O$ on the Structure of Cement Paste, Proceedings of 7th Int. Chem. of Cem. Congress, Vol. 4, Paris, France, 1980, pp 487-492

5. LONGUET, P., Reactivity of Calcareous Filler, Ciments-Betons-Platrex-Chaux, Vol. 23, No. 742, 1983, pp 24-28

6. RAMACHANDRAN,V.S., CHUN-MEI, Z., Influence of CaCO3 on Hydration and Microstructural Characteristics of Tricalcium Silicate, Il Cemento, Vol. 3, 1986, pp 129-152

7. DIDAMONY H.EL, EL-ALEI, E., Addition of Limestone in Low Heat Portland Cement, Ceramics Silicaty, Vol. 4, 2000, pp 146-150

NUMERICAL AND EXPERIMENTAL STUDIES OF PERCOLATION PHENOMENA OF HARDENING CEMENT PASTES

G Ye

K van Breugel **A L A Fraaij**

Delft University of Technology

Netherlands

ABSTRACT. This paper presents the results of studies of the pore structure and the permeability for several hardening cement pastes. . Cement pastes with water/cement ratio of 0.40, 0.50 and 0.60 from 1 day to 28 days were tested with mercury intrusion porosimetry. In parallel, the water permeability was measured. With the help of a numerical simulation model the correlation between the pore structure and permeability was found. The water permeability is more related to the pore size distribution and the effective porosity. The de-percolation of capillary porosity was found neither from experiments nor numerical simulations during the first 28 days after casting.

Keywords: Cement paste, Porosity, Permeability, Modelling.

G Ye Msc is a researcher at the Delft University of Technology. His research topic is experimental study and numerical simulation of the microstructure, porosity and permeability of cement-based materials.

K van Breugel PhD civil engineering, is a professor at Delft University of Technology. His research interests are early age concrete, design for imposed deformations and concrete structures for environmental protection.

A L A Fraaij PhD civil engineering, is an associate professor at the Delft University of Technology, his research interest is fly ash in concrete, using of recycled material as an aggregate in concrete and durability of concrete.

INTRODUCTION

It is widely accepted that the permeability of cement paste is governed by the microstructure, i.e. the volume, size distribution, connectivity and shape of the pores. In order to make a good permeability model based on the microstructure parameters, it is also necessary to investigate experimentally the correlation between the pore structure and permeability. These experiments help us to identify the influence of the w/c ratio and curing age on changes of both the pore structure and water permeability. In parallel with the experimental studies numerical simulations of the microstructure were performed. Of particular interest were the percolation phenomena. For the numerical simulation of the microstructure the HYMO-STRUC program [1] was applied in this study.

Systematic experiments, both on the pore structure and the water permeability of cement pastes, were conducted for cement paste samples with water/cement ratio of 0.4, 0.5, 0.6, at curing age of 1, 3, 7, 14 and 28days. The influence of water/cement (w/c) ratio and curing age on the pore structure parameters are discussed in detail. The results from simulations and from experiments were compared with each other and discussed in detail.

EXPERIMENTAL PROCEDURES

The experiments include two parts. The first part concerns pore structure measurements using mercury intrusion porosimetry (MIP). The second part deals with the percolation phenomenon as determined from permeability tests.

The cement paste specimens were prepared with CEM I 32.5R Portland cement. The chemical composition of cement is listed in Table 1. Cement pastes with three different w/c ratios, i.e. w/c 0.40, 0.50 and 0.60 were used for both MIP measurements and permeability tests. The degree of hydration of cement paste was calculated with the HYMOSTRUC model and was confirmed by heat release measurements of cement pastes using an isothermal calorimeter.

Table 1 Chemical compositions for CEMI 32.5R cement

ANALYTE (OXIDE)	MASS (%)
CaO	63.4
SiO_2	21.0
Al_2O_3	5.03
SO_3	3.0
Fe_2O_3	2.83
MgO	2.0
Na_2O	0.24
K_2O	0.65
TiO_2	0.30
P_2O_5	0.16
Mn_2O_3	0.06
f_{CaO}	0.5
Total	99.17

Mercury Intrusion Porosimetry Measurements

The specimens were mixed with de-aired distilled water. A commercial machine was used for mixing. After 1 minute of mixing at a low speed and two minutes mixing at high speed, the pastes were immediately poured into 500 ml glass bottles. The bottles were shaken continuously in order to remove air bubbles and then sealed with a plastic cover. The samples were then rotated at a speed of 5 revolutions per minute at the room temperature of 20°C for 24 hours. This procedure minimizes the possible effects of bleeding and segregation in the cement paste.

After rotation, the bottles were moved into a climate box and exposed to a temperature of 20°C until the age of testing. Curing periods of 1, 3, 7, 14 and 28 days were considered. At the end of the curing period, the glass bottle was broken and the sample was removed from the bottle and split into small pieces of around 1 cm^3. The split samples were exposed to freeze-drying with liquid nitrogen. The whole drying procedure was the same as suggested by Gallé [10].

Mercury intrusion measurements were performed with a Micromeritics PoreSizer 9320. The PoreSizer 9320 is a 207 MPa mercury intrusion porosimeter, which determines pore sizes in the range of 7 nm to 500 μm. The apparatus has two low-pressure chambers and one high-pressure chamber. A measurement is conducted in two stages: a manual low-pressure run from 0 to 0.170 MPa and an automated high-pressure run from 0.170 to 205 MPa. Data is collected and handled by a computer acting as a control module.

By extended capacity of MIP, the whole MIP test procedure, including the first intrusion, extrusion and following the second intrusion, were performed on the same sample. The curve of mercury intrusion and extrusion hysteresis can be obtained. From the hysteresis curve, the continuous pore and inkbottle pore can be detected from the first intrusion. During first intrusion, all pore space will be filled with mercury. When the pressure is released, the mercury is sucked from pore space except from the inkbottle pores and the dead-end pores. The amount of removed mercury is defined as effective porosity. If the mercury is re-pressurized, the pore size distribution of effective porosity can be obtained from the second intrusion curve.

Permeability Measurements

The permeability apparatus is shown schematically in Figure 1. The system includes regulated gas pressure source, a gas/water reservoir, three parallel permeability cells, computer controlling system and appropriate valves and tubing.

1. Permeability Cells

The specimens are disks with a diameter of 95 mm. The thickness varied from a maximum of 50 mm to a minimum of 10 mm. A ring of Room Temperature Vulcanised (RTV) silicone rubber has an outer diameter of 150 mm and inner diameter of 95 mm. The ring is confined by a steel confining cylinder with in inside diameter 200 mm and 80 mm long.

Two fitted end plates are provided to prevent axial length changes of the RTV so that it expands only radially in order to seal the annular space between the specimen and outer steel confining cylinder. The specimen itself is not subjected to mechanical load. The top plate is connected with the pressure source supply. The bottom is connected with a needle as outlet. A steel porous disk as a support was placed between specimen and bottom plate to prevent cracking of the samples.

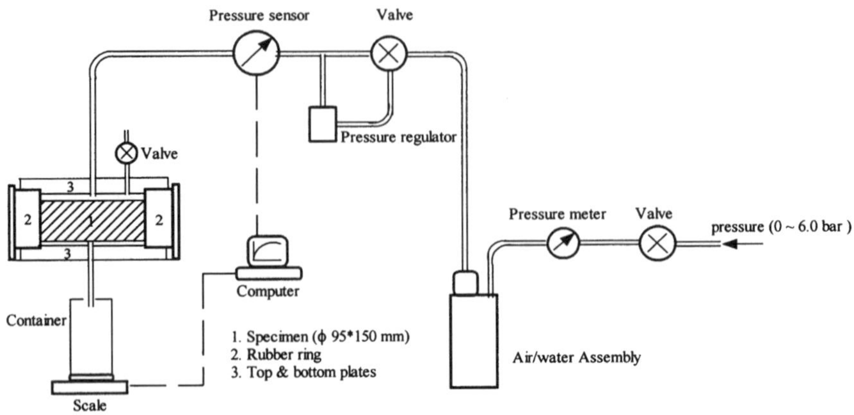

Figure 1 Permeability testing system

2. Data Acquisition System

The data acquisition system consists of the electronic balances for weighing the penetrated water in the container with an accuracy of 0.1 g and is connected to an industrial computer. The system is programmed to allow independent start and stop for each cell. The scanning sequence and timing is set at the start of a test by defining the interval of the time or the change of water weight in the container. The graphs showing the cumulative collected water versus time are plotted on the computer. The steady flow state can be observed directly from the plots.

3. Sample Preparation and Curing

Sample preparation and curing were almost the same as for the MIP test. In this way, it was possible to correlate the pore structure data and permeability results of the cement paste. However, small changes were adopted for the specific requirement for permeability test, viz

Instead of cement paste cast in the glass bottle, the samples for the permeability test were cast into a PVC cylinder. The length of cylinder was 800 mm and the inner diameter was 96 mm.

After casting and vibration, the cylinder was sealed in order to prevent moisture loss. It was rotated for 24 hours to avoid segregation of the cement paste. The speed of the rotation machine was set at 5 rpm. After 24 hours rotation, the specimens were cured in a climate box at a temperature of 20 °C. At each curing age four disks were sawn from the whole sample with a thickness of 15 mm. The samples were then immersed into the water and vacuumed for 4 hours to saturate the samples. The state of sample saturation was monitored as well.

3D microstructural Simulation of Cement Paste.

The 3D microstructure of cement paste was simulated and analyzed with the HYMO-STRUC-3D model [1,2,3]. In the HYMOSTRUC-3D model cement hydration is simulated as a function of the particle size distribution and the chemical composition of the cement, the w/c ratio and the reaction temperature. The model starts from the cement particle size distribution in a 3D body, the cement particles being considered as sphere. Based on an interaction-expansion mechanism, the cement grains gradually grow. With progress of the reaction process, the growing particles become more and more connected. In this way, a simulated porous microstructure is formed.

Figure 2 shows an example of simulated cement paste and its pore structure with w/c 0.40, at a degree of hydration (α) 0.75. The geometrical and topological parameters of the simulated microstructure were obtained with the algorithm described in [3], in which the percolation theory was applied to explain the connectivity of the solid phase and the pore phase.

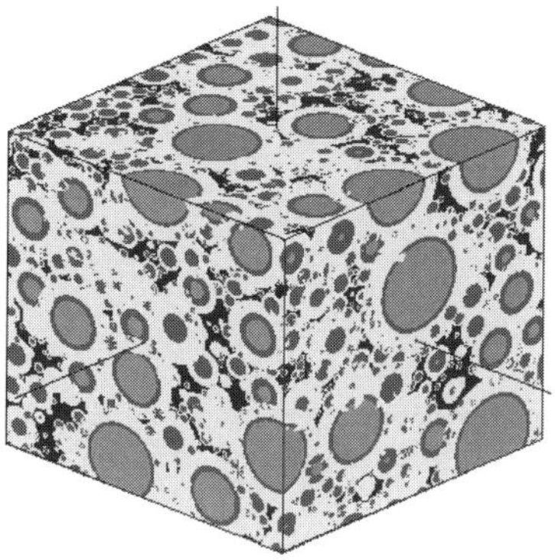

a). Cement paste

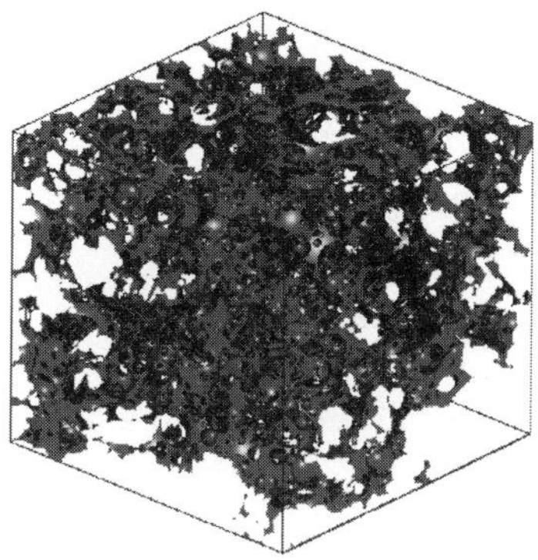

b). Pore structure

Figure 2 3D simulated cement paste and its pore structure (w/c=0.4, α=0.75)

RESULTS

Parts of the simulation and the experimental results on the pore structure and measured permeability results are summarized in table 2. The simulation results are in good agreement with the experiments.

Table 2 Characterization of cement paste and permeability

W/C RATIO	AGE (DAYS)	DEGREE OF HYDRATION (%)		TOTAL POROSITY (ml/ml %)		EFFECTIVE POROSITY (ml/ml %)		PERMEABILITY MEASURED (m/s)
		Measured	Simulated	Measured	Simulated	Measured	Simulated	
0.40	3	45.1	45.6	33.2	32.11	22.2	29.1	6.42×10^{-12}
	14	64.1	64.8	29.1	26.01	19.3	23.5	4.48×10^{-13}
	28	-	72.1	22.1	21.02	11.3	16.8	6.20×10^{-14}
0.50	3	46.2	46.4	40.5	41.4	27.3	40.8	1.54×10^{-10}
	14	68.9	69.2	35.0	32.1	20.9	30.9	2.14×10^{-11}
	28	-	77.2	31.1	28.9	17.0	25.8	1.33×10^{-12}
0.60	3	47.3	48.2	45.7	46.0	33.7	45.5	4.61×10^{-10}
	14	73.6	73.8	41.5	40.2	26.0	39.1	7.61×10^{-11}
	28	-	84	39.5	36.5	23.0	34.2	1.82×10^{-11}

1 Total porosity and Permeability

It is logical that the sample with higher porosity gives a higher permeability. However, the relationship between porosity and permeability is not unique. This can be illustrated with the porosity-permeability curves shown in Figure 3, which show the results for the samples with different water/cement ratio. For example, for a given MIP porosity of 39% of the samples with w/c ratio 0.40 and 0.60, the permeability of the sample with w/c 0.40 is one order of magnitude higher than the sample with w/c 0.60. It can be concluded that the porosity is not only the factor that governs the permeability. This is in agreement with the finding of Mehta and Manmohan [9].

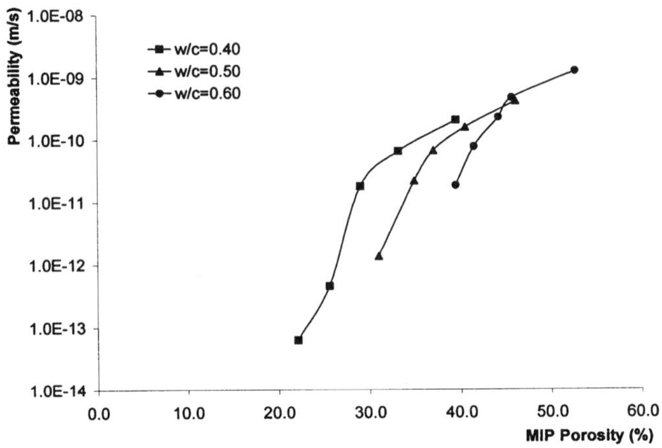

Figure 3 Relationship between permeability and total porosity measured by MIP

Table 3 lists the pore structure properties measured by MIP and the measured permeability of cement pastes with w/c 0.40 at 1 day, w/c 0.5 at 3 days and w/c 0.60 at 28 days, respectively. They have the same porosity of around 40%. However, if we compare the pore size distribution for the sample of w/c 0.40 with the sample of w/c 0.6, Figure 4 shows that the sample with w/c 0.60 at 28 days has much more smaller pores. This can also be observed from the critical pore diameter (The definition of critical pore diameter was given by Gallè [10]). Thus, though the total porosity is the same, the pore size distribution can differ dramatically. The fact that the w/c 0.60 sample has a much lower permeability implies that the fine pore system in the cement paste leads to a lower permeation even though the porosity is same.

Table 3 Pore structure and permeability of three different cement pastes with the same porosity

SAMPLE	POROSITY (%)	AVERAGE PORE DIAMETER (μm)	CRITICAL PORE DIAMETER (μm)	PERMEABILITY (m/s)
w/c 0.40, 1 days	39.52	0.0560	0.54	2.00×10^{-10}
w/c 0.50 3 days	40.54	0.0564	0.38	1.54×10^{-10}
w/c 0.60 28 days	39.50	0.0406	0.035	1.82×10^{-11}

If we compare the pore structure parameters and permeability of cement pastes with w/c 0.40 at 1 day with the cement pastes with w/c 0.50 at 3 days from Table 3 and Figure 5, both samples show a similar pore size distribution, and similar value of the average pore diameter and the critical pore diameter. Most importantly, the permeability value is found for both samples nearly same. Thus, regardless the water/cement ratio and curing age, the water permeability of cement paste mainly depends upon the pore size distribution and connectivity of pores.

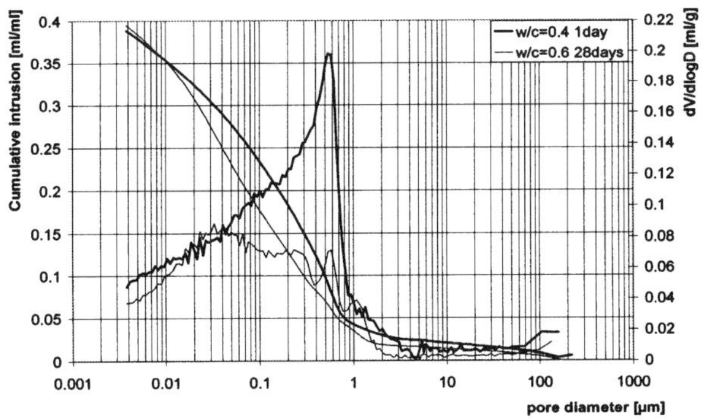

Figure 4 Comparison of the pore structure for the cement pastes with w/c ratio 0.40 at 1day and w/c ratio 0.60 at 28 days.

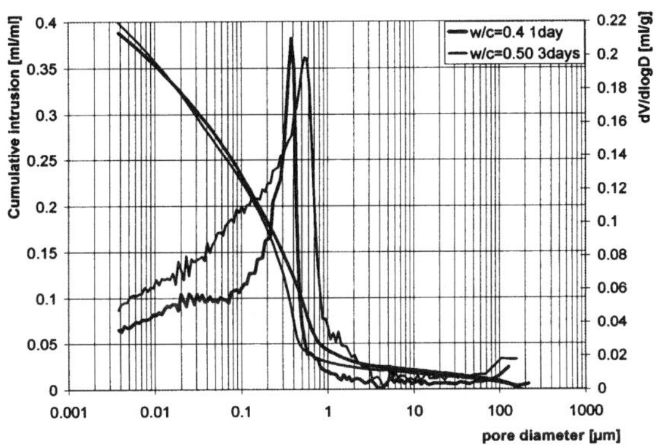

Figure 5 Comparison of the pore structure for the cement pastes with w/c ratio 0.40 at 1 day and w/c ratio 0.50 at 3 days.

2. *Effective Porosity and Permeability*

As defined previously, the effective porosity measured from MIP is the total porosity minus the inkbottle porosity. The effective porosity represents the path through which water can penetrate the sample. Figure 6 plots the water permeability as function of effective porosity.

From the trend line, a good relationship can be found between the water permeability and the effective porosity.

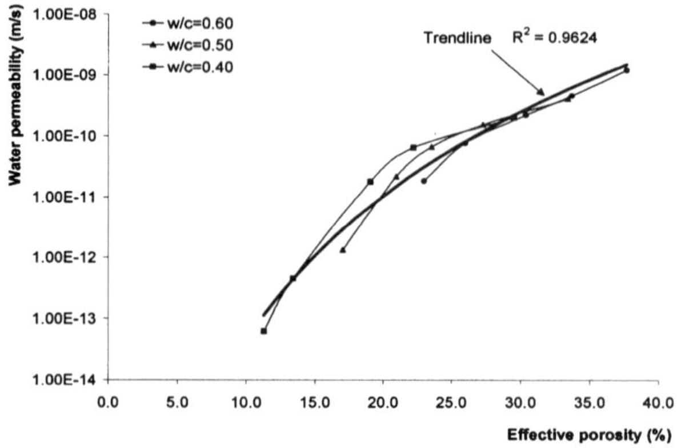

Figure 6 The relationship between permeability and effective porosity measured by MIP

3. Percolation of Capillary Porosity and Permeability

The percolation theory deals with disordered media in which the disorder is characterized by a random variation in the degree of connectivity. The microstructure of cement-based materials provides numerous examples of percolation phenomena [4]. In percolation studies one is often interested in the fraction of a phase, which is connected across the microstructure as a function of the total volume fraction of the phase. In particular, the de-percolation of capillary porosity is of great interest for studying the transport phenomena in cement-based materials.

The ideal idea of application of percolation theory for explaining the transport properties in cement-based materials can be summarized as follows: in the initial stage when cement is mixed with water, the capillary pores are all connected with each other and the cement matrix has a high permeability. With the increase of the degree of hydration, the volume and the size of capillary pores decrease and capillary pores start to become disconnected and the permeability decreases as well. At a certain degree of hydration, the capillary pores are not connected with each other any more and a significant (sudden) reduction in permeability is observed. The porosity of this point is called de-percolation threshold.

There are three questions brought to us when we apply the percolation theory in explaining the capillary transport in cement-based materials. Firstly, does a real de-percolation threshold exist? Secondly, if a real de-percolation threshold exists, what is the de-percolation threshold for a given cement paste? This also links to the third question: what is the permeability when the cement matrix is considered as impermeable. A certain answer seems logical for the first question since the cement paste itself is a random porous material. In theory [5], a de-percolation threshold always exists if the porosity continuously decreases to a certain point. The numerical simulations of cement paste by Bentz and Garboczi [4] and Ye et al. [3] have demonstrated the 'numerical' existence of de-percolation of capillary porosity. However, the important issue is what is the 'actual' de-percolation threshold value in practice and how to define it from permeability tests. These questions are crucial for developing a reliable permeability model. In fact, most researchers [6,7] have followed Powers's conclusion [8] stating "the capillary porosity in cement paste exhibits a percolation transition (from connected to disconnected) at a volume fraction of about 20 % porosity". This value, in fact, can be deduced from permeability test. Since the capillary pores are much bigger than the gel pores, the water transport is dominated by the connected capillary porosity and expresses a higher permeability. When the capillary pores become disconnected, the gel pores will play a key role for water penetration. The Minimum capillary permeability of cement paste according to Powers is around 2.54×10^{-14} to 2.54×10^{-15} m/s [8]. From experiments, once the permeability value drops to this value, the cement matrix can be treated as capillary de-percolated. The porosity at this moment is the de-percolation threshold.

Now, if we look at the permeability experiments, the lowest permeability is 9×10^{-14} m/s for the sample with w/c 0.40 at 28 days. This value is higher than reported by Powers, no de-percolation of capillary porosity was found. This can also be confirmed from the results reported by Metha and Manmohan [9]. They found the de-percolation of capillary pores close to Powers's results only for the samples with w/c ratio 0.30 at an age of 1 year.

Numerical simulation of the pore structure with HYMOSTRUC-3D also provided similar results. Figure 7 shows the simulation results. The capillary de-percolation in the simulation for the sample with w/c 0.30 was found at a degree of hydration around 68%, at which the de-percolation threshold porosity is around 5%. This is close to observations in the practice.

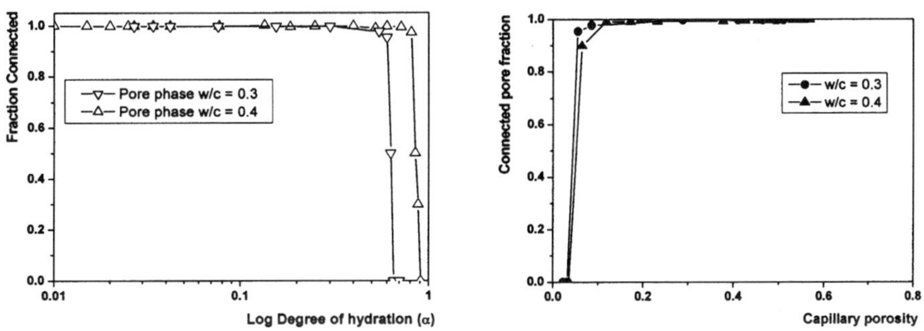

Figure 7 Development of microstructure in cement paste with two different w/c ratios, after [3].

CONCLUSIONS

From this study, it was found that the total porosity is not a unique parameter that governs the permeability, whereas the pore size distribution, critical pore diameter and effective porosity are more crucial for water permeability. No de-percolation was found in the experiments for the samples with w/c ratio 0.40 to 0.60 at curing ages up to 28 days. This agrees with the results by Metha and Manmohan. The numerical simulated pore structure from HYMOSTRUC3D has shown that the de-percolation of capillary porosity occurs only in the sample with very low water/cement ratio at a high degree of hydration. These results are important for developing and tuning permeability models.

ACKNOWLEDGEMENTS

The authors would like to thank Mr. E.M. Horeweg for his assistance in conducting the permeability tests. We would also like to acknowledge the Microlab at Delft University of Technology, where the mercury intrusion porosimetry measurements were performed. The research was financially supported by the Dutch Technology Foundation (STW), which is gratefully acknowledged.

REFERENCES

1. BREUGEL, K. van., Simulation of hydration and formation of structure in hardening cement-based materials, Dissertation, Delft University of Technology, The Netherlands, 1991

2. KOENDERS, E.A.B., Simulation of volume changes in hardening cement-based materials, Dissertation, Delft University of Technology, The Netherlands, 1997

3. YE, G., BREUGEL, K. van., FRAAIJ, A.L.A., Three-dimensional microstructure analysis of numerically simulated cementitious materials, Cement and Concrete Research, Vol.33, No. 2, February 2003, pp 215 – 222

4. BENTZ, D.P., GARBOCZI, E.J., MARTYS, N.S., Application of Digital-Image-Based models to microstructure, transport properties, and degradation of cement-based materials, in H. Jennings et al. (eds.), The modelling of microstructure and its potential for studying transport properties and durability, 1996, pp 157-165

5. STAUFFER, D., AHARONY, A., Introduction to percolation theory, 2nd ed., Taylor and Francis, London 1994

6. GARBOCZI, E.J., BENTZ, D.P., Computer Simulation and Percolation Theory Applied to Concrete, in R. Edge et al. (eds.), Annual Reviews of Computational Physics. Volume VII, World Scientific Publishing Co., 1999, pp 85-123

7. COOK, R.A., HOVER, K.C., Mercury porosimetry of hardened cement pastes, Cement and Concrete Research, Vol.29, No. 6, June 1999, pp 933-943

8. POWERS, T. C., COPELAND, L.E., MANN, H.M., Capillary continuity or discontinuityin cement pastes," PCA Bulletin, Vol. 1, No. 110, May 1959, pp 2-12

9. MEHTA, P.K., MANMOHAN, C., Pore size distribution and permeability of hardened cement paste, Seventh International congress on the chemistry of cement, Paris, France, Vol. III, Theme VII, 1980, pp VII-1/VII-5

10. GALLÉ, C., Effect of drying on cement-based materials pore structure as identified by mercury intrusion porosimetry: A comparative study between oven-, vacuum-, and freeze-drying, Cement and Concrete Research, Vol. 31, No. 10, October 2001, pp 1467-1477.

ESTIMATION OF THE POROSITY OF PORTLAND CEMENT PASTES USING BACKSCATTERED ELECTRON IMAGES

M Mellas **B Mezghiche**

University of Biskra

Algeria

J E Ash

University of Birmingham

United Kingdom

ABSTRACT. Microscopy plays an important role in the examination of cementitious materials. Electron-optical techniques allow examination of microstructural details. The increased application of scanning electron microscopy in cement and concrete investigations has brought attention to the estimation of the porosity. Backscattered electron image has been established as a method for quantitative analysis of hardened cement paste. This paper describes work and its application to Ordinary Portland cement paste. Results show that the technique can be used to determine the porosity of Portland cement pastes. The success of these investigations is, influenced by the type and quality of specimen preparation. In particular, backscattered electron and X-ray imaging modes are influenced by the specimen surface characteristics, with the ideal surface being highly polished. The technique can be a powerful tool to determine this essential factor of durability in concrete materials.

Keywords: Porosities, Cement paste, Strength, Scanning electron microscope, Methanol exchange.

M Mellas is the Head of Department of Civil Engineering at the University of Biskra.

B Mezghiche is a researcher in the Department of Civil Engineering at the University of Biskra

J E Ash is a researcher at the University of Birmingham.

INTRODUCTION

It is widely recognized that the porosity of a material exerts an enormous influence on its physical properties. For hardened cement paste a large volume of pores is inherent in the set structure. This porosity is derived mainly from excess water required to ensure cement hydration and provide workability of cement paste, but can also be present due to inadequate compaction. The residue of water filled space in fresh cement paste become voids in hardened cement paste. These voids are divided in two classes, capillary pores and gel pores. The former represents the volume of the capillary pores and it depends mainly on both cement-water ratio of the mix and the degree of hydration. The latter represents the gel pores. As hydration progresses, the amount and distribution of porosity between capillary and gel pores will change. Initially all the pores are capillary pores. As hydration precedes the capillary pore volume is reduced because the capillary space becomes filled with hydration products, and the gel porosity increases. There is a net reduction in total porosity.

During the last fifteen years Mercury intrusion porosimetry which involve forcing mercury into pores of a body by application of pressure and assuming that all pores have a simple shape, has been used for determining the pore structures sizes of hydrated cement paste. This method enables a wide range of pore-size distribution to be measured between 1000μm and 30 A° depending on the pressure [1]. However, the application of great pressure on hydrated cement paste may cause destruction in the structure and pores below 30 A° cannot be considered [2]. In addition to this method there are other investigations involving fluid replacement as methanol and helium pycnometry. The use of methanol as displacement medium for porosity measurement [3], have indicated that the porosity values were equivalent to those when helium was used as displacement method. Parrot [3] reported on the effect of drying upon the exchange of pore water with methanol. He found that the water/methanol exchange results of 0.6 water/cement ratio by weight pastes indicated that the pore volume penetration by methanol was only on average 0.7 per cent greater than that occupied by original pore water and the exchange provided reliable pore structure information.

With the application of backscattered imaging of hydrated Portland cement paste [4], in a polished cement surface the anhydrous grains appear brighter than the hydration products, which have a lower atomic number due to the presence of water. It was found that in hydrated cement paste the unhydrated cement appeared bright, the area of calcium hydroxide appeared slightly darker, other hydrated products still darker, and the pores black. The use of scanning electron microscope (SEM) for cement pastes, with the application of and programming of an image store for the SEM has been applied in previous work [5]. Some of the work described in this paper deals with an approaches to the porosity determined by SEM and Methanol exchange.

MATERIALS

The cement was an Ordinary (typical) Portland cement supplied by Blue Circle Industries (Table 1). The composition was determined accurately by Gutteridge at the British Cement association (UK), and the results are quoted in Table 2.

Table 1 Oxide composition of cement

Oxide	CaO	SiO	Al_2O_3	Fe_2O_3	SO_3	MgO_2	K_2O	TiO_2	Na_2O	P_2O_5	Mn_2O	Loss On ignition	Insoluble residue	Free CaO
Content (%)	64.50	20.00	5.30	3.40	3.00	1.20	0.78	0.26	0.10	0.08	0.06	0.90	0.20	2.20

Table 2 Phase composition by QXRD

Compound	C_3S	C_2S	C_4AF	C_3A
QXRD (%)	53.0	25.0	7.0	4.0

METHODS OF INVESTIGATION

Determination of the porosity has been measured by two methods firstly by methanol exchange and then by a back scattered electron image.

Methanol Exchange

Determination of the porosity has been achieved by method of solvent replacement. Analar grade methanol was used which can penetrate into the microstructure without damage. The unbound water in the specimen was replaced by methanol, thus the hydration was arrested and all water eventually replaced by methanol; this is shown by a constant weight of the specimen between 3 and 14 days. Specimens were then dried over silica gel for 24 hours at 20 °C. The porosity was measured using methanol exchange [4].

Backscattered Electron Image (BEI)

Polished specimens of hardened cement paste samples was a technique developed to provide qualitative information by Backscattered Electron Image (BEI). With the application of BEI of hydrated Ordinary cement pastes, anhydrous grains appear brighter than the hydrated products, because of differences in atomic numbers [4]. The unhydrated grains appear bright, area of calcium hydroxide slightly darker, other hydrated products darker and the pores black. Although quantitative measurement of BEI has been possible some time, the use of TV rate scanning without image storage usually prevents the SEM being used with the best performance in terms of contrast discrimination and beam settings, the application and programming of an image store for SEM has been used [6].

SPECIMEN PREPARATIONS

Several small prisms (125 x 35 x35 mm) and cubes (50 mm sides) were cast with cement ratios by weight of 0.40 and 0.45. Mixing was prolonged and intermitted to minimize the separation of water from the setting paste. Careful curing ensured no loss of water from surfaces for the first 24 hours. Following demoulding, samples were kept under wet and dry curing conditions and chosen for test at ages which provided a variation in the porosity. When selected for analysis, cubes were crushed. Prisms were cut perpendicular to the longitudinal axis with a diamond saw to yield a 3 mm thick slice. Samples were then immersed in methanol immediately in methanol to prevent carbonation and stop hydration prior to the determination of porosity.

A resin vacuum impregnation technique was developed as a means of supporting the paste microstructure during the specimen preparation, which was particularly important in the case of the weaker early age's pastes. An Araldite combination resin system consisting of 10 parts of AY18 resin and 1 part of HY 18 hardener was of low viscosity and proved to be suitable. For ease of handling during process of polishing, specimens were contained in a fast setting metaset resin. Specimen preparation is important in any microscopical technique with proper preparation methods facilitating examination and interpretation of microstructural features. Improper preparation methods may obscure features, and even create artifacts that may be misinterpreted. Scanning electron microscope (SEM) analysis using backscattered electron and X-ray imaging requires a highly polished surface for optimum imaging. Rough-textured surfaces, such as those produced using only saw-cutting diminish the image quality by reducing contrast and loss of feature definition. Specimens investigated by BEI must be plane; otherwise reflected shadows will adversely affect the image. Samples were automatically using a sequence of sand paper sizes, varying from grade 250 to Grade 1200. Satisfactory surfaces were produced when the final polishing was carried out initially with 6 μm diamond paste followed by 1 micron paste. In order to prevent the build of electron, the final surface was coated with carbon (7, 8). Proper sample preparation is absolutely vital in the study by SEM. A summary of the time required to produce one sample of hydrated cement paste is given below, where the term in brackets the time is undertaken for each operation.

- Cutting the sample to yield 3 mm thick slices, (15 mn).
- Drying the sample by methanol exchange, (15 days).
- Resin impregnation of the sample, (1hours). Then the impreganation sample was left to be hardened, (1 day).
- Cutting the impregnated sample to yield 15 mm square, (15 mn).
- The sample was then placed into a 30 mm diameter plastic tube mould fast-setting resin and hardener, (20 mn)
- Manual polishing on sand paper through the grades 25. 400.800.1000.1200.(60mn).
- Automatic polishing on a rotating wheel using Metaserv cloths and diamound paste respectively 6 and 1 um (12 hours).
- Coating with carbon
- Study by SEM (4 hours).

RESULTS AND DISCUSSION

Table 3 shows results for a cement paste with a water cement ratio by weight of 0.4 and 0.45, cured under water and in air, and tested at range of ages to provide information about the porosity determined by methanol exchange and BEI.

Table 3 results of compressive strength and porosity of cement pastes

W/C	Ages (days)	COMPRESSIVE STRENGTH (MPa)	METHANOL POROSITY (%)	BEI POROSITY* (%)
Water cured				
0.4	7	48.4	29.4	14.3
	14	57.9	27.6	13.1
	28	74.0	26.2	10.8
	65	77.2	24.9	7.7
0.45	7	41.7	35.8	15.3
	14	48.2	32.3	13.4
	28	53.8	31.1	9 .7
	65	57.8	29.4	8.8
Air Cured				
0.4	7	41.0	30.1	14.9
	14	45.4	29.1	14.4
	28	46.7	28.5	13.4
	65	47.6	28.2	9.2
0.45	7	33.4	37.1	15.1
	14	39.1	35.6	13.7
	28	41.0	32.2	12.7
	65	46.5	30.7	10.5

*each porosity is the average range of a ten random observations within +-1 standard deviations

The porosity by BEI is determined at 200 magnification (X200), this magnification was found adequate in the previous work (5), for the determination of the degree of hydration. Figure 1 shows an image taken from a sample of ordinary Portland cement paste where the unhydrated product are bright, area of calcium hydroxide slightly darker, other hydrated products darker and the pores black. Figures 3, 4 and 5 show respectively the same image where the pores, the uhydrated products and the unhydrated cement are selected, thus quantified as a percentage.

Figure 5 shows porosity based upon the methanol exchange method and BEI of hydrated cement pastes as a function of the compressive strength . This illustrate clearly the increase of strength while the porosity decrease, the results are for pastes with water cement ratio of 0.4 and 0.45, tested at various ages. Figure 6 shows the compressive strength of hydrated cement paste as function of age. While Figure 7 shows the porosity of OPC pastes based upon methanol exchange as function of age.

A correlation can be made between the two methods, (Figure 8). This relation shows that the porosity determined by BEI is proportional to that obtained by Methanol Exchange method, thus giving a difference between the different curing condition and various age. The porosity obtained by BEI is less than that obtained by methanol exchange and this is due essentially that the first one represents the capillary porosity but the second represents the hole porosity with even cracks.

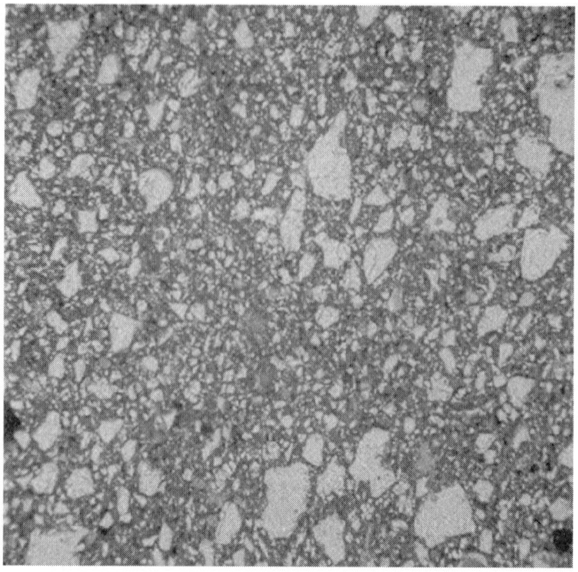

Figure 1 BEI of hydrated ordinary Portland cement paste

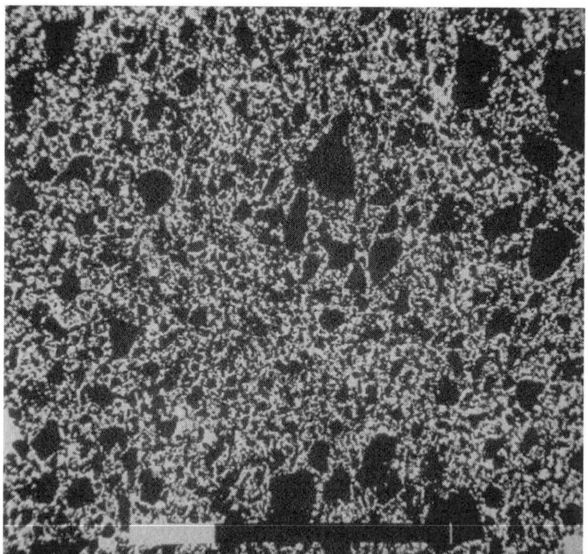

Figure 2 BEI of hydrated ordinary Portland cement paste, where the pores are selected (white)

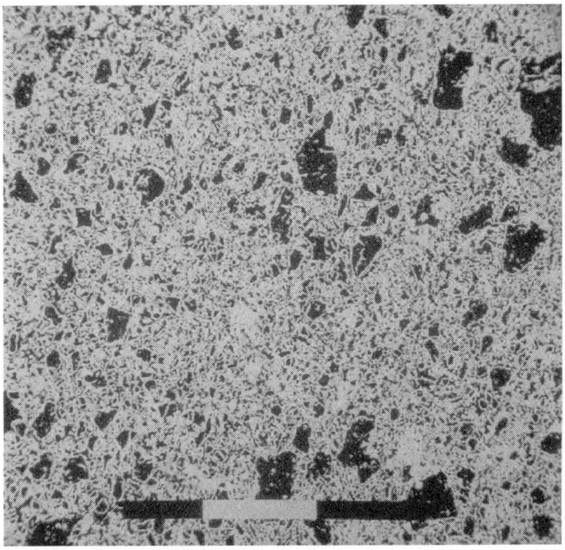

Figure 3 BEI of hydrated ordinary Portland cement paste, where the hydrated products are selected (white)

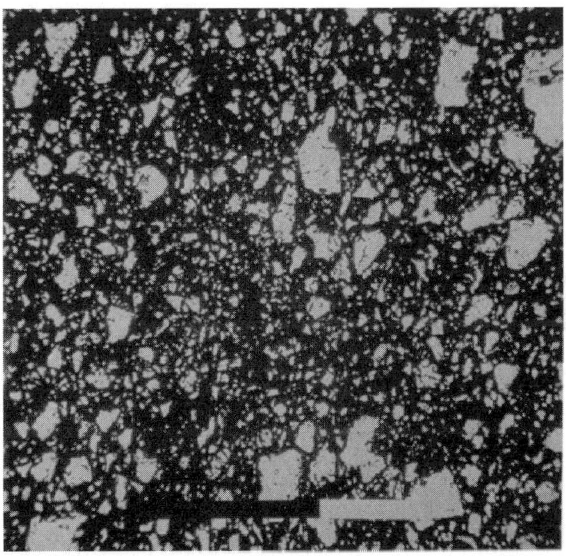

Figure 4 BEI of hydrated ordinary Portland cement paste, where the unhydrated cement is selected (white)

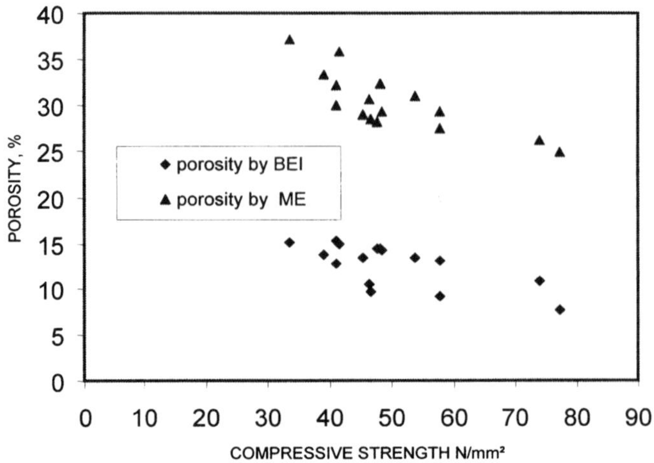

Figure 5 porosity based upon methanol exchange and BEI of hydrated cement paste as function Compressive strength

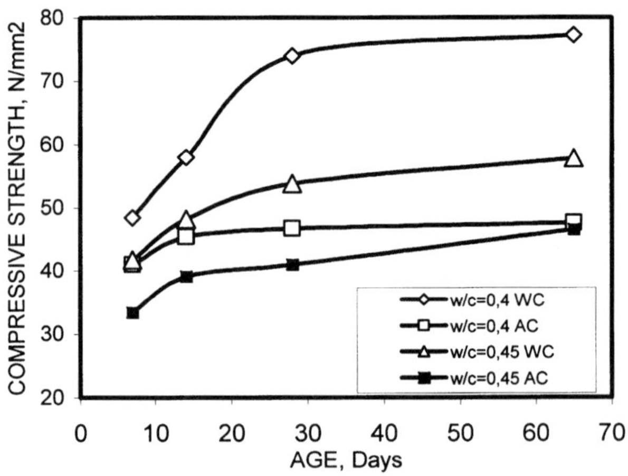

Figure 6 Compressive strength of hydrated cement paste as function of age

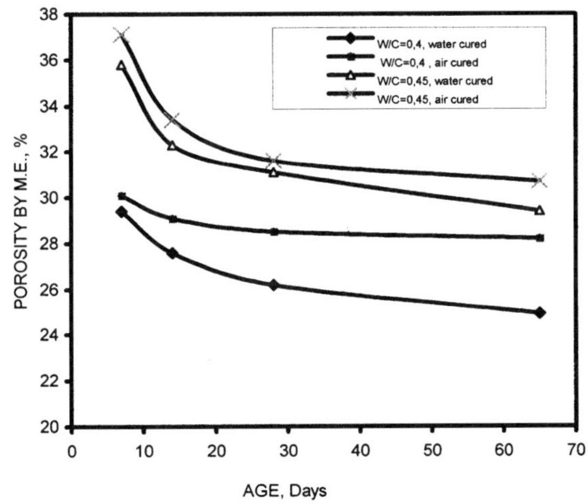

Figure 7 Porosity of cement pastes by Methanol exchange as function of age

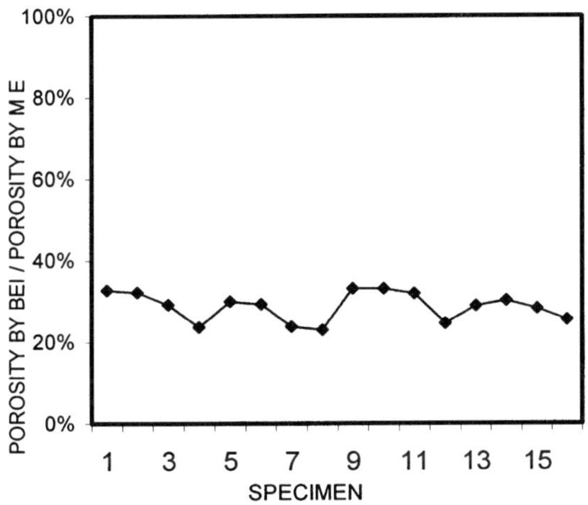

Figure 8 Relation between the porosity obtained by BEI and that obtained by Methanol Exchange (ME)

The significant differences between the two methods is also subjected to a different factors such as:

- Difficulties in sample preparation for the in the SEM
- Effect of magnification, the calculation of the porosity under 200 magnifications, and the area shown on the SEM represents an area of 0.5X0.5 mm^2 of the sample considered.
- Random selection of a number of small area; It depends on the selection of different parts of the sample and the boundary between the compounds
- Subjective analysis of the images

Rough-textured surfaces, such as those produced using only saw-cutting diminish the image quality by reducing contrast and loss of feature definition. Additionally, the lack of a polished specimen makes quantitative estimates arduous, as the surface is no longer planar. Nonetheless this method remains and can be applied for concrete materials to show the porosity of cement paste in concrete materials. Even thought the porosity obtained by the scanning electron microscope is less than that obtained by methanol, it remains the only one tools to observe and determine the porosity of cement pastes. The results obtained show clearly the variation of the porosity of different curing condition and water cement ratios.

Durability of concrete is dependant upon the porosity. Figure 5 demonstrate a wide spread of results in comparing compressive strength to the porosity. This shows that the water cement ratio has a marked and interesting effect worthy of further research. Various authors have suggested that curing, age, degree of hydration, porosity and cement constituents will also have an effect It is clear that, although cube compressive strengths remain as an easy, practical measure of compliance of site of concrete, adequate strength are not absolute guarantee of satisfactory durability

REFERENCES

1. ORR C.: Application of mercury penetration to material analysis. Powder Technology, vol 3, 1970

2. SHI D. and WINSLOW D N. : Contact angle and damage during mercury intrusion into cement paste. Cement and Concrete Research, vol 15, 1985.

3. PARROT, L. J., "Effect of drying history upon the exchange of pore water with methanol and upon subsequent methanol sorption behaviour in hydrated alite paste", Cement and Concrete Research, Vol 11, 1981, p 651 – 658.

4. SCRIVENER , K L and PRATT P L; Backscattered Electron Images of Polished Cement Section in the Scanning Electron Microscope. Proc. 6th int. Conf. On Cement Microscopy, 26 – 29,March 1984, Albubkerque , New Mexico, pp 145 – 155.

5. MELLAS M, ASH E J, HALL M G & LANGFORD I ,L; Estimations of degree of hydration of Portland cement pastes. Cement and Concrete Research, vol 23, pp 399-405, 1993

6. Hall M G ; The application and programming of an image store for the Scanning Electron Microscope; Scanning, vol 13 N° 3, 1991

7. BENTZ D P. and. STUTZMAN P E; SEM Analysis and Computer Modelling of Hydration of Portland Cement Particles. Petrography of Cementitious Materials, ASTM STP 1215, Sharon M. DeHayes and David Stark, Eds., American Society for Testing and Materials, Philadelphia, pp. 60-73, 1994

8. STUTZMAN P E & CLIFTON J R; Specimen preparation for Scanning Electron Microscopy. Proceedings of the twenty-First international conference on cement microscopy. April 25-29,1999, Las vegas, Nevada , pp 10-22,1999

A DIGITAL METHOD TO SIMULATE THE MICROSTRUCTURE OF THE ITZ IN CONCRETE

R Gao

Northern Jiaotong University

China

P Stroeven

Delft University of Technology

Netherlands

ABSTRACT. The ITZ between aggregate and bulk has been a subject of considerable research interest in recent years. Cement particle packing is considered to be one of the main factors influencing the features of the ITZ. The description of the particle packing mechanism in the ITZ to simulate the microstructure, therefore, becomes essential for the assessment of its micro-mechanical behaviour. A digital method which is called non-placement modal, based on "arch structures", is proposed in this paper to simulate the micro-structure of the ITZ. The arch structure, together with aggregate edge, can be used to form an enclosed zone where no particles could be located. The enclosed zone is used to simulate the pores near the aggregate edge in the ITZ. The size and the stability of the arch structure depends on the diameter of the cement particles and the distance between the particles. Several basic connects of particle packing near the edge are discussed. The principles for forming microstructure in the ITZ are presented. The advantage of the method is the easy transfer of a two-dimensional particle packing problem in the ITZ into one-dimensional one, and a three-dimensional problem into a two-dimensional one. Further, it offers a convenient way to describe the formation of microstructure and to study the variation of porosity in the ITZ. An example for microstructure formation of mono-size particle packing near the aggregate edge in two dimensions is illustrated and the simulation results are discussed

Keywords: Concrete, Interfacial transition zone, Microstructure, Particle packing.

Ri Gao received his Master degree in civil engineering from Northern Jiaotong University, and now is a professor of Northern Jiaotong University, Beijing, China.

Piet Stroeven, received his Doctor degree in civil engineering from Delft University of Technology, and now is a professor of Delft University of Technology, Delft, the Netherlands.

INTRODUCTION

Generally, concrete is considered to be a composite material and can be divided into three main parts according to the different material properties. They are the bulk paste, the ITZ and the aggregate. The ITZ is defined as the region between the aggregate surface and the bulk paste.

The Interfacial transition zone (ITZ) between aggregate and bulk has been a subject of considerable research interest in recent years because it constitutes a weak zone in concrete. Both qualitative and quantitative characterizations of ITZ have been conducted[1-6]. The microstructure of the ITZ is different from that of the bulk in the same concrete. The most evident micro-feature of the ITZ is its higher porosity and larger pores. The mechanical behaviour of concrete is significantly affected by the properties of this zone.

Concrete is a granular composite material of which the microstructure is build up of particular elements ranging in size from nanometers via micrometers to millimeters. It is a very difficult to try to theoretically describe the ITZ. An excellent approach to this subject is to use computer simulation.

There are many digital models for simulating and evaluating the ITZ[4,7,8,9]. It is found that particle packing and hydration of cement particles have great effects on the microstructure formation and development of the ITZ. On the one hand an important part of the material properties of concrete are linked to the packing characteristics of the particulate phase; on the other hand, microstructure development of concrete during hydration is based on the initial structure formed by particle packing. Obviously, the investigation on the microstructure formation and development starts with the understanding of the particle packing.

Particle packing effect arises from particles not packing together as well near a flat surface as in free space. A major concern is for the influence of the particle packing on the microstructure and development of the ITZ. The desire may be to attain also a high packing density in the ITZ and reduce thereby greatly the weakness of the ITZ in cement-based materials. Container wall or wall effect[10] is a main factor affecting particle packing. Other factors, such as particle size, particle shape and surface texture, agglomeration, surface active agents, segregation, bridging and vibration, as well as particle size distribution, etc., contribute to the particle packing effect also.

German[11] discussed in detail the factors affecting the particle packing and described the mechanisms underlying the container wall effect. The container in which particles are poured will induce a local area of "order" at the container wall in an otherwise "random" packing of particles. The effect is more pronounced for flat, smooth container walls. Stroeven[8] simulated the effect of a rigid interface on the particle structure by the SPACE system. He outlined that the interface causes local segregation in the particle packing. Furthermore, he discusses the structural development in the ITZ and the packing characteristics of fresh cement.

It this paper, a new model called non-placement model will be presented to simulate the cement particle packing effect near an aggregate surface. On the basis of cumulative and fractional weight distributions of cements, the microstructure of multi-size particle packing is formed in ITZ.

NON-PLACEMENT MODEL

Description of Non-Placement Model

In a non-placement model, it is assumed that near the wall a "particle arch structure" may be frequently formed by particle contact (see Figure 1). Because of the wall, no cement particles can be placed into the space between the wall and the arch. In Figure 1, zone I and zone II are enclosed by the wall and the arch structures, so it is impossible for cement particles to get into the zones. The result is some free spaces are formed near the wall. Zone III and zone IV may also take the form of "arch structures", but unlike zone I and zone II, they will be easily filled by cement particles because there is no wall to form an enclosed area with an arch structure. This description, herein, is called non-placement mechanism.

In the non-placement model the basic element is the particle arch structure which is formed by particles. The particle arch must be a stable system so that the forces between cement particles can be transferred. Several basic forms of particle arch structures are discussed in the following.

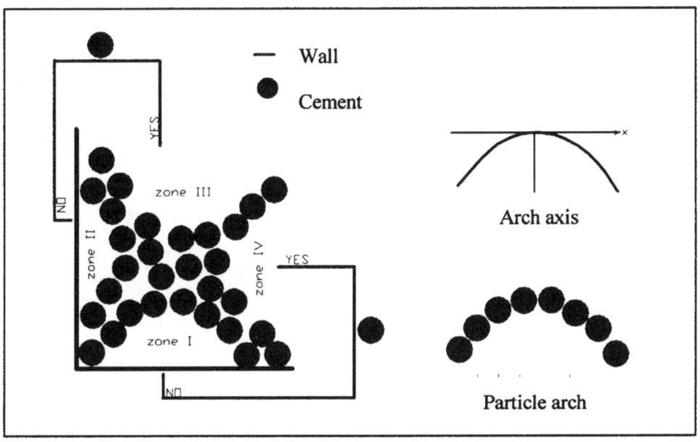

Figure 1 Description of non-placement mechanism and arch structure.

Basic Forms of Particle Arch Structure

As a first step towards the further analysis of cement particle packing near an aggregate, it is necessary to consider the basic forms of a particle arch structure. Several basic forms of arch structures of mono-size sphere packings in two dimensions are shown in Figure 2.

In the basic forms of arch structures, d_p is the diameter of the mono-size cement particle; d_i (i=3,4,5,6), means the distance between two particles in contact with the wall. The geometric conditions for the existences of the above structures are:

case (a): $0 < d_3 < d_p$;
case (b): $0 < d_4 < 2d_p$ and further stability analysis;
case (c): $0 < d_5 < 3d_p$ and further stability analysis;
case (d): $0 < d_6 < 4d_p$ and further stability analysis.

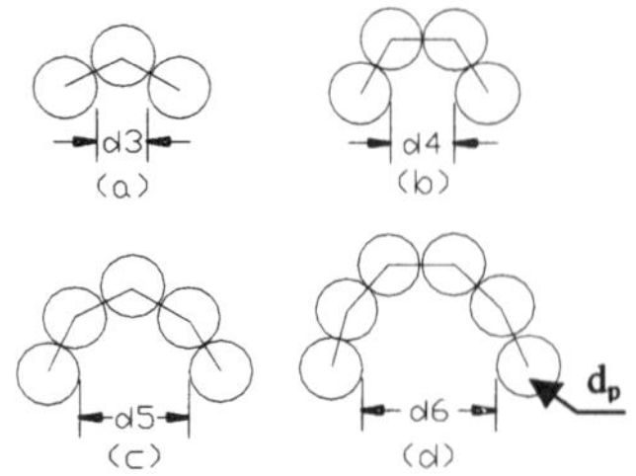

Figure 2 Four basic forms of the arch structure

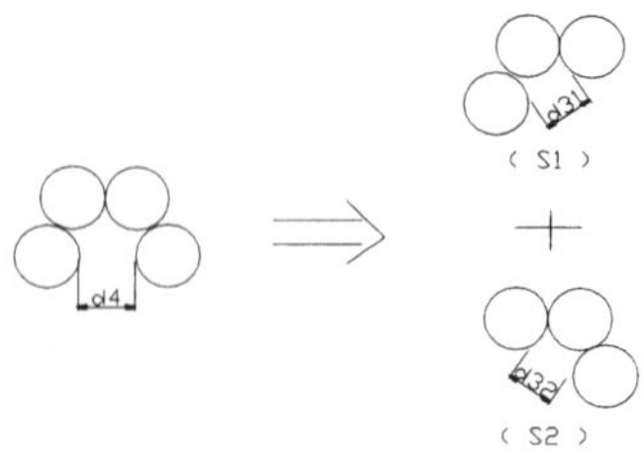

Figure 3 Dividing process of a four-particle system.

CEMENT PARTICLE PLACEMENT NEAR THE WALL

In the simulation, a random size and a random location are attributed to cement particles. Herein, it is assumed that a particle string is placed near the wall. If some conditions are satisfied, as described later, a particle arch will be formed.

The parameters used in the simulation procedure have following meaning:

M- the maximum number of the particles of the string;
L_w-the length of an aggregate (or the wall);
D_{min}-the minimum diameter of the cement particles;
D_{max}-the maximum diameter of the cement particles;
N_p-particle number of a particle string;

In the procedure, the following steps can be distinguished.

Generating N_p randomly

A particle string is composed of N_p particles. N_p is between 1 and M, where M is given by

$$M = \frac{L_w}{D_{\min}}$$

A random number, α, in the interval between 0 and 1, is generated by computer. Then, N_p is determined by

$$N_p = M \times \alpha$$

If $N_p<1$, let $N_p=1$.

Fix the Diameter of Each Particle in the String

$G(d)$ is defined as a cumulative distribution function of cement particle sizes. If α_g is a random number, it can be generated by computer in the interval [α_{min}, α_{max}] with uniform distribution, where α_{min} and α_{max} can be obtained by:

$$\alpha_{\min} = G(D_{\min}) \tag{1}$$
$$\alpha_{\max} = G(D_{\max}) \tag{2}$$

The diameter, *d*, of a cement particle will be determined by solving the following equation.

$$\alpha_g = G(d) \tag{3}$$

When $G(d)$ is associated with the Rosin-Rammler distribution function, equations (1), (2) and (3) will become

$$\alpha_{\min} = 1 - e^{-bD_{\min}^n}$$
$$\alpha_{\min} = 1 - e^{-bD_{\max}^n}$$
$$\alpha_g = 1 - e^{-bd^n}$$

where *b* and *n* are constants. For different values of fineness ratio of the cement, *b* and *n* are different.

Determine the Length of the String

It is supposed that all spherical centers in the string are located at same line, and d_s is the constant free spacing of the spheres, the length of the string can be obtained by: (see Figure 4).

$$L_s = (N_p - 1)d_s + \sum_{j=1}^{N_p} d_j \tag{4}$$

where L_s is the length of the string, and d_j is the diameter of the *j*th sphere.

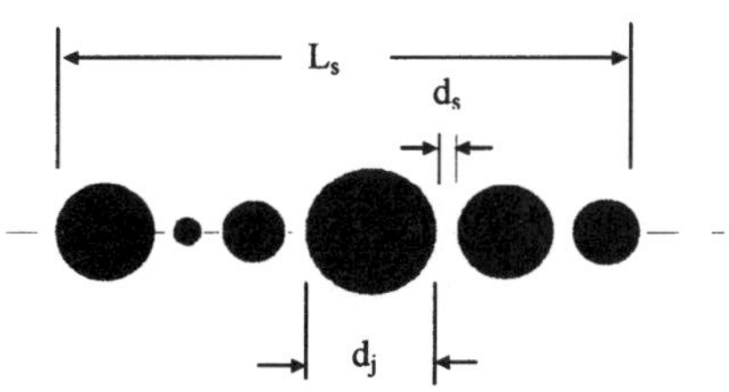

Figure 4 A cement particle string

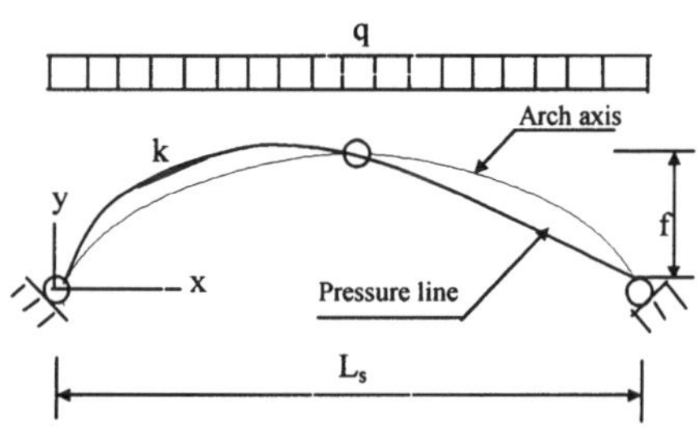

Figure 5 A three-hinged arch under uniform distribution load

Determine Reasonable Shape for the Arch Axis

When the axis (center line) of an arch coincides with the line of pressure of all the dead loads acting on the arch, the arch will be in maximum economy, or its axis is reasonable shape for the arch axis. In this case, the loads will produce no bending moment and no shear force in the structure. Let us consider the case of a three-hinged arch (see Figure 5).

If y and η are defined as the ordinates of the arch axis and of the line of pressure, respectively, these ordinates are a certain function of x, i.e.

$$y = f(x) \quad \text{and} \quad \eta = \varphi(x)$$

If the arch axis were reasonable as defined above, we should have

$$y = \eta$$

In this case, the arch axis will follow a second order parabola given by

$$y = \frac{4f}{L_s^2}(L_s - x)x \tag{5}$$

in which f is the rise of the arch.

Assess the Conditions for Particle Placement at the Wall

Consider the case where a space d_{ij} exists between two generated particles (see Figure 6). The diameters of the particles are d_i and d_j, respectively. The following conditions pertain to the particle placement near the wall.

If $d_{ij} > L_s$, a particle string is placed in the space;

If $d_{ij} < L_s < d_{ij} + \frac{1}{2}(d_i + d_j)$, a particle arch is formed;

If $L_s > d_{ij} + \frac{1}{2}(d_i + d_j)$, a particle string will be regenerated.

SIMULATED MICROSTRUCTURES OF THE ITZ

Microstructures formed by cement particle packing in the ITZ are simulated with the non-placement model (NPM). The cement particle size distribution is in agreement with the Rosin-Rammler function, in which the values of constants b and n are set to 0.038 and 0.98, respectively. Particle sizes larger than 13μm and smaller than 1μm will be neglected. Other parameters are given below. To compare the results, microstructures simulated by random packing method (RPM, particles are randomly placed one by one) are also presented.

Wall thickness: $L_t = 4\mu m$;
Wall length: $L_w = 200\mu m$;
Size of digital model: $L \times L = 200\mu m \times 200\mu m$.

The simulated results of a cement particle packing are shown in Figure 6, in which there are three parts: part A, part B and the wall (aggregate edge). The microstructure shown in part A is simulated by using NPM, while that in part B is simulated by RPM. By comparison, we can find that there are more particle arches (larger pores) near the wall in part A than those in part B. This is because in non-placement model, not only the wall effect, but also particle bridging is considered.

Figure 7 shows the fractional density ($L_v = V_v$, in 2-D) in correspondence with the microstructure in Figure 6. In Figure 7, the fractional density, as a function of distance to the wall, is plotted. The curve with solid line is the fractional density of the particle packing in part A. The curve with dashed line is the fractional density of the particle packing in part B.

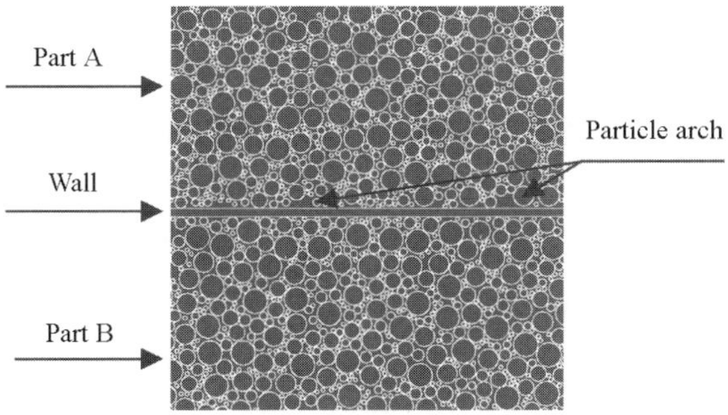

Figure 6 Simulated microstructure of particle packing

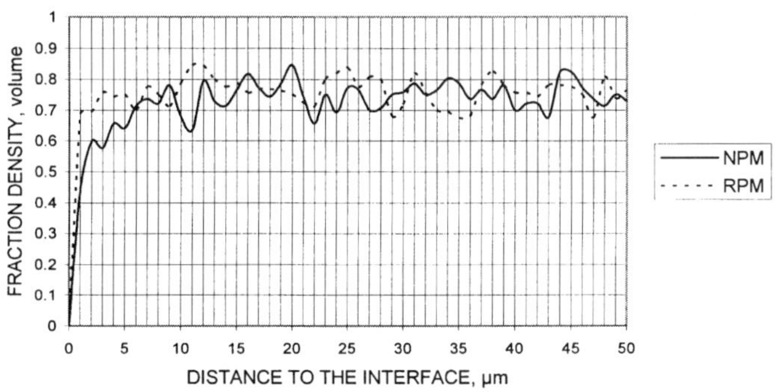

Figure 7 Fractional densities of particle packing

The lowest density in both parts can be seen close to the aggregate edge, due to the fact that particles can not be packed as efficiently near the wall as in the bulk. The density then increases to its bulk value, with some oscillations because of the discrete nature of the cement particles. It is evident, by comparison of two curves, that the solid curve extends more deeply into the bulk than the dashed does. This means that the thickness of the ITZ simulated by non-placement model is larger than that by the random packing.

SUMMARY AND DISCUSSIONS

A non-placement model for simulation of the microstructure in cement particle packing has been established. The simulated results have been presented. It can be clearly shown that larger pores occur near the aggregate surface. The features of the ITZ can be reproduced with this new model.

The model is based on the idea that *no particles can fill the space formed by particle arches and the wall. These voids result in larger pore space and thus in lower fractional density in the ITZ.* The particle arch is used to approach the geometrical shapes of larger pores generated by the wall effect, particle bridging and water bulbs during cement particle packing. The question can be raised why an arch structure is used to simulate the pores in the ITZ, instead of other structural forms. The main consideration is that an arch structure, generally speaking, is one of the most reasonable structural forms in nature. Under gravitational condition, a second order parabolic arch will only subjected to axial forces. In particle packing, the contact force between any two particles is in normal line direction. Therefore, a particle arch will be statically stable and will approach to the real conditions in particle packing.

However, the particle arch is not always geometrically stable, as its shape can be altered by deformations caused by external forces, such as by vibration excitations. The stability of a particle arch also depends on the particle sizes, the numbers of particles, etc.

REFERENCES

1. DIAMOND, S., HUANG, J., The Interfacial Transition Zone: Reality or Myth, Proceedings of the International RILEM Conf., 1998, pp3~42.

2. MINDESS, S., Interfaces in Concrete, Material Science of Concrete I. Jan P.Skalny (ed.), The American Ceramic Society, 1989, pp163~180.

3. BENTUR, A., ITZ Structure and Its Influence on Engineering and Transport Properties, Proceedings of the International RILEM Conf., 1998, pp335~337.

4. BENTZ, D. P., GARBOCZI, E. J., STUTZMAN, P.E., Computer Modelling of the Interfacial Zone in Concrete, Interfaces in Cementitious Composites, Proceedings of the International RILEM Conf., J.C.Maso (ed.), E&FN Spon, 1992, pp107~116.

5. STROEVEN, P., Some Mechanical Effects of Interface Debonding in Plain Concrete, Interfaces in Cementitious Composites, Proceedings of the International RILEM Conf., J.C.Maso (ed.), E&FN Spon, 1992, pp187~196.

6. ZHANG, M.H., GJORV, O.E., Microstructure of the Interfacial Zone Between Lightweight Aggregate and Cement Paste, Cem. Conc. Res. Vol.20, 1990, pp610~618.

7. STROEVEN, P., STROEVEN, M., Impact of Packing Phenomena on the Extent of the ITZ in Plain and Fibre Concrete, Proceedings of the Third International RILEM Workshop, 1999, pp131~142.

8. STROEVEN, M., Discrete Numerical Modelling of Composite Materials, Ph.D thesis, TU Delft, The Netherlands. 1999.

9. JENNINGS, H.M., JOHNSON, S.K., Simulation of Microstructure Development During the Hydration of a Cement Compound, Journal of the American Ceramic Society, Vol.69, 1986, pp790-795.

10. VERMAN, L.C., BANERJEE, S., "Effect of Container Walls on Packing Density", Nature, Vol.157, 1946, pp 584-585.

11. GERMAN, R.M., Particle packing characteristics, Printed in the U.S.A., 1989.

EFFECTS OF CEMENT TYPE ON CONCRETE MATURITY

O Kasap

M Tokyay

Middle East Technical University

Turkey

ABSTRACT. Maturity is a concept that accounts for the combined effects of temperature and time on the properties of concrete. Since 1940's, many maturity functions have been proposed. However, most of the functions were only tested for ordinary Portland cements. The investigations on maturity for different types of cements are inconclusive. For this purpose, different concrete classes (C20, C30 and C40) were prepared with four different types of cements which were ordinary Portland cement, Portland composite cement, sulfate resisting cement and blast furnace slag cement. The specimens were isothermally cured at 5, 20 and 40°C. Results revealed that Nurse-Saul and equivalent age functions were also applicable to cement types other than ordinary Portland cement used in this investigation. As a new method, the apparent activation energy for equivalent age function was determined from the relation between the degree of hydration and the rate of hydration, which was obtained from the experimental calorimetric hydration studies.

Keywords: Maturity, Cement type, Heat of hydration, Degree of hydration, Strength development

O Kasap, is a researcher within the Middle East Technical University, Turkey.

M Tokyay, is a Professor within the Middle East Technical University, Turkey

INTRODUCTION

For safe and economic construction, application and scheduling of various critical processes like form removal, post-tensioning, transportation of prefabricated elements, etc. prediction of concrete strength becomes important. One of the common methods for this makes use of the maturity concept, which was first developed in the early 1950's. Maturity is a concept that accounts for the combined effects of time and temperature on concrete properties. Among the several maturity functions that have been proposed since the 1950's, the following ones have received general acceptance:

The Nurse-Saul maturity function, which defines maturity as the product of time and temperature [2,3]:

$$M = \sum_{0}^{t} (T - T_0)\Delta t \qquad (1)$$

where M is maturity, T is the average temperature during the time interval Δt and T_0 is the datum temperature. Datum temperature is the temperature below which concrete is assumed not to be able to gain strength.

Although the Nurse-Saul maturity function has a wide usage due to its simplicity, it has some shortcomings. It is based on the assumption of linear relationship between the initial rate of strength gain and the temperature. Bergstrom showed that this linear relationship could be applied to the concretes cured at normal temperatures with a datum temperature of -10°C [4]. However, it may not be valid when the temperature varies over a wide range [5,6].

In order to overcome the deficiencies of the Nurse-Saul maturity function, a new function based on Arrhenius law that defines the effect of temperature on the rate of chemical reactions was proposed [7]:

$$t_e = \sum_{0}^{t} e^{-\frac{E}{R}\left(\frac{1}{T}-\frac{1}{T_r}\right)} \Delta t \qquad (2)$$

where t_e is the equivalent age, E is the apparent activation energy (J/mol), R is the universal gas constant (8.314J/mol.K), T_r is the reference temperature (°K) and T is the average temperature during the time interval Δt.

Equivalent age is the curing age at a constant reference temperature T_r that results in the same maturity and the same strength as cured under the actual temperature history [8]. Although it defines the behaviour of concrete against temperature better, especially at early ages, the equivalent age function may overestimate the effect at high temperatures [9,10]. The accuracy of this method depends on the accuracy of the value of apparent activation energy, which is the key parameter of the function. Therefore, recent studies about maturity mainly focus on the determination apparent activation energy of concrete [10-14]. The purpose of this paper is to investigate the effects of cement type on concrete maturity through:

- Checking the validity of the common maturity functions for different types of the cements
- Relating the maturity of concrete to the heat of hydration of cement

EXPERIMENTAL STUDY

Cements

Ordinary Portland Cement (PC), Portland Composite Cement (PCC), Sulfate Resisting Cement (SRC) and Blast Furnace Slag Cement (BFSC) produced in Turkey were used. Their properties are given in Table 1.

Aggregates

Crushed limestone was used in the concrete mixes. The properties of the aggregates are given in Table 2.

Table 1 Properties of the cements used

	PC	SRC	PCC	BFSC
%		CHEMICAL COMPOSITION		
SiO_2	19.3	20.9	30.0	23.8
Al_2O_3	5.7	4.5	8.2	10.7
Fe_2O_3	2.4	4.2	2.8	3.3
CaO	63.3	63.2	49.7	53.4
MgO	3.0	1.9	2.7	5.0
SO_3	3.0	2.7	2.1	2.5
LOI	2.7	2.4	3.4	1.4
IR	0.4	0.5	19.2	2.7
		PHYSICAL AND MECHANICAL PROPERTIES		
Density (g/m^3)	3.01	3.03	2.90	2.97
Blaine Fineness (cm^2/g)	3015	2808	3633	3970
Compr. str.(MPa) 3day	28.7	13.2	11.2	24.8
7day	34.6	20.7	18.9	32.1
28day	49.9	39.9	32.5	51.9

Table 2 Properties of the Aggregates

SIZE FRACTION	0-7mm	5-10mm	7-15mm	15-25mm
Specific Gravity				
Dry	2.71	2.71	2.71	2.69
SSD	2.72	2.72	2.72	2.70
Absorption, %	0.42	0.40	0.48	0.39

Concrete Mixture Proportions

Three strength classes as C20, C30 and C40 were aimed in preparing the concrete mixes. However, due to the differences in cement types, the aimed 28-day strength values were not achieved in some cases. The slump value was taken as the primary criterion in the mix designs. It was of 8-10cm for all mixes prepared. The mix proportions are given in Table 3.

Table 3 Mix Proportions of Concretes

CEMENT	MATERIALS	C20	C30	C40	CEMENT	MATERIALS	C20	C30	C40
PC	Cement	280	330	380	SRC	Cement	350	400	450
	Water	189	209	209		Water	206	213	211
	0-7mm	1120	1060	983		0-7mm	1047	987	898
	5-10mm	200	193	185		5-10mm	190	177	169
	7-15mm	163	149	142		7-15mm	146	141	131
	15-25mm	530	501	479		15-25mm	492	474	437
BFSC	Cement	280	330	380	PCC	Cement	350	400	450
	Water	203	212	228		Water	208	219	225
	0-7mm	1160	1060	983		0-7mm	1032	971	898
	5-10mm	200	193	185		5-10mm	188	163	163
	7-15mm	163	149	142		7-15mm	144	144	126
	15-25mm	530	501	479		15-25mm	486	486	437

Test Methods and Test Results

Heat of hydration determination

For each cement type, the heat of hydration of cement pastes cured isothermally at 5, 20 and 40°C are determined according to ASTM C186 (Standard Test Method for Heat of Hydration of Hydraulic Cement) at the ages of 3, 7, 14 and 28 days. For each type of cement, curing temperature and age, three specimens were tested. Results of heat of hydration tests are summarised in Table 4.

Table 4 Heat of Hydration Test Results

CEMENT TYPE	CURING TEMPERATURE	HYDRATION HEAT (cal/g)				
		1-day	3-day	7-day	14-day	28-day
PC	5 °C	44.7	49.4	58.9	62.9	78.4
	20 °C		67.5	74.9	85.2	92.1
	40 °C		78.2	80.8	82.8	95.5
PCC	5 °C	10.5	30.9	39.9	51.6	65.8
	20 °C		37.1	55.6	61.5	73.6
	40 °C		43.9	59.2	68.4	84.6
SRC	5 °C	23.7	22.6	48.0	61.1	71.9
	20 °C		43.5	58.9	69.6	78.1
	40 °C		58.6	75.1	82.1	87.3
BFSC	5 °C	55.1	66.9	73.1	82.3	91.4
	20 °C		77.9	84.4	88.6	95.8
	40 °C		78.2	96.9	102.1	107.2

Compressive strength determination

For the determination of compressive strength of concrete samples, 15cm cube specimens were moist-cured at 5, 20 and 40°C. The standard compressive strength tests were applied at 1, 3, 7, 14 and 28 days. For each cement type, concrete class, curing condition and age, three specimens were tested. Test results are shown in Table 5.

Table 5 Compressive Strength Test Results

CEMENT TYPE	AGE (days)	COMPRESSIVE STRENGTH (MPa)								
		C20			C30			C40		
		5°C	20°C	40°C	5°C	20°C	40°C	5°C	20°C	40°C
PC	1	4.0	10.5	15.5	8.4	17.1	21.0	10.4	19.5	23.7
	3	8.1	13.8	18.0	17.1	24.4	25.3	20.8	22.6	27.2
	7	12.6	18.6	20.7	24.7	26.3	27.7	27.3	28.2	32.4
	14	16.7	19.4	22.4	26.1	26.9	29.2	31.9	33.3	35.9
	28	18.1	20.3	23.1	26.8	30.8	32.1	34.2	40.8	43.3
PCC	1	1.3	3.8	7.7	2.7	6.6	10.5	2.8	7.3	12.7
	3	4.6	9.2	12.2	7.7	13.2	17.5	7.9	13.8	18.3
	7	8.3	12.1	15.9	12.5	16.9	20.0	13.6	18.2	23.9
	14	13.1	17.6	18.1	16.3	23.2	23.9	18.3	24.7	28.6
	28	18.5	19.9	21.1	24.6	27.1	28.4	27.6	31.9	34.7
SRC	1	1.8	6.9	13.9	2.0	7.9	14.5	2.5	9.0	13.8
	3	8.1	15.1	20.6	9.0	16.0	19.4	9.3	18.1	24.7
	7	12.9	22.2	28.3	13.0	23.6	26.3	16.1	26.4	34.2
	14	18.3	26.7	30.2	18.7	27.8	33.6	21.4	33.5	35.0
	28	24.2	29.4	34.8	25.8	31.7	39.1	29.0	37.0	38.1
BFSC	1	1.8	6.1	9.2	2.4	8.5	11.8	3.0	10.0	14.1
	3	7.2	11.3	14.6	7.9	13.5	18.4	10.3	17.2	23.1
	7	10.4	16.5	21.1	12.2	19.7	22.6	16.1	22.6	29.5
	14	14.4	22.6	24.4	17.6	25.6	27.1	20.4	30.2	34.7
	28	18.7	25.3	26.2	21.7	29.8	31.7	25.0	35.8	39.7

DISCUSSION OF RESULTS

Applicability of the Common Maturity Functions

Nurse-Saul maturity function

The compressive strength values obtained during the investigation for different ages and curing temperatures are used to calculate maturity indices according to ASTM C1074. The determined datum temperature values according to ASTM C1074 for each cement type are given in Table 6.

Table 6 Datum Temperatures of Cements

CEMENT TYPE	DATUM TEMPERATURE, T_0 (°C)
PC	-3.99
PCC	0.35
SRC	3.27
BFSC	-2.57

Using the datum temperature values for the related cement type, strength-maturity relationships are obtained for each concrete class. The plots are shown in Figures 1-4.

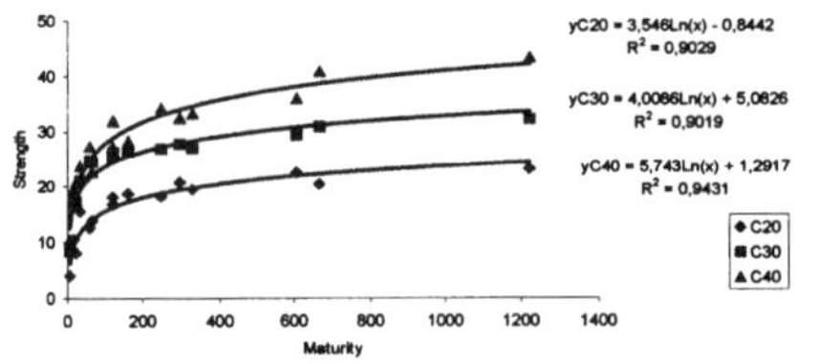

Figure 1 Strength-Nurse-Saul maturity for PC

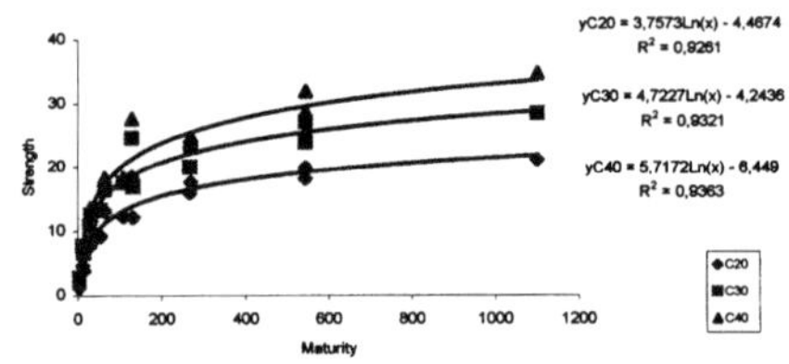

Figure 2 Strength-Nurse-Saul maturity for PCC

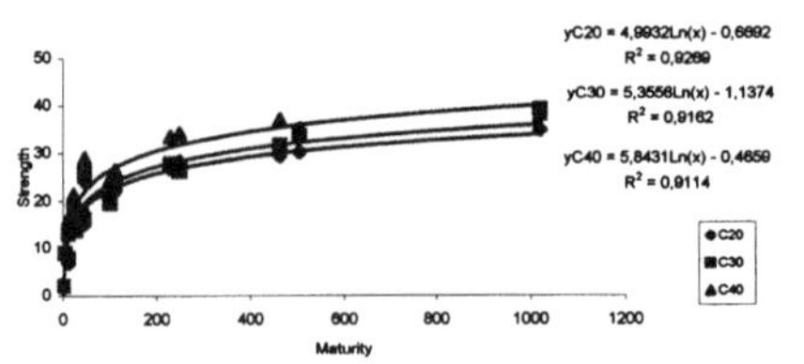

Figure 3 Strength-Nurse-Saul maturity for SRC

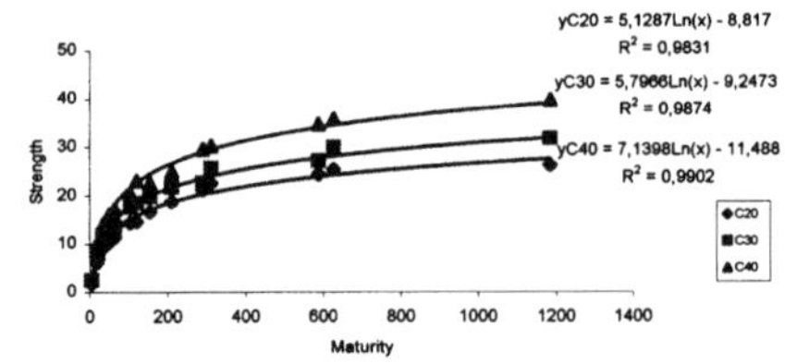

Figure 4 Strength-Nurse-Saul maturity forBFSC

It is obvious from Figures 1-4 that the logarithmic curves correlate well with the experimental data. Therefore, the Nurse-Saul maturity function is appropriate for the cement types used in this investigation when the ASTM C1074 procedure is applied.

Equivalent age function

In order to calculate equivalent ages at a specified reference temperature, the apparent activation energy values of each cement type are determined according to ASTM C1074. The results are shown in Table 7.

Table 7 Apparent Activation Energy of Cements

CEMENT TYPE	APPARENT ACTIVATION ENERGY, E (J/mol)
PC	35474
PCC	41459
SRC	60737
BFSC	43514

The apparent activation energy values obtained in this study according to ASTM C1074 are similar to the values reported in the literature [7,8,15,16]. However, the apparent activation energy obtained for PC is a little lower than the value reported in ASTM C1074. The main reason for this difference may be the higher C_3S content (60,7%) of the PC cement used in this research than that of a typical ordinary Portland cement (50%).

The equivalent age-strength relations are obtained with a reference temperature of 20°C for each concrete and given through Figures 5-8.

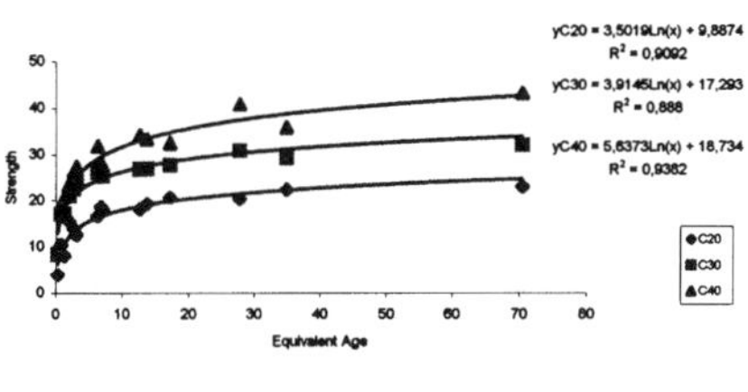

Figure 5 Equivalent age-strength for PC

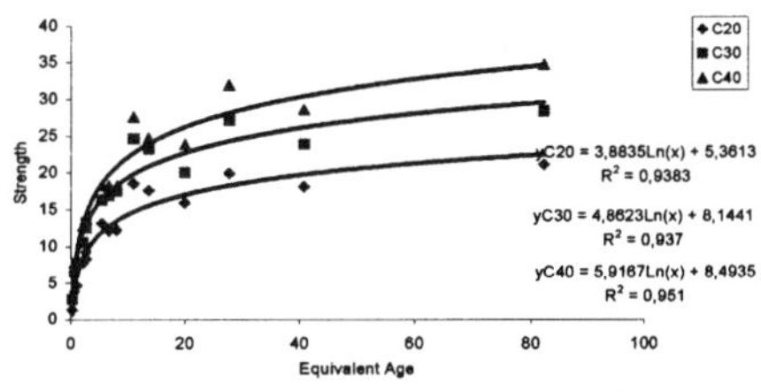

Figure 6 Equivalent age-strength for PCC

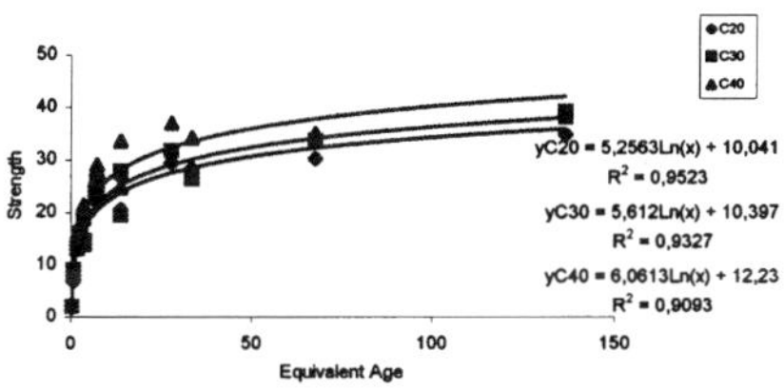

Figure 7 Equivalent age-strength for SRC

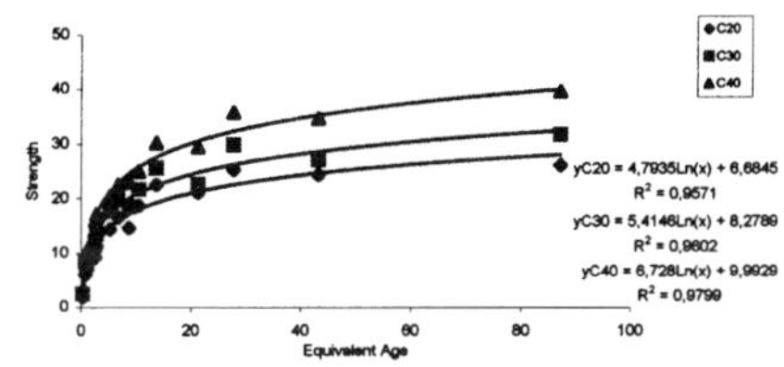

Figure 8 Equivalent age-strength for BFSC

Development of a New Method for Determination of Apparent Activation Energy

In a broad sense, the maturity function of the concrete can be written as:

$$M(t,T) = \int_0^t K[T(\tau)]d\tau \qquad (3)$$

where M (t; T) is the maturity of concrete cured at T for a time period of t, K is the kinetics constant and T (τ) is the temperature at time τ.

Arrhenius law is the most appropriate method in the determination of the kinetics constant for concrete [11]:

$$K(T) = Ae^{-\frac{E}{RT}} \tag{4}$$

where A is a constant.

In this approach, the effects of cement hydration on the development of concrete properties are reflected by the term E. The rate of hydration differs from one cement type to another and it is related to the degree of hydration. The degree of hydration is a criterion that shows the extent of the cement hydration. It can be calculated as:

$$\alpha(t) = \frac{X(t)}{X(\infty)} \tag{5}$$

where α(t) is the degree of hydration at time t, X(t) is the value of a mechanical or physical property at time t and X(∞) is the maximum value of the same property.
The rate of hydration can be defined for two isothermal curing conditions of the same cement as in Equations (6.a) and (6.b).

$$(d\alpha_1/dt)_{T_1} = K(T_1)f(\alpha) \tag{6.a}$$
$$(d\alpha_2/dt)_{T_2} = K(T_2)f(\alpha) \tag{6.b}$$

where $(d\alpha/dt)_T$ is the rate of hydration for isothermal curing at temperature T, and f(α) is a function related to the degree of hydration. Since f (α)'s are the same for the same degree of hydration, the following Equation is obtained when the rates of hydration at different temperature are proportioned:

$$\frac{(d\alpha_1/dt)_{T_1}}{(d\alpha_2/dt)_{T_2}} = \frac{K(T_1)}{K(T_2)} \tag{7}$$

As mentioned above, Arrhenius law describes the rate of hydration best:

$$\frac{(d\alpha_1/dt)_{T_1}}{(d\alpha_2/dt)_{T_2}} = e^{\frac{E}{R}\left(\frac{1}{T_2}-\frac{1}{T_1}\right)} \tag{8}$$

From the above relationship, the apparent activation energy is derived as:

$$E = R\left(\frac{T_1 \cdot T_2}{T_1 - T_2}\right)\left[\ln\frac{(d\alpha_1/dt)_{T_1}}{(d\alpha_2/dt)_{T_2}}\right] = R\left(\frac{T_1 \cdot T_2}{T_1 - T_2}\right)\left[\ln\frac{K(T_1)}{K(T_2)}\right] \tag{9}$$

Determination of apparent activation energy from heat of hydration studies

The theoretical approach [11] explained in the previous section was applied to each type of cement separately. The physical property that was chosen to determine the degree of hydration is the heat of hydration.

Since cements evolve most of their total heat of hydration during the first 3 months, 90-day heat of hydration values of cement pastes cured at 20°C is assumed to be a good approximation as the maximum value.

The method of calculation is summarised below:

1. Measurement of the heat of hydration values for the different ages and temperatures according to ASTM C186,
2. Plotting heat of hydration-age curves for each curing temperature,
3. Calculating the heat of hydration of cement pastes cured at 20°C for 90 days,
4. Defining the degree of hydration for each temperature and age as H/H_{90},
5. Plotting the degree of hydration-age graphs,
6. Determining the rate of hydration from the degree of hydration curves,
7. Defining the relationship between the rate of hydration and degree of hydration,
8. Determination of the kinetics constants for each temperature from the Arrhenius Equation,
9. Calculating the apparent activation energy according to Equation (9)

Applicability of the Proposed Method

The kinetics constants for PC were determined as 1.3943 and 3.4427 for 5 and 20°C respectively. Inserting these values into Equation (9), the apparent activation energy for PC is calculated as E=40806.9J/mol. Equivalent age-strength relation for PC obtained by using this value of activation energy is shown in Figure 9.

The same procedure was applied to SRC. The kinetics constants for 5 and 20°C were found as 0.2225 and 0.7376 respectively. This led to an apparent activation energy value of E=54107.8J/mol. The eqqagshghdguivalent age-strength relationships for the three concrete classes produced from SRC obtained with the new calculated activation energy are shown in Figure 10.

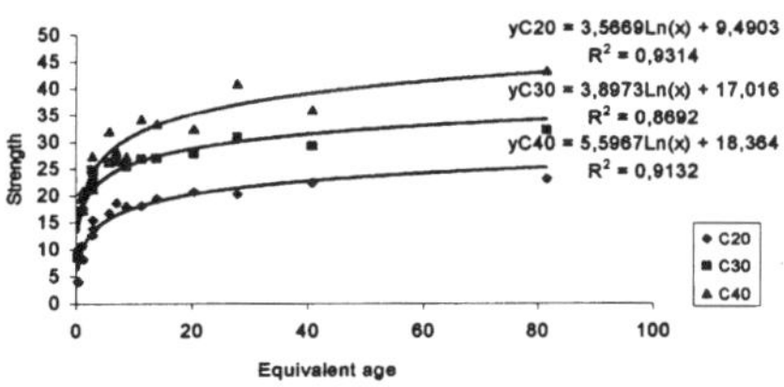

Figure 9 Equivalent age-strength relations for PC with the new method

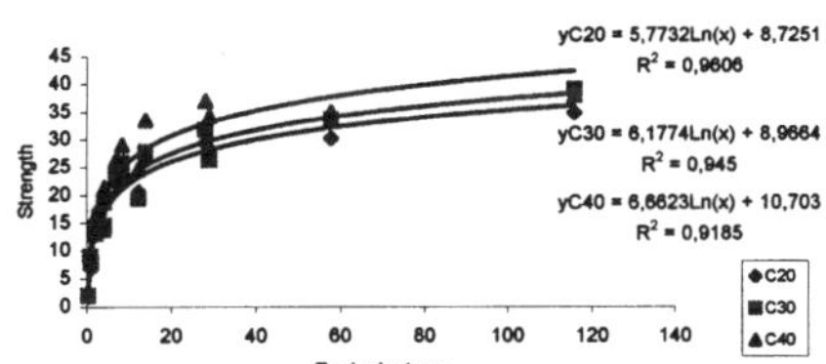

Figure 10 Equivalent age-strength relations for SRC with the new method

After substituting the kinetics constants obtained for PCC in Equation (9), a value of 8078 J/mol is obtained as apparent activation energy, which is meaningless for any type of cement. So Equation (9) is modified so that the amount of admixture is taken into consideration for cements containing admixtures:

$$E = R\left(\frac{T_1 \cdot T_2}{T_1 - T_2}\right)\left[\ln\left[(1-a)\frac{K(T_1)}{K(T_2)}\right]\right] \tag{10}$$

where, a is the amount of admixture.

The mineral admixture content of PCC was determined as 32.48%. When this value is substituted in Equation (10), the activation energy is calculated as E=25818.4J/mol, which is much more meaningful. The equivalent age-strength relations for PCC are given in Figure 11.

Since BFSC contains chemical admixtures, again only the clinker part is considered and Equation (10) is applied. It is determined that BFSC contains 57% slag. After substituting the values into Equation (10) the apparent activation energy of BFSC is obtained as E=41344.5 J/mol. The equivalent age-strength relations for concretes produced with BFSC is shown in Figure 12.

Comparison of ASTM C1074 and the New Method

The correlation coefficients of equivalent age-strength curves obtained according to the proposed calorimetric method are compared with those obtained according to ASTM C1074 for each type of cement. The coefficients of correlation values are summarised in Table 8.

Table 8 Comparison of ASTM C1074 and the New Method

CEMENT	CORRELATION COEFFICIENTS FOR DIFFERENT CONCRETE CLASSES, R^2					
	C20		C30		C40	
	ASTM	New M.	ASTM	New M.	ASTM	New M.
PC	0.9092	0.9314	0.8880	0.8692	0.9382	0.9131
SRC	0.9523	0.9606	0.9327	0.9450	0.9093	0.9185
PCC	0.9383	0.9696	0.9370	0.9685	0.9510	0.9739
BFSC	0.9602	0.9579	0.9571	0.9625	0.9799	0.9808

As seen from Table 8, both methods have similarly high correlation coefficients which means both methods are applicable to the cement types considered.

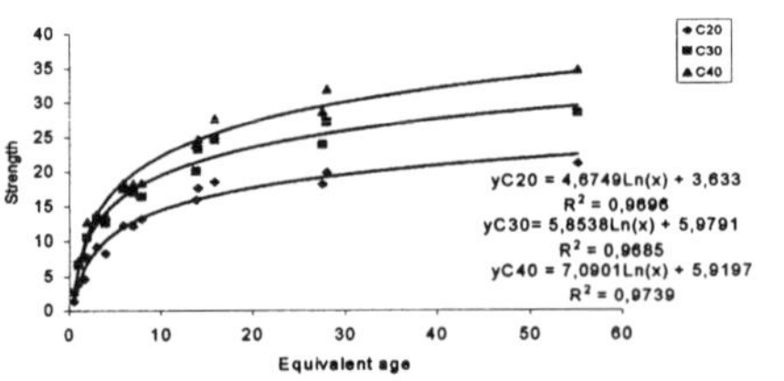

Figure 11 Equivalent age-strength relations for PCC with the new method

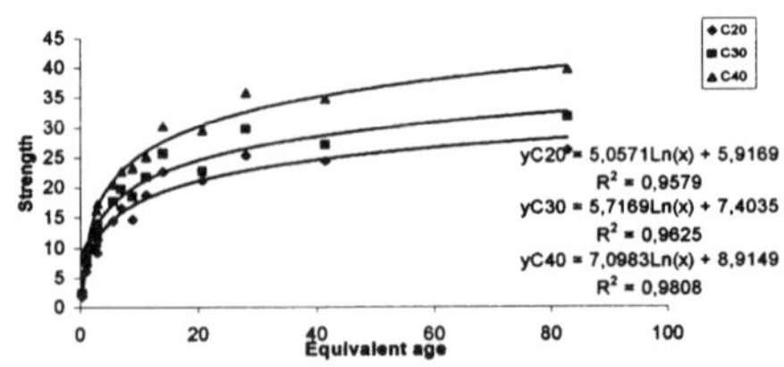

Figure 12 Equivalent age-strength relations for BFSC with the new method

CONCLUSIONS

The following conclusions were drawn from this experimental study:

1. The Nurse-Saul maturity function, as described in ASTM C1074, is also applicable to the cement types used in this study as well as for ordinary Portland cement.
2. The equivalent age function as described in ASTM C1074 can also be used for the strength prediction of Portland composite cements, sulfate resisting cements and blast furnace slag cements as well as normal Portland cements.
3. Equivalent ages can be calculated by using the apparent activation energy obtained from the degree of hydration and the rate of hydration relations for Portland cements. So, it can be concluded that Equation (9) defines a practical and reliable method for Portland cements.
4. The proposed method needs to be modified by a constant related to the amount of admixture for cements containing admixtures.
5. The proposed method becomes practical once the heat of hydration values of cement to be used and concrete strength class are known.
6. Similar investigations are necessary to end up with more conclusive results.

REFERENCES

1. McINTOSH, J.D., "Electrical Curing of Concrete", Magazine of Concrete Research, V.1, 1949, pp.21-28.

2. NURSE, R.W., "Steam Curing of Concrete", Magazine of Concrete Research, V.2, 1949, pp.79-88.

3. SAUL, A.G.A., "Principles Underlying the Steam Curing of Concrete at Atmospheric Pressure", Magazine of Concrete Research, V.2, 1951, pp.127-140.

4. BERGSTROM, S.G., "Curing Temperature, Age and Strength of Concrete", Magazine of Concrete Research, V.5, 1953, pp.61-66.

5. McINTOSH, J.D., "The Effects of Low-Temperature Curing on the Compressive Strength of Concrete", RILEM Symposium on Winter Concreting, Copenhagen, Danish Institute for Building Research, 1956, Session B-II.

6. ALEXANDER, K.M. and TAPLIN, J.H., "Concrete Strength, Cement Hydration and the Maturity Rule", Australian Journal of Applied Science, V.13, 1962, pp277-284.

7. FREISLEBEN HANSEN, P. and PEDERSEN, E.J., "Maturity Computer for Controlled Curing and Hardening of Concrete", Nordisk Betong, V.1, 1977, pp19-34.

8. CARINO, N.J., "The Maturity Method: Theory to Application", Cement, Concrete and Aggregates, CCAGDP, V.6, Winter 1984, pp.61-73.

9. CARINO, N.J. and LEW, H.S., "The Maturity Method: From Theory to Application", Proceedings of 2001 Structures Congress & Exposition, Washington D.C., ASCE, P.C. Chang (Ed.), 2001, 19p.

10. KJELLSEN, K.O. and DETWILLER, R.J., "Later-Age Strength Prediction by a Modified Maturity Model", ACI Materials Journal, V.9, 1993, pp.220-227.

11. KADA-BENAMEUR, H., WIRQUIN, E. and DUTHOIT, B., "Determination of Apparent Activation Energy of Concrete by Isothermal Calorimetry", Cement and Concrete Research, V.30, 2000, pp.301-305.

12. KIM, J.K., HAN S.H. and LEE K.M., "Estimation of Compressive Strength by a new Apparent Activation Energy Function", Cement and Concrete Research, V.31, 2002, pp.217-225.

13. WIRQUIN, B., BRODA M. and DUTHOIT B., "determination of the Apparent Activation Energy of One Concrete by Calorimetric and Mechanical Means: Influence of Superplasticizer", Cement and Concrete Research, V.32, 2002, pp.1207-1213.

14. D'ALAIO, L. and CHANVILLARD, G., "Determining the Apparent Activation Energy of Concrete E_a-numerical Simulations of the Heat of Hydration of Cement", Cement and Concrete Research, V.32, 2002, pp. 1277-1289.

15. GUO CHENGJU, "Maturity of Concrete: Method for Predicting Early-Stage Strength", ACI Materials Journal, V.86, 1989, pp.341-353.

16. CARINO, N.J. and TANK R.C., "Maturity Functions for Concretes Made With Various Cements and Admixtures", ACI Materials Journal, V.89, 1992, pp.188-196.

THEME THREE:

SUPPLEMENTARY CEMENTING MATERIALS, ACTIVATORS AND INHIBITORS

Keynote Paper

PORTLAND CEMENT BINDERS: SCIENCE AND SUSTAINABILITY

S Wild

University of Glamorgan

United Kingdom

ABSTRACT. Portland cement (PC) is the key component of concrete, an essential and indispensable component of today's world. However, changes in the composition of Portland cement binders and the exponential growth in Portland cement production are creating a whole range of potential future problems. The success of Portland cement concrete and its universal acceptance and use has been achieved largely as a result of advances in cement science, which have played a leading role in its development and in solving many of the problems that have arisen in its use. As the world moves further towards establishing more sustainable systems for the production, application and use of Portland cement concrete, the diversity of Portland cement binders will continue to increase. That is, cement compositions will be increasingly tailored both to specific applications and to the more sustainable use of available materials, the latter including utilisation of waste materials and industrial by-products as cement replacement materials. The development of an increasing diversity of cement types and the problems that will undoubtedly arise from their application will generate an even greater demand for cement science to provide answers and solutions. This paper looks at some of the successes and failures of Portland cement based binders, and highlights the role of cement science in their development and in providing solutions to current and possible future problems.

Keywords: Cement science, Supplementary cementing materials, Waste, Durability, Sustainability

Professor S Wild is Professor of Construction Materials and Head of the Building Materials Research Unit in the School of Technology at the University of Glamorgan. His current research involves the development of novel sustainable cementitious materials and particularly the utilisation of waste materials in construction, with the object of minimising environmental impact and reducing energy demand and raw materials extraction. He has published over one hundred papers in the materials field and has also carried out numerous consultancies principally in the general area of physical and chemical testing of cements, mortars and concretes.

INTRODUCTION

Concrete is a very ancient invention [1]. One of the earliest known examples is a hut floor in Yugoslavia dating from 5600BC. The concrete consisted of a mixture of red lime as binder, plus sand and gravel. The Egyptians also used cement in construction. The Great Pyramid at Giza contains mortar for binding some of the stone masonry. The Egyptians however tended to use burnt gypsum rather than lime, as the cementing material. From about 500BC concrete and mortar were used extensively, first by the Greeks and particularly by the Romans. Concrete during this period was used mainly as an in-fill material between masonry walls as it had insufficient strength as a stand-alone material. One very important discovery first made by the Greeks but attributed generally to the Romans was that certain sands and rock fragments, when mixed with lime and water, produced a cement of much superior hardening power and strength to that of a normal sand-lime-water mix. The Greeks first produced this type of concrete, at around 1300BC, from a mixture of water, lime and volcanic tuff. The tuff, which is fragmentary volcanic rock and ash, came from the island of Thera (now called Santorini). The Greeks obtained the ash from the cliffs on the south coast of Therasia, which is a small island off Greece and which forms part of the caldera of an extinct volcano which erupted in about 1500 BC. This same volcanic ash was used during the 1860s, by the Suez Canal Company, in the construction of the Suez Canal. The Romans perfected the development of this type of concrete to a fine art. They discovered that pink sand, from a source near Pozzuoli in Italy, produced exceptionally strong concrete when combined with lime and water. The sand was again a fine volcanic ash. Normally, when lime is mixed with sand and stone, the resultant concrete hardens very slowly by carbonation of the lime and attains a very low ultimate strength. If however the sand is a volcanic ash the oxides of silica and alumina from the glassy phase go into solution in the highly alkaline environment created by the lime to produce calcium silicate hydrate gel. This reaction is now termed a pozzolanic reaction and the cement is known as pozzolanic cement.

The numerous Roman structures still in existence today are an enduring testament to the durability of this material. There are some outstanding examples. The Theatre at Pompeii was built in 75 BC, and pozzolanic cement was used in the concrete to in-fill the walls. The Colosseum in Rome, was built in 80 AD. Here dense concrete was used in foundations and column bases, and light-weight concrete was used in some of the arches. The light-weight concrete was produced by using pumice (porous volcanic rock), for the coarse aggregate. The Pantheon in Rome was built in 127 AD. This is one of the few buildings to survive intact after the decline of the Roman Empire. Its magnificent domed roof, constructed of lightweight concrete, is 50 metres in diameter and is said to have inspired Sir Christopher Wren in his design of St Paul's Cathedral. In the UK, where no volcanic ash was available, the Romans produced another form of pozzolanic concrete, in which the binder consisted of lime, sand and crushed clay brick or powdered pot fragments. A good example of this is Hadrian's Wall, built between 122 and 130 AD, the core of which is a lime concrete of enduring strength. Whether this was an intelligent association between volcanic activity and burnt earth, or a result of trial and error will never be known. In the UK much of the knowledge gained by the Romans appears to have been lost after the fall of the Roman Empire, and where concrete was used the binder appears to have been solely lime and not lime-pozzolan. It would appear that an innovative approach towards cementitious binders did not arise again until the middle of the eighteenth century when there was a renewed activity in the development of cementitious binders. One example is William Champion, a metal manufacturer, who in 1750 built a giant statue of Neptune in the grounds of his new mansion at Warmley near Bristol. The cement consisted of a mixture of lime and copper slag, which produced very durable cement and after 250 years the statue is still in a good state of preservation. Thus using waste materials such as slag and crushed brick, or natural pozzolanic materials such as volcanic ash and pumice to produce cement is not a new concept and goes far back into history, and the structures still standing demonstrate the durability of these materials.

PORTLAND CEMENT

On the 21st October 1824 a patent was filed by an inventor, one Joseph Aspdin[1] from Wakefield, for a new type of cement, called Portland cement as it was thought to resemble the colour of Portland stone. It was produced by mixing together clay and lime, which were then heated to very high temperatures and the resultant product was ground to a fine powder. This powder required only the addition of water for it to set and harden, no added lime was needed. Also its hardening rate was much more rapid than all previous cements. The novelty of the material was such that Rafaelle Monti, a famous Victorian sculptor was commissioned to create a concrete statue for the Great Exhibition of 1851. The statue, of Old Father Thames, survived the Crystal Palace fire in 1936 and is now re-sited, near the source of the Thames in Lechlade, Gloucestershire. This cement rapidly replaced the older pozzolanic cements. The use of Portland cement concrete has accelerated at an exponential rate from the 1850s to the present day. Portland cement has, however, evolved and developed over the years and Portland cements used today are significantly different in composition and properties from those that were used 179 years ago and even those that were used 30 years ago.

The major changes in Portland cement over time are in the balance of compounds making up the cement and the fineness to which the cement is ground. Emphasis in the past has been very much on the strength of Portland cement concrete and over time architects have designed much more slender and increasingly exotic structures, which have required concrete of increasingly higher performance. Also commercial and financial demands for ever-increasing efficiency, have dictated that structures be completed in ever-shorter time periods. This increase in performance has been achieved principally by increasing the proportion of C_3S and decreasing the proportion of C_2S in the Portland cement to produce faster hardening and strength gain. Over time the proportion of C_3S, which reacts much more rapidly with water than C_2S, has increased progressively [2, 3] Table 1.

Thus the early strength of standard Portland cement concrete has increased progressively with time (Figure 1). Strength is however not the only measure of high performance. Over-emphasis on strength gain and increased efficiency, has often led to the neglect of durability. The acceleration in the hydration reactions produces increased temperatures and as heating is subsequently followed by cooling the differential shrinkage of the concrete results in the formation of micro-cracks, particularly where high cement contents are used in high performance concrete[4]. These micro-cracks increase the permeability of concrete and destroy its water tightness. In addition the increase in the proportion of C_3S relative to C_2S results in higher CH production during hydration. All of these factors can lead to a reduction in the ability of concrete to resist attack by aggressive agents.

Table 1 Typical Portland cement compositions over time#

CEMENT COMPOUND	1905 %	1918 %	1929 %	1960 %	1970 %	1980 %	1992 %	1998* %	REACTION RATE WITH WATER
C_3S	25	31	39	47	51	53	55	63	fast
C_2S	45	39	34	23	19	18	18	11	slow
C_3A	13	13	13	11.5	10.5	9.7	8.5	7.9	controlled by gypsum
C_4AF	10	10	10	9	7	7	6.5	6.75 (est.)	limited effect
Fineness m^2/kg	?	?	?	343	332	337	355	350	

data after [3]
*average of values from Rugby Cement and Blue Circle Cement

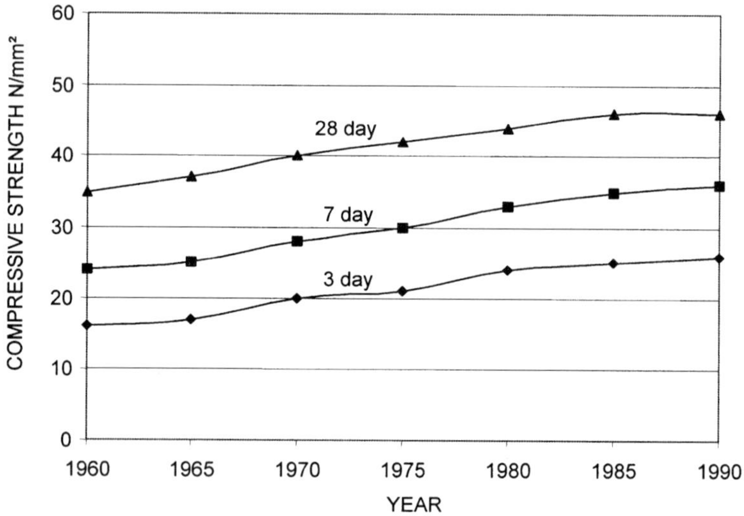

Figure 1 Change in average compressive strength of standard concrete from 1960 to 1990 [3]

CONCRETE DURABILITY

To be durable, concrete must have two principal characteristics:

It must be able to resist penetration by aggressive liquids and gases (i.e. it must be impermeable or have low permeability). Finely divided solids, including pozzolans and ground slags blended with PC assist in reducing permeability by acting as fillers and producing a denser cement paste and also generating additional C-S-H which reduces pore interconnectivity and produces a finer pore structure.

It must be chemically stable (i.e. the chemical compounds formed during cement hydration must not react readily with components in the immediate environment). The CH formed on PC hydration acts as a buffer maintaining a high pH and ensuring the stability of the C-S-H and the passivation of steel reinforcement. Even greater pH levels are achieved when alkali metals are present. However CH is also implicated in damaging reactions including sulphate attack and alkali silica reaction. Dilution of the PC by blending it with pozzolans and ground slags can reduce the CH levels and provide some chemical protection whilst still maintaining pH and the stability of the C-S-H.

There are numerous examples of serious failures of concrete structures over the past few decades, where these two principal characteristics have not been achieved. Our understanding of the poor performance and failure of the concrete and the formulation of solutions to resolve the problems have been derived primarily from cement science. A number of leading examples include corrosion of steel reinforcement, alkali silica reaction (ASR), delayed ettringite formation (DEF) and thaumasite attack (TSA).

Alkali Silica Reaction

In ASR the hydroxyl ions in the pore solution react with particular forms of silica in aggregate such as opal to produce an expansive alkali silica gel containing from 28 - 86% SiO_2, 0.1 - 60% CaO, 0.4 - 19% K_2O and 0 - 20% Na_2O. The sodium and potassium alkalis from the cement concentrate in the pore fluid, suppress the solubility of $Ca(OH)_2$ and enhance the pH of the pore solution promoting attack of active siliceous material. The presence of $Ca(OH)_2$ is essential for alkali gel formation although high and low calcium gels are none expansive. An external source of water is required to induce expansion and the swelling mechanism is attributed to either osmosis or imbibition. Both require a variation in chemical potential and an absence of free mixing, the former requiring a semi-permeable membrane and the latter a solid sufficiently rigid to prevent free mixing, sufficiently flexible to expand, sufficiently porous to imbibe water, sufficiently polar to attract water molecules and relatively insoluble. However the distinction between imbibition and osmosis in relation to colloidal materials is not clearly differentiated and they are both manifestations of the same effect. Pozzolans reduce the C/S ratio of the paste, inhibit ASR by rapidly depleting the pore solution of alkali and reducing the hydroxyl concentration, and by depletion of CH. High silica and low alkali content pozzolans are most effective in reducing expansion due to ASR, the replacement level required being dependent on the alkali content of the concrete. Active pozzolans such as silica fume are highly effective in reducing the pore fluid alkali levels in cement paste. This is because the C-S-H that develops has a low C/S ratio, which enables it to more easily absorb alkali ions. PFA however is much less effective as it consumes less CH and also tends to release alkali into the system. Hong and Glasser[5] have demonstrated, using synthesised alkali-free, homogenous C-S-H, prepared from CH - silica gel mixtures, that alkali binding in the solid, relative to alkali concentration in solution (distribution ratio R_d), improves as the C/S ratio decreases (Figure 2, after[5]) and is little influenced by alkali concentration between 1 and 300 mM. Hong and Glasser[6] have also demonstrated that the introduction of Al into C-S-H to form C-A-S-H gel markedly increases the distribution ratio R_d indicating further enhancement in alkali binding. A relatively small level of substitution of Si by Al (e.g. 1 in 16 to 1 in 13) markedly enhances the alkali sorption of the gel and enhancement is greater when the substitution by Al, is in gel of lower C/S ratio.

The increased sorption power is attributed to valence compensation (i.e. $M^+ + Al^{3+}$ for Si^{4+}) in addition to the silanol (Si-OH) binding capacity. The results confirm that positive benefits will be achieved by adding aluminosilicate rich pozzolans such as PFA, GGBS and MK to Portland cement in sufficient quantities to lower the C/S ratio of the gel to values well below its "normal" value of 1.7.

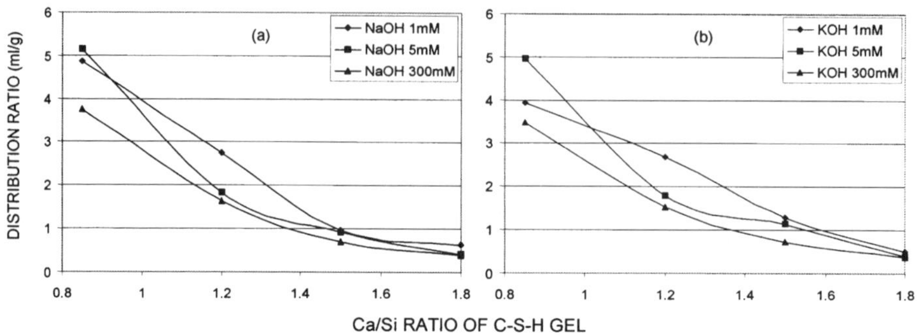

Figure 2 Solid aqueous distribution ratios for (a) sodium and (b) potassium versus C/S ratio of the C-S-H gel fro different concentrations of aqueous solutions [5]

Corrosion Of Steel Reinforcement

Many thousands of concrete structures including bridges and fly-overs, multi-storey car parks, tower blocks and sea defences, have suffered from this problem at a cost of many billions of pounds. In fact modern slender highly designed structures such as bridges require high-strength concrete, which has required typically high cement contents. Therefore high internal temperatures are generated during hardening and shrinkage and the likelihood of micro-cracking is increased. The steel, whilst giving the material high tensile strength also restrains thermal contraction and drying shrinkage and contributes to micro-cracking. However supplementary cementing materials can significantly reduce the heat output, reduce the permeability and in some cases enhance the strength. For example the Great Belt Crossing, linking the Danish islands of Funen and Zealand, is constructed from concrete with a binder comprising supplementary cementing materials (FA, SF) and is designed to have very low permeability to chloride ingress and a lifetime of at least 100 years.

The diffusion of chloride ions is directly influenced by the microstructure of the cement paste. The microstructure develops principally by deposition of C-S-H gel in the capillary pore system, which reduces pore interconnectivity and inhibits diffusion of chloride ions through the pore solution. Both an increase in hydration time and the partial replacement of cement by mineral admixtures including FA and GGBS increase the amount of C-S-H gel deposited. Also some cement replacement materials (e.g. GGBS and MK) result in the development of a very fine pore structure further inhibiting diffusion. However, modelling of chloride ingress in order to predict chloride diffusion is highly problematic because of the numerous chemical and physical processes involved[7]. Modelling is normally based upon Fick's law but this has to be modified to take account of the fact that the concrete comprises a dynamic system.

The apparent diffusion coefficients obtained vary with exposure time (particularly for concrete cured for only a short period of time before exposure), and with chloride binding capacity of the binder (due to formation of chloro-aluminate phases and chloride binding by the C-S-H which increases with increasing C/S ratio and increasing chloride concentration). Considerable research is required in this area to develop predictive and reliable models.

Sulphate Attack

One form of attack that has recently caused a significant degree of controversy is delayed ettringite formation. Here the problem arises mainly in precast products such as railway sleepers that have been subjected to high curing temperatures (80-100°C) although mass concrete may also be susceptible to this form of attack particularly where high cement contents are employed and where the heat of hydration causes substantial temperature rise. The sulphate derives from the cement paste and damage occurs only in materials that have been subjected to temperatures in excess of 70°C. In many cases ASR is also present and provides initial cracking, which then facilitates expansion from DEF. The expansion is attributed to expansion of the paste although the often-observed crystallisation of ettringite in cracks in the concrete is not in itself the cause of the cracking. A number of authors have shown that during high temperature curing initially formed ettringite disappears, in agreement with Glasser's [8] observations that ettringite solubility increases markedly with increase in temperature (and alkali content). Numerous authors including Taylor[9] have suggested that the sulphate, which does not remain in the pore solution or in any residual monosulphate and ettringite, is contained in the C-S-H gel. This sulphate together with calcium is then released during subsequent storage under moist conditions at normal temperatures and reacts with any available alumina (possibly from monosulphate) to form ettringite. The available evidence suggests that expansion occurs primarily as a result of ettringite growth within the paste. Taylor[9] has observed that for mortar that subsequently expands after initial heat treatment, the outer product C-S-H gel exhibits higher S/Ca ratios during heat treatment than in mortar that does not subsequently expand. Also after long-term moist storage, although both expanding and non-expanding mortars exhibit outer product C-S-H with reduced S/Ca ratios, the C-S-H in the former is intimately mixed with ettringite whereas in the latter it is not. This clearly shows that expansion is due to delayed ettringite formation in the paste. There have been other terms used for this phenomenon including *late ettringite formation* and *secondary ettringite formation* but these do not uniquely define the process. The reason for the confusion arose simply as a result of a lack of understanding of the mechanism and as this is now better understood possibly the term thermally delayed ettringite formation should be used (Glasser[10]). Another form of sulphate attack that has recently been brought to the attention of the cement science community is the thaumasite form of sulphate attack. Ettringite ($Ca_3Al_2(SO_4)_3(OH)_{12}.26H_2O$) and thaumasite ($Ca_3SiSO_4CO_3(OH)_6.12H_2O$) are closely related structurally[11] and solid solutions occur between them separated by a miscibility gap. Over the past five years there has been heightened interest and concern regarding thaumasite attack of buried concrete, which was initiated by the discovery in March 1998 of thaumasite sulphate attack (TSA) below ground level on the columns of the Tredington - Ashchurch road-bridge on the M5 motorway in Gloucestershire. The sulphates, dissolved in the ground water, penetrate the concrete and in the presence of carbonate and particularly at low temperatures (<15°C) cause decalcification of the C-S-H gel and formation of thaumasite, which destroys the cement bond and reduces the concrete to a soft mushy mass. Numerous other examples of this form of attack have since been reported. Current evidence suggests that significant resistance to TSA can be achieved by use of supplementary cementing materials

Cement science has made an enormous contribution to our understanding not only of the above high-profile problems but also to many others and the application of this knowledge is contributing substantially to the prevention and future occurrence of these phenomena.

CEMENT PRODUCTION AND ENVIRONMENTAL DAMAGE

Recently society has become aware of even longer-term destructive effects associated with the universal, extensive and large-scale use of cement-based concrete. This is the damage which cement production is doing to our environment. The production of cement is making a significant and increasing contribution to world CO_2 emissions and hence to global warming. The manufacture of 1 tonne of cement results in the emission of approximately 1 tonne of CO_2 into the atmosphere. Between 0.92 and 0.7 tonnes of CO_2 is emitted per tonne of cement produced (depending on the efficiency of the kiln and the fuel used). Over half of this (0.5 tonnes/tonne) derives from the dissociation of calcium carbonate, and cannot be reduced significantly without changing the basic chemistry of the cement[12]. In addition the quarrying of vast quantities of limestone also produces considerable environmental damage to the countryside.

Table 2 Estimated annual world cement consumption

WORLD REGION	ESTIMATED AND PROJECTED WORLD CEMENT CONSUMPTION (m.t./a.)		
Year	1994	2000	2005
Europe including Eastern Europe	313	393	432
Asia	680	853	1000
Middle East	65	79	82
Africa	63	71	77
North America	90	92	92
South and Central America	92	118	142
Other regions	7	9	10
Total	1310	1625	1835

It is however not feasible to continue to advance and develop the type of infrastructure required by our modern technological post industrial society (i.e. mass transport systems, reservoirs, hospitals, residential and industrial buildings, leisure complexes) without the consumption of vast quantities of cement. For example the construction of the M25 around London (circa 1990) consumed 600,000 tonnes of Portland cement and the construction of the channel tunnel between England and France (circa 1993) consumed an estimated 400,000 tonnes of Portland cement. If we extend this order of cement consumption by major construction projects (Table 2, after Mehta[13]) to all the developed nations and then extrapolate this to future consumption by the developing nations of the third world the potential environmental damage is considerable. Much of the predicted increase is in developing third world countries where there is rapid transition from rural agricultural based society to urban city based society and where populations are increasing most rapidly. However it is important to get this into perspective. For example in the UK concrete production accounts for only 2.5% of total UK CO_2 emissions, and cement only 1.1%[14].

The Approach To Reducing Environmental Damage

The approach to overcoming the problems of producing increasing and irreversible environmental damage is to move to more sustainable development. Two possible ways in which this may be achieved are (i) reduce the amount of cement required to produce concrete by blending it with other materials, especially waste or recycled materials, which can be produced or acquired with no, or limited, environmental damage. This is already standard practice but much more has yet to be done; and (ii) produce improved high-performance cemented construction materials, which will be highly durable, have increased life-spans and will not require frequent repair and replacement. Most, if not all, supplementary cementing materials currently available, can (and normally do) improve the performance and durability of concrete.

In blended cements the clinker/cement ratio (C/C ratio) can be reduced, by blending the clinker with industrial wastes and by-products. These currently include FA, GGBS, SF (Table 3a), together with other more recently used products such as MK[15], RHA[16] and limestone filler[17, 18] (Table 3b). There are also other industrial waste products that show potential as supplementary cementing materials, examples being wastepaper sludge ash (WSA)[19-23], municipal solid waste (MSW)[26], ground waste glass, concrete demolition waste (CDW)[27, 28] and fluidised bed waste cracking catalyst[29, 30] (Table 3c). Some European countries, for example Belgium and France, have low C/C ratios whereas the UK operates with relatively high C/C ratios indicating slow adoption of blended cements. The major barrier to further application of blended cements appears to be a natural conservatism reflected in the standards and specifications and building codes. Therefore the potential for CO_2 reduction by the greater adoption of blended cement is considerable. If more of the waste materials generated by developed technological societies are used as supplementary cementing materials it would be possible to meet the projected demands for cement well into the next century, without any further increase in Portland cement production. For example, fifty percent of the FA produced throughout the world is stockpiled/landfilled as waste. In the UK 50% is placed in long-term storage (~ 3 million tonnes) and 41% is used for cementitious purposes including 14% as direct cement replacement in concrete[31]. Parrott [32] has suggested that blending PC with 30% FA would, in comparison with an equivalent strength PC concrete, reduce CO_2 emission by 17% and primary energy requirement by 14%. There are also numerous other wastes/byproducts that could contribute in the same manner (Table 3c). Making use of these would also reduce some of the problems associated with modern Portland cement, because many of these supplementary cementing materials also reduce temperature rise and the tendency towards thermal cracking, reduce permeability and increase resistance to chemical attack.

In order to establish the viability of these newer wastes for use as supplementary cementing materials considerable cement science needs to be carried out. The different wastes/by-products react in different ways. Some are highly reactive pozzolans (e.g. SF, MK and RHA); some are relatively low reactivity pozzolans (e.g. FA and ground waste glass); some are latently hydraulic materials (e.g. GGBS); some act principally as fillers (e.g. limestone-dust); and some are both hydraulic and latently hydraulic (e.g. WSA). Although, with the exception of some fillers, the wastes are typically high in silica, which in general results in additional formation of C-S-H gel and reduction in CH, many also contain significant levels of alumina, which results in formation of calcium aluminate and silico-aluminate hydrates.

Table 3a Wastes and byproducts as supplementary cementing materials/fillers to Portland cement (PC) a) well established products; b) more recently used products;

a) WELL ESTABLISHED PRODUCTS

Waste/byproduct	Characteristics	Actual/potential drawbacks
Fly ash (FA)	Low-reactivity, low-cost pozzolan consisting mainly of alumino-silicate glassy spheres, improves concrete workability and long-term strength and durability, abundant supply.	Reduces early strength. High calcium FA can impair sulphate resistance.
Ground granulated blastfurnace slag GGBS	Latently hydraulic, glassy calcium alumino-silicate activated by PC, reduces concrete permeability and improves durability, abundant supply.	Lower reactivity than PC, reactivity dependent on composition and fineness. Strength development significantly retarded at low temperatures.
Silica fume (SF)	Ultra-fine (0.03-0.3 μm) high reactivity pozzolan consisting principally of amorphous silica. Produces highly durable, high strength concrete with low permeability and a densified interfacial zone.	Expensive, difficult to handle, usually used as slurry or pre-blended. Has a high water demand, superplasticisers are required and concrete can suffer from differential drying shrinkage. Supply limited to output of silicon/ferro-silicon industries.

b) MORE RECENTLY USED PRODUCTS

Metakaolin (MK) [15]	High reactivity pozzolan, derived from heating kaolin (700-800°C), consisting of an amorphous alumino-silicate. Produces highly durable, high strength concrete, with low permeability.	Expensive manufactured product, made from high purity Kaolin. However much less pure, low-cost kaolins are available (for example waste oil sand fine tailings) which could be used, but impurities may adversely influence reactivity and performance. Has a high water demand.
Rice husk ash[16] (RHA)	High reactivity pozzolan consisting of amorphous micro-porous silica with a high specific surface (20-100 m^2/g). Produces pore refinement, densification of the interfacial zone, permeability reduction and improved durability. Potential current world supply 20 m.t./annum.	Quality of RHA highly dependent on heat treatment of rice husks, to produce a highly porous low-carbon product. Increases water requirement, reduces setting time
Limestone [17, 18]	Limestone ($CaCO_3$) interground with PC clinker acts as a filler. Blended with PC at levels up to 35%. Causes acceleration of PC hydration, improves particle packing, reacts with C_3A and C_4AF to form carbo-aluminate hydrate. Can have beneficial effects on workability and bleeding. Potential for use of waste limestone fines.	PC-limestone blends more susceptible to sulphate attack and less durable than equivalent PC binder. More susceptible to TSA.

Table 3b Wastes and byproducts as supplementary cementing materials/fillers to Portland cement (PC): products with potential

c) PRODUCTS WITH POTENTIAL		
Wastepaper sludge ash(WSA) [19-23]	Complex combination of crystalline and amorphous calcium silicate/alumino-silicate phases derived from high temperature combustion of wastepaper sludge. Contains both hydraulic and latently hydraulic components. Can, when blended with GGBS/PC produce concrete with improved durability and enhanced strength.	Promotes rapid setting. Contains free lime and can promote unsoundness. High water demand and high heat of hydration.
Ground waste glass [24, 25]	Pozzolan comprising ground soda glass of cement fineness consisting principally of amorphous silica but high in alkalis. Strength development superior to FA concrete Evidence indicates alkalis are incorporated in C-S-H gel.	Waste glass (e.g. soda glass) contains alkali metal oxides (typically 15% Na_2O) and silica (typically 70% SiO_2) and could therefore promote ASR in concrete.
Municipal solid waste(MSW) [26]	MSW heated with ~5wt% limestone at >1300°C and water-cooled to produce a latently hydraulic calcium alumino-silicate glass slag. The slag is low in chlorides and elutes no heavy metals. Performance similar to a typical GGBS.	High-temperature process and a small amount of limestone needs to be added in order to achieve required hydraulicity.
Concrete demolition waste (CDW)[27, 28]	In the UK 90 m.t. of construction waste are generated annually of which an estimated 50% is concrete. Heating concrete rubble at 700°C separates the cement stone, which can be used to partially replace Portland cement clinker. Alternatively the cement rich fine fraction from crushed concrete could be used as a supplementary cementing material.	Research is at a very preliminary stage. Processing costs to produce a usable cementitious material may prove prohibitive.
Fluidised bed cracking catalyst[29, 30]	High surface area (50-250 m^2/g) zeolite catalyst, comprising about 52% SiO_2 and 41% Al_2O_3, with pozzolanic properties. Can produce strength enhacement (up to 15% PC replacement and at low w/b ratios) particularly if superplasticisers are used. Abundant supply (~400,000 tonnes/annum worldwide).	Reduces setting time and increases water demand. Increasing cement replacement leads to increasing water absorption.

Also other oxides are commonly present, for example waste soda glass has high levels of Na_2O, GGBS has significant levels of MgO. In addition MSW can contain significant levels of chloride and heavy metals, FA from flue gas de-sulfurisation (FGD) plants can contain very high levels of sulphate. Many of these additional components can have very deleterious effects upon the performance of concrete. Thus, in order to achieve the desired properties and performance, it may be necessary not only to blend PC with the waste but also to blend with other materials (e.g. other wastes/by-products). Clearly the more complex the system the greater the likelihood of potentially damaging reactions occurring, which makes it essential to understand the science of such systems. One example of such a new system under investigation is WSA-GGBS-PC for use as a binder for concrete [19-23].

WASTEPAPER SLUDGE ASH AS A BINDER FOR CONCRETE

The oxide composition of WSA (CaO - 42.61% ± 2.49; SiO_2 - 27.81% ± 6.12; Al_2O_3 - 18.05% ± 2.76) is similar to the composition of ground granulated blastfurnace slag (GGBS) but it is lower in silica and higher in alumina than a typical slag. However the phase composition of WSA is very complex[23], mainly because it has undergone non-equilibrium heat treatment. The sludge (consisting of kaolin and calcite) from which the ash is derived is combusted in a fluidised bed by heating up to about 1200°C in a few seconds and cooling to 200°C within a few seconds. TEM analysis shows the WSA to consist principally of 'small' crystals (~0.1μm), 'large' crystals (~1μm) and 'glassy' globules (~ 1-10 μm). Estimates from SEM suggest that the WSA comprises up to 20% of the glassy phase. The small crystals consist of two different cation substituted crystalline calcium silicate type phases having two distinct compositions, one high in Mg and low in Al (bredigite) and one low in Mg and high in Al (an Al substituted α'-C_2S). The large crystals consist of gehlenite (C_2AS) with some substitution of silicon by aluminium. The glassy phase has a mean composition (35% CaO, 36% SiO_2, 25% Al_2O_3 plus minor oxides) within the specified range of commercial slags but at the weaker end of the hydraulicity scale. It therefore has the potential for activation and cementation. In addition to the phases mentioned above WSA also contains unreacted free lime (CaO) some of which may hydrate and carbonate during storage. Typical values would be 5% free lime, 5% calcite and 1% portlandite. There is also a small amount of anorthite (CAS_2) in the WSA and traces of quartz. Figure 3 shows the composition of the principal phases within the SiO_2 - Al_2O_3 - CaO system.

Due to the presence of free lime in WSA the amount of heat evolved on the addition of water[34] is substantial (~100 J/g in 3 hours compared to 20 J/g for a typical PC) and unsoundness and cracking occur. The WSA paste sets very rapidly[34]. Initial set at standard consistence (water/solid ratio 0.65) is 5.9 minutes compared to 110 minutes at standard consistence (water/solid ratio 0.27) for a typical Portland cement. The principal hydration products of WSA are CH, C_4AH_{13}, $C_4A.0.5CO_2.H_{12}$ and C-S-H gel. The CH is produced from hydration of the free lime and hydration of the α'-$C_2(S,A)$. The C_4AH_{13}, $C_4A.0.5CO_2.H_{12}$ and C-S-H gel are produced principally from α'-$C_2(S,A)$ hydration, but in the alkaline environment produced by the CH the glass (with a low-Ca slag composition) will also hydrate and contribute to these hydration products. Thus the CH content does not increase significantly with curing time and remains at a few percent. Because of the poor soundness of WSA paste it also shows poor strength development and cannot be used alone as a cementitious binder. For example the 28-day strength of a WSA paste with w/s ratio of 0.5 is about 4 N/mm^2. However, when combined with ground granulated blastfurnace slag (GGBS), and where necessary a small amount of Portland cement (PC), it produces a very effective binder.

Combining WSA with GGBS dilutes the WSA, thus reducing the amount of expansive product(s) per unit volume of pore space. Also for a given w/s ratio the water initially available for WSA hydration is increased enabling a greater degree of CaO hydration to take place prior to setting. This substantially reduces expansion and improves soundness. In addition separation of the WSA particles by the GGBS particles increases setting times. Also the alkaline environment provided by the hydrated CH not only produces conditions favourable for hydration of the glassy phase in the WSA, but also for hydration of the GGBS.

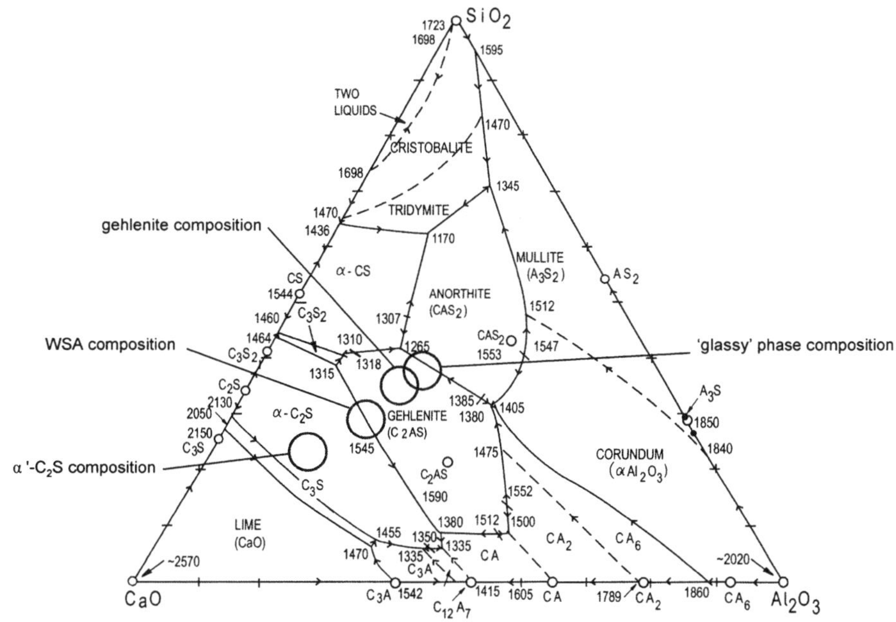

Figure 3 The system $CaO-Al_2O_3-SiO_2$ (after Taylor [33])

Thus much greater strengths and a much finer pore structure can be achieved although the rate of strength development is less than that of Portland cement, particularly at 1 and 7 days. However, further blending of the WSA/GGBS blend with a small amount of PC (10 or 20%) can overcome this problem.

Application of WSA-GGBS binders has to take account of the relatively rapid setting of WSA-GGBS pastes, the significantly higher rate of heat evolution of WSA-GGBS pastes relative to that of PC paste and the smaller strength attained by WSA-GGBS pastes relative to PC paste (about 25% at 7 days and 48% at 90 days). One ideal application of the WSA-GGBS binder is the production of concrete blocks[22].

During the automated production of concrete blocks, quick setting times are desirable to enable subsequent handling and throughput of blocks, concrete volumes (per block) are very small therefore heat evolution is not a problem and may in fact be an advantage, and only low strength concrete is required. Blocks made from WSA binder have already been produced on a commercial scale and current trials are underway to effect full commercial production.

THE WAY AHEAD

Glasser[35], in a keynote lecture on waste utilisation in cement stated that the keys to successful reuse are *characterisation, performance* and *education.*

Characterisation is defined as the quantitative determination of the physico-chemical characteristics of the waste relative to its fitness for use or for disposal.

Performance applies to suitability for use and implies a quantification of the reactions between the waste material and the blended cementitious components and the implications of the reactions in terms of resultant mechanical and chemical behaviour.

Education is necessary to create awareness of codes, specifications and site practice particularly with respect to using new products, where there is no reservoir of acquired knowledge and experience. It should also encompass awareness of environmental impacts.

These objectives should provide the principal thrust for future research work. This century will undoubtedly be the century of sustainable development, in which protection of the environment is pre-eminent. The construction industry (in meeting the infra-structural requirements for increasing industrialisation and urbanisation of the world) is ideally placed to take a leading role in the provision of a built environment, the construction and operation of which is in balance with the natural world rather than destroying it. One way in which the construction industry, which is a very conservative industry, can contribute to this is in strongly supporting future research in this field and lobbying for the much wider application of supplementary cementing materials in construction, including processed municipal wastes, processed industrial wastes and processed clays. The UK national waste strategy requires the reduction of commercial waste sent to landfill (much of which is construction and demolition waste currently running at 90 million tonnes per annum) to 80% of the 1998 levels. This translates to 36 million tonnes by 2005 and some of this could be utilised as supplementary cementing materials (e.g. WSA). For municipal and domestic waste, the landfill directive and current UK legislation requires the UK to recover 40% of the waste stream by 2005, 45% by 2010 and 67% by 2015. 'Recover' is defined as recycling, composting or other forms of recovery. This represents an enormous tonnage to recover, and utilisation of some of this material as supplementary cementing products could make a significant contribution to this recovery. For example, it is likely that to achieve these targets in the limited time period available many local authorities will choose combined heat and power plants that will produce substantial quantities of waste ash with the potential to be incorporated in cement. Clearly however, much cement science needs to be carried out to produce good quality reliable products. We have gone a long way along the 'green road' but there is still a long way to go.

CONCLUSIONS

It is evident that in attempting to solve some of the current problems of our post-industrial society we have to pay due attention to history in addition to utilising the most recent technological developments. The Greeks were one of the first peoples to demonstrate that burnt earth in the form of volcanic ash combines with lime to produce durable concrete. Structures built using this form of cement as binder (pozzolanic cement) can be very durable as evidenced by the numerous 2000 year-old Roman structures still standing. The advent of Portland cement in the 19th century, which could set and harden rapidly on addition of water, and the harnessing of cement science in its subsequent development has provided the world with a universal and highly adaptable construction material. However it has also created many problems, the principal ones being low durability and increased environmental pollution.

Cement science has made an invaluable contribution to both understanding these problems and providing solutions to them. The problems associated both with the durability of modern Portland cement and the environmental damage associated with its production, can in part be alleviated by partial replacement of cement with supplementary cementing materials many of which can be derived from waste products. However in order not to compound existing problems by introducing additional problems much scientific investigation, testing and education and training is needed.

In celebrating the immense contribution that Professor Fred Glasser has made to this area, the current research community should be working in collaboration with industry and government, to produce a climate and environment in which new cement scientists can emerge. People whom, being inspired and guided by the work of the past, will develop new cementing materials and find solutions to the problems of the future. That surely is the best possible legacy that a scientist can leave behind.

ACKNOWLEDGEMENTS

The author would like to thank members of the Building Materials Research Unit at the University of Glamorgan for their contribution, and the EPSRC and the School of Technology for their support.

REFERENCES

1. STANLEY C. C., Highlights in the History of concrete, Cement and Concrete Association publication, 1979.
2. TECHNICAL REPORT NO. 29, Report of a Concrete Society Working Party, Changes in the properties of ordinary Portland cement and their effects on concrete. October 1987, pp 1-30.
3. CORISH A., Portland cement properties - updated. Concrete, Jan-Feb 1994, pp 25-30.
4. MATHER B., Use less cement. Concrete International, Nov 2000, pp 55-56.
5. HONG S-Y., GLASSER F. P., Alkali binding in cement pastes: Part I. The C-S-H phase, Cement and Concrete Research 29 (1999) 1893-1903.
6. HONG S-Y., GLASSER F. P., Alkali sorption by C-S-H and C-A-S-H gels: Part II. Role of alumina, Cement and Concrete Research 32 (2002) 1101-1111.
7. JENSEN O. M., HANSEN P. F., COATS A. M., GLASSER F. P., Chloride ingress in cement paste and mortar, Cement and Concrete Research 29 (1999) 1497-1504.
8. GLASSER F. P., The role of sulphate mineralogy and cure temperature in delayed ettringite formation. Cement and Concrete Composites, 18 (1996) 187-193.
9. TAYLOR H. F. W., Delayed ettringite formation. Cement and Concrete research. 31 (2001) 683-693.
10. GLASSER F. P., Private communication.

11. BARNETT S. J., MCPHEE D. E., CRAMMOND N. J., Extent of immiscibility in the ettringite-thaumasite system. Proceedings of the First International Conference on Thaumasite in Cementitious Materials, 20-21st June, 2002, BRE, UK., Paper 4, 8 pages, 2002, Publ. Construction Research Communications Ltd., Available on CD, ISBN 1 86081 557 X.

12. WORRELL E., PRICE L., MARTIN N., HENDRIKS C., MEIDA O., Carbon dioxide emissions from the global cement industry, Annu. Rev. Energy Environ. 2001, 26, 303-329.

13. MEHTA P. K., Role of pozzolanic and cementitious material in sustainable development of the concrete industry. Proc. 6th CANMET/ACI International Conference on Fly Ash, Silica Fume, Slag and Natural Pozzolans in Concrete, May 1988, Vol. 1, SP-178, pp. 1-20, Bangkok, Thailand, Edit. V. M. Malhotra.

14. WORKING PARTY OF THE CONCRETE SOCIETY'S MATERIALS GROUP. Concrete and the Environment, Concrete, September 2001, 8pp.

15. SABIR, B. B., WILD S., BAI J., Metakaolin and calcined clays as pozzolans for concrete – A review, Cement and Concrete Composites, 23, 2001, 441-454.

16. HWANG C. L., CHANDRA S., The use of rice husk ash in concrete, Waste Materials Used in Concrete, 1997, pp. 184-231, Edit Satish Chandra, Noyes Publications, New Jersey, USA.

17. BONAVETTI V. L., DONZA H., MENEDEZ G., CABRERA O., IRASSAR E. F., Limestone filler cement in low w/c ratio concrete: a rational use of energy. Cement and Concrete Research, 33, (2003)

18. IRASSAR E. F., BONAVETTI V. L., GONZALEZ M., Microstructural study of sulphate attack on ordinary and limestone Portland cements at ambient temperature, Cement and Concrete Research, 33, (2003), 31-41.

19. KINUTHIA J. M., O'FARRELL M., SABIR, B. B., WILD S., A preliminary study of the cementitious properties of wastepaper sludge ash – ground granulated blastfurnace slag (WSA - GGBS) blends, Proceedings of International Symposium on Recovery and Recycling of Paper, Edit. R. K. Dhir, M. C. Limbachiya and M. D. Newlands, 2001, 93-104, University of Dundee, 19th March 2001, Thomas Telford Ltd., London,

20. O'FARRELL M., CHAIPANICH A., KINUTHIA J. M., SABIR B. B., WILD S., A new concrete incorporating wastepaper sludge ash (WSA). International Congress "Challenges of Concrete Construction", University of Dundee, 5-11th September 2002, Innovations and Developments in Concrete Materials & Construction, Theme 1 – Developments in Concrete Constituent Materials, 2002, 149-158, Thomas Telford Ltd, London.

21. VEERAPPAN G., The development of a low-cost, low-energy cement from waste, PhD transfer report, University of Glamorgan, December 2001.

22. VEERAPPAN G., KINUTHIA J. M., O'FARRELL M., SABIR B. B., WILD S., Compressive strength of concrete block using wastepaper sludge ash. International Symposium on Advances in Waste Management and Recycling, Symposium W2, Theme W23, Subject area W234, 10 pages, Dundee University, September 2003.

23. BAI J., CHAIPANICH A., KINUTHIA J. M., LEWIS M. H., O'FARRELL M., SABIR B. B. WILD S., Compressive strength and hydration of wastepaper sludge ash – ground granulated blastfurnace slag (WSA – GGBS) blended pastes. Cement and Concrete Research, at press.

24. DYER T. D., DHIR R. K., Chemical reactions of glass cullet used as a cement component. Journal of Materials in Civil Engineering, Nov/Dec 2001, pp 412-417.

25. SHAO Y., LEFORT T., MORAS S., RODRIGUEZ D., Studies on concrete containing ground waste glass. Cement and Concrete Research, 30, (2000), pp 91-100.

26. KITATSUJI M., ENDO T., ISHIDA H., FUJII K., Study on the utilisation of melting slag from municipal solid waste as concrete materials, International Congress "Challenges of Concrete Construction", University of Dundee, 5-11th September 2002, Sustainable Concrete Construction, Theme 2 - Reuse of Waste Materials, 2002, 323-333, Thomas Telford Ltd, London. ISBN 0 7277 3179 3.

27. MULDER E., BLAAKMEER J., NIJLAND T., TAMBOER L., A closed material cycle as part of an integrated process for the reuse of the total flow of C&D waste. International Congress "Challenges of Concrete Construction", University of Dundee, 5-11th September 2002, Sustainable Concrete Construction, Theme 4-Total recycling opportunities, 2002, 555-562, Thomas Telford Ltd, London.

28. GRANT N., GLASSER F. P. Recycling the cement-rich fraction of waste concrete: A status report. International Congress "Challenges of Concrete Construction", University of Dundee, 5-11th September 2002, Sustainable Concrete Construction, Theme 4 -Total recycling opportunities, 2002, 797-804, Thomas Telford Ltd, London.

29. SU N., FANG H. Y., CHEN Z. H., LIU F. S., Reuse of waste catalysts from petrochemical industries for cement substitution. Cement and Concrete Research, 30, (2000) 1773-1783.

30. HSU K., TSENG Y., KU F., SU N., Oil cracking waste catalyst as an active pozzolanic material for superplasticised mortars. Cement and Concrete Research, 31, (2001) 1815-1820.

31. SEAR L.K.A., The environmental benefits of using PFA in cementitious systems, Proceedings of the International Congress Challenges of Concrete Construction, University of Dundee, 9-11 Sept. 2002, Sustainable Concrete Construction, Theme 2: Reuse of waste materials, 2002, pp 313-322, Edit. R. K. Dhir, T. D. Dyer & J. E. Halliday, Thomas Telford Ltd. London.

32. PARROTT L., HIGGINS D. D., SEAR L. K., A. Defining and improving the environmental performance of concrete. Concrete Industry Alliance Project Report, January 2000, DETR project reference 39/3/487, CC1553.

33. TAYLOR H. F. W., Cement Chemistry, 2nd edition, Thomas Telford, London, 1997.

34. CHAIPANICH A., O'FARRELL M., KINUTHIA J. M., SABIR B. B., WILD S., Setting time and heat of hydration of wastepaper sludge ash - ground granulated blastfurnace slag (WSA-GGBS) blended pastes. Cem. and Concr. Res., under review.

35. GLASSER F. P. Properties of cement waste composites, Waste Management, Vol. 16, Nos 1-3, pp 159-168, 1996.

INFLUENCE OF DIFFERENT CEMENT AND CHEMICAL ADMIXTURE TYPES ON STRENGTH DEVELOPMENT

N U Kockal

Bogazici University

F Turker

Akdeniz University

Turkey

ABSTRACT. In the study, specimens from fluid and highly fluid concrete specimens containing different kinds of chemical admixtures and cement types for various strength concrete classes were prepared and tested for compressive strength and modulus of elasticity. Concrete specimens were produced with ordinary Portland cement (PC 42,5) and two different types of Portland composite cements (PCC-A 32,5 and PCC-B 32,5R). Super-plasticizer, set retarder, set accelerator, air entraining agent and hiper-plasticizer were used as chemical admixtures in concrete mixtures. Test results of concrete specimens were experimentally analyzed and compared. The results have indicated that strength values of concrete containing air entraining agent were higher than those of concretes containing set accelerating and set retarding admixtures. PCC-B 32,5R concrete having highly fluid consistency showed the highest strength development. High-strength concretes exprienced less strength development than lower strength concretes.

Keywords: Chemical admixture, Cement types, Compressive strength, Modulus of elasticity, Portland composite cement, Air entraining agent, Set accelerator, Set-retarding admixture.

N U Kockal, is a research assistant in the Department of Civil Engineering at Bogazici University. His research focuses on concrete durability.

F Turker, is a Professor in the Department of Civil Engineering at Akdeniz University. His research focuses on concrete durability and the use of waste materials in concrete manufacturing.

INTRODUCTION

Normal and high-strength concretes with different kinds of chemical admixtures and cement types are widely used in the construction industry. It is clear that compressive strength of concrete has a vital importance for quality control in concrete production. Therefore, it is necessary to understand the effectiveness of these different kinds of chemical admixtures and cement types in improving concrete strength.

The use of mineral additives in cement and concrete can achieve not only economical and ecological benefits, but technical benefits as well. It is generally agreed that, with proper selection of admixture, mixture proportioning and curing technique, mineral additives can greatly improve the durability of concrete [1]. The materials with the best pozzolanic characteristics have not always been used in a country, where there are natural pozzolans of volcanic origin like tuff. The importance of using natural pozzolans in the cement industry requires a complete evaluation of their effects on concrete [2].

It is important to realise that all plasticising chemical admixtures do not behave similarly in concrete, with performance levels often depending on Portland cement and admixture compatibility [3]. The influence of the superplasticizer on the characteristics of the hardening and hardened concrete is largely unclear. A better understanding of these aspects is, therefore, essential for further improvement of these admixtures and for analyzing the long-term properties of the concrete [4].

The scope of this study is to investigate the influence of different cement and chemical admixture types on strength development of concretes with fluid and highly fluid consistency.

MATERIALS

Aggregates

Four different aggregates were used in the concrete mixtures. Natural river sand (A1) and crushed stone sand (A2) as fine aggregate, natural gravel (A3) and coarse crushed stone (A4) as coarse aggregate. Maximum particle size was 25 mm. The density of aggregates in saturated surface dry condition and the water absorption of the aggregates are given in Table 1. The aggregate gradation is shown in Table 2. Aggregate proportions were changed in order to improve grading and obtain ultimate desired strength and Table 3 shows the percentage of aggregates used in each concrete strength class.

Table 1 Some physical properties of aggregates used

	A1	A2	A3	A4
Density, (kg/dm^3)	2.64	2.63	2.66	2.70
Water absorption, (%)	2.0	2.0	0.8	0.7

Table 2 The aggregate gradation

AGGREGATE TYPE	SIEVE SIZE, (mm)							
	0.25	0.5	1	2	4	8	16	31.5
	PERCENTAGE PASSING							
A1	12	23	37	59	92	100	100	100
A2	10	22	41	65	91	100	100	100
A3	0	0	0	0	2	47	100	100
A4	0	0	0	0	0	0	23	100

Table 3 The percentage of aggregates used in each concrete strength class

AGGREGATE TYPE	CONCRETE STRENGTH CLASS		
	C 20	C 30 – C 40	C 60
A1	20%	20%	7%
A2	40%	32%	42%
A3	22%	27%	22%
A4	18%	21%	29%

Cement

Ordinary Portland cement (PC 42,5) and two different types of Portland composite cements (PCC-A 32,5 and PCC-B 32,5R) were used. Their chemical analysis, physical and mechanical properties are given in Table 4, Table 5 and Table 6, respectively. Table 7 shows permitted constituents of Portland composite cements according to TS 12143 [5].

Table 4 Chemical analysis of cements

	PC 42.5	PCC-A 32.5	PCC-B 32.5R
SiO_2, %	21.10	22.12	-
Al_2O_3, %	3.97	4.70	-
Fe_2O_3, %	3.49	3.32	-
CaO, %	64.29	59.23	-
MgO, %	2.52	2.48	-
SO_3, %	2.89	2.97	1.86
Loss on ignition, %	1.39	3.80	-
Insoluble residue, %	0.50	-	-
Free CaO, %	0.78	0.49	-
Cl^-, %	0.003	0.003	0.005

Table 5 Physical properties of cements

	PC 42.5	PCC-A 32.5	PCC-B 32.5R
Specific surface, Blaine, (cm^2/g)	2916	3326	3750
Specific gravity, (g/cm^3)	3.11	3.03	2.94
Retained on 200 μm sieve,(%)	0.1	0.1	-
Retained on 90 μm sieve, (%)	1.5	3.7	-
Initial time of setting, (min)	165	175	165
Final time of setting, (min)	245	245	195
Volume expansion, (mm)	1.0	1.0	1.0

Table 6 Mechanical properties of cements

COMPRESSIVE STRENGTH, (MPa)	PC 42.5	PCC-A 32.5	PCC-B 32.5R
2 Days	22.7	16.1	17.4
7 Days	41.2	31.4	-
28 Days	54.8	43.5	37.2

Table 7 Permitted constituents of Portland composite cements (%)

CEMENT	CLINKER	GRAN. BLAST. SLAG (S)	SILICA FUME* (D)	NATURAL POZZOLANA (P)	INDUSTRIAL POZZOLANA (Q)	SILICEOUS FLY ASH (V)	CALCAREOUS FLY ASH (W)	BURNT SHALE (T)	LIME STONE (L)
PCC-A	80-94	← 6 – 20 →							
PCC-B	65-79	← 21 – 35 →							

* In the case of using silica fume, it is limited to 10% on cement.

Chemical Admixtures

Superplasticizer (SP), set retarder (SR), set accelerator (SA), air entraining agent (AE) and hiper plasticizer (HP) were used in concrete mixtures. Air entraining agent was added to the mix with superplasticizer, but retarder and accelerator were used singly due to their superplasticizing ability. Table 8 gives the content of chemical admixtures in the concrete mixtures.

METHOD

Mix Proportions

The investigation includes four different concrete strength classes: C 20, C 30, C 40 and C 60. 15 mixture proportions were prepared. In the strength class C 20, three different cement types were used with SP in order to have fluid and high fluid concretes. SA, SR, SP-AE admixtures were used with PC 42,5 and SP-AE admixtures were used with PCC-A 32,5 and PCC-B 32,5R in class C 30. The slump of all concrete mixtures were maintained within the range of 14 to 17 cm in this strength class. HP admixture was used with PC 42,5 and PCC-A 32,5 in order to obtain highly fluid consistency in the strength class C 40 and C 60. Mix proportions and some properties of concretes are indicated in Table 8.

Table 8 Mix proportions and some properties of concretes

MIXTURE	ADM. TYPE	ADM. PROP. (%)	W/C	SLUMP (cm)	AIR CONT. (%)	CONSTITUENTS CONT. (kg/m^3)						FRESH CONCRETE DENSITY (kg/m^3)
						C	A1	A2	A3	A4	w	
PC 42.5-20F	-	-	0.66	12	0.6	301	370	740	406	334	199	2350
PC 42.5-20HF	SP	1.0	0.65	21	1.8	298	367	733	402	331	194	2325
PC 42.5-30SA	SA	0.8	0.52	15	1.4	339	373	594	502	390	177	2375
PC 42.5-30SR	SR	1.3	0.53	15	1.4	339	369	593	501	389	180	2371
PC 42.5-30AE	SP/AE	1.2/ 0.022	0.47	17	4.3	333	369	589	496	386	157	2330
PC 42.5-40	HP	0.8	0.45	18	1.2	343	384	615	519	405	154	2420
PC 42.5-60	HP	0.9	0.28	19	1.4	453	132	798	417	549	127	2476
PCC-A 32.5-20F	SP	0.75	0.58	13	1.6	301	376	754	415	340	174	2360
PCC-A 32.5-20HF	SP	1.6	0.57	19	1.9	300	378	753	414	339	171	2353
PCC-A 32.5-30AE	SP/AE	0.9/ 0.026	0.44	16	4.6	370	356	568	483	372	161	2310
PCC-A 32.5-40	HP	0.8	0.40	19	1.3	383	375	602	509	394	153	2416
PCC-A 32.5-60	HP	1.0	0.26	19	1.1	454	136	804	419	554	118	2485
PCC-B 32.5-20F	SP	1.4	0.52	11	2.8	309	376	750	413	336	161	2345
PCC-B 32.5-20HF	SP	1.7	0.52	21	2.5	345	359	719	397	325	180	2325
PCC-B 32.5-30AE	SP/AE	1.6/ 0.036	0.38	14	5.4	370	362	576	487	379	141	2315

Preparation of Specimens and Experimental Procedure

The concrete cylinders used for compresive strength and modulus of elasticity tests were 150x300 mm. The compressive strength and modulus of elasticity tests were performed at 28, 56, 90, 180 days for the strength class C 20; at 28, 90 and 180 days for other concrete strength classes C 30, C 40 and C 60 in accordance with TS 3502 [6]. The concrete cylinders were demoulded after 24 hours and put into a water curing tank, saturated with lime at 23°C for 28 days. After 28 days, they were taken from the tank and stored under constant environmental condition; in the curing room at 23±2°C and 60±5 % RH. Three concrete specimens of each mixture were prepared for the tests.

RESULTS AND DISCUSSION

Compressive Strength and Modulus of Elasticity

The relative compressive strength and relative modulus of elasticity values are illustrated in figures from Figure 1 to Figure 8.

Highly fluid concretes exhibited higher strength development as compared to the fluid concretes except for specimens prepared from PCC-B 32,5R in C20, Figure 1. The possibility of strength reduction in highly fluid concrete mixtures was considered, therefore as shown in Table 8 cement dosages were kept constant, w-c ratios were decreased in PC 42,5 and PCC-A 32,5 concrete mixtures, conversely w/c was kept constant, cement dosage was increased in PCC-B 32,5R concrete mixtures in order to prevent this strength reduction. As can be seen in Figure 1, the differences in strength development of highly fluid concretes in C 20 seem to be related to decreasing w/c and increasing cement dosage besides consistency differences in the mixtures. This result emphasizes the importance of changing w/c values rather than changing cement dosage values so as to obtain higher strength development. In addition to these factors mentioned above, superplasticizer type and its content can have an impact on strength development.

The concretes containing set accelerator gained lower strength than concretes containing air entraining agent and set retarder at the age of 90 days. Concretes with set accelerator and concretes with set retarder showed almost equal development at 180 days. PC 42,5 and PCC-A 32,5 concretes exhibited similar behaviour in strength development among the concretes having air entraining agents. PCC-B 32,5R concrete containing AE showed the highest strength development at the age of 180 days, Figure 2. A similar tendency was also observed in the case of modulus of elasticity.

Although the strength development of PCC-A 32,5 concretes was higher than that of PC 42,5 concretes at 90 days, 180-day strength values of PCC-A 32,5 fell slightly in both classes C40 and C60. However, PC 42,5 concretes exhibited a steady rise at all ages, Figure 3 and Figure 4.

As compared to the PC 42,5 concretes, modulus of elasticity values of PCC-A 32,5 concretes were higher at all ages in C40 and C 60, Figure 7 and Figure 8.

It can be seen from the figures that the strength development of high-strength concrete specimens was lower than that of lower strength concrete specimens.

Disregarding cement types, differences in w/c and cement content values decreased continuously and became smaller in the higher strength classes.

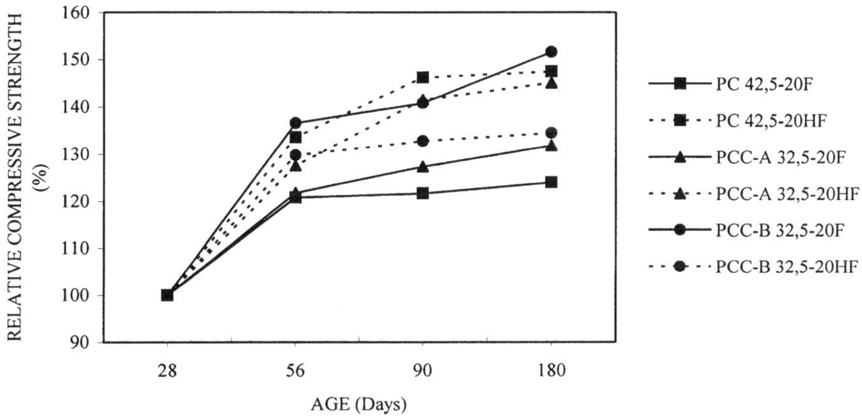

Figure 1 Relative compressive strength versus age for C 20

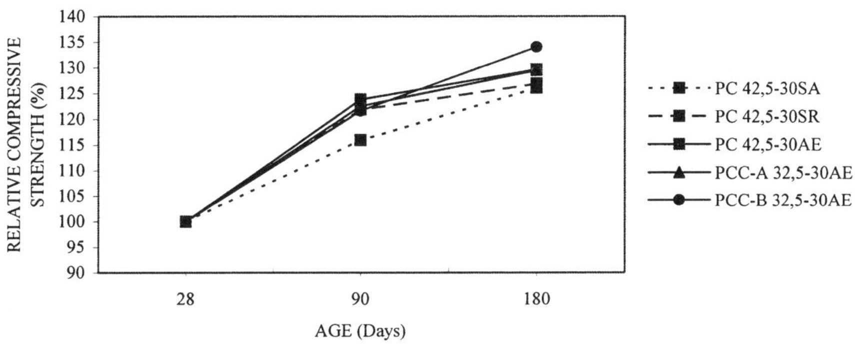

Figure 2 Relative compressive strength versus age for C 30

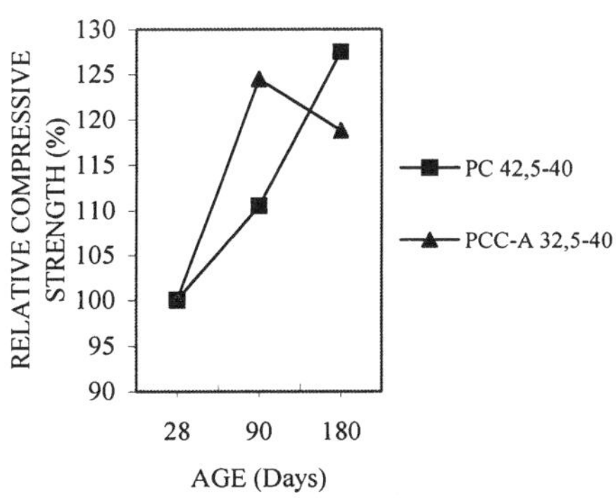

Figure 3 Relative compressive strength versus age for C 40

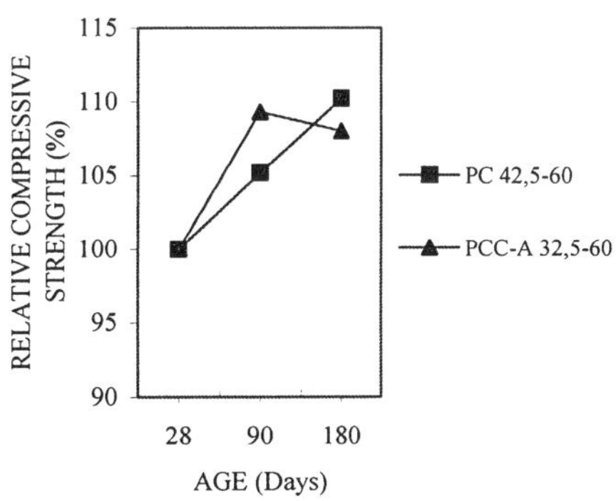

Figure 4 Relative compressive strength versus age for C 60

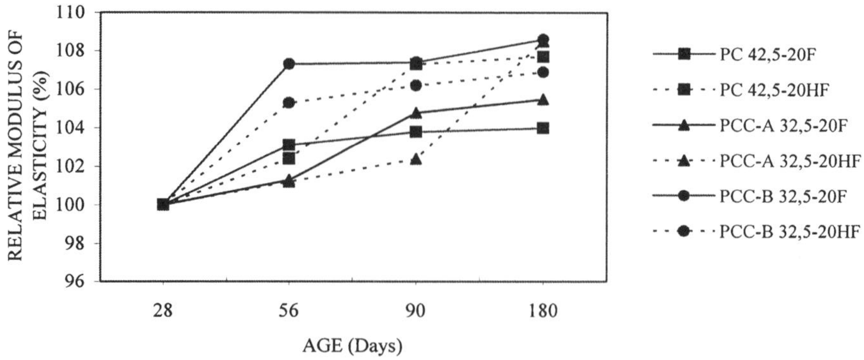

Figure 5 Relative modulus of elasticity versus age for C 20

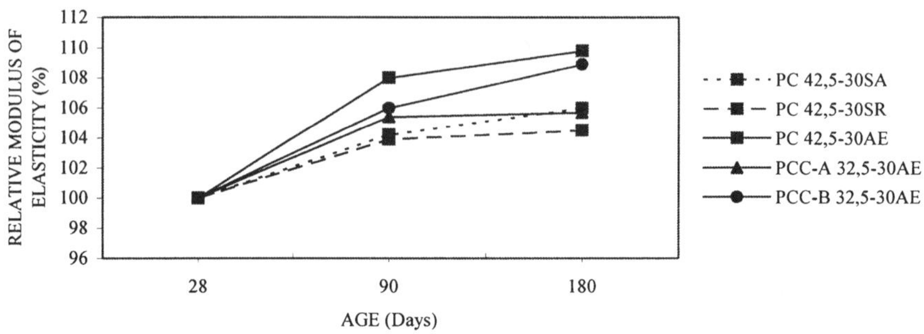

Figure 6 Relative modulus of elasticity versus age for C 30

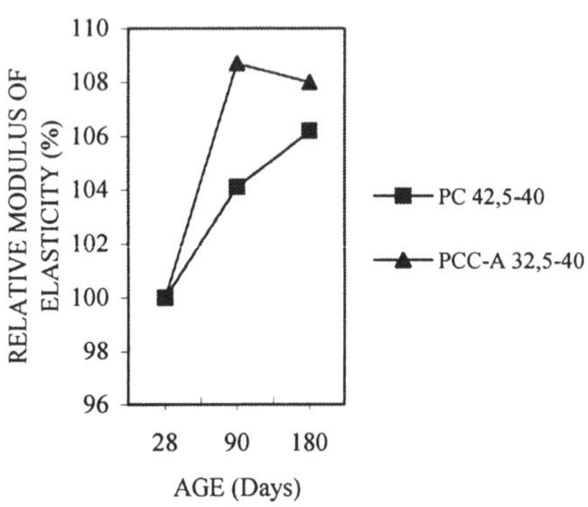

Figure 7 Relative modulus of elasticity versus age for C 40

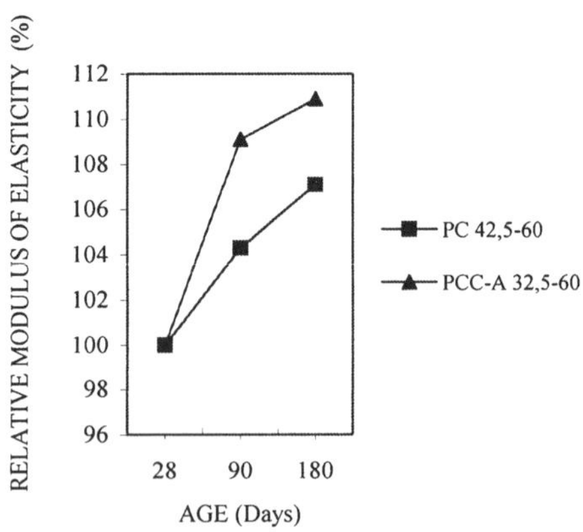

Figure 8 Relative modulus of elasticity versus age for C 60

CONCLUSIONS

According to the investigation results, the following conclusions can be drawn;

- Highly fluid concretes exhibited higher strength development as compared to the fluid concretes except for specimens made with PCC-B 32,5R. Concrete specimens prepared with PCC-B 32,5R having highly fluid consistency showed the highest strength development. It is noteworthy that the concretes produced with high cement content have considerably small increase rates in strength. Apparently, the influence of cement content was more pronounced than that of cement type for such behaviour of PCC-B 32,5R concretes in strength class C 20.

- As compared to the concretes containing SA and SR admixtures, the higher strength values were obtained by the concretes containing AE.

- A similar tendency was also observed in the case of modulus of elasticity in both concrete classes (C 20 and C 30).

- PC 42,5 and PCC-A 32,5 concretes exhibited similar behaviour in strength development among the concretes having air entraining agents. PCC-B 32,5R concrete containing AE showed the highest strength development.

- PCC-A 32,5 concretes showed higher strength development than PC 42,5 concretes at 90 days, there was a slight drop in 180-day strength values of PCC-A 32,5 in both classes C40 and C60. Contrary to PCC-A 32,5 concretes, PC 42,5 concretes exhibited a steady rise at all ages.

- The modulus of elasticity values of PCC-A 32,5 concretes were higher than those of PC 42,5 concretes at all ages in C40 and C 60.

- High-strength concrete specimens experienced lower strength development than lower strength concrete specimens.

- Differences in w-c ratio and cement content values in compositions of PC 42,5 and PCC-A 32,5 concretes declined continuously in the higher strength classes regardless of cement types.

REFERENCES

1. WU, Z., NAIK, T.R., Properties of concrete produced from multicomponent blended cements, Cement and Concrete Research, Vol. 32, 2002, pp 1937-1942.

2. RODRIGUEZ-CAMACHO, R.E., URIBE-AFIF, R., Importance of using the natural pozzolans on concrete durability, Cement and Concrete Research, Vol. 32, 2002, pp 1851-1858.

3. MAGEE, B.J., ALEXANDER, M.G., Simple test method to assess the relative effectiveness of plasticising chemical admixtures, Cement and Concrete Research, Vol. 31, 2001, pp 303-307.

4. RONCERO, J., VALLS, S., GETTU, R., Study of the influence of superplasticizers on the hydration of cement paste using nuclear magnetic resonance and X-ray diffraction techniques, Cement and Concrete Research, Vol. 32, 2002, pp 103-108.

5. TS 12143, Cement – Portland Composite, T.S.E., Ankara, 1977.

6. TS 3502, Test Method for Static Modulus of Elasticity and Poisson's Ratio of Concrete in Compression, T.S.E., Ankara, 1981.

DURABILITY OF CEMENT AND CEMENT PLUS RESIN STABILISED EARTH BLOCKS

A Guettala **A Abibsi**

Biskra University

H Houari

Constantine University

Algeria

ABSTRACT. The main drawback of earth construction is the rapid deterioration of the material under severe weather conditions. The objective of this work is to improve the behaviour of stabilized blocks of earth blocks against water attacks. The blocks manufactured with one type of earth were tested in compressive strength as dry blocks and after immersion, in intensive sprinkling and in absorption. Tests of wetting-drying and tests of freeze- thaw were also carried out. The results show the influence of the different manufacturing parameters: compacting intensity, sand, cement and cement plus resin content on the mechanical strength in the dry state as well as in the wet state, water resistance coefficient, weight loss and absorption.

Keywords: Stabilized, Cement, Resin, Earth, Compacting stresses, Sand content, Compressive strength, Water resistance coefficient, Wetting-drying, Freeze-thaw, Absorption, Durability.

A Guettala is a lecturer in the Department of Civil Engineering and member of a Research Laboratory, Biskra University – Algeria. He specializes in concrete properties and has carried out research on the durability of construction materials.

H Houari is a lecturer in the Department of Civil Engineering and a director of a Research Laboratory, Constantine University – Algeria. He specializes in concrete properties and has carried out research on the durability of construction materials.

A Abibsi is a senior lecturer in the Department of Mechanical Engineering, Biskra University. He has been involved for the past 4 years in Research in the field of construction materials.

INTRODUCTION

The use of earth as a building material dates back to the period of the ancient Mesopotamia (5000-4000 B.C). For economical reasons and by studying what already has been done until now, scientists and builders consider that it is judicious to try to improve the life span of construction materials [1], [2]. Large research programs have been undertaken all over the world into the durability of earth walled buildings [2], [3], [4]. The durability prevision of stabilized earth blocks is still a controversial matter amongst construction actors. In order to know the limits of this kind of material destined to construction, it is intended to find solutions that can improve its life span by the know how of its general use as well as its mass treatment (additions of binders, compacting energies,...). Obtaining a durable material would need a treatment which would result in sufficient mechanical strength as well as low sensitivity to water attacks [5].

These two main conditions should be preceded by a very precise study of parameters related to the grading and mineralogy of these materials. The type and the content of the binders, aggregates grading, compacting stresses and water content would be adapted as conditions of making of these materials [6], [7]. The durability can be improved by other additions such as lime [5], cement and lime or cement and microsilica [8]. In this present work, we have tried to improve the durability of earth blocks by several methods: by the additions of cement (5, and 8 %), cement and resin (5 % cement + 50 % resin; 8 % cement + 50 % resin(1)), sand content (0, 10, 20, 30 and 40%) and the compacting stresses (5, 7.5, 10, 12.5, 15, 17.5 and 20 MPa).

(1) The resin content is relative to compacting water.

PHYSICAL CHARACTERISTICS OF SOIL

Soil samples of the region of Biskra (south east of Algeria) have been taken as reference samples and subjected to several laboratory tests as specified by ASTM standards [9].

Atterberg Limits

According to [11], the best earth soils for stabilization are those with low plasticity index (P.I) and the product (P.I x M) in the vicinity of 500 to 800, where M is the percentage of mortar, in this case P.I x M = 644, see Table 1.

Table 1 Atterberg's limits.

SOIL CHARACTERISTICS							
Sample	W_L	W_P	P_I	W_S	Wa	Ca	PIxM
No 1	31	17	14	10	9.5	0.77	644
Biskra	P.Z(1)	P.Z	P.Z	P.Z		A.A(2)	

(1) Preference Zone.
(2) Average Activity.

Grading Aggregate Analysis

In Figure 1, the grading curves of the soils as well as the corrected soils with sand and limits of the recommended zone for compressed earth blocks are represented [10]. It is noted through these curves that soil and corrected soil with contents of 10, 20 and 30 % of sand are very close to the lower limit of the recommended zone; whereas the corrected soil with 40 % sand content is out of the recommended limit zone.

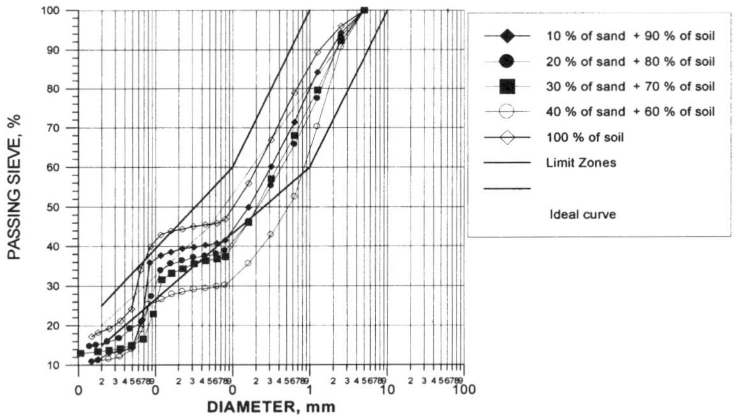

Figure 1 Grading curves aggregate analysis of used soil, corrected soil and the recommended limit zone of stabilized earth concrete.

Chemical Analysis

Clay analysis has been accomplished in the cement factory of Hamma Bouziane (Constantine, East of Algeria) using X ray Fluorescence, in accordance to NF6 P 15-467. The obtained results, showing the constituents of the soil, are presented in Table 2.

Table 2 Chemical composition of the soil.

SAMPLE	CONTENT, %					
	SiO_2	AL_2O_3	Fe_2O_3	MgO	CaO	SO_3
No 1	32.22	2.24	0.53	0.03	31.8	5.81

SAMPLE	CONTENT, %					
	K_2O	Na_2O	Cl	TiO_2	MnO	F.W(1)
No 1	0.15	0.03	0.005	0.2	0.02	26.9

(1) loss due to fire.

Mineralogical Analysis

To differentiate the clay soils, a mineralogical analysis by X rays is important. The analyses have been carried out in the geology laboratory of Boumerdes (Algiers, Algeria) using a SIEMENS 500 diffractometer, interfaced to a computer for data collecting.

Tests have been conducted on aggregates passed on 80 microns sieves. The obtained results in Table 3 show that the soil is composed mainly of kaolin (non-expansive and non-absorbent) and illites.

Table 3 Mineralogical constituents of the soil

Sample	CLAYEY MINERALS, %			NON CLAYEY MINERALS, %	
	Kaolin	Illites	I. M[(1)]	Quartz	Calcite
N 1	45	40	15	5	10

[(1)] Interstratifiers.

Organic Matter

During the treatment of soil with oxygenated water, it was noticed that the soil-water reaction was very slow and the organic mater was essentially free of vegetable fragment (0.15%).

Measurement of pH

The analysis shows that the tested sample was almost neutral, pH = 7.1.

Mechanical Characteristics (Proctor test)

The results are shown in Table 4.

Table 4 Proctor test

	OPTIMAL (Wc), %	MAX. DRY DENSITY (γ), kg/m^3
	11.75	1877
Appreciation	Excellent	Satisfactory

The Proctor test shows that the water content (Wc) of the studied sample is excellent and the dry density is satisfactory.

Physical Characteristics of Sand

Using AFNOR [12] regulations, the sand samples have been tested and found the following results :

- Disturbed apparent density (ρo) = 1520 kg/m^3
- Specific mass (γ) = 2640 kg/m^3
- Fineness modulus (F.M) = 2.33
- Sand equivalence value by sight (SE) = 70
- Sand equivalence value by test (SE_t) = 64

Cement

The cement used was manufactured in Algeria, under the commercial label C.P.J 45 and has been tested following the AFNOR [12] regulations in order to determine its real class. Tests carried out on mortar cubes have shown that the strength at 28 days was 46 MPa.

Resin

The resin used for this work has a commercial name of 'Medalatex'; supplied by Granitex; a private Algerian company of additives making. Medalatex is an aqueous dispersion of resin of white colour. It's compatible with most cements as well as lime.

In general, the latex content varies between 10 and 20% in respect to the cement mass. The latex addition gives a good adherence to the support. It gives also the impermeability, the durability and the improvement in protection of the reinforcement, thus resistance to chemical attacks.

INFLUENCE OF SAND CONTENT

In order to determine the influence of sand content on the mechanical strength, durability and the optimal quantity of soil-sand mix, several blends have been used (0 – 40%) with cement contents of 5% and 5% + 50% resin and a compacting stress of 10 MPa. Samples have been stored in a humid environment.

Compressive Strength

Figure 2a shows that the compressive strength of dry and humid sand-soil samples increases with increasing sand content. However, in percentage terms, the compressive strength evolution is 27.5% for dry samples and 30 % for humid samples, when the concentration of sand is 30%. For the same sand content, the addition of the resin has resulted in the increase in strength of the order of 11% in the dry state and 29% in the wet state.

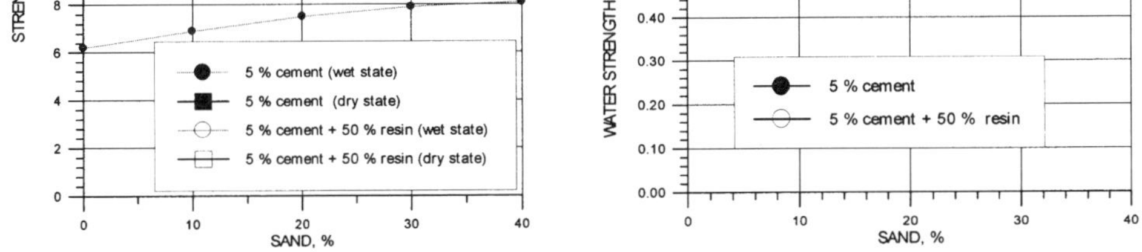

a) Mechanical strength b) Water strength coefficient

Figure 2 Sand content influence on compressive strength and water strength coefficient with 5 % cement and 5 % cement + 50 % resin stabilizer, using 10 MPa compacting stress.

Water Strength Coefficient

The water strength coefficient is determined from the compressive strength ratio for dry and humid states. Figure 2b shows that the sand content does not affect the water strength coefficient which varies between 0.53 and 0.54 when the sand content varies between 0 and 40%. However, the addition of the resin has resulted in an increase of the coefficient of the order of 17 %.

INFLUENCE OF THE COMPACTING STRESS AND THE CEMENT AND CEMENT PLUS RESIN CONTENT

In the following section, the effect of the compacting stresses (5, 7.5, 10, 12.5, 15, 17.5 and 20 MPa), the cement content (5 and 8 %) and cement plus resin (5 % + 50 % and 8 % + 50 %) on the compressive strength on dry and humid sand samples is studied. In addition, durability tests of wetting and drying, freeze-thaw, water absorption (total and capillary) tests with the optimal sand content of 30% are carried out.

Mechanical Compressive Strength in Dry Samples

Figure 3a clearly shows that the compressive strength evolution is the same for the different cement and cement plus resin content: the compressive strength increases with increasing the compacting stress until 17.5 MPa which is the optimal compacting stress. Again, the addition of the resin had a great effect on the strength of the samples in the dry state. As can be seen in Figure 3, the compressive strength increases with the addition of resin content for both cement content cases.

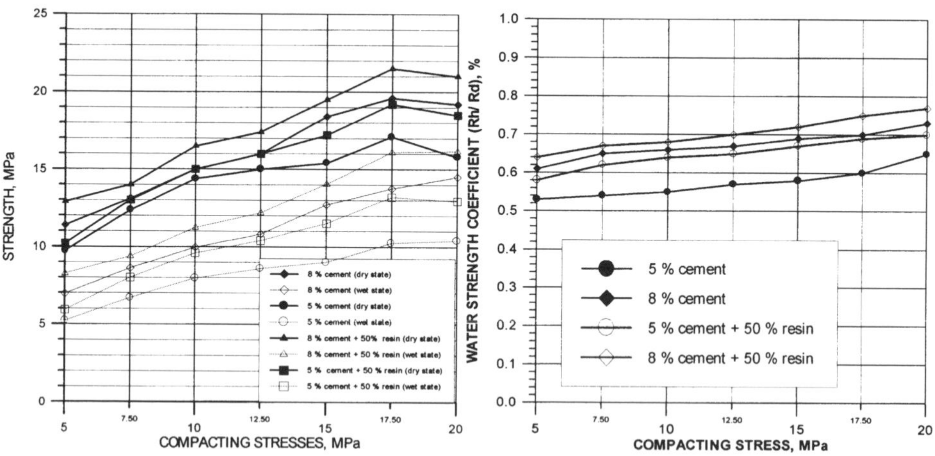

a) Compressive strength b) Water strength coefficient

Figure 3 Influence of compacting stresses and cement and cement plus resin content on the compressive strength.

Compressive Strength in Humid Samples

The mechanical strength of humid soil sample increases with increasing the compacting stresses, Figure 3a. The compressive strength also increases with the addition of resin. Like in the case of dry samples, the compressive strength evolution is not regular. The effect of the resin addition is more important in the case of the 5 % sample.

Water Strength Coefficient

The water strength coefficient evolution depends on the cement and cement plus resin percentage and on the compacting stresses, Figure 4b. It increases with the increase of cement content, the compacting stresses as well as the addition of resin. For example, the addition of the resin by 50 % as is the case in this work had resulted in variation of the water strength coefficient by 15 % and 7.5 % for both cases of 5 % and 8 % cement respectively, with a compacting stress of 15 MPa.

Water Absorption

The absorption capacity of earth stabilized blocks gives a general idea on the presence and importance of voids. When a volume of soil is subjected to the action of a stress, the material is compressed and the voids ratio decreases. As the density of soil is increased, its porosity is reduced and less water can penetrate it [13].

Total Absorption

The present test consists of immersing the soil samples in water and measuring the increase in weight during 24 hours. The absorption is evaluated in dry weight percentage. Table 5 shows that the absorption decreases when increasing the compacting stresses. Up to a certain value of 15 MPa and above; it has a minor effect. We also notice that the increase in cement content lowers the water absorption factor. The resin addition lowers considerably the water absorption factor for both cement contents. The total absorption varies between 18 to 33 % and between 18 to 39 % for 5 and 8 % cement respectively; for the compacting stresses varying from 5 to 20 MPa.

Table 5 Influence of the compacting stresses and the cement and cement plus resin content on the total absorption

	TOTAL ABSORPTION, %			
Compacting stress (MPa)	5% cement	5% cement + 50 % resin	8% cement	8% cement + 50 % resin
5	10.12	8.2	9.17	7.5
7.5	9.78	7.5	9.23	6.6
10	9.12	6.1	8.26	5.5
12.5	8.33	6.0	7.84	5.6
15	8.27	5.9	7.35	5.3
17.5	7.54	5.8	7.25	5.2
20	8.71	5.8	8.59	5.2

Capillary Absorption

Capillary absorption test consists of placing the soil sample on a humid surface with voids, constantly water saturated, and measuring its weight after 7 days. Absorption is evaluated in percentage of dry weight. Table 6 shows that the capillary absorption decreases when increasing the compacting stresses and the cement as well as cement plus resin content. For instance, it varies from 2.15 to 2.02% when the cement content varies from 5 to 8% with a compacting stress of 17.5 MPa. Again, the addition of the resin decreases considerably the water capillary absorption.

Table 6 Influence of the compacting stresses and the cement and cement plus resin on the capillary absorption

	CAPILLARY ABSORPTION, %			
Compacting stress (MPa)	5% cement	5% cement + 50 % resin	8% cement	8% cement + 50 % resin
5	3.99	3.8	3.52	3.1
7.5	3.19	3.7	2.48	2.2
10	3.90	3.1	2.30	2.1
12.5	2.74	2.5	2.24	1.9
15	2.35	2.3	2.23	1.8
17.5	2.15	2.1	2.02	1.2
20	1.17	2.0	1.19	1.1

Wetting and Drying Test

This test is carried out according to the ASTM D 559-57; it consists of immersing soil samples in water for a period of 5 hours and then removing to be dried in an oven at 71 °C for a period of 42 hours. The procedure is repeated for 12 cycles, samples are brushed every cycle to remove fragments of the material affected by the wetting and drying cycles. For every sample, the variation in weight is computed after the 12 cycles [13]. Table 7 shows that the loss in weight diminishes when increasing the compacting stress and the cement and cement plus resin content. For the cases of 5 and 8% cement, the effect of cement on the weight loss is important for the compacting stresses up to 12.5 MPa. Above this value the addition of cement is less significant. The resin addition has a slight effect on the weight loss.

Table 7 Influence of the compacting stresses and the cement and cement plus resin on the weight loss.

	WEIGHT LOSS, %			
Compacting stress (MPa)	5% cement	5% cement + 50 % resin	8% cement	8% cement + 50 % resin
5	2.65	2.1	2.07	1.8
7.5	2.28	1.8	1.86	1.7
10	1.9	1.5	1.4	1.2
12.5	1.34	1.2	1.35	1.3
15	1.27	0.9	1.17	0.9
17.5	0.31	0.25	0.25	0.22
20	0.13	0.10	0.05	0.05

Freeze - Thaw

Following the procedure described by ASTM D560, the freeze-thaw test consists of placing a soil sample on an absorbent water saturated material in a refrigerator at a temperature of -23°C for a period of 24 hours and then removed. The sample is then thawed in a moist environment at a temperature of 21°C for a period of 23 hours and then removed and brushed. The test is repeated for 12 freeze-thaw cycles and then dried in an oven to obtain a constant weight [13]. Table 8 shows that the weight loss diminishes when increasing the compacting stress and the cement content as in the case of wetting and drying test discussed previously. For the 5% cement sample, the weight loss changes from 17 to 3.3% when the compacting stress varies from 5 to 20 MPa. And the weight loss is very important with low compacting stresses. The effect of the resin is more pronounced at lower compacting stresses.

Table 8 Influence of the compacting stresses and the cement and cement plus resin on the weight loss

	WEIGHT LOSS, %			
Compacting stress (MPa)	5% cement	5% cement + 50 % resin	8% cement	8% cement + 50 % resin
5.0	17.0	3.8	5.28	4.1
7.5	12.0	3.7	4.67	3.2
10.0	4.5	3.1	2.4	2.1
12.5	4.2	2.5	1.3	1.1
15.0	3.9	2.3	1.02	1.0
17.5	3.6	2.1	0.24	0.2
20.0	3.3	2.0	0.14	0.1

CONCLUSIONS

The main objective of this work was to investigate the different factors affecting the durability of cement stabilized earth blocks as well as the importance of the resin addition.
The work showed the importance of the sand content, the compacting stress and the cement and the cement plus resin contents on the behaviour of stabilized earth blocks with respect to water attacks as well as elucidating certain points:

- The principle effect of the stabilization with the cement is to prevent water attacks. We would achieve good stabilization if we could obtain a durable material with a limited loss in mechanical strength in a wet state [14].
- The sand content does not affect considerably the compressive strength and the water strength coefficient.
- Increasing the compacting stress from 5 to 20 MPa and the cement content from 5 to 8% improves the compressive strength in dry as well as wet state and the water strength coefficient. We notice also that the increase of these two parameters decrease the weight loss and the water absorption.
- The latex addition is shown also to be beneficial concerning the durability in general. This is due to the fact that such additions consolidate the cementitious matrix and play a role of a co-matrix. However, when considering the economical factor, mainly the resin price (4 to 6 times more expensive than cement), this addition is not recommended.

REFERENCES

1. GHOUMARI, F., Matériau en Terre Crue Compactée: Amélioration de sa Durabilité à l'Eau; Thèse de Doctorat, INSA de Lyon, 1989.

2. GREGORY, M., KEVAN, H., "Earth Building in Australia - Durability Research", Proceedings of Modern Earth Building 2002, 19 to 21 April 2002 in Berlin, ISBN 3-8167-6118-6 Germany, 2002, pp. 129-139.

3. HEATHCOTE, K.A., "Durability of earthwall buildings" Construction and Building Materials Volume 9, Number 3, 1995, pp 185-189.

4. GUETTALA, A., HOUARI H., MEZGHICHE, B., CHEBILI, R., "Durability of Lime stabilized Earth Blocks", Proceedings of an international conference University of Dundee, Scotland UK 9 /11 Sep 2002, pp. 145-654

5. GUETTALA, A., GUENFOUD, M., Influence des Types d'Argiles sur les Propriétés Physico-mécaniques du Béton de Terre Stabilisée au Ciment, ANNALES du Bâtiment et des Travaux Publics N° 1 Février, Paris, 1998, pp. 15-25.

6. GUETTALA, A., GUENFOUD, M., Béton de Terre Stabilisée Propriétés Physico-Mécaniques et Influence des Types d'Argiles, LA TECHNIQUE MODERNE N°1-2, Paris, 1997, pp. 21-26.

7. KERAALI, A.G., "Durability of compressed and cement-stabilized building blocks" PhD Thesis, University of Warwick, school of Engineering, UK, September 2001.

8. American Society for Testing and Materials. Annual Book of ASTM Standards, V.04.01, Philadelphia, 1993.

9. DOAT, P., HAYS, A., HOUBEN, H., Construire en Terre, Collection en Architecture, Paris, 1981.

10. MICHEl, J., Etude sur la Stabilisation et la Compression des Terres Pour leur Utilisation dans la Construction, Annales de l'Institut Technique de Bâtiment et des Travaux Publics, Série Matériaux N° 339 Mai, 1976, pp. 22-35.

11. AFNOR 1984, Recueil de Norme Françaises. Bâtiment Béton et Constituants du Béton. Paris.

12. HOUBEN H., GUILLAUD, H., (CRATerre), Earth Construction, Primer Brussels, CRATerre / PGC / CRA / UNCHS / AGCD, 1984.

13. GUETTALA, A., MEZGHICHE, B., CHEBILI R., HOUARI H., Durability of Blocks of Earth Concrete, Proceedings of II International Symposium Cement and Concrete Technology in the 2000, September 6-10 2000 Istanbul, Turkey, pp. 273-281.

HEAT EVOLUTION AND HYDRATION MODELLING OF GGBS CEMENT

L Zheng

K A Paine

R K Dhir

University of Dundee

United Kingdom

ABSTRACT. The advisability of controlling peak temperature of concrete at early-age is well recognised, and the choice of appropriate low heat cement with suitable heat of hydration characteristics can assist in this control. This is particularly pertinent with respect to water-retaining and massive concrete structures where the need to prevent early-age thermal cracking is paramount. PC/ggbs combinations are often used in these structures since they have low heat of hydration but achieve strength rapidly enough to meet most strength classes. The paper describes tests carried out to assist in the modelling of the hydration of PC/ggbs combinations for use in estimating the early-age temperature of concrete structures. Two sources of ggbs were used; one regarded as a standard ggbs typical of most use in UK, and the other specifically selected as a high fineness ggbs. Systematic isothermal conduction calorimetry tests for PC/ggbs combinations were carried out over a range up to 90% ggbs content by mass in the combination, at temperatures of 5°C, 20°C, 40°C and 60°C. The curves of rate of heat evolution against time and cumulative heat were analysed and a mathematical model based on the test results was proposed. In general, the relationships appeared to be double peak curves, which were considered to result from an initial Portland cement reaction, followed by a later ggbs reaction activated by the lime made available during hydration of the Portland cement. A further refinement can be achieved by considering co-reactivity effects of PC and additions.

Keywords: Heat of hydration, PC/ggbs combination, Modelling, Peak temperature

Dr Li Zheng is a Research Fellow in the Concrete Technology Unit in the Division of Civil Engineering, School of Engineering, at the University of Dundee, UK. His research focuses mainly on concrete durability, mix optimisation, cement hydration and modelling.

Dr Kevin A Paine is a Lecturer in the Concrete Technology Unit at the University of Dundee. His research focuses mainly on the use of recycled materials and industrial by-products as cement and aggregate components in concrete.

Professor Ravindra K Dhir is Director and Professor of the Concrete Technology Unit at the University of Dundee. His is a member of numerous national and international technical committees and has published extensively on many aspects of concrete technology, cement science, durability and construction methods.

INTRODUCTION

The advisability of controlling peak temperature of concrete at early-age is well recognised [1, 2]. When the peak temperature of concrete exceeds a certain value, a temperature gradient generated by the hydration of cement can initiate microscopic cracks that may seriously affect the durability performance of concrete. The peak temperature of a mass concrete structure can be controlled by various means and the use of low heat cement with suitable heat of hydration characteristics is one of the most effective ways [3, 4]. Indeed, combinations consisting of Portland cement (PC) and ground granulated blastfurnace slag (ggbs) are widely used all over the world, since they have low heat of hydration but achieve strength rapidly enough to meet most strength classes [1, 4, 5, 6].

In order to simulate the heat evolution inside a concrete structure and predict the peak temperature, it is vital to establish a proper hydration model of the cement for the temperature prediction. To-date, there have been a number of cement hydration models published [7-13]. These models not only play an important role in prediction of the temperature distribution inside the concrete, but can also be applied to predict concrete formwork striking times [14], simulating microstructure formation and degradation processes [9, 12, 13, 15]. However, from the view of concrete temperature simulation, some models are over-complicated, requiring a large number of parameters, many of which are not available in general engineering practice, for example cement mineral compositions, i.e. C_3S, C_2S, C_3A and C_4AF content, and their particle sizes. On the other hand, some models are over-simplified and cannot deal with the variation of ggbs content in cement combinations, thus cannot be used for designing an optimum PC/ggbs combination for a particular application in order to achieve both an optimal point for temperature control and satisfactory concrete performance. This paper describes a study to fill this gap and to propose an appropriate hydration model for PC/ggbs combination for use in estimating the early-age temperature of concrete structures.

In this study, systematic isothermal conduction calorimetry tests for PC/ggbs combinations were carried out over a range of ggbs content up to 90% by mass in the combination, at temperatures of 5°C, 20°C, 40°C and 60°C. The curves of rate of heat evolution against time and cumulative heat were analysed and a mathematical model based on the test results was proposed. In general, the relationships appeared to be double peak curves, which were considered to result from an initial Portland cement reaction, followed by a later ggbs reaction activated by the lime made available during hydration of the Portland cement. A further refinement can be achieved by considering co-reactivity effects of PC and additions. The effect of ggbs content on the hydration and heat evolution performance has been examined. Based on test results and modelling, a computer program was developed for thermal simulation of concrete to provide design data for thermal crack management.

EXPERIMENTAL PROGRAMME

Materials

A single source of cement strength class 42.5N Portland cement (PC) conforming to BS12 and BS EN 197 -1 was used throughout, and two sources of ggbs: one regarded as a standard ggbs typical of most use in UK (GB), and the other specifically selected as a high fineness ggbs (GA). Both conformed to BS 6699:1992 and BS EN 197-1. The physical and chemical properties are shown in Table 1.

Table 1 Properties of cements and additions used in study

PROPERTY	PC	GA	GB
Relative Density	3.14	2.86	2.88
Fineness, m^2/kg	405	602	466
Bulk Oxide Composition			
SiO_2	21.5	35.2	36.3
Al_2O_3	5.4	13.1	12.6
Fe_2O_3	2.6	0.2	0.5
CaO	64.2	41.0	42.1
MgO	2.6	8.1	6.9
P_2O_5	0.1	0.0	0.0
TiO_2	0.3	0.7	0.6
SO_3	2.8	1.3	1.1
K_2O	0.7	0.5	0.3
Na_2O	0.3	0.2	0.2
MnO	0.0	0.5	0.3

Cement Combinations and Test Programme

Isothermal conduction calorimetry was used to measure the heat evolution processes of PC/ggbs combination. To judge the "apparent activity energy" of the cement combinations with different ggbs contents, each cement combination was tested at four different temperatures of 5°C, 20°C, 40°C and 60°C.

The isothermal conduction calorimetry used is JAF conduction calorimeter manufactured by Wexham Development, UK. The heat evolution process up to 72 hours was recorded and calculated using software supplied by the manufacturer.

To examine the effect of ggbs content, tests were carried out over a range combination contents including 100% PC, and four different ggbs contents, i.e. 35%, 50%, 70% and 90% by mass. For each test, 30 grams of cement were used and w/c was fixed at 0.5.

EXPERIMENTAL RESULTS

Portland Cement

Table 2 gives (i) the time to maximum rate of heat evolution, (ii) maximum rate of heat evolution, (iii) total heat at maximum rate of heat evolution and (iv) total heat after 72 hours of Portland Cement (PC) at 5°C, 20°C, 40°C and 60°C.

The rate of heat evolution, q, against time, t, and cumulative heat, Q_{cc}, are shown in Figure 1. The figure is characterised by the smooth curves and the clear single peak of maximum rate of heat evolution. It can be seen that the maximum rate of heat evolution, q_{max}, increases with an increase in hydration temperature and that the time to reach the maximum rate is shorter for the higher temperatures (Figure 1a).

Table 2 Results of isothermal conduction calorimetry tests for PC

	TIME TO MAX. RATE OF HEAT EVOLUTION, t_{max}, hrs	MAX. RATE OF HEAT EVOLUTION, q_{max}, W/kg	HEAT OUTPUT at q_{max}, Q_{tm}, kJ/kg	HEAT OUTPUT at 72 hrs, Q_{t72}, kJ/kg
5°C	21.5	1.24	55.8	189.1
20°C	8.90	3.30	59.1	266.1
40°C	3.80	9.16	53.9	305.9
60°C	1.21	22.0	56.0	347.5

In Figure 1b, it can be seen that the cumulative heat, Q_{tm}, at q_{max} is similar for different hydration temperatures, and ranges from 50 to 60kJ/kg for the PC tested over the temperature range from 5 to 60°C. This result is similar to results published by De Schutter [11, 16], in which the measured temperature range was 5 to 35°C.

The measured cumulative heat after 72 hours hydration was greater for the higher temperatures. This was anticipated, since at a given time, the higher the temperature, the higher the degree of hydration of the cement.

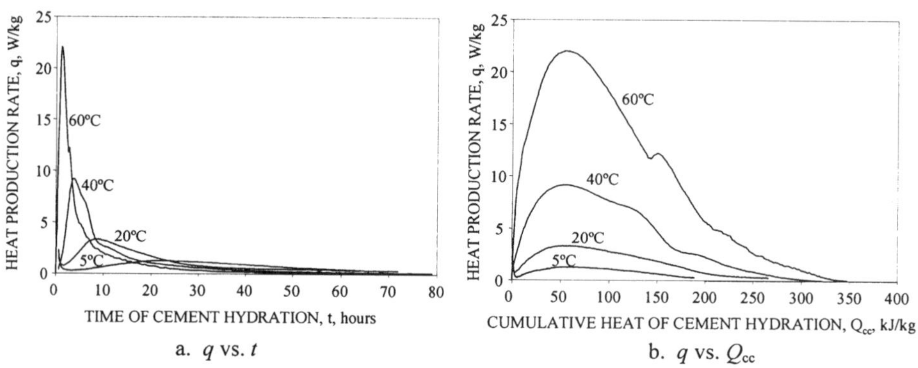

Figure 1 Rate of heat evolution of Portland cement

PC/ggbs combinations

Some typical curves of the rate of heat evolution, q, against time, t, and cumulative heat, Q_{cc}, for PC/ggbs combinations are shown in Figures 2-3. In general, the relationships are characterised by the typical double peak curves, which are considered to result from an initial Portland cement reaction, followed by a later ggbs reaction activated by the lime made available during hydration of the Portland cement. However, for a temperature of 60°C, the second peak tends to become coincident with the first peak.

From comparison of the results for the two types of ggbs, it is clear that the curves results for the same temperature conditions were quite similar. However, it can be seen that the rate of heat evolution and total heat output for GA is greater than that of GB, reflecting the higher fineness and greater reactivity of this material. Comparisons of curves of rate of heat evolution against cumulative heat are shown in Figure 4. It can be seen that the difference in heat evolution rate become more apparent at higher hydration temperatures.

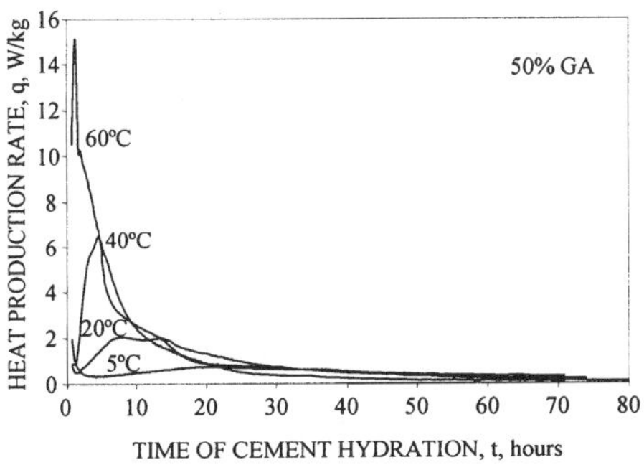

a. q vs. t

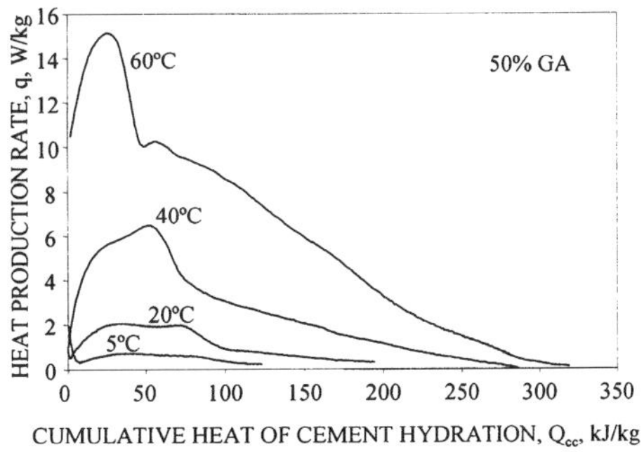

b. q vs. Q_{cc}

Figure 2 Rate of heat evolution of 50%PC/50%GA cement combination

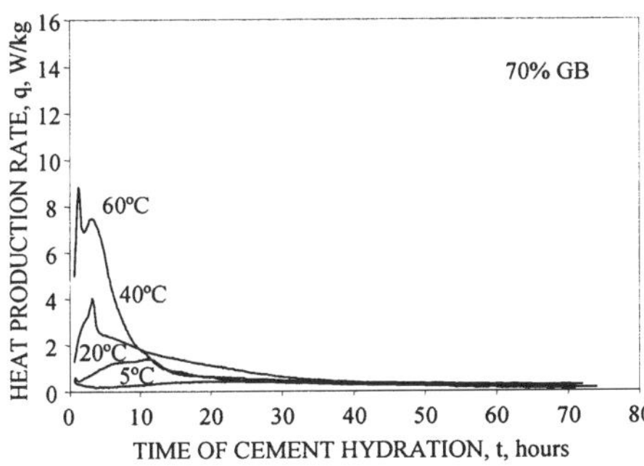

a. q vs. t

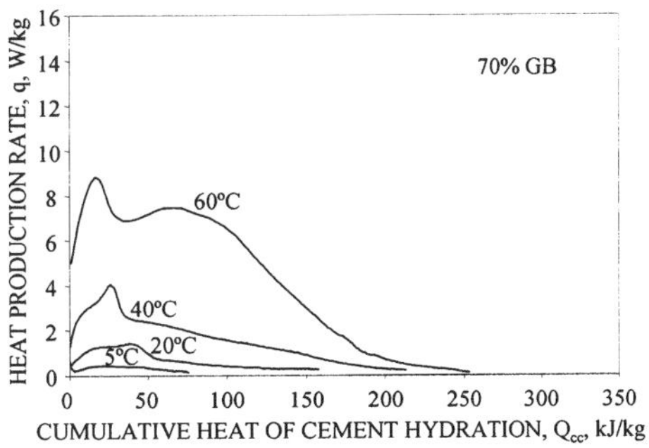

b. q vs. Q_{cc}

Figure 3 Rate of heat evolution of 30%PC/70%GB cement combination

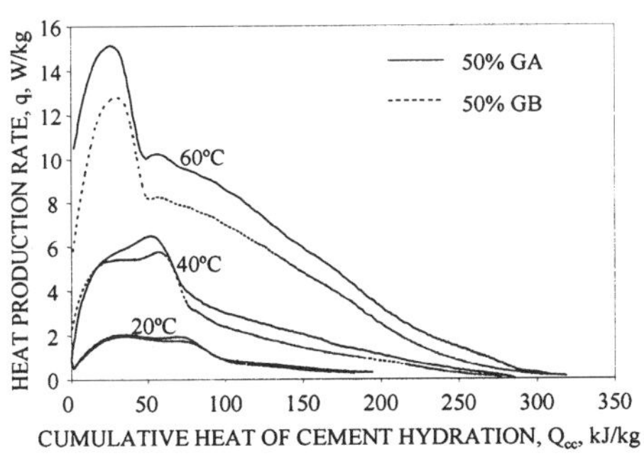

a. 50% ggbs cements

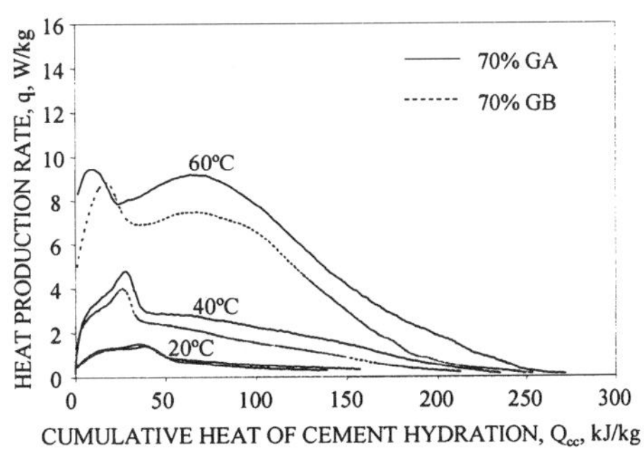

b. 70% ggbs cements

Figure 4 Comparison of q vs. Q_{cc} curves between the two ggbs cements

The effects of ggbs content on the maximum rate of heat evolution are shown in Figure 5. For all temperatures, a nearly linear relationship between the two parameters is observed for both types of ggbs.

Relationships between Q (after 72 hours) and ggbs content are shown in Figure 6. At lower temperatures, 5-20°C, the relationship appears linear. However, curved relationships are observed at higher temperatures, 40-60°C.

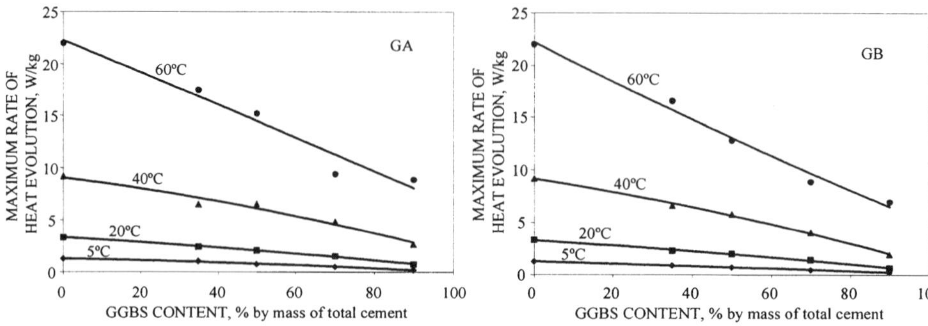

a. ggbs A cements

b. ggbs B cements

Figure 5 Relationship between maximum rate of heat evolution and ggbs content

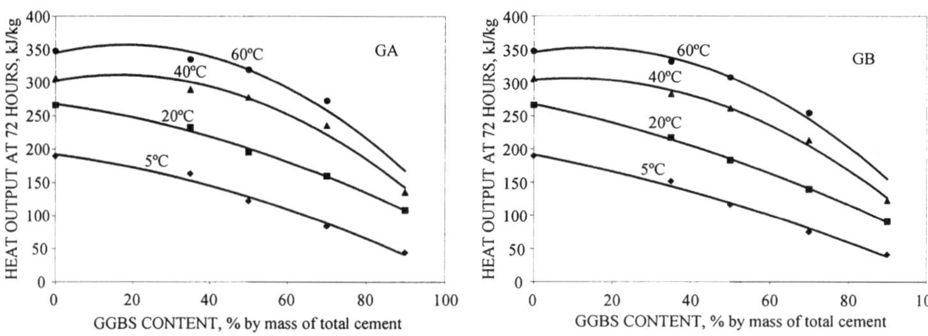

a. ggbs A cements

b. ggbs B cements

Figure 6 Relationship between total heat after 72 hours and ggbs content

HYDRATION MODELLING

Portland cement

Examination of several cement hydration models and comparison with the results obtained in this study found that the De Schutter model [11, 16] is accurate in describing the hydration of Portland cement. A relationship between rate of hydration heat and degree of hydration at 20°C for PC proposed by De Schutter is given as follows:

$$q_{20} = q_{\max 20} \cdot f(r)$$
$$f(r) = \mathrm{c} \cdot [\sin(r\pi)]^{\mathrm{a}} \cdot \exp(-\mathrm{b} \cdot r) \qquad (1)$$
$$r \approx Q_{cc}(t) / Q_{\max}$$

Where: q_{max20} = maximum heat hydration rate of PC at 20°C
r = degree of hydration, $\approx Q_{cc}(t)/Q_{max}$
$Q_{cc}(t)$ = cumulated heat of hydration at t
Q_{max} = the total heat of hydration released at the end of reaction
a, b, c = constants

In considering the effect of temperature on the hydration, the Arrhenius Function has been chosen. According to Arrhenius, when heat is produced by a zero-order chemical reaction, the rate of heat production may be represented by the following function [17]:

$$q = q_0 \times \exp\left[\frac{-E}{R \cdot (273 + T)}\right] \qquad (2)$$

Where, q_0, k = constants
E = apparent activation energy
R = universal gas constant

When T_{ref} = 20°C (293 K) and q_{20} are used as a reference, it can be derived that:

$$q = q_{20} \times \exp\left[\frac{E}{R}\left(\frac{1}{293} - \frac{1}{273 + T}\right)\right] \qquad (3)$$

Based on the test data, parameters in the formulae above can be calibrated as given in Table 3 together with De Schutter's original model parameters. The differences between the two sets of parameters reflect the different hydration characteristics of cements used in this study and by De Schutter. The curves from the model simulation are shown in Figure 7 and compared with the measured data. It can be seen that the simulated curves agree well with the measured data.

Table 3 Modelling parameters for PC hydration

	MODELLING PARAMETERS FOR PC HYDRATION	
	(calibrated by the data from this study)	(Schutter's original published parameters)
a	0.875	0.667
b	5.0	3.0
c	4.22	2.60
q_{max20} (W/kg)	3.30	2.16
Q_{max} (kJ/kg)	350.0	270.0
E (kJ/mol)	38.5	33.5

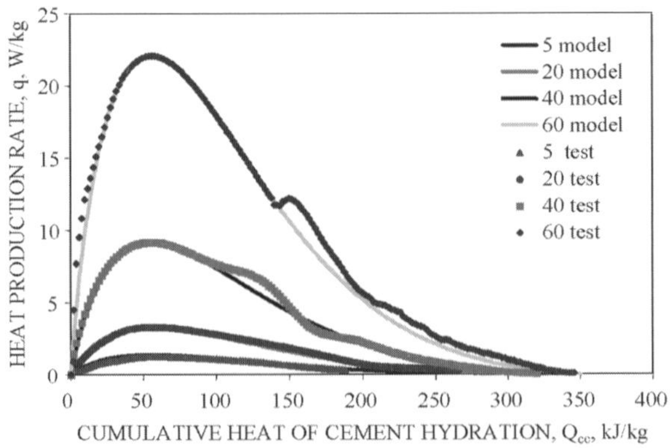

Figure 7 Comparison between model curves and test data of rate of heat evolution and cumulative heat for Portland cement

PC/ggbs combinations

Although De Schutter has proposed a model for ggbs cement hydration [11, 16], this model did not fully link the heat production rate to the ggbs content. When the model was applied to the results obtained in this study, no meaningful relationships were found between the model parameters and ggbs content. Therefore, further development was required to deal with the problems. A refined model was proposed for this study and the new developed model has the characteristics as follows:

(1) Heat production rate, q, of the cement hydration was divided into three portions: (i) from PC hydration, q_p; (ii) from ggbs hydration, q_s and (iii) co-reactivity effect of PC and additions, q_r. So $q = q_p + q_s + q_r$

(2) Similar model patterns to Equation 1 and 3 were used for both PC and ggbs hydrations:

$$q_{p,20} = q_{p,\max 20} \cdot f_p(r)$$
$$f_p(r) = c_p \cdot [\sin(r\pi)]^{a_p} \cdot \exp(-b_p \cdot r) \qquad (5)$$
$$r \approx Q_{cc}(t) / Q_{\max}$$

and

$$q_{s,20} = q_{s,\max 20} \cdot f_s(r)$$
$$f_s(r) = c_s \cdot [\sin(r\pi)]^{a_s} \cdot \exp(-b_s \cdot r) \qquad (6)$$
$$r \approx Q_{cc}(t) / Q_{\max}$$

Where:

$q_{p,\max 20}$	= maximum heat hydration rate of PC at 20°C
$q_{s,\max 20}$	= maximum heat hydration rate of ggbs at 20°C
r	= degree of hydration, $\approx Q_{cc}(t)/Q_{\max}$
$Q_{cc}(t)$	= cumulated heat of hydration at t
$Q_{\max}$	= the total heat of hydration released at the end of reaction
a_p, b_p, c_p	= constants for PC hydration
a_s, b_s, c_s	= constants for ggbs hydration

The Arrhenius Functions are expressed as:

$$q_p = q_{p20} \times \exp\left[\frac{E_p}{R}\left(\frac{1}{293} - \frac{1}{273+T}\right)\right] \tag{7}$$

and

$$q_s = q_{s20} \times \exp\left[\frac{E_s}{R}\left(\frac{1}{293} - \frac{1}{273+T}\right)\right] \tag{8}$$

Where, E_p = apparent activation energy for PC
E_s = apparent activation energy for ggbs

Both $q_{p,max20}$ and $q_{s,max20}$ are functions of PC/ggbs proportions, however, it is assumed that both E_p and E_s are constants only associated with the chemical reactivity of PC and ggbs respectively, independent to the PC/ggbs proportions. In this way, the effect of ggbs content on cement hydration can be modelled. Relationships of $q_{p,max20}$ and $q_{s,max20}$ to ggbs contents are shown in Figure 8. For PC hydration, the relationship is linear. The maximum heat hydration rate for ggbs increases linearly, however, since the hydration of ggbs requires lime made available during the PC hydration, this linear relationship only exists in a certain range, in which the release of $Ca(OH)_2$ matches the requirement of ggbs reaction. Clearly, the results show that for ggbs content larger than 70%, the potential $q_{s,max20}$ cannot be achieved due to the lack of $Ca(OH)_2$, and there is a $q_{s,max20}$ plateau.

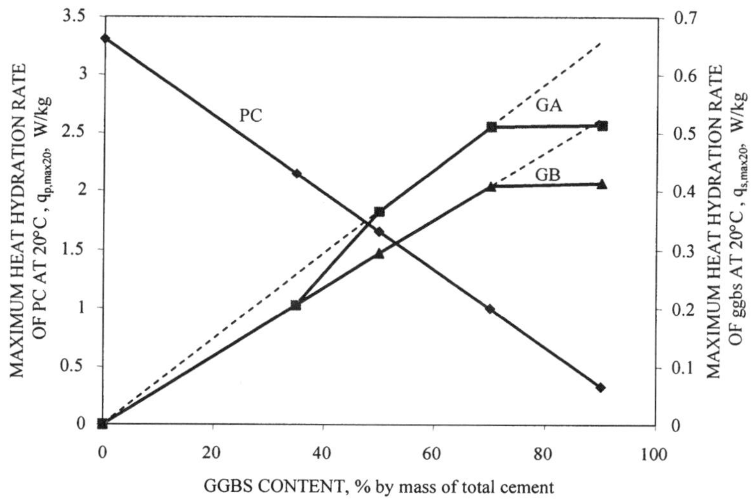

Figure 8 Relationships of $q_{p,max20}$ and $q_{s,max20}$ to ggbs content

(3) Due to the consideration that the hydration of a blended cement (PC + ggbs for this study) may not be just the simple summary of the hydration of the two individual components, co-reactivity effect of PC and additions, q_r, is taken into consideration. The following equation is used to modelling the co-reactivity effect:

$$q_r = \mathrm{a}_r \exp\left[-\frac{(Q_{cc} - \mathrm{b}_r)^2}{\mathrm{c}_r}\right] \tag{9}$$

where, a_r, b_r and c_r are constants for co-reactivity effects of PC and additions, and they will depend on temperature and also on PC/ggbs combinations.

Since so called PC reaction and ggbs reaction occurred coincidently and the exist of co-reactivity effect in a PC/ggbs combination system, further division of the total heat of hydration into heat from PC and heat from ggbs seems difficulty. It was also not necessary in the modelling process proposed in this study. Figure 9 shows a set of full modelling curves in comparison with test results for 70% GB cement. It can also be seen that the modelling curves agree well with the measured data.

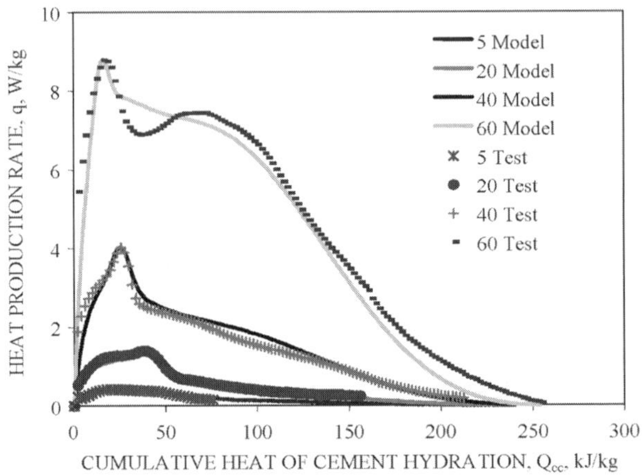

Figure 9 Comparison between model curves and test data of rate of heat evolution and cumulative heat for 70% GB cement

CONCLUSIONS

Systematic isothermal conduction calorimetry tests for PC/ggbs cement combinations were carried out over a range of cement combinations up to 90% ggbs content by mass, at different temperatures. Based on test results, a mathematical model to simulate the hydration of PC/ggbs cement combinations was proposed. The model comprises three parts which were considered to result from an initial Portland cement reaction, followed by a later GGBS reaction activated by the lime made available during hydration of the Portland cement and a further refinement by considering co-reactivity effects of PC and additions. De Schutter's PC hydration model is used for PC reaction part. However, for the ggbs reaction and co-reactivity effects of PC and ggbs, new model was developed. With the model proposed, the effect of ggbs content can be taken into account and it is possible to design an optimum PC/ggbs cement combination to achieve an optimal point for both temperature control and satisfactory concrete performance. A computer program was developed for thermal simulation of concrete to provide design data for thermal crack management based on the proposed model.

ACKNOWLEDGEMENTS

The authors would like to acknowledge the support provided by the UK government Department of Trade and Industry, Babtie Group, Castle Cement Ltd, Cementitious Slag Makers Association, Concrete Structures Group, John Doyle Construction Limited, Mott MacDonald, North of Scotland Water Authority, Quarry Products Association, Rugby Cement, and ScotAsh Ltd. Thanks are also given to Professor D C Spooner for supplying his prototype computer programs for concrete temperature simulation, and to Mr Oliver Hart for his assistance with the experimental work.

REFERENCES

1. HARRISON T.A., Early-age thermal crack control in concrete, CIRIA Report 91, 1992, 57pp

2. LIU C., MELINE R., LEE J.N., Heat removal from mass concrete footing, CONCRETE, 2002, No.1798, pp39-42

3. FURUYA N., TATSUMI M., Technical innovation for realization of Akashi-Kaikyo Bridge, Proceedings of the Institution of Civil Engineers-Structures and Buildings, 1994, Vol.104, No.3, pp285-296

4. TOLLOW, A.J., Durbans newest water-resource - the Inanda Dam, Journal of the Institution of Water and Environmental Management, 1991, Vol.5, No.5, pp519-528

5. DE SCHUTTER G., Hydration and temperature development of concrete made with blast-furnace slag cement, Cement and Concrete Research, 1999, Vol. 29, pp143-149

6. XIONG X., VAN BREUGEL K., Apparent activation energy and its application in the numerical simulation of blended cement hydration process, Innovations and Developments in Concrete Materials and Construction, Edited by R K Dhir, P C Hewlett and L J Csetenyi, Proceedings of the International Conference, University of Dundee, Scotland, UK, 2002, Thomas Telford, pp91-100

7. PITKÄNEN P., Prediction of temperature fields of massive concrete structures during hardening, Nordic Concrete Research, 1984, Pub. No. 3, pp183-190

8. SPOONER D.C., BCA computer model for temperature rise prediction in fresh concrete, British Cement Association, TDH 51062, undated, CIRIA 1992

9. VAN BREUGEL K., HYMOSTRUC: A computer based simulation model for hydration and formation of structure in cement based materials, Hydration and Setting of Cements, Edited by A Nonat and J C Mutin, 1992 RILEM, E & FN Spon, London, pp361-368

10. WANG C., DILGER, W.H., Prediction of temperature distribution in hardening concrete. Thermal Cracking in Concrete at Early Ages, Proceedings of the International RILEM Symposium, edited by R. Springenschmid, Institute of Building Materials, Technical University of Munich, Germany, E & FN Spon, London 1994, pp21-28

11. DE SCHUTTER G., Hydration and temperature development of concrete made with blast-furnace slag cement, Cement and Concrete Research, 1999, Vol. 29, pp143-149

12. MAEKAWA K., CHAUBE R., KISHI T., Modelling of concrete performance: hydration, microstructure formation and mass transport, 1999, E &FN Spon, London, 308pp.

13. BENTZ D.P., Three-dimensional computer simulation of Portland cement hydration and microstructure development, Journal of the American Ceramic Society, 1997, Vol.80, No.1, pp3-21

14. HARRISON T.A., Formwork striking times – criteria, prediction and methods of assessment, CIRIA Report 136, 1995, 71pp

15. BENTZ D.P., GARBOCZI E.J., MARTYS N.S, Application of digital-image-based models to microstructure, transport properties, and degradation of cement-based materials, The Modelling of Microstructure and Its Potential for Studying Transport Properties and Durability, 1996, Kluwer Academic Publishers, Dordrecht, The Netherlands, pp167-185

16. DE SCHUTTER G., TAERWE L., General hydration model for Portland cement and blast-furnace slag cement, Cement and Concrete Research, 1995, Vol. 25, pp593-604

17. CARSLAW H.S., JAEGER J.C., Conduction of Heat in Solids. 2nd Edition, Clarendon Press, Oxford, 1959, reprinted 1992, 510pp

NEW ASPECTS OF BLAST FURNACE SLAG CEMENT HYDRATION

T Adam
Darmstadt University of Technology
Germany

ABSTRACT. Blast furnace slag cements are qualified for applications in nearly all fields of construction. Their properties are similar to pure Portland cements and often beyond this. Anyhow, these cements still meets resistance, because of their reputation for requiring an extended curing duration. Due to the self commitment of the German cement industry to lower its fuel energy consumption, the share of blast furnace slag cements in production will increase. As a logical consequence, the application of blast furnace slag cements will be intensified in fields of construction, which were reserved for Portland cement up to now. The purpose of this work is to describe the course of hydration of blast furnace slag cement for varying slag contents. The specimens were exposed to different relative humidities. The content of non-evaporable water, relative to that in a fully hydrated paste of the same cement, was used as a measure of the degree of hydration. The content of non-evaporable water was analysed by the Carbon-Water-Method. The results so far indicate that the late hydration has been underestimated up to now. With growing slag content, a considerable hydration occurs after the 28th day. This happens regardless of the preceding moisture conditions. Thereby, a much higher degree of hydration was reached than could have been expected from blast furnace slag cements.

Keywords: Hydration, Blast furnace slag cement, Curing, Non-evaporable water

Thomas Adam is a research assistant of construction materials engineering at Darmstadt University of Technology. The focus of his research is on investigating the effect of different curing environments on the hydration and durability of concrete and the use of pozzolanic materials as replacement for cement. Mr. Adam received the Diploma (5-year degree) in civil engineering from Weimar Bauhaus University in 2000.

INTRODUCTION

Blast furnace slag cements are standardized in Germany since 1917. From that time on, more than 200 millions tons of these cements were used. Although, there are long experiences with the application of those cements, they still meet on resistance. They are said to be slow in their strength development, need too much time for curing and are less durable under frost-thaw-salt impact. Therefore, their use was often limited to fields of construction where those properties were less important or where their advantages like a slower heat liberation and improved resistance against chemical impact were demanded. Typical fields of applications are hydraulic structures, monolithic foundations and precast elements in connection with steam curing.

In the year 1995 the German cement industry, together with other energy-intensive industries, committed itself to make its contribution to global warming prevention and lower its specific fuel energy consumption by 20 % between 1987 and 2005. This commitment has been updated and, since 2000, has provided for a 28 % reduction in energy-related specific CO_2 emissions from 1990 to 2012. To reach this target, the German cement industry will continue to increase the share of unburnt cement constituents. Due to the rising output of blast furnace slag from the iron production, the cement industry will primarily increase the share of blast furnace slag cement production. As a result, it is expected that those cements will invade in fields of application, which are dominated by Portland-cement up to now.

With the increasing utilisation of blast furnace slag cements, and especially those containing high loadings of slag, the cement industry has to meet the objections against blast furnace slag cements. Hence, the investigations in incorporating large volumes of blast furnace slag have attracted intensive research attentions. First results already led to slag cements with improved strength development [4].

On the other hand, an understanding of the similarities and differences of the hydration kinetics between Portland cement and blast furnace slag cement is important. Especially the investigation of the hydration of the slag fraction under different moisture conditions has to proceed. This may provides a means for estimating the real curing demand of concretes made with blast furnace slag cements. Ultimately, such data allows estimates of the relationship between the degree of hydration and the performance of the concretes.

MATERIAL AND METHODS

The aim of the work was to achieve results as close as possible to those in practical use. On that score, it was refrained from blending the cement in the laboratory. Instead, three cements of the same strength class were used in commercial quality. The cements were made in the same plant, contained the same Portland clinker but varying slag contents. The chemical composition and physical properties are shown in Table 1.

Due to the combined grinding of clinker and blast furnace slag, the particle size distribution of the slag was not determinable. The chemical composition of the blast furnace slag was examined on a separate sand sample and may differ from the actual composition in the cement ,due to the tolerance of production.

Table 1: Physical and chemical characteristics of the cements and slag

	CEM I 32,5 R	CEM III/A 32,5 N	CEM III/B 32,5 N-NW	PURE BLAST FURNACE SLAG
Blast furnace slag content	-	49 %	71 %	(100%)
Blaine Specific Surface [cm²/g]	2959	3239	3725	-
SiO_2	19.72 %	25.69 %	28.53 %	35.13 %
Al_2O_3	5.20 %	11.44 %	8.90 %	11.43 %
Fe_2O_3	3.14 %	2.02 %	1.39 %	0.65 %
CaO	64.13 %	49.16 %	48.35 %	40.56 %
MgO	1.68 %	4.87 %	6.32 %	8.24 %
SO_3	2.42 %	2.76 %	2.48 %	2.53 %
K_2O	0.96 %	0.75 %	0.65 %	0.52 %
Na_2O	0.19	0.22 %	0.23 %	0.25 %
Ignition loss	2.60 %	1.30 %	1.10 %	0.31 %
Insoluble residue	0.30 %	0.30 %	0.40 %	-
Cl^-	0.020 %	0.013 %	0.011 %	0.007 %

Six concrete mixes were used. Mix details are shown in Table 2. For fine and coarse aggregates a fractured quartz was used. The 28th-day-compressive strengths of the CEM I and CEM III/A were about the same, while those of the CEM III/B-mixes were about 15 % higher.

Table 2 Mix proportions (kg/m³)

	CEM I 32,5 R		CEM III/A 32,5 N		CEM III/B 32,5 N-NW	
w/c-ratio	0.40	0.60	0.40	0.60	0.40	0.60
Cement	357.5	537.5	357.5	537.5	357.5	537.5
Water	215	215	215	215	215	215
Aggregates	1705	1505	1690	1560	1690	1560
0/2	975	840	962.5	890	962.5	890
2/8	730	665	727.5	670	727.5	670

The specimens were produced as follows: First, 100-mm cubes were cast, table vibrated and covered with a foil for overnight curing. Then, the cubes were sawed into small slices of 10 mm -thickness and moist cured at 20 °C and 100 % relative air humidity. After this curing period, the specimens were stored in a controlled climate at the following relative humidities:

- 92 % r. h.
- 85 % r. h.
- 75 % r. h.
- 50 % r. h.
- 0 % r. h.

The specimens were exposed to the storage conditions on the first, second and third day. A control group was stored at 20 °C and 100 % r. h. continuously.

The measurements took place on day 1, 2, 3, 7, 14, 28 and 91.

Having reached the testing day, the specimens were taken from the storage climate. The weights were measured and the hydration of the cement was stopped by evaporating the water at 105 °C. After heating to constant mass, the specimens were weighed again. The difference of both values is the content of evaporable pore water. Before the specimens were further worked on, they were cooled in a desiccator.

The CWA-measurement was used to determine the non-evaporable water. In order to accomplish this, the completely dried specimen was milled to a fineness less than 0,125 mm. Next, a few grams of the powder were brought to a red-hot glow in the CWA-unit at about 1000 °C. The percentages of carbon dioxide, water and the total weight loss were displayed. The evaporable and non-evaporable water content was calculated as follows:

$$w_{105°} = \frac{m_0 - m_{105°}}{m_Z} \cdot 100 \qquad (1)$$

$$w_{1000°} = \frac{l_{w,CWA}}{100} \cdot \frac{m_{CWA}}{\left(\frac{m_{CWA} \cdot m_Z}{m_{105°}} \right)} \cdot 100 - K \qquad (2)$$

$$m_Z = \frac{z}{100} \cdot m_{105°} \cdot \left(1 - \frac{l_{total,CWA}}{100} \right) \qquad (3)$$

$w_{105°}$	mass-%	content of evaporable water
$w_{1000°}$	mass-%	content of non-evaporable water
m_Z	g	weight of the cement in the specimen
m_0	g	weight of the specimen before evaporation
$m_{105°}$	g	weight of the specimen after evaporation at 105 °C
m_{CWA}	g	weight of the specimen before carbon-water-analysis
z	mass-%	content of cement in the mixture
$l_{total,CWA}$	mass-%	total weight loss after glow (output CWA-unit)
$l_{w,CWA}$	mass-%	water loss after glow (output CWA-unit)
K	mass-%	reduction factor - glow loss of the pure cement

Although, an unknown content of interlayer water and water from the AFt-phases disappears during the drying at 105 °C, the non-evaporable water can nevertheless be used as an empirical measure of the degree of hydration $\alpha(t)$. The degree of hydration is defined as the quotient of the chemically bound water at the time t (w_t) and the maximum capacity of the cement to bind water chemically (w_{max})(4).

$$\alpha(t) = \frac{w_t}{w_{max}} \cdot 100 \tag{4}$$

The maximum water binding capacity of Portland cements can be estimated by stoichiometric calculations. Typically it rangesfrom 23 % to 26 % [7]. Due to the difficulties to detect or to calculate the maximum water binding capacity of blast furnace slag cements, the content of non-evaporable water, relative to that in a moist cured specimen of the same cement on the 91st day, was used as a measure of the degree of hydration.

RESULTS AND DISCUSSION

Figure 1 gives curves showing the non-evaporable water content of three different cements as functions of the time.

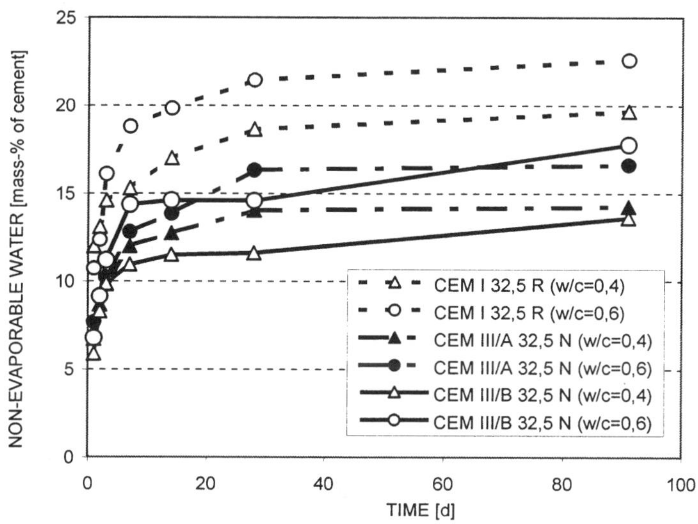

Figure 1 Non-evaporable water content of different types of cement and varying w/c-ratio (saturated storage)

It is evident that the Portland-cement reacts fastest. It was expected that with growing slag content the speed of hydration as well as the maximum non-evaporable water content was going to decrease. However, the opposite was found. On the one hand, mixes with CEM III/B exhibit more or less the same content of non-evaporable water as mixes with CEM III/A on the 91st day. On the other hand, the speed of hydration between the 1st and 7th day depends more on the w/c-ratio than on the slag content. The curves of mixes with a w/c-ratio of 0.60 show an steep rise while the mixes with w/c of 0.40 obviously run below them. The consequence is, that the CEM III/B with a w/c-ratio of 0.60 reacts faster in this stage of the hardening than the CEM I, with w/c=0.40. The lower non-evaporable water content of the CEM III/B-mix is mainly caused by a retarded hydration within the 1st day. Beyond this, it can be seen from Figure 1, that CEM III/B in contrast to CEM III/A and CEM I do not exhibit significant changes in non-evaporable water between the 7th and 28th day. That doesn't mean, that no hydration occurs. It is probable, that in this stage the blast furnace slag favoured the reaction with the CH. The water of the CH-phases will appropriate for the new C-S-H-phases [8].

After the 28th day, the differences of the three cements, regarding their course of hydration, continue. While the CEM I exhibits a slight progress in hydration, the hydration of the CEM III/A is paused and shows similarities to the interruption of the CEM III/B before. The CEM III/B's hydration proceeds again and exhibits a fair rise of the non-evaporable water. However, it could be, that this increase does not start exactly on the 28th day. It is more probable, that the upward trend is initiated later. Surprisingly, this growth in the late stage of hydration occurs independently of the storage conditions. It can be seen in Figure 2 that the gradients of the curves hardly differ from each other.

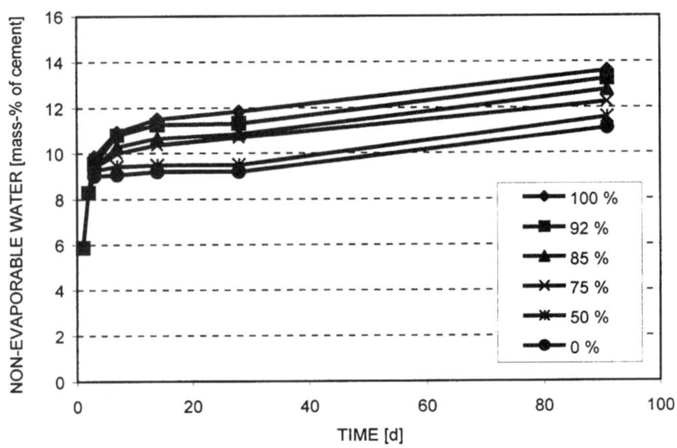

Figure 2 Non-evaporable water content of CEM III/B 32,5 N NW/HS with w/c-ratio of 0.40 (cured for 2 days under saturated conditions)

Both the relative humidity and the preceding curing under saturated conditions have no vital influence on the late hydration of the CEM III/B. According to Figure 3, the duration of curing has its effect above all on the hydration in the early stage.

In Figure 4, a balance of all measured water contents in a specimen of a CEM III/A-mix is given. Because of the storage under saturated conditions the changes in the non-evaporable water comply with the evaporable water. However, this proposition doesn't fit CEM III/B-mixes.

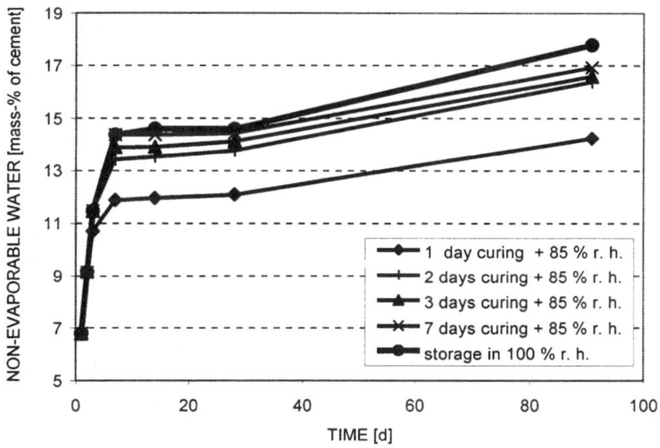

Figure 3 Non-evaporable water of a mix with CEM III/B 32,5 N NW/HS (w/c=0.60) stored at 85 % relative humidity (r. h.) after curing under saturated conditions

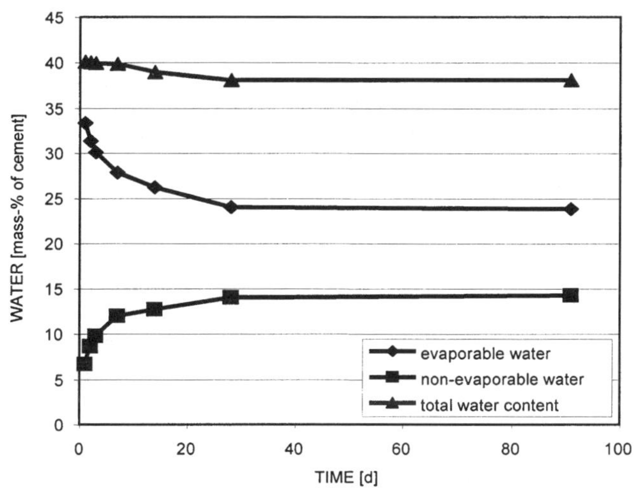

Figure 4 Balance of water of a mix with CEM III/A 32,5 N (w/c=0.40) stored under saturated conditions

Although, it was expected that the addition of the single water content results in the initial w/c-ratio, CEM III/B-mixes exhibit an anomalous behaviour. The total water content neither hit the initial w/c-ratio nor exceeded it. The latter would occur, if the specimens took up additional water during the storage under saturated conditions. But, Figure 5 and 6 contradict this assumption. Not even, that the present total water content is not higher than the initial w/c-ratio, the curve is dropping notably below it.

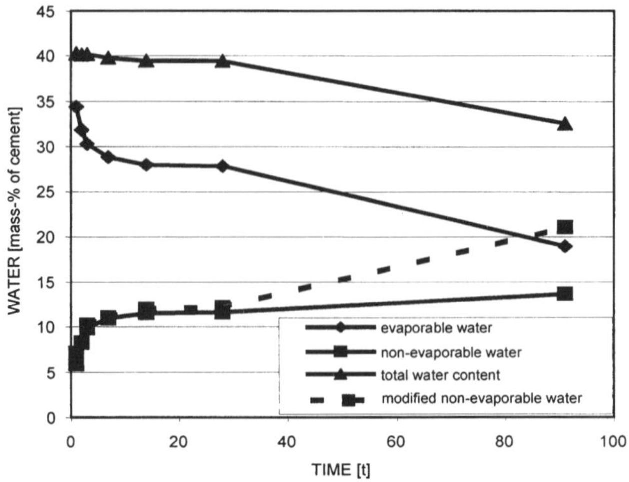

Figure 5 Balance of water of a mix with CEM III/B 32,5 N NW/HS (w/c=0.40) stored under saturated conditions

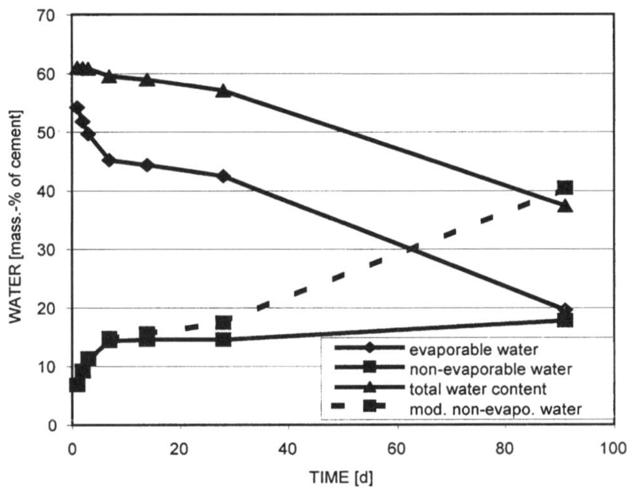

Figure 6 Balance of water of a mix with CEM III/B 32,5 N NW/HS (w/c=0.60) stored under saturated conditions

Assuming that no water was able to disappear in an uncontrolled way, the missing fraction of water must still be in the specimen. If this turns out to be true, hydrated phases were formed, which are stable to up to 1000 °C. From this it could be followed, that the assumed degree of hydration of blast furnace slag cements was probably too low.

In addition to the determined water contents, Figures 5 and 6 show a modified curve. These values result from the measured non-evaporable water and the difference between the total water content and the initial w/c-ratio. It can be seen from these modified curves, that the mixes have 21 % and 40 %, respectively, bound water. As a consequence, the CEM III/B has a higher degree of hydration on the 91st day than a comparable CEM I and CEM III/A.

The quantity of non-evaporable water depends strongly on the initial w/c-ratio. Therefore, it will be difficult to estimate the maximum capacity of the cement to bind water. The idea, that complete hydration occurs if the w/c-ratio is above 0.40, has to be query in the case of blast furnace slag cements. Assuming that the hydration will proceed with the same rate beyond the 91st day, values not much lower as the initial w/c-ratio will be reached.

However, an increase of strength in this height was never observed, although blast furnace slag cements are known for their excellent late strength development. This led to the conclusion, that the contribution on the strength of the phases, formed in this stage of hardening, is not too large. But they may have an influence on the resistance against the impact of fire. Due to the different influences on the fire resistance, such as type and size of aggregates, the data about the fire resistance of concretes made with blast furnace slag cements are not uniform. There are hints [7] that the strength loss of those concretes after the impact of high temperatures is significantly lower than that of Portland-cement concrete.

It has been recognized for a long time that concretes made by blast furnace slag cement show a denser microstructure and a lower capillary-porosity than Portland-cements of the same w/c-ratio [8]. The formation of additional phases may explain this phenomenon.

CONCLUSIONS

The effect of the evaporation on the hydration of blast furnace slag cements was studied. The kinetics of the hydration were analysed by the measurement of the non-evaporable water. To do this, the Carbon-Water-Analysis was used. Results indicated that the degree of hydration of cements with large amounts of blast furnace slag was underestimated up to now. This is due to the misjudgement that all gel-phases evaporate at 1000 °C as the reaction products of Portland cement generally do.

In comparison with it, this study showed that a considerable amount of water was still bound after the red-hot glow. This led to the assumption that hydrate phases were formed, which are resistant up to 1000 °C. Thereby, the amount of blast furnace slag in the mix had a relevant effect in the process. A significant rise of the non-evaporable water was monitored only on mixes with high slag loadings. In these cases the progress of the hydration was quite independent of the preceding curing conditions. From the results it can be followed that moist curing has no effect on the hydration of blast furnace slag in the late stage of the hardening.

This study reflected the strong influence of the slag content on the durability of the concrete. With rising slag content the water binding capability altered as well as the composition of the cement gel. Therefore, the outstanding resistance to chemical impact and fire is a result of an increased degree of hydration in connection with the formation of very stable phases.

REFERENCES

1. ADAM, TH., GRÜBL, P.: The influence of the humidity on the hydration of concrete with high contents of fly ash and blast furnace slag; Darmstadt-Concrete; Vol. 17 (2002) http://www.darmstadt-concrete.de/dacon/2002/humidity.pdf [PDF-file]

2. JIANG, W.; SILSBEE, M.R.; ROY, D.M.: Similarities and differences of Microstructure and macro properties between Portland and blended cements; Cement and Concrete Research 27 (1997); pp. 1501-1511

3. PAQUÉE, T.: Charakterisierung der Hydratation von Betonen mit hohem Hüttensand- oder Steinkohleflugascheanteil im frühen Stadium der Erhärtung; Diplom Thesis; TU Darmstadt 2002

4. RENDCHEN, K.: Hochofenzement – Charakteristische Merkmale und Eigenschaften im Beton; Hüttensandhaltiger Zement; Edited by RENDCHEN, K.; VBT Verlag GmbH, Düsseldorf 2002; pp. 23-31

5. RICHARDSON, I.G.: The nature of C-S-H in hardened cements; Cement and Concrete Research 29 (1999); pp. 1131-1147

6. ROY, D.M.: Alkali-activated cements / Opportunities and challenges; Cement and Concrete Research 29 (1999); pp. 249-254

7. STARK, J.; WICHT, B.: Zement und Kalk; Birkhäuser Verlag; Basel 2000

8. TAYLOR, H.F.W.: Cement chemistry; 2nd edition; Thomas Telford; London 1997

EFFECT OF FLY ASH AND SILICA FUME ON THE STATIC MODULUS OF ELASTICITY OF A CONCRETE MIX

V Kumar

M M Prasad

National Institute of Technology

India

ABSTRACT. An experimental investigation of M20-grade of concrete for evaluating the effect of flyash and silica fume on the static modulus of elasticity has been performed and presented in the present paper. A design mix of M20-grade of concrete having proportions of 1:1.40:3.40, w/c = 0.52 has been chosen as a reference mix. Partial replacement of Ordinary Portland Cement with a constant 7% silica fume and with a varying 10%, 15%, 20% and 25% flyash has been incorporated. In this mix proportion, blast furnace slag has been utilized as coarse aggregates in place of conventional stone chips. The stress-strain curves for flyash-silica fume concrete at 56-day maturity have been plotted and analysed. The test results show that the static modulus of elasticity of concrete mix containing up to 22% flyash plus silica fume is comparable to that of the designed mix of M20-grade of concrete..

Keywords: Ordinary Portland Cement, Flyash, Silica-fume, Industrial Waste, Pozzolanic Reaction, Stress, Strain , Modulus of Elasticity.

Virendra Kumar is a lecturer in the Department of Civil Engineering at the National Institute of Technology, Jamshedpur, India. His research interests include performance evaluation of new materials, analysis, design and detailing of reinforced concrete structures.

Dr M M Prasad is a Professor and Head of the Civil Engineering Department at the National Institute of Technology, Jamshedpur, India. He obtained his Ph.D. degree from Indian Institute of Technology, Roorkee, India in 1990. His research field includes low cost building technology, analysis and design of reinforced concrete structures and performance evaluation of new building materials in different environments.

INTRODUCTION

Energy plays a crucial role in growth of developing countries like India. In context of low availability of non-recoverable energy sources coupled with the requirements of large quantities of energy for providing materials like cement, steel, etc., the importance of utilization of industrial wastes as building materials cannot be underestimated. About 110 million tonnes of flyash have been produced by 68 major thermal power stations in India and are likely to be doubled within the next 10 years. It has been established fact from research that waste materials like flyash, silica fume, etc. through their use as construction materials can be converted into meaningful wealth. Also, the partial replacement of cement with flyash is desirable, and indeed essential due to a verity of technical, economical and ecological reasons. An attempt to solve these problems, the mineral admixture may often be used as partial replacement of cement. In the present investigation, flyash and silica fume have been chosen as mineral admixtures, which not only improve the properties of concrete but also result in cost reduction. The flyash in concrete makes efficient use of product of hydration of cement, such as, calcium hydroxide (C-H) which is otherwise a source of weakness in normal cement concrete and converts it into denser[1] and stronger C-S-H compounds by pozzolanic reaction[2]. The heat generated during hydration initiates the pozzolanic reaction of flyash. When concrete containing flyash cured properly, the reaction products fill in the space between hydrating cement particles. This reduces the permeability of concrete to water and aggressive chemicals.

Extremely small size of silica fume may be beneficial for replacement of cement in conventional concrete. It results in densification of the fresh cement paste[3]. Researchers[4] have reported that silica fume having smaller in size and round in shape fills the voids between the coarser cement particles otherwise the same may be occupied by water. A properly proportioned flyash and silica fume in concrete mix improves properties of concrete that may not be achievable through the use of Portland Cement alone. The resulting concrete mix becomes strong, durable and economical, and also eco-friendly as it utilizes an ecologically hazardous material. Herein, the blast furnace slag has been used as coarse aggregate in place of conventional coarse aggregates like stone chips to cast test specimens, and the effect of 17%, 22%, 27% and 32% replacement of cement by flyash and silica fume in design concrete mix has been studied on the static modulus of elasticity at 56 days of maturity. The water binder (i.e. cement plus flyash and silica fume) ratio was kept constant as 0.52 for all mixes.

EXPERIMENTAL PROGRAMME AND MATERIALS

The effect of flyash and silica fume as partial replacement of Ordinary Portland Cement (OPC) of 43-grade on the static modulus of elasticity is reported here. The experimental programme consists of casting of cube and cylinder specimens with mix proportion 1:1.40:3.40 (w/c=0.52) for M20 grade of concrete. All specimens were normally cured for 56 days and then tested to failure.

Cement

Ordinary Portland Cement satisfying the requirements of IS 8112-1989[5] has been used in the present investigation.

Flyash

The Fly ash used was locally available and collected from Tata Power Plant, Jojobera, Jamshedpur, India. Its physical and chemical properties of flyash are as under.

Table 1 Physical and chemical properties of flyash

PARTICULARS	IS: 3812-1981 Specification[6]	FLY ASH USED
Physical Properties		
Fineness, m^2/kg	320 (minimum)	410
Compressive strength at 28 days, N/mm^2	Not less than 80% of the strength of cement corresponding to plain cement mortar cube	83.33% to 137.15% corresponding to plain cement mortar cube
Chemcial Properties		(In percent)
Silica	$SiO_2+Al_2O_3+Fe_2O_3$=70 %(min)	60.10
SiO_2 (alone)	35 %(Min)	45.00
Aluminium	--	14.77
Iron	--	2.84
Calcium	--	0.60
Sulphur (SO_3)	2.75 (max)	1.20
Loss on ignition	12% (max)	3.40

Silica Fume

The Silica fume used in this investigation was collected from the ferro-silicon alloy factory situated at Bhutan. Its physical properties are given below:

- + 90 micron fraction – 0.25 percent by weight
- - 90 micron fraction – 99.75 percent by weight
- Colour of silica fume – array (light)

Fine Aggregate

Locally available sand from the bed of the River Koel has been used as fine aggregate. Its physical properties are given in Table 2.

Table 2 Physical properties of sand

PHYSICAL PROPERTIES	VALUE OBSERVED
Free moisture content	0.15%
Specific gravity	2.55
Water absorption	1.20%
Fineness modulus	2.08

Coarse Aggregate

Blast furnace slag of size passing the 20 mm down and the 10 mm sieves retaining, the by-product obtained from Tata Iron and Steel Company (TISCO), Jamshedpur, India was used as coarse aggregate. Its physical properties are given in Table 3.

Table 3 Physical properties of blast furnace slag

PHYSICAL PROPERTIES	VALUE OBSERVED
Aggregate impact value	11.70%
Aggregate crushing value	20.65%
Los Angles abrasion value	23.0%
Specific gravity	2.69
Water absorption	0.84%
Free moisture content	0.25%
Fineness modulus	6.93

Water

Normal tap water was used for mixing and curing of all specimens.

Mix Proportion

The procedure outlined in IS:10262-1982[7] has been adopted for proportioning of mix to prepare M20-grade concrete. Cylinders of size 150 mm diameter and 300 mm high have been cast, cured in water for 56 days and then tested under a 200-tonne compression testing machine for determining the static modulus of elasticity. Table 4 shows the mix proportion of flyash-silica fume concrete.

Table 4 Mix proportion of flyash-silica fume concrete

MIX	W/ (C+F+SF)	CEMENT	FLYASH	SILICA FUME	FINE AGGREGATE	COARSE AGGREGATE	CEMENT REPLACEMENT	
							Flyash (%)	Silica fume (%)
M_1 *	0.52	1.0	--	--	1.40	3.40	0	0
$M_{1.1}$	0.52	0.83	0.10	0.07	1.40	3.40	10	7
$M_{1.2}$	0.52	0.78	0.15	0.07	1.40	3.40	15	7
$M_{1.3}$	0.52	0.73	0.20	0.07	1.40	3.40	20	7
$M_{1.4}$	0.52	0.68	0.25	0.07	1.40	3.40	25	7

* reference mix

Table 6 Static modulus of elasticity of ordinary portland cement concrete

Mix (M_I) – 1:1.40:3.40 (reference mix) W/C=0.52

STATIC MODULUS OF ELASTICITY AT 56 DAYS		
Stress (N/mm^2)	Strain, (x10^{-5})	E (kN/mm^2)
0	0	
1.132	0.75	
2.264	2.0	
3.395	2.5	
4.526	9.5	
5.658	10.0	41.38
6.79	10.5	
7.921	11.25	
9.053	11.75	
10.185	21.25	
11.316	23.25	
12.448	31.25	

Table 7 Static modulus of elasticity of fly ash-silica fume concrete.

Mix ($M_{I.1}$) – 0.83:0.10:0.07:1.40:3.40, W/C=0.52

STATIC MODULUS OF ELASTICITY AT 56 DAYS		
Stress (N/mm^2)	Strain, (x10^{-5})	E (kN/mm^2)
0	0	
1.132	1.25	
2.264	3.20	
3.395	4.50	
4.526	8.00	
5.658	10.00	34.67
6.79	12.00	
7.921	14.75	
9.053	19.25	
10.185	21.75	
11.316	26.50	
12.448	27.50	
13.62	32.50	

Table 8 Static modulus of elasticity of fly ash-silica fume concrete.

Mix ($M_{I.2}$) – 0.78:0.15:0.07:1.40:3.40, W/C=0.52

STATIC MODULUS OF ELASTICITY AT 56 DAYS		
Stress (N/mm^2)	Strain, (x10^{-5})	E (kN/mm^2)
0	0	
1.132	2.0	
2.264	5.0	
3.395	8.5	
4.526	9.5	
5.658	12.0	35.17
6.79	15.0	
7.921	18.5	
9.053	24.5	
10.185	32.0	
11.32	35.0	
12.48	40.0	

Table 9 Static modulus of elasticity of fly ash-silica fume concrete.

Mix ($M_{I.3}$) – 0.73:0.20:0.07:1.40:3.40, W/C=0.52

STATIC MODULUS OF ELASTICITY AT 56 DAYS		
Stress (N/mm^2)	Strain, (x10^{-5})	E (kN/mm^2)
0	0	
1.132	2.0	
2.264	6.5	
3.395	7.0	
4.526	15.0	
5.658	17.0	29.25
6.79	20.0	
7.921	25.0	
9.053	28.0	
10.185	38.5	

Table 10 Static modulus of elasticity of fly ash-silica fume concrete.

Mix ($M_{I.4}$) – 0.68:0.25:0.07:1.40:3.40, W/C=0.52

STATIC MODULUS OF ELASTICITY AT 56 DAYS		
Stress (N/mm^2)	Strain, (x10^{-5})	E (kN/mm^2)
0	0	
1.132	3.4	
2.264	9.0	
3.395	14.0	23.23
5.658	18.5	
6.79	23.0	
7.921	30.0	
9.053	36.0	
10.185	40.5	
11.316	50.0	

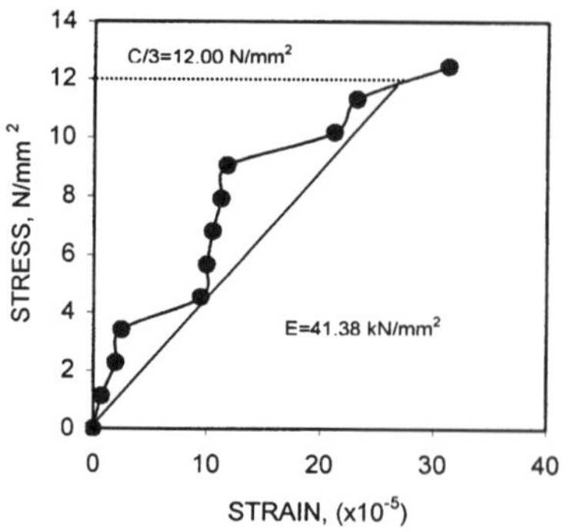

Figure 2 Stress vs strain curve for mix M_1.

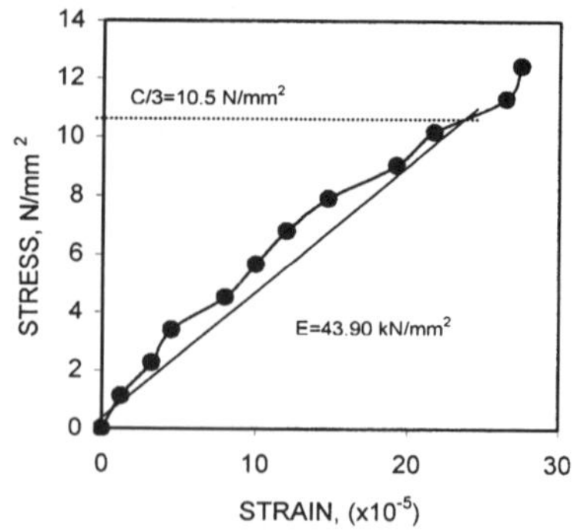

Figure 3 Stress vs strain curve for mix $M_{1.1}$.

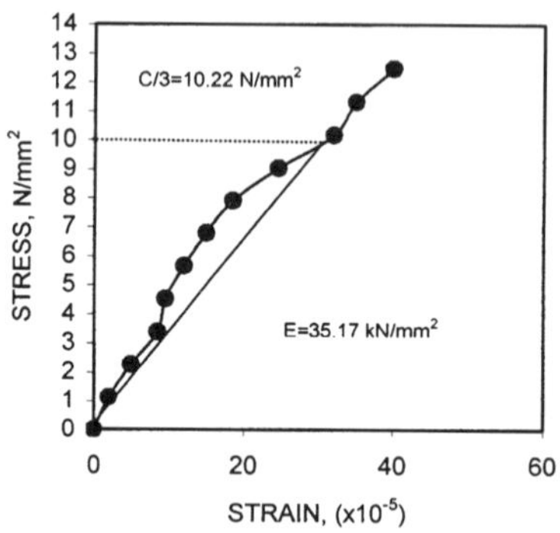

Figure 4 Stress vs strain curve for mix $M_{1.2}$.

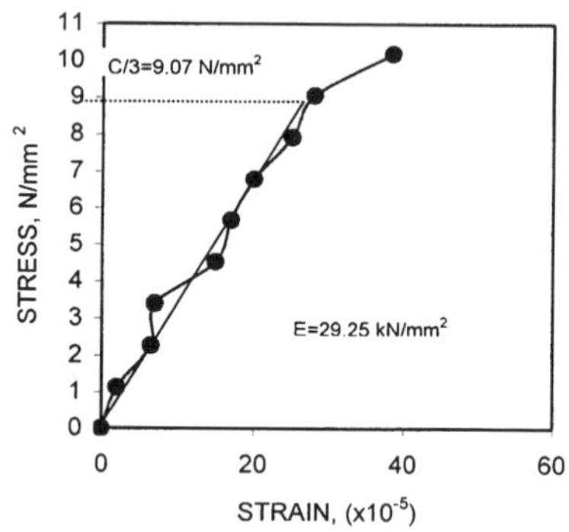

Figure 5 Stress vs strain curve for mix $M_{1.3}$.

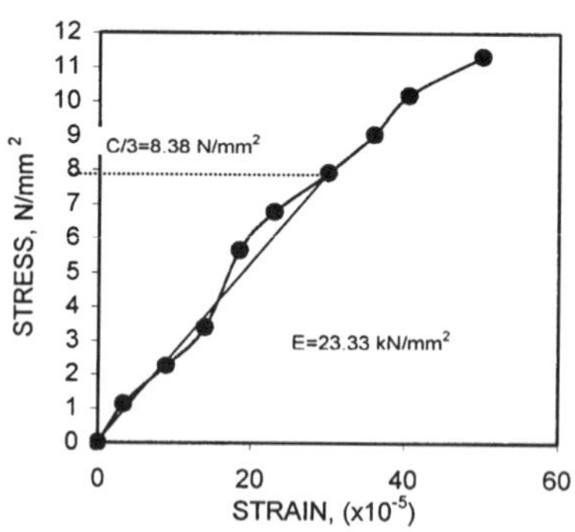

Figure 6 Stress vs strain curve for mix $M_{1.4}$.

TESTING AND TEST RESULTS

The static modulus of elasticity has been determined at 56-day of maturity on cylindrical specimen under uniaxial compression as per IS 516-1956[8], and the deformation measured by means of dial gauge fixed between certain gauge length. The load was applied continuously and without shock at a rate of 14 N/mm^2/minute until an average stress of (C+0.5) N/mm^2 is reached, where C is one-third of the average compressive strength of three cubes calculated to the nearest 0.5 N/mm^2. The compressive strengths of cubes at 56-day of maturity have been given in table 5 and the stress-strain values are shown in Tables 6,7,8,9 and 10. These results have been reproduced in graphical form through Figures 2,3,4,5, and 6 respectively for various concrete mixes. The ratio of applied load to the area of cross section of the specimen indicates the stress whereas the strain is obtained from the ratio of dial gauge reading to gauge length.

Table 5 Compressive strength of flyash-silica fume concrete at 56-day

MIX DESIGNATION	COMPRESSIVE STRENGTH *, N/mm^2
M_1	36.00
$M_{1.1}$	33.48
$M_{1.2}$	30.66
$M_{1.3}$	27.22
$M_{1.4}$	25.14

* average of three cubes

ANALYSIS AND DISCUSSION

As defined, the static modulus of elasticity of a material under tension or compression is obtained by the slope of the stress-strain curve for a concrete under uniaxial loading. Since the stress-strain curve of concrete is non linear[9], the secant modulus of elasticity of any mix is considered by the slope of a line drawn from the origin to a point on the curve corresponding to 33 percent stress of the failure load of the cube compressive strength of the same concrete mix. In the present investigation the stress, at which the secant modulus determined is shown in the figure of stress-strain curve for various mixes of flyash-silica fume concrete.

Stress-strain curves for flyash-silica fume concrete for mixes M_1, $M_{1.1}$, $M_{1.2}$, $M_{1.3}$, and $M_{1.4}$ containing 0%, 15%, 22%, 27% and 32% flyash and silica fume respectively at 56-day of maturity have been shown in figures 2 to 6. The corresponding values of static modulus of elasticity obtained from these curves are 41.38 kN/mm^2, 43.90 kN/mm^2, 35.17 kN/mm^2, 29.25 kN/mm^2 and 23.33 kN/mm^2 respectively and all the stress-strain relationship of FSC concrete are found to be nonlinear. This may be due to the less development of micro cracks at the interface of the aggregate and paste[10]. It is clear from Figures 2 and 3 that the static modulus of elasticity obtained for FSC having constant 7 percent silica fume, with 10% and 15% flyash have shown comparable results with respect to reference concrete mix. This may be due to the fact that there is a reduction in pore size (the filler effect) by silica fume and pozzolanic activity by flyash and silica fume. Also the gel formation like C-S-H/C-A-H fills in the pore generated by liberation of Ca $(OH)_2$ during hydration of OPC concrete. All these mechanisms enhance the bonding inside the concrete matrix and thus the strength increases. This may be the possible reason that the FSC experiences less strain[11] for a given stress and gives higher modulus of elasticity as compared to the conventional OPC concrete of the same mix proportion with stone chips as coarse aggregate. It has become obvious that the concrete mix with BFS as coarse aggregate has higher compressive strength in respect of concrete mix with stone chips as coarse aggregate. However, a decrease in the value of static modulus of elasticity for other concrete mix has been found due to increase in percentage replacement of OPC by silica fume and flyash. The similar behaviour has already been observed in respect of static modulus of elasticity at 28 days of maturity[12]. It is to be pointed out that more extensive study is required to understand the effect of particular non-conventional constituents (either flyash or silica fume as cementitious materials, and BFS as coarse aggregate) on static modulus of elasticity of FSC concrete. The results of static modulus of elasticity obtained for FSC concrete containing more than 22 percent flyash and silica fume are not comparable to the value obtained for reference mix due to higher percentage replacement of cement.

CONCLUDING REMARKS

The investigation on flyash-silica fume concrete has been carried out with the objective of exploring the large-scale use of these waste material coming out from different industires as a supplementary cementitious material. The effect of flyash and silica fume as partial replacement of Ordinary Portland Cement on M20-grade of concrete (mix proportion 1:1.40:3.40, w/c = 0.52) has been studied in respect of the static modulus of elasticity. The 56-day static modulus of elasticity for different mixes are 41.38 kN/mm^2, 43.90 kN/mm^2, 35.17 kN/mm^2, 29.25 kN/mm^2 and 23.33 kN/mm^2 respectively for flyash silica fume concrete, containing 0%, 17%, 22%, 27% and 32% of flyash including a constant 7% silica fume.

FSC concrete containing up to 22% percent (15% flyash and 7% silica fume) gives encouraging result for static modulus of elasticity at 56 days of maturity when compared to designed mix of M20-grade of concrete. Cement replaced by higher percentage of flyash and silica fume indicates decrease in static modulus of elasticity of FSC concrete. The stress-strain relationship of FSC concrete is nonlinear and similar in nature to the stress-strain characteristics of conventional OPC concrete. More extensive study will reveal the effect of particular non-conventional constituents on static modulus of elasticity of flyash-silica fume concrete.

ACKNOWLEDGEMENTS

Authors are thankful to Director, National Institute of Technology, Jamshedpur, India for his permission to publish the paper. The contribution rendered by Mr. Rajeev Kumar and Miss Snigdha and et. al., final year Students of Civil Engineering Department is highly acknowledged.

REFERENCES

1. KUMAR, VIRENDRA; PRASAD, M.M. & JAIN, K.K. "Ultrasonic Pulse Velocity for High Volume Flyash Concrete" Prof. Int. Conf. on Advances in Civil Engg. (ACE-2002) Jan 3-5, 2002, India, pp. 1444-1450.

2. MALHOTRA, V.M. "Performance of High Volume Flyash Concrete in Large Experimental Monoliths", ACI Material Journal, vol.-19, NO.2, March-April 1994.

3. K.H. KHAYAT & P.C. AITCIN- "Silica Fume in Concrete – An Overview" SP 132-46, 1992, Istanbul Conf. pp. 836-855.

4. M. BAALBAKI, S.L. SARKAR, P.C. AITCIN & H. ISABELLE "Properties and Microstructure of High Performance Concrete Containing Silica Fume, Slag and Flyash" SP 132-50, 1992 Istanbul Conf. pp. 921-942.

5. I.S. 8112-1989, 43-Grade Orindary Portland Cement – Specification (1st Revision) Bureau of Indian Standard, New Delhi.

6. I.S. 3812-1981, Specification for use as Pozzolana and Admixture (1st Revision) Bureau of Indian Standard, New Delhi.

7. I.S. 10262-1981, Recommended Guidelines for Concrete Mix Design, Bureau of Indian Standard, New Delhi.

8. I.S. 516-1959, Methods of Tests for Strength of Concrete, Bureau of Indian Standard, New Delhi.

9. A.M. NEVILLE, Properties of Concrete, ELBS Pub.

10. P. KUMAR MEHTA & PANLO, J.M. MONTERIRO "Concrete Microstructure, Properties, and Material" Indian edition (1st).

11. KUMAR, VIRENDRA; PRASAD, M.M. & JAIN K.K., "An Investigation on Surface Drying Shrinkage of High Volume Flyash Concrete" Int. Conf. on "Concrete technology for developing countries" 21-23 Oct. 2002, Amman, Jordan.

12. KUMAR, VIRENDRA & PRASAD, M.M. "Static Modulus of Elasticity of Flyash-Silica Fume Concrete" Proc. Construction Management and Materials (CONMAT-2003), 9-11 Jan 2003, I.I.T., Kharagpur, India, pp. 549-558.

INFLUENCE OF FINENESS LEVELS ON THE SULFATE RESISTANCE OF CEMENT MATRIX WITH GGBS

H Y Moon **S T Lee**

H S Jung **J P Kim**

Hanyang University

S S Kim

Daejin University

Korea

ABSTRACT. This paper describes the sulfate resistance of cement matrix with or without ground granulated blast furnace slag (GGBS). Sulfate attack was performed on the cement matrix, which had been prepared by using water-binder ratio of 0.45. Variables were the fineness levels of GGBS and two sulfate solution. Microstructural analyses such as X-ray diffraction (XRD) and scanning electron microscopy (SEM) as well as determination of physical properties, compressive strength and length change were carried out to evaluate the sulfate resistance of GGBS with various fineness levels

Keywords: Ground granulated blast furnace slag, Sulfate resistance, Fineness levels, Compressive strength, Length change, XRD, SEM.

H Y Moon is Professor of Civil Engineering at Hanyang University in Korea. His research focuses mainly on binder technology, concrete durability and high strength concrete.

S S Kim is Professor of Civil Engineering at Daejin University in Korea. His main interest is durability and repair of reinforced concrete structures.

S T Lee is Researcher of Civil Engineering at Hanyang University in Korea. His main interest is durability and attack of reinforced concrete structures.

H S Jung and **J P Kim** are Ph. D. candidates of Civil Engineering at Hanyang University in Korea. Their main interest is durability and attack of reinforced concrete structures.

INTRODUCTION

Recently, there has been a growing trend for use of supplementary cementitious materials, whether natural, waste, or by-products, in the production of composite cements because of ecological, economical, and diversified product quality reasons. GGBS, a by-product of the transformation of iron ore into pig-iron in a blast furnace, has been increasingly used for additive material in concrete [1].

On the other hand, as with all cementing materials, the reactivity of slag is greatly influenced by its surface area. In general, increased fineness results in better strength development, but in practice, fineness is limited by economic and performance considerations and factors such as setting time and shrinkage. The fineness of GGBS is a very important parameter, which is dependent on energy-saving and economic considerations, influences the reactivity of GGBS in concrete, early strength development and water requirement of concrete. Swamy [2] reported that an increase in fineness of two to three times that of OPC can preserve the benefits of materials fineness on a variety of engineering properties such as bleeding, time of setting, heat evolution, high strength and excellent durability. Thus, for better performance, the fineness of GGBS must be greater than that of cement.

In particular, sulfate attack on concrete is a complex process as was reported by Cohen and Mather [3-4]. Many factors, such as cement type, sulfate cation type, sulfate concentration and exposure period, may affect the sulfate resistance. However, not much research has been conducted to evaluate the effect of sulfate concentration on the performance of OPC and blended cements. In the present paper, the sulfate resistance of the cement matrix with or without consideration of the fineness levels of GGBS was evaluated by XRD and SEM, and by measurement of SDF and length change after various periods of exposure.

EXPERIMENTAL

Materials

Cement and mineral admixture

ASTM C 150 Type I cement was used in preparing the OPC paste/mortar specimens. The GGBS blend was made with 50% GGBS and 50% Type I cement. GGBS with three different nominal fineness levels of 4,450 (SG4), 6,000 (SG6) and 8,000 (SG8) cm^2/g was adopted, respectively. The mixture proportion of all mortars was binder :sand = 1:2. Mortar and paste specimens were made at a water-binder ratio (w/b) of 0.45. The chemical composition and physical properties of OPC and GGBS used in this experiment are also detailed in Table 1.

Table 1 Chemical composition and physical properties of OPC and GGBS

	CHEMICAL COMPOSITION (%)							PHYSICAL PROPERTIES	
	SiO_2	Al_2O_3	Fe_2O_3	CaO	MgO	SO_3	Ig. loss	Specific gravity	Specific surface area (cm^2/g)
OPC	20.2	5.8	3.0	63.3	3.4	2.1	1.2	3.13	3,120
SG4	31.7	14.5	0.7	41.7	5.4	2.1	2.6	2.92	4,450
SG6	32.2	14.6	0.4	44.0	5.6	1.1	1.2	2.92	6,000
SG8	32.3	14.6	0.4	44.0	5.6	1.1	1.2	2.92	8,000

Aggregate

River sand, being a fine aggregate inert to most chemical agents and having little organic compound content was employed for manufacturing of mortar specimens. The specific gravity, absorption and fineness modulus of fine aggregate were 2.60, 0.80% and 2.80, respectively.

Test solution

The exposure solution used to provide sulfate attack to the specimens, was by dissolving reagent grade chemical in water. The chemical used was 5% of sodium sulfate solution (NS solution) and 4.24% magnesium sulfate solution (MS solution). All specimens were initially cured in water for 7 days and then immersed in test solutions for periods of exposure. Water and solutions temperature were kept 20 ± 1°C Before immersion, mortar specimens were tested to determine the initial compressive strength and length change.

Test Methods

Compressive strength

Compressive strength measurements were performed on 50mm cube mortar specimens. All specimens, mortar specimens with or without GGBS, were demoulded after 1 day of mixing and cured in water at room temperature for an additional 6 days. Some of them were moved to test solution and were continuously immersed for predetermined periods. Compressive strength tests were performed on each mixture at 0, 28, 91 and 180 days of immersion. At each interval, the compressive strength of four specimens drawn from water and test solutions were measured and then their values were averaged.

Length change

Based on ASTM C 1012, length change was measured at each immersion period. All length change values were compared with the initial length of mortar prisms (after 7 days of curing in tap water). The length change is calculated by the following Equation.

$$\Delta L_i = (L_i - L_o) / L_o \times 100(\%)$$

Where, ΔL_i = length change at (i) immersion period (%), L_i = length of mortar prism specimen measured at (i) immersion period and L_o = initial length of mortar prism, which is measured before the mortar specimen is exposed to the attacking solution.

X-ray diffraction (XRD)

In order to determine the solid phases in the paste specimens qualitatively, X-ray diffraction analysis technique is used, The XRD is conducted using the RINT D/max 2500 (Rigaku) X-ray diffractometer. The source of radiation is CuKα With a wavelength of 1.54 Å at a voltage of 30 kV. The 2θ angle range covered for each paste specimen is from 5° to 40°. The scanning speed is 2° /min.

Scanning electron microscope (SEM)

SEM investigation is used to analyze the microstructure of the paste specimen attacked by sulfate solution. An XL30ESEM (Philips) equipped with EDAX Falcon Energy System 60SEM was used for microscopic examination. SEM is a direct observation of deteriorated surface using a stub sample. Small pieces of paste were dried in a vacuum desiccator for 1 day. They were sputter coated with a gold coating. The accelerating voltage is 20 kv.

RESULTS AND DISCUSSION

Compressive Strength

The test results of the compressive strength of OPC mortar specimens made with a w/b ratio of 0.45, immersed in water, NS and MS solutions are shown in Figure 1. This figure indicates that mortars immersed in NS and MS solutions begin to reduce in compressive strength after 91 days of exposure. Whereas mortar specimen immersed in water didn't decrease in compressive strength even up to 180 days of exposure. At 180 days of exposure, the compressive strength of mortar specimens immersed in MS and NS solution was about 42MPa and 47MPa, respectively.

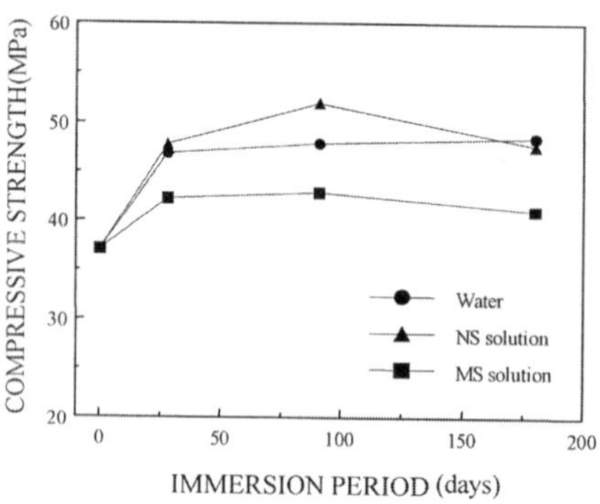

Figure 1 Compressive strength of OPC mortar specimens immersed in water and sulfate solutions

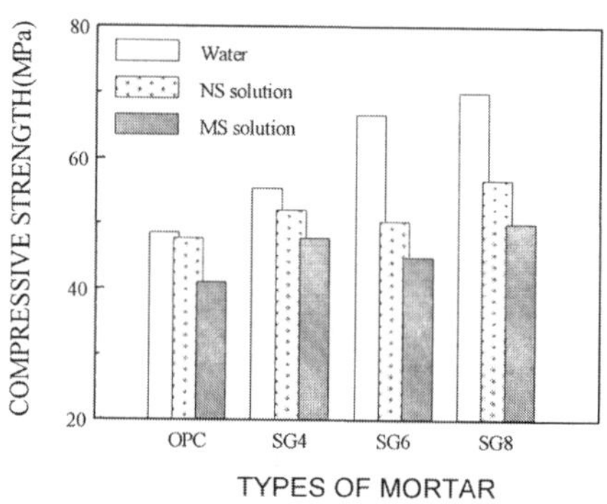

Figure 2 Compressive strength of mortar specimens with different fineness levels of GGBS

On the other hand, Figure 2 illustrates compressive strengths of GGBS mortar specimens with different fineness levels, subject to sulfate attack at 180 days. The compressive strengths of mortar specimens immersed in NS and MS solutions were significantly lower than those of mortar specimens immersed in water. In the mortar specimens with three different fineness GGBS immersed in NS solution, the compressive strengths were 52MPa, 50MPa and 58MPa, respectively. For the mortar specimens immersed in MS solution, the compressive strength was 48MPa, 46MPa and 50MPa, respectively. The reduction in compressive strength was observed to be higher in both SG6 and SG8 mortar specimens compared to SG4 mortar specimen immersed to the sulfate solutions. In particular, the reduction in compressive strength for mortar specimen immersed MS solution was higher than in NS solution. Therefore, the deterioration modes of cement matrix with GGBS were dependent on the exposure solutions.

Length Change

The length change of OPC and GGBS mortar specimens immersed in NS and MS solution are shown in Figure 3 and Figure 4, respectively. The average length change values at 180 days of immersed in NS solution were about 0.087%, while for SG4, SG6 and SG8 mortar specimens it was 0.039%, 0.028% and 0.027%, respectively. In the case of OPC mortar specimens, the length change was higher than for GGBS by about 40%. The length changes of OPC mortar specimens in both NS and MS solution were similar to each other up to 56 days of immersion. However, after 91 days, OPC mortar specimens exposed to NS solution experienced much larger length change than in the MS solution.

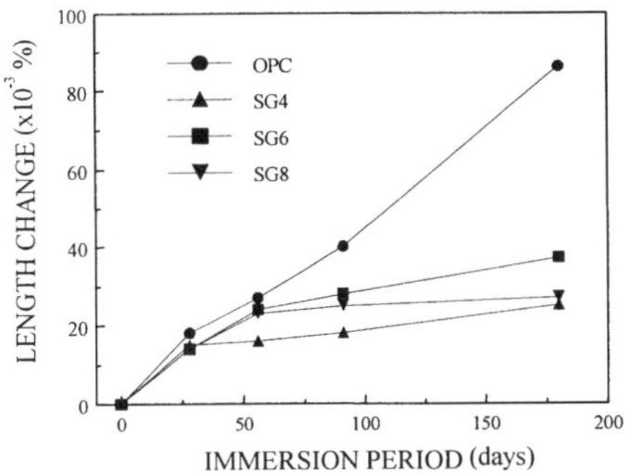

Figure 3 Length change of mortar specimens (NS solution)

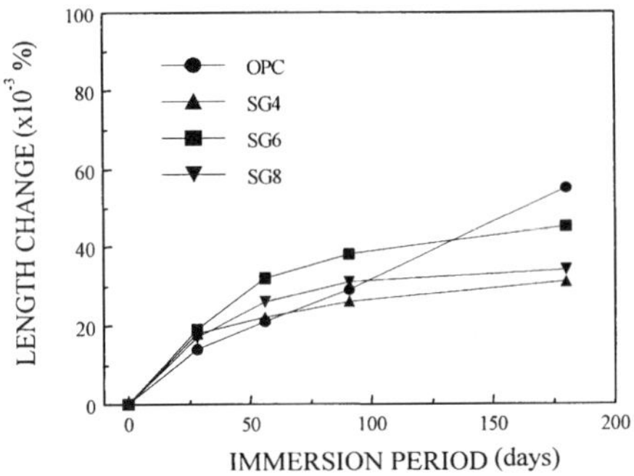

Figure 4 Length change of mortar specimens (MS solution)

Meanwhile, the average length change in OPC mortar specimens immersed in MS solution was 0.055%. In the case of specimens immersed in MS solution, the length change values of SG4, SG6 and SG8 were 0.045%, 0.033% and 0.030%, respectively. Therefore, in the OPC mortar specimens immersed in NS solution, the length change had some increase in comparison with the MS solution. The length change of GGBS mortar specimens of various fineness indicates similar a trend as immersed in the two sulfate solutions. Therefore, the influence of fineness levels of GGBS on the sulfate resistance was somewhat little because of a relatively short exposure period.

X-ray Diffraction Analysis

XRD analysis gives clear information on the crystalline phase formed in the cement matrix due to sulfate attack. In this study, change in mineralogy with 240 days of sulfate exposure was monitored by XRD. The x-ray diffraction patterns on the deteriorated surface of the OPC and GGBS paste specimens subjected to sodium sulfate and magnesium sulfate attack for 240 days are presented in Figure 5 and Figure 6, respectively. The X-ray diffraction patterns of the OPC and GGBS paste specimens stored in NS solution are presented in Figure 5. For XRD pattern of OPC paste specimen (Figure 5 (a)), the presence of peaks for ettringite, thaumasite and gypsum were observed. Similarly, the SG8 paste specimen showed a similarity with the OPC paste specimen in terms of the formed products. However, no gypsum and thaumasite peaks were detected in SG4 and SG6 paste specimens. This trend may explain that the compressive strength of SG8 mortar specimen was similar to that of SG4 mortar specimen, although SG8 had significantly higher compressive strength than SG4 in water. The intensive peaks for portlandite at 18.1° and 34.1° 2θ in OPC system was observed to be similar to those in the SG6 system. As commonly known, it is difficult, in particular at around 9.1° and 16.0° 2θ, to distinguish thaumasite peaks from ettringite peaks. However, the thaumasite peaks at around 19.5, 23.4, 26.0 and 28.4° 2θ are absent from the ettringite peaks, as shown in Figure 5. Calcite peaks were present in all paste samples. Figure 6 shows the XRD patterns to identify the reaction products in powders drawn from OPC and GGBS paste specimens placed in MS solution. XRD trace (Figure 6 (a)) on the surface part of OPC paste specimen indicates the presence of thaumasite, ettringite, portlandite, calcite, and gypsum. Additionally, weak peaks at around 18.6 and 38.0° 2θ confirmed the presence of brucite. Another important observation in XRD traces of SG4 and SG6 paste specimens, subjected to magnesium sulfate attack, was the eminent presence of gypsum peaks at around 11.7 and 20.7° 2θ (see Figure 6 (b) and (c)). Similar to XRD patterns of OPC, the XRD pattern of SG8 paste specimen showed the maximum intensity peak for thaumasite, which represents following equation [5].

$$3Ca^{+} + SiO_3^{2-} + CO_3^{2-} + 2(OH)^{-} + Mg^{2+} + SO_4^{2-} \rightarrow CaSiO_3{\cdot}CaSO_4{\cdot}CaCO_3{\cdot}15H_2O + Mg(OH)_2$$

However, there was no evidence for the existence of thaumasite in the SG4 and SG6 systems.

Scanning Electron Microscopy

The microstructural images of paste samples continuously stored in NS and MS solutions for 240 days were shown in Figure 7 to 10, respectively. For the OPC paste sample, a thick thaumasite phase, which was confirmed by EDS analysis, can be seen in the center, as shown in Figure 7. On the other hand, Figure 8 corresponding to the image of SG8 paste specimen exhibits the gypsum crystals were prevailing in the image. These images were consistent with the results of XRD analysis.

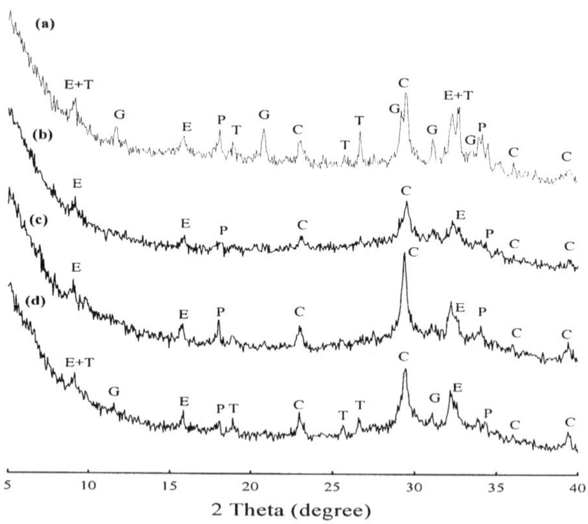

Figure 5 XRD pattern of paste specimens stored in NS solution for 240 days, (a = OPC, b = SG4, c = SG6, d = SG8), (Note, C = calcite, E = ettringite, G = gypsum, P = portlandite, T = thaumasite)

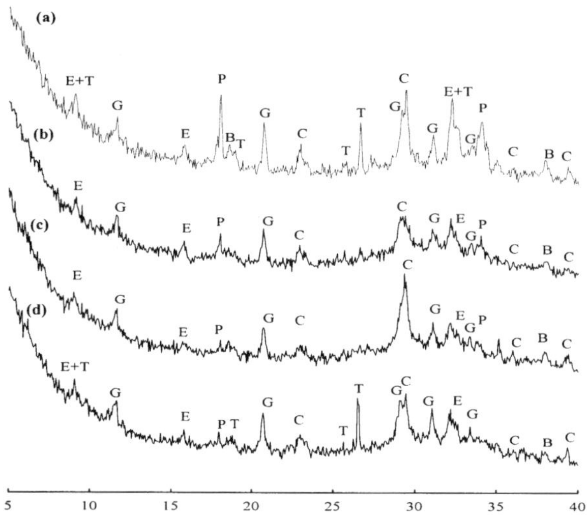

Figure 6 XRD patterns of paste specimens stored in MS solution for 240 days, (a = OPC, b = SG4, c = SG6, d = SG8), (Note, B = brucite, C = calcite, E = ettringite, G = gypsum, P = portlandite, T = thaumasite)

Figure 9 shows the brucite phase formed in the OPC paste specimen subjected to MS solution. This image also shows a round crack along with different brucite formations, and a large amount of the phases were observed elsewhere in the sample. At a high magnification (× 10,000), the microstructural image related to the SG8 paste specimen subjected to MS solution, as shown in Figure 10, indicated that many subhedral gypsum crystals formation were formed. This suggested that the main deterioration mode of SG8 cement matrix was the formation of gypsum.

Figure 7 SEM image of OPC paste specimen stored in NS solution for 240 days

Figure 8 SEM image of SG8 paste specimen stored in NS solution for 240 days

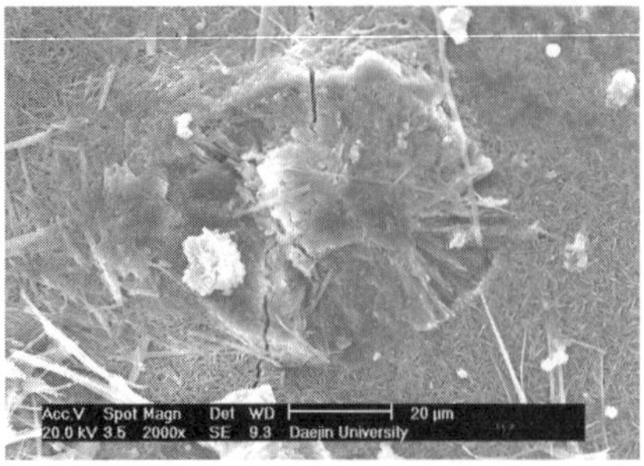

Figure 9 SEM image of OPC paste specimen stored in MS solution for 240 days

Figure 10 SEM image of SG8 paste specimen stored in MS solution for 240 days

CONCLUSIONS

1. All mortar specimens immersed in MS solution had lower compressive strength than those in NS solution. Therefore, the deterioration degrees of cement matrices with GGBS were dependent on the exposure solutions.

2. The results of length change of GGBS mortar specimens immersed in NS and MS solutions were similar irrespective of fineness levels of GGBS. The influence of fineness levels of GGBS on the sulfate resistance was somewhat little, because of the relatively short exposure period.

3. It can be seen that the deterioration modes of cement matrices with or without GGBS were dependent on their exposure environments. A GGBS cement matrix with a higher fineness level showed a poor sulfate resistance. This was obviously confirmed by XRD and SEM analyses. .

REFERENCES

1. HIGGINS, D. D., Ground Granulated Blast Furnace Slag, World cement. 6, 1996, pp 51-52.

2. SWAMY, R. N., Design for Durability and Strength Through the Use of Fly Ash and Slag in Concrete, CANMET/ACI International Workshop on supplementary Cementing Materials, Superplasticizers and Other Chemical Admixtures in Concrete, Toronto, Canada, 1998.

3. COHEN, M. D., MATHER B., Sulfate attack on concrete-research needs., ACI Materials Journal, Vol. 88, No 1, January-February 1991, pp 62-69.

4. IRASSAR, E. F., MAIO, A. DI, BATIC, O. R., SULFATE ATTACK ON CONCRETE WITH MINERAL ADMIXTURE, Cement and Concrete Research, Vol. 26, No 1, 1996, pp 113-123.

5. HARTSHORN, S. A., SHARP, J. H., SWAMY, R. N., The Thaumasite Form of Sulfate Attack in Porland-Limestone Cement Mortars Stored in Magnesium Sulfate Solution, Cement Concrete Composites, Vol. 24, 2002, pp 351-359.

FLY ASH INTERMIXES: A PATH TOWARDS SUSTAINABLE CEMENT PRODUCTION

S Antiohos **V Kasselouri-Rigopoulou**

A Karamberi **A Moutsatsou**

S Tsimas

National Technical University of Athens

Greece

ABSTRACT. The introduction of industrial solid by-products possessing cementitious and pozzolanic properties into the construction sector has nowadays become a worldwide trend. Fly ash is one of the main solid wastes being utilized in the cement industry. However, problems associated with the use of each type of fly ash (high and low calcium) in cementitious systems (low early strength, undesired expansion, etc.) contribute to the skepticism with which this material is still being handled by a significant part of the industrial world. The present study presents an innovative way to overcome the aforementioned problems, that is, by developing several intermixtures consisting of a high and a low calcium fly ash, both generated in great quantities globally. Their efficacy was examined in terms of pozzolanic potential, compressive strength development, *k*-values and CH-depletion. It was found that for certain substitution ratios, the new ternary blends outperformed the corresponding binary mixtures, as a result of a synergistic action that took place between the fly ashes used.

Keywords: Ternary blend, Active silica, Efficiency factor, Fixed lime, Synergistic effect.

S Antiohos, is a Chemical Engineer holding an MSc in Environmental Management and a PhD Candidate at the NTUA with a special interest in interpreting the reactivity of fly ashes.

V Kasselouri-Rigopoulou, is a Professor at the Chemical Engineering School of NTUA. Her teaching and research fields include, High Temperature Chemistry and Technology.

A Karamberi, is a Chemical Engineer, awarded MSc in 'Materials and Conservation Interventions' and a PhD candidate in the field of cullet utilization in construction materials.

A Moutsatsou, is an Associate Professor with great experience in chemical analysis. She specializes in Chemistry and Technology of metals, recycling of materials and methods for the treatment of solid and liquid waste.

S Tsimas, is an Associate Professor at the NTU of Athens. His major interests are in chemistry and technology of aluminosilicates, specializing in cement technology and upgrade techniques of industrial minerals and by products.

INTRODUCTION

In the production of cement and concrete, more and more supplementary cementing materials (such as fly ash, slag and silica fume), deriving as by-products from several industrial activities, are being incorporated. This is not only due to the fact that sustainability in the construction industry can be achieved through the increased usage of such materials [1], but also because their employment as admixtures can lower the cost and impart certain technological benefits to the final product. For fly ash in particular, it has been repeatedly proved that despite its relatively slow rate of reaction [2], it induces improved workability [3], higher later age strength [3, 4] and superior resistance towards aggressive media [4, 5] when is added in cement and concrete.

The literature is rich with papers that highlight the differences between the two main types of fly ashes. Apart from critical differences in their mineralogical structure, variations in their chemical composition are also accountable for the uneven effect that these have on cementitious systems. It is low-calcium fly ashes for example that react slower, especially during the early stages of hardening, mainly due to the higher presence of crystalline phases, which are considered chemically inert in concrete [2]. High calcium fly ashes on the other hand provide better early age strengths as a result of the cementitious compounds they contain. It is believed [3] that calcium substitution in the glassy phase is generally increasing the reactivity of high lime ashes providing for the formation of the calcium-silicate and calcium-aluminate phases in the absence of an external source of lime. On the other hand, the dissociation of the firmly chained glass of the low lime ashes is far more depended on the calcium hydroxide that is liberated during the hydration of cement.

Despite the clear advantages that the two main types of fly ashes bring about on cementitious systems, certain drawbacks associated with each of them are held responsible for their rather limited use. In Greece, for example, a recent report [6] has revealed that during the last two decades, the utilization rate of this material is stuck to 10%, indicating the skepticism with which this material is still being handled with by a significant part of the industry. In order to deal with the mutual shortcomings of the two types of fly ashes, the majority of the scientific community has proposed the construction of ternary or even quaternary blends by introducing into the binary FC system a third (or fourth) highly reactive pozzolan. Materials such as silica fume [7] and metakaolin [8], possessing significant quantities of silica in active form have been employed to compensate for the early strength loss and slow rates of reactions associated with both types of fly ashes. Although these tactics have been proved to be very effective in that aspect, they often result in substantial cost increases, a fact that inhibits their wider acceptance as day-to-day routine methods. To deal with that, a number of researchers have attempted to produce ternary blends by using the two types of fly ashes aiming at taking advantage of their assets and a possible synergistic action, taking place between them. Naik et al. [9] prepared several mixtures of blends of class F and class C ashes. The constructed blends, that occupied 40% of the total cementitious material, showed either comparable or better results (in terms of both mechanical and durability properties) than either the reference mixture or the mixture containing the class C fly ash solely. Antiohos et al. [10], prepared intermixtures, using a low-lime ash and a highly siliceous active high lime fly ash, and replaced up to 30% of cement. The encouraging results obtained were attributed to the synergy between the ashes used, given that no serious alterations in the nature of hydration products were detected.

In the present study, new supplementary cementing materials (SCM) were prepared from mixing two types of fly ashes, of typical composition, in several proportions. The efficiency of the new products was examined in terms of pozzolanic activity, compressive strength development and CH depletion rate. Comparisons with both the reference mixture and the corresponding binary mixtures were performed in order to detect if a combination of those ashes can compensate for the handicaps connected with the incorporation of each of them in a cementitious matrix. As long as the two types of fly ashes examined here can assist each other, or even better interact synergistically, they may constitute an excellent solution for producing viable, superior cementitious systems of relatively low cost.

EXPERIMENTAL PROCEDURES

Raw Materials and Preparation of Fly Ash Intermixtures

Two different fly ashes, one with high calcium content (designated here as T_k) and one with lower calcium content (T_m) were used as the basis for producing the several intermixtures. For preparing mortar and paste specimens, a normal setting cement (CEM I 42.5 according to European Standard EN 197-1) was used. The chemical composition of the ashes and cement are given in Table 1. Due to differences observed in the particle size distribution of the two utilized ashes, they were both ground in a lab ball mill in order to obtain ashes of similar fineness. When the ashes were brought to the required fineness, they were mixed in a rotating blender until homogeneity of the new blend was achieved. The blended fly ash mixtures were prepared by using various dosages of the high (T_k) and low (T_m) fly ashes.

Table 1 Chemical composition of the raw materials and fly ash intermixtures

	PERCENT, % by mass								
	SiO_2	Si_{re}*	CaO	Al_2O_3	Fe_2O_3	MgO	SO_3	Na_2O	CaO_f
C	20.28	-	65.01	4.75	3.76	1.61	2.55	0.17	-
T_k	31.54	24.75	33.90	14.00	6.25	3.29	6.31	0.13	6.96
T_m	51.36	31.36	13.80	16.73	8.75	2.26	1.49	0.77	0.95
T_1	37.62	27.46	27.92	14.63	6.52	3.13	5.11	0.17	4.68
T_2	42.16	29.16	23.93	15.70	7.14	2.80	3.52	0.35	3.76
T_3	46.55	29.25	19.28	15.92	7.77	3.01	2.55	0.56	2.23

* *The method specified in European Standard EN 196-2 was followed for estimating the reactive silica content (Si_{re}) of fly ashes.*

The proportions applied were 75% T_k and 25% T_m (blend designated as T_1), 50% T_k and 50% T_m (for blend T_2) and 25% T_k and 75% T_m for the mix designated as T_3. The chemical composition of the new blends is given also in Table 1. An initial observation is that the blending procedure had a beneficial effect on the sulfate and free lime (CaO_f) contents of the blended fly ashes, as these either met the requirements stated in EN 450-1 'Fly ash for concrete' or in the worst case approached conformation with those.

Testing Procedures

Pozzolanic activity potential

Before introducing the new blended materials into mortars and pastes, their pozzolanic reactivity was determined using the Chapelle test. According to this test, dilute slurry of the pozzolan reacts with calcium hydroxide at 100°C for 18 hours. The remaining quantity of lime in the suspension is then determined by titration and the results are expressed in grams of lime reacted per grams of each pozzolan tested. The same test was also applied to the initial fly ashes for comparison purposes.

Compressive strength development

For monitoring the compressive strength of the new blends, mortar mixes were prepared by adopting cementitious material-to-sand ratio of 1:3, water to binder ratio (w/b) of 0.5 and 20 and 30% by weight cement replacement. Keeping the w/b ratio constant, cement mortar without any fly ash was prepared as the control specimen. Additionally, two more mortars incorporating the two initial fly ashes (also replacing 20 and 30% by weight cement) were prepared for comparison purposes.

The dry mortars were mixed and then water was added. After sufficient mixing, they were cast in 4x4x16 cm mortar prisms and vibrated for 20 s on a vibration table. The molds were stripped after 24 h, and the specimens were immersed in lime-saturated water at 20°C, until testing. The testing ages were 2, 7, 28, and 90 days. For each age, two specimens of each mixture were tested for compressive strength and the mean value of these measurements is reported. Based on the compressive strength measurements, efficiency factors (simpler, *k*-values) were determined, following the procedure described elsewhere [6].

CH depletion rate

To simulate mortar in paste and monitor the evolution of the hydration process, pastes were prepared both with the initial ashes and their intermixtures, adopting a representative 20% cement replacement in all cases. During preparation, the same w/b ratio with the mortar case was used and curing under water at 23°C followed. At the dates of testing, the samples were fractured into pieces. Hydration was terminated with the addition of acetone and diethylethere, followed by drying until constant weight in a vacuum chamber. Finally, the specimens were placed into polypropylene bags and stored in a dryer until they were tested.

Thermogravimetry and Differential Thermal Analysis (DTA) were carried out to monitor the progress of the pozzolanic reaction in each blend. This can be accomplished by measuring the amount of calcium hydroxide contained in each paste. The measurements were performed in a Mettler STARe 851/LF/1600 TG/SDTA. A platinum crucible with 70μL capacity was used and mass of the examined samples ranged from 45 to 55 mg.

The samples were tested in a nitrogen atmosphere (50 ml min^{-1}) at a heating rate of 10°C min^{-1} from ambient temperature to 1000°C. The weight of calcium hydroxide in the samples (expressed through the abrupt weight loss occurring in the temperature region of 400-550°C) and the quantity of free-CH transformed into calcium carbonate (due to possible carbonation during handling) were estimated and afterwards added to determine the total CH percentage in all specimens.

RESULTS AND CONSIDERATIONS

Chapelle Test

The results of the Chapelle test are plotted in Figure 1, against the active silica content of each of the ashes (raw and blended) tested. It can be seen that the low calcium fly ash (T_m) exhibits the greatest pozzolanic potential, combining a significant amount of lime. On the contrary, high calcium ash (T_k), presents the worst behavior, in terms of grams of combined lime, as a result of its low active silica content. The activity of the constructed intermixtures fluctuates between the activities of the initial ashes, confirming the positive influence of the T_m addition to T_k. From the same figure it can be also observed that the role of active silica of the tested specimens is determinative on their pozzolanic activity, a fact that was somehow expected since soluble silica is actually the fraction that participates in the pozzolanic reactions [11]. It should be noted however, that despite the fact that the Chapelle test can provide an indication of the reactivity of several pozzolanic materials, it cannot provide any direct information about the rate of reaction of the same material under ambient temperatures in a cementitious matrix.

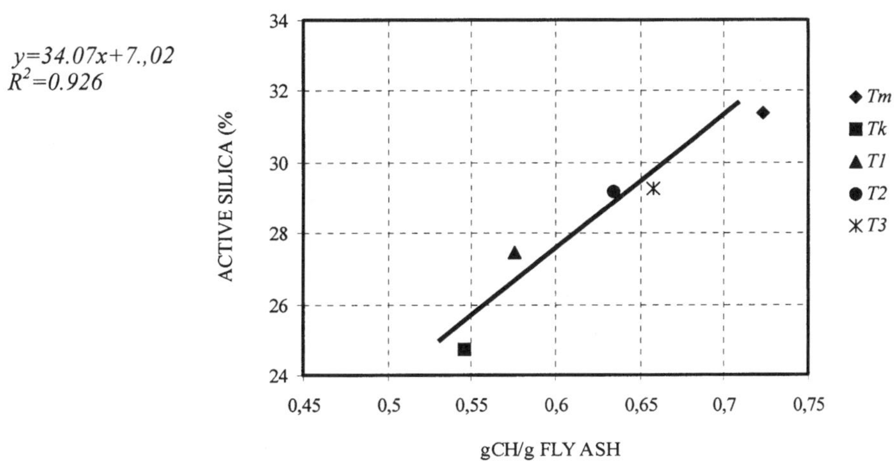

Figure 1 Pozzolanic potential (expressed as g of CH per g of pozzolan) in relation to the active silica content of each pozzolan

Compressive Strength Test

The results of the compression test for the control sample and all the mixtures that replaced 20% by weight cement are given in Figure 2 as a function of the hydration age. It can be seen that throughout the first month of the hydration process, the control specimen (no fly ash addition) outperforms both the systems with each raw fly ash as well as the systems incorporating their intermixtures. This is mainly due to the slow evolution of the fly ash pozzolanic reaction, which determines to a large extent the strength development of blended mixtures. Generally T_k ash performs slightly better than T_m especially during the early stages of hardening, as a result of its higher active calcium oxide content. Up to the first 28 days of hydration, the ash intermixtures are performing *better* than slowly reacting T_m and slightly worse than T_k. During the same period, blend T_1 outperforms the rest blended ash mixtures, probably due to the significant presence of T_k ash in it.

The situation differs substantially after 90 days of hydration, where all the intermixtures exceed the value of the control specimen, whereas blends T_2 and T_3 exhibit the maximum strength values. Somewhat surprisingly at the same age the strength values of the ashes that participate in those blends are lower. This could be an indication that a synergistic effect between the two ashes has taken place. Especially noticeable is the fact that during the last two months of the examined hydration period, T_3 intermixture presents a dramatic strength improvement. This could be additionally attributed to its high active silica, which it has been found [12] to hold a critical role after the first month of hydration in systems that incorporate high calcium fly ashes.

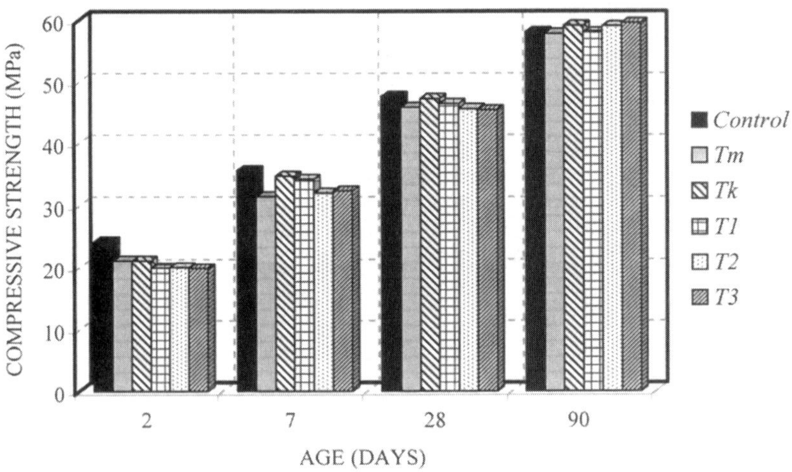

Figure 2 Compressive strength development raw fly ashes and their intermixtures for 20% cement replacement

At increased cement replacement (Figure 3) the situation is different. In this case, the raw ashes as well as their intermixtures significantly retard the strength development as a result of the increased cement content substitution and the well-established incompetence of fly ashes to act drastically from the beginning of the hydration period.

However, synergistic effects occurring between the different ashes are more clearly detected even from the beginning of hydration. The initial ashes (T_k and T_m) are presenting a similar behavior to the previous replacement dosage with high-lime fly ash T_k exhibiting higher strength values throughout the hydration period. On the other hand, the apodosis of the tested intermixtures varies with the hydration stage. In particular, blends T_1 and T_2 (with substantial participation of high calcium T_k) are the most efficient after 7 and 28 days of hydration respectively, whilst after 90 days the blend T_3 (containing mainly low-lime T_m) exhibits the highest strength value among all the blended systems. It is possible that a small participation of high-lime ash T_k reduced the period before the onset of the pozzolanic reaction leading to a notable strength increase after 28 days [10]. In general it can be noted that regardless of the hydration stage and the cement replacement level (20 or 30% by weight), almost none of the intermixtures presents the worst behavior among the pozzolanic samples tested.

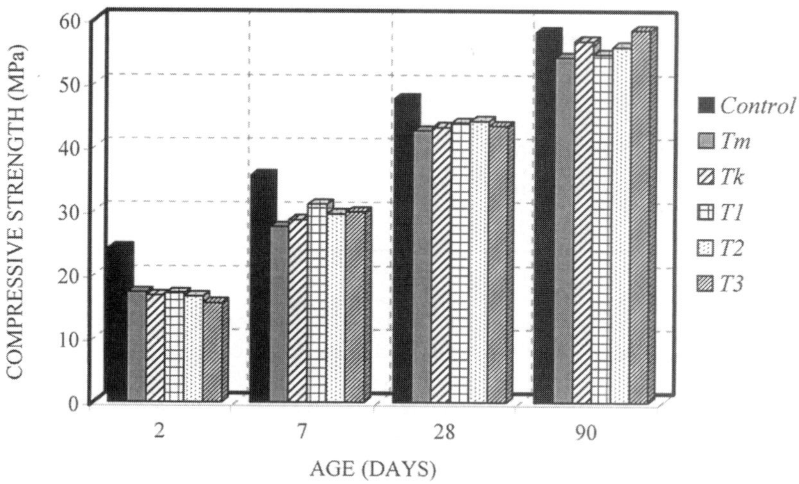

Figure 3 Compressive strength development of raw fly ashes and their intermixtures for 30% cement replacement

Efficiency Factors

The encouraging performance of the intermixtures is also confirmed via the concept of the efficiency factor. In the case of mortars and concrete that incorporate supplementary cementing materials, the *k*-value derives from the following expression for the measured compressive strength (f_c) [6, 10]:

$$f_c = K \left(\frac{1}{W / (C + kP)} - a \right) \quad (1)$$

where K is a parameter depending on the cement type (here 38.8MPa), C and P are the cement and fly ash contents respectively, in the mortar (kg/m^3), W is the water content (kg/m^3) kept constant in all the mixes and α a parameter depending mainly on time and curing. Using this equation, the measured values of the compressive strength given in Figures 2 and 3, the *k*-values for the intermixtures of the present work were calculated and are presented in Table 2.

In previous attempts [6] dealing with Hellenic fly ashes that replaced 20% by weight cement, the reported *k*-values were around unity during the early ages and they progressively exceeded it as the hydration evolved. A similar image is obtained here for all the blends tested, indicating that they could substitute equivalently for cement.

Table 2 Efficiency factors (*k*-values) for raw and blended ashes

	k-VALUES									
	SCM addition, % by cement mass									
	20					30				
Age (days)	T_k	T_m	T_1	T_2	T_3	T_k	T_m	T_1	T_2	T_3
2	*0.81*	*0.81*	0.73	0.73	0.73	*0.68*	*0.71*	0.70	0.68	0.63
7	*0.94*	*0.72*	0.91	0.76	0.78	*0.70*	*0.65*	0.80	0.73	0.75
28	*0.97*	*0.88*	0.92	0.87	0.86	*0.80*	*0.78*	0.83	0.84	0.81
90	*1.06*	*0.97*	0.99	1.05	1.09	*0.94*	*0.82*	0.85	0.90	1.01

The beneficial role of the mixing procedure is validated here, as the *k*-values of all the blended fly ashes are either slightly smaller (at the early stage of hydration) or bigger (with progressing hardening) than the corresponding ones of the initial ashes. This is possibly the outcome of the evolution of their pozzolanic reaction and the synergy detected between them. It becomes evident that in both cases of cement replacement, a small incorporation of a class C fly ash in a matrix redundant with class F ash (T_3 blend) can assist the strength development of such a system especially after the first month of hydration. This becomes clear from the emphatic increase observed in the k-values of the aforementioned blend during the last two months of the hydration period. The authors have observed a similar, but certainly more pronounced, trend in a previous study [10] dealing with fly ash intermixtures consisting of a high active siliceous high lime ash and a low lime ash.

CH Consumption and Fixed Lime Capability

Monitoring the amount of calcium hydroxide existing in a Fly Ash-Cement (FC) system during hydration, can provide information on the ability of each pozzolan (and fly ash intermixture for the work presented herein) to bind CH towards the formation of second generation hydration products. In the frame of this study, the CH measurements were used to compute the percentage of fixed lime in the constructed pastes. For this purpose, the following equation proposed by Paya et. al. [13] was applied for determining the percentage of fixed lime in each case.

$$\text{Fixed Lime (\%)} = \frac{(CH_c \cdot C_{\%}) - CH_P}{CH_c \cdot C_{\%}} \times 100 \qquad (2)$$

where CH_c is the CH content of the control paste for a given curing time, CH_P is the CH content of the FC paste at the same age and $C_{\%}$ is the proportion of cement in the examined paste (obviously in this case where a 20% cement was substituted, $C_{\%}$ is 0.8). The fixed lime values of all the specimens containing fly ash (raw and intermixed) are plotted against hydration age in Figure 4. In general it can be observed that fixed lime values increase with curing time for all samples due to the evolution of the pozzolanic reaction.

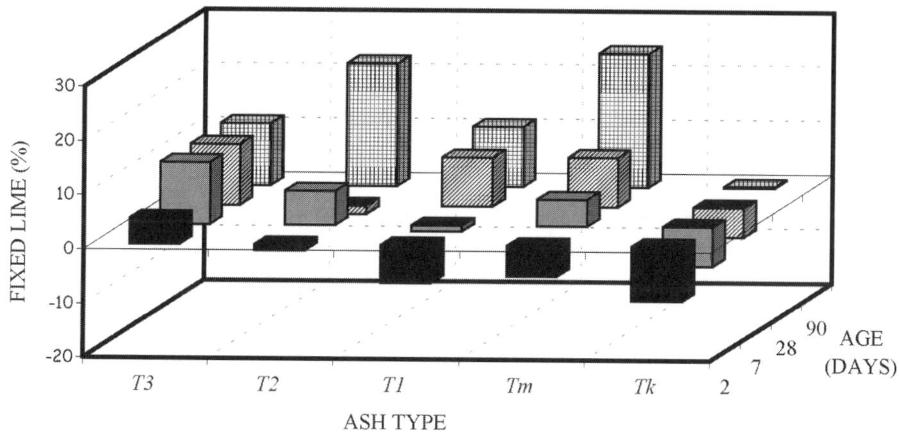

Figure 4 Fixed lime capabilities of fly ash specimens with hydration age

Negative values observed for almost all specimens tested during the early stages of hardening are attributed to their inability to consume significant amounts of CH from the beginning of the hydration process and also to the instant hydration of the free lime present in all blends (especially those with a significant participation of high-lime T_k) which creates an excess of CH into the matrices. Intermixture T_3 is the only that binds lime at a constant rate throughout the testing period, whilst T_2 and T_1 are fixing notable amounts of lime after the first month of hydration, with the former exhibiting an impressive increase in its fixing lime capability during the last two curing months. This provides an additional explanation of the previously presented strength supremacy of the T_2 and T_3 intermixtures at the end of the hydration period.

CONCLUDING REMARKS

In the frame of this study, it was demonstrated that using intermixtures based on two types of fly ash is an effective means for dealing with mutual shortcomings related with each type of fly ash. It was not only shown that unique assets of the one ash can compensate for the problems of the other, but most importantly, synergistic effects between them can occur that may further improve the performance of the final product. Although the production of such blended cements is not a cure for all construction threats, it certainly constitutes an alternative path towards upgrading the quality of concrete structures without adopting environmental baleful tactics of great cost.

REFERENCES

1. MEHTA, P.K., Role of Pozzolanic and Cementitious Materials in Sustainable Development of the Concrete Industry, Proc. 6th CANMET/ACI Int. Conf. Fly Ash, Silica Fume, Slag & Natural Pozzolans in Concrete, Ed. V. M. Malhotra, ACI SP-178, 1998, Vol 1, pp.1-20.

2. HEMMINGS, R.T. and BERRY, E.E., On the Glass in Coal Fly Ashes: Recent Advances. Presented at the MRS Symposium, Pittsburgh, United States, 1988, vol. 113, pp. 3.

3. THOMAS, M.D.A, SHEHATA, M.H., and SHASHIPRAKASH, S.G. The Use of Fly Ash in Concrete: Classification by Composition, Cement and Concrete Aggregates, 1999, Vol. 21, pp 105-110.

4. GHOSH, S.N., SARKAR, L.S. Mineral Admixtures in Cement and Concrete. In: Progress in Cement and Concrete, 1st ed., ABI Books, New Delhi, 1993, p. 565.

5. DHIR, R.K. and JONES, M.R., Development of Chloride-Resisting Concrete Using Fly Ash, Fuel, 1999, Vol. 78, pp 137-142.

6. PAPADAKIS, V.G., TSIMAS, S. Supplementary Cementing Materials for Sustainable Building-Sector Growth, European Commission DGXII, Marie Curie Fellowship, Final Report, Project No HPMF-CT-1999-00370, NTU of Athens, Greece, 2001, p. 74.

7. LAM, L., WONG, Y.L., and POON, C.S., Effect of Fly Ash and Silica Fume on Compressive and Fracture behaviors of Concrete, Cement and Concrete Research, 1998, Vol. 28, pp 271-283.

8. BAI, J., SABIR, B.B., and KINUTHIA, J.M., Strength Development in Concrete Incorporating PFA and Metakaolin, Magazine of Concrete Research, 2000, Vol. 52, pp 153-162.

9. NAIK, T.R., SINGH, S., and RAMME, B., Mechanical Properties and Durability of Concrete Made with Blended Fly Ash, ACI Materials Journal, 1998, Vol. 95, pp 454-462.

10. ANTIOHOS, S., PAPADAKIS, V.G., MAGANARI, K., and TSIMAS, S., The Development of Blended Supplementary Cementing Materials Consisting of High and Low Calcium Fly Ashes, Accepted for Publication in the Proceedings of the 11th ICCC, Durban, South Africa, May 2003.

11. RICHARTZ, W., Composition and Properties of Fly Ashes, ZKG, Translation 2/84, 1984, pp 62-71.

12. ANTIOHOS, S. and TSIMAS, S. Investigating the Role of Reactive Silica in the Hydration Mechanisms of High-calcium Fly Ash/Cement Systems. Cement and Concrete Composites (Accepted for Publication), 2003.

13. PAYA, J., MONZO, J., BORRACHERO, M.V., VELASQUEZ, S., Evaluation of the Pozzolanic Activity of Fluid Catalytic Cracking Catalyst Residue (FC3R). Thermogravimetric Analysis Studies on FC3R-Portland Cement Pastes, Cement and Concrete Research, 2002, Vol. 33, pp 603-609.

STRENGTH BEHAVIOUR OF CONCRETES CONTAINING METAKAOLIN

K Ganesh Babu

Ch V Apparao

IIT Chennai

India

ABSTRACT. The use of mineral admixtures in concrete is well accepted today. Though there are several methods for the design of normal concretes of a specified strength grade, the design of concretes containing the newer mineral admixtures like metakaolin is not yet well understood. Also the information available on the behaviour of metakaolin in concrete is very little. This paper is directed towards an understanding of the strength behaviour of concretes containing metakaolin, through a systematic evaluation of the most recent information available. These evaluations are based on the efficiency concepts that were successfully utilized earlier by the author for other pozzolans like fly ash, silica fume, blast furnace slag etc. It was observed that metakaolin is a high efficiency pozzolan and can exhibit efficiencies comparable to that of silica fume.

Keywords: Metakaolin, Pozzolan, Efficiency, Compressive strength, Mix design, Replacement percentage, Fineness.

Professor K Ganesh Babu, after obtaining his PhD in structural engineering has been a Scientist in the SERC, Chennai, and later joined the Department of Ocean Engineering in 1980. He is in charge of the Structures and Materials laboratory and is active in the areas of "effective utilization of mineral admixtures, HPC, concrete in marine environment, corrosion of steel in concrete, repair and rehabilitation of structures".

Ch V Apparao, obtained his B. Tech degree from JNTU, Hyderabad, was a student of MS by research program at IIT, Chennai and worked in the area of "behaviour of concretes with metakaolin".

INTRODUCTION

Cement and concrete composites are accepted as materials of construction even for structures that have to perform under very severe environmental conditions. The adverse nature of such an environment leads to the deterioration of the concrete and to the corrosion of reinforcement, two of the major problems of structural performance. Efforts towards an improvement of the characteristics of concrete in the recent past suggest that cement replacement materials along with chemical admixtures can improve the durability and corrosion characteristics when appropriately utilised. Metakaolin (MK), produced through the calcination of kaolinite and having a very high pozzolanic reactivity, is one such material that can be utilised to produce highly durable concrete composites. However, the information available on the behaviour of metakaolin in concrete is very little and the present study is an effort to understand the compressive strength characteristics of metakaolin in concrete.

METAKAOLIN IN CONCRETE

Though there are several methods for the design of normal concretes of a specified strength grade, the design of concretes containing mineral admixtures is not yet well understood. In general, concretes containing mineral admixtures are produced through methods like simple addition, direct replacement or modified replacement. In recent times the cementitious efficiency of the mineral admixture is used in the modified replacement methods for achieving a specific strength. This methodology was successfully implemented earlier in this laboratory for the design of concretes containing mineral admixtures like fly ash, silica fume, blast furnace slag etc., (Babu et al, 1993, 1994, 1996, 2000). The present effort is directed towards an understanding of the contribution of metakaolin to the strength of concrete. In specific, the cementitious efficiency factor for metakaolin in concrete is established through a systematic evaluation of the available results from literature, at the different percentages of replacement. A mix design methodology was also suggested using these efficiency factors.

PRESENT STATUS

To understand the effectiveness of mineral admixtures in cementitious composites, some of the parameters like chemical composition, hydraulic reactivity and fineness have been carefully examined by many earlier. However, there is no specific correlation that is obvious with either the physical or chemical composition, though certain specifications regarding the minimum compositions and fineness could be given for these mineral admixtures.

It was reported that metakaolin, in general, is a poorly crystallized white powder and has a specific surface of 12,000 m^2/kg with the average particle size varying from 1.5 – 2.5 μm. It comprises of nearly 95 percent (SiO_2 + Fe_2O_3 + Al_2O_3) as elongated particles. The glass content, considered to be the most significant variable and certainly the most critical to hydraulicity, is generally in the range of 71 - 88% depending on the source. Metakaolin has been studied by a few for its compliance with these minimum specifications along with its behaviour in concrete. At this stage it was decided that the reactivity of metakaolin in concrete and its cementitious efficiency should be evaluated through the compressive strength results reported in the literature. However, it was observed that the information available was rather small for a comprehensive analysis.

It is, however, to be noted that Caldarone et al (1994) reported that High Reactivity Metakaolin (HRM) has shown promise as a product capable of expanding the applications of high performance concrete. Based on an investigation directed towards a comparison of the mixtures produced with 5 and 10% HRM and silica fume addition, he observed that the compressive and flexural strength development of HRM mixtures was significantly greater than that of non-pozzolanic control mixture and slightly higher than the silica fume mixtures.

In conclusion, he reported that HRM has extremely high reactivity and that the strength and performance of 10% HRM concrete was similar to, or better than the 10% silica fume concrete. In their later publication (Caldarone and Gruber, 1995), based on the concretes containing 10% addition and 8.5% replacement of both metakaolin and silica fume, reported similar conclusions.

Wild et al. (1996) presented the role of metakaolin (MK) in enhancing the strength through concretes containing 0-30 replacement. It is to be noted that as the replacement level increased the superplasticiser requirement also increased and was observed to be 0.6% for every 5% increase in metakaolin content. They also presented a relative strength comparison of the concretes containing metakaolin at 1 day and 90 days. They opine that, at 1 day where the contribution due to pozzolanic activity is expected to be small, relative strength increases with increase in MK content up to a maximum of 1.18 at 10% replacement, and then falls linearly to 0.76. The increase in relative strength at 1 day was interpreted as being principally due to the increased acceleration of OPC hydration combined with the contribution from the filler effect and the fall in relative strength at high MK contents must clearly be a result of the dilution effect. At 90 days all the cementitious reactions are close to completion and the short term effect of strength enhancement due to acceleration of OPC hydration has been lost. Therefore the 90 day figure represents the long term effect that pozzolanic activity and the filler effect have in enhancing the concrete strength. The above facts by themselves show that the strength efficiency of metakaolin is dependent upon both the age and the percentage of replacement.

Dubay and Banthia (1998) studied concretes containing 10% replacement of metakaolin and silica fume individually and in combination with and without steel fiber. They observed only a marginal increase in the compressive strengths due to silica fume and HRM. They also reported that although the air content of hardened concrete with HRM was higher than that of concrete with equal amount of silica fume, the strength and performance of HRM concrete was similar to, or better than, concrete with silica fume.

Aldea et al. (1998) studied the behaviour of metakaolin in comparison to the to silica fume in extruded pressure pipes. However, they found that the pipe containing metakaolin in place of silica fume did not perform well in comparison. It is to be noted that a comparison of the mix proportions shows that the silica fume and metakaolin contents (30% and 20%), the water contents in these two mixes (28% and 40%), as well as the fibers added (PVA and glass) were all different in the mixes making it difficult for the results to be understood. They have attributed the poor behaviour of metakaolin mix to the dimensional variability of the extruded pipe (may be due to the higher water percentage) and to the glass fiber inefficiency or breaking during the extrusion.

Bai et al. (2000) studied the strength development of concretes containing metakaolin replacements varying from 0-15% and at varying water to cementitious material ratios of 0.4, 0.5 and 0.6 and compared these with mixtures also containing fly ash. It was reported that up to 15% partial replacement by metakaolin resulted in a considerable enhancement in strength in both the short and medium term at all the water to binder ratios investigated. It is obvious from even the few investigations reported in literature that the performance of metakaolin can be similar to that of silica fume.

EVALUATION OF EFFICIENCY

The efficiency of mineral admixtures was expressed by some in terms of the pozzolanic activity. The cementing efficiency of a mineral admixture is defined as the number of parts of cement that could be replaced by one part of the admixture without changing the property being studied. The present investigation is an effort towards a specific understanding of the strength efficiency of metakaolin in concrete. The basic methodology in this is to bring together the water to cementitious material ratio to strength relations for both normal and metakaolin concretes. This was done by using the "Δw" concept, which attempts to bring the water cementitious materials ratio [w/(c+m)] nearer to the water cement ratio of the control concrete (w/c_o) by applying the cementitious efficiency of metakaolin "k" and ensuring that (w/c_o) = [w/(c+km)] at any particular strength.

A detailed presentation of this method for evaluation of efficiency of mineral admixtures in concrete was discussed in detail earlier (Babu et al 1993). Thus, in this method the water cement ratio (w/c_o) to strength relation of normal concrete will also be valid for the metakaolin concretes, by considering the "water to effective cementitious materials ratio, [w/(c+km)]".

EVALUATION OF AVAILABLE DATA

For a specific understanding of the behaviour of metakaolin in concrete, to start with, it was felt essential to study the results available from literature. In view of this, the reactivity of metakaolin in concrete and its cementitious efficiency is verified through the compressive strength results reported in the literature. For this, the data available from research efforts in the recent past were collected and presented in Table 1.

Table 1 Range of Metakaolin Concretes Evaluated

No.	YEAR	REFERENCE	REPLACEMENT, %	CEMENT CONTENT, kg/m^3	W / (C+K)	28 DAY COMPRESSIVE STRENGTH RANGE, N/mm^2
1	1994	Caldarone	5, 10	656 - 660	0.38	58.48 – 63.10
2	1995	Caldarone	8.5, 10	353 - 390	0.36 – 0.40	55.0 – 85.80
3	1996	Wild	5 - 30	243 - 332	0.38 – 0.45	63.5 – 82.47
4	2000	Bai	2.5 - 15	331 - 381	0.40 – 0.60	48.0 – 92.0

It was made sure that these will form a fairly representative group governing all the major parameters that influence the behavior of metakaolin in concrete and present the complete information required for such an evaluation. During this evaluation it was seen that some of these mixes do not form a part of investigations on metakaolin concretes, due to the addition of other cementitious materials like fly ash or silica fume etc. It was also noted that the total number of metakaolin concretes that could be used in this assessment were far too small for a complete and comprehensive investigation of even the compressive strength efficiencies of the material at different ages.

It was also difficult to look at the specific contribution of the age (general efficiency factor, k_e) and the effect of percentage of replacement (percentage efficiency factor, k_p), as was reported for the other pozzolans earlier. In view of this only the 28 day overall strength efficiency of metakaolin (k, where $k = k_e + k_p$) at the different percentages was attempted.

The data suitable for an effective evaluation of the strength behaviour of conventional concretes from these metakaolin studies was also studied. The water to cement ratio to strength relations of these control concretes at the 7, 28 and 90 days, for which the information is available, have been presented in Figure 1. However, it was observed that even these conventional concretes were highly modified through one or more superplaticizers. In view of this, the 28 day water cement ratio to strength relation presented by Raja (1993), by combining the normal and high strength relationships recommended by ACI, was considered to be more appropriate for depicting the behaviour of normal concretes.

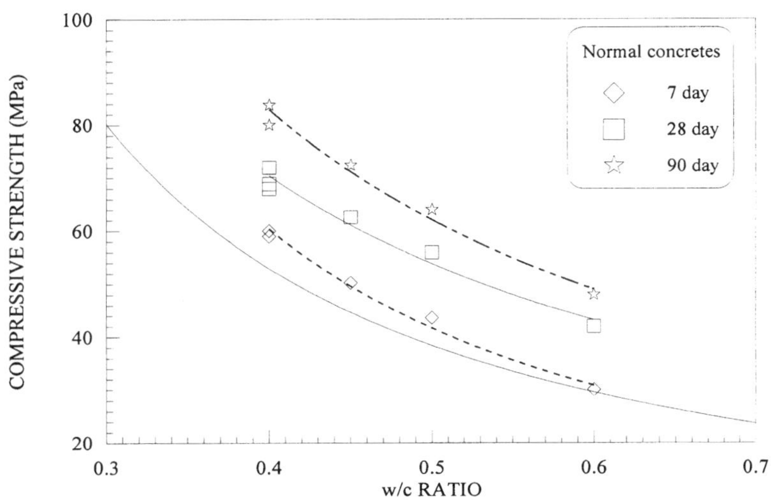

Figure 1 Water cement ratio to strength relations of normal concretes

The data available for the metakaolin concretes was now reevaluated through the proposed methodology. The water cementitious materials ratio [w/(c+m)] to compressive strength relations at different percentages of replacement were plotted for all the concretes at 28, days (Figure 2). It can be seen from this that the 28 day compressive strengths of concretes containing metakaolin were all above that for normal concretes. The above figure by itself shows the strength efficiency of metakaolin at the different percentages of replacement. Attempts to go through the proposed methodology from this point were not very successful because of the limited data as already stated. However, it was previously reported that the strength behaviour of concretes with metakaolin was similar to (or even better than) silica fume.

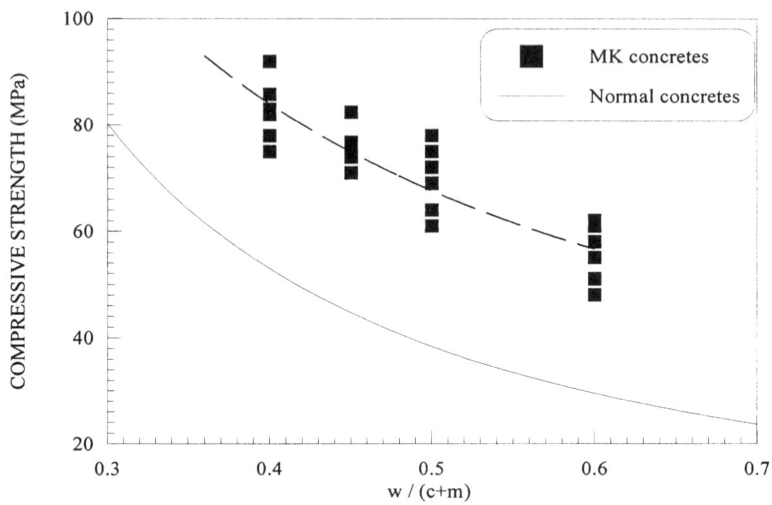

Figure 2 Variation of the 28 day compressive strength with [w/(c+m)] ratio

At this stage the overall efficiency factor "k" was directly assessed considering the water to cement ratio to strength relation for normal concrete proposed above. The variation of these "k" values with the percentage replacement was plotted in Figure. 3. The variation of the "overall efficiency factors" obtained for silica fume at 28 days as already reported earlier (Babu, 1996), is also presented in this figure. This relationship, superimposed in Figure 3 above to compare the relative efficiencies of the two materials, clearly shows that the overall efficiency of silica fume and metakaolin are almost the same.

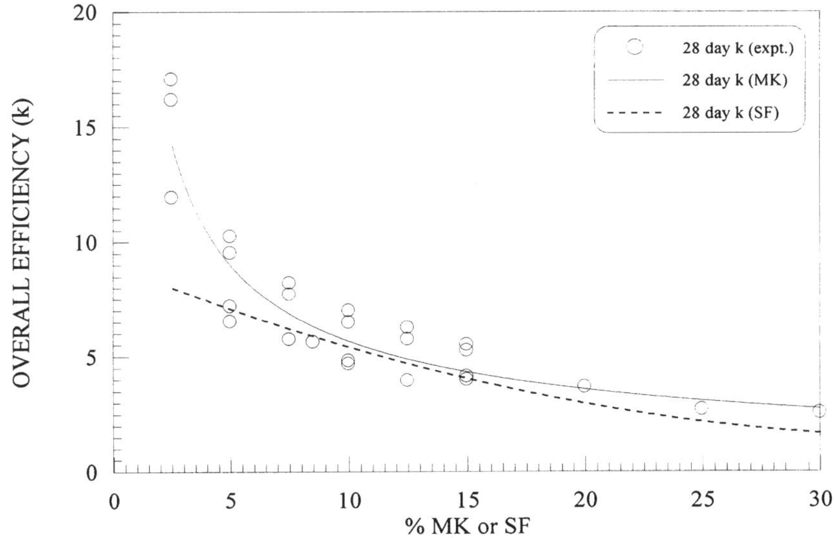

Figure 3 Variation of the overall efficiencies of MK and SF concretes

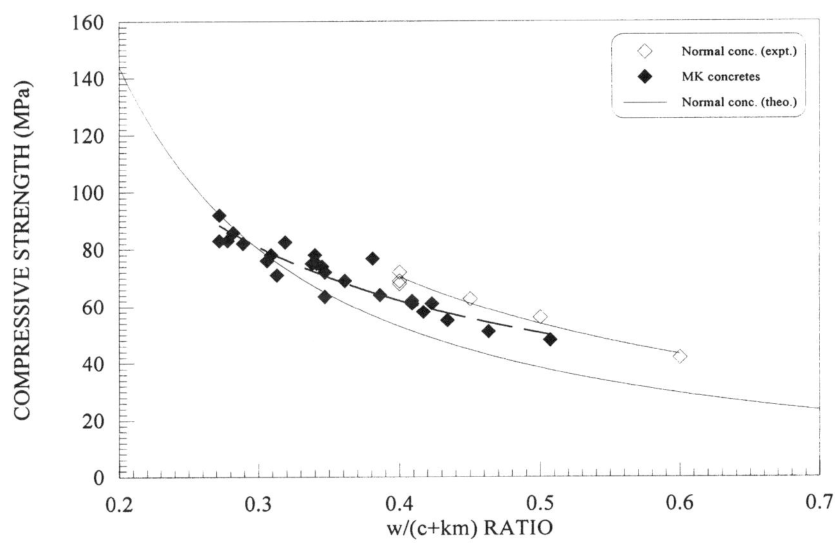

Figure 4 Comparison of the strength variation with water to effective cementitious materials ratio for MK concretes

PROPOSED MIX DESIGN METHODOLOGY

By using the already established efficiency values of metakaolin in concretes, at the various percentage replacements, a mix design methodology was proposed. This was essentially similar to the one proposed earlier from this laboratory for pozzolans like fly ash and silica fume (Babu et al. 1993). The water to cement ratio (w/c_0) was chosen for the strength required as per any of the recognized mix design methodologies for normal concrete. The water content required from the workability considerations was also chosen from the same procedure. This basically means that the mix design procedure for concretes with pozzolans is only a simple, yet significant, extension of the normal concrete mix design and at the zero replacement level will result in the normal concrete. The cement content (c_0) required for normal concrete was then estimated. The efficiency factor (k) corresponding to the required replacement is now taken from Figure 4. The quantities of cement (c) and metakaolin content (m), to have a cementitious efficiency of the equivalent of c_0, is obtained from the relation [$c_0 = c + km$]. The total aggregate content is then determined by the absolute volume method. The coarse aggregate to fine aggregate ratio is then chosen from the earlier adopted normal mix design procedure for the total aggregate content, remembering the slump, the size and grading of aggregates. This method was found to result in concretes with the desired quality of high performance in the limited laboratory experimentation under taken.

CONCLUSIONS

- The limited information available on the behaviour of matakaolin in concrete was utilised towards an understanding of the compressive strength characteristics of concretes containing matakaolin, through a systematic evaluation of the strength efficiency.

- The 28 day water to cement ratio to strength relation for normal concrete obtained by combining the normal and high strength relationships recommended by ACI was used for the evaluation of the efficiency of MK concretes.

- A comparison of the efficiency values obtained for the matakaolin concretes in the present investigation with that for the silica fume concretes established earlier shows that there is a reasonably good comparison between the two and that metakaolin concretes can be designed with confidence using the method.

- As a check, the strength obtained for the concretes investigated at different water to effective cementitious materials ratio [w/(c+km)], using "k" values of silica fume as already reported, shows that concretes can be satisfactorily obtained using the proposed methodology.

- A mix proportioning method was also proposed for metakaolin concretes based on this efficiency concept.

REFERENCES

1. ALDEA, C., MARIKUNTE, S., SHAH, S.P., "Extruded Fiber Reinforced Cement Pressure Pipe", Advanced Cement Based Materials V 8, (1998), pp. 47-55.

2. BABU, K.G., RAO, G.S.N., PRAKASH, P.V.S., "Efficiency of Pozzolans in Cement Composites", CONCRETE - 2000, Dundee, U.K., (1993), pp. 497-509.

3. BABU, K.G., PRAKASH, P.V.S., "Efficiency of Silica Fume in Concretes", Cement and Concrete Research, V 25, N 6, (1995), pp. 1273-1283.

4. BABU, K.G., RAO, G.S.N., "Efficiency of Fly Ash in Concrete with Age", Concrete and Concrete Research, V 26, No.3, (1996)pp. 465 - 474.

5. BABU, K.G., KUMAR, V.S.R., "Efficiency of GGBS in Concrete", Cement and Concrete Research, V 30, (2000), pp. 1031-1036.

6. BAI, J., SABIR, B.B., WILD, S., KINUTHIA, J.M., "Strength Development in Concrete Incorporating PFA and Metakaolin", Magazine of Concrete Research, V 52, No 3, (2000), pp. 153-162.

7. CALDARONE, M.A., GRUBER, K.A., "High Reactivity Metakaolin (HRM) for High Performance Concrete", SP 153-43 ACI, Detroit, V 2, (1995), pp. 815-827.

8. CALDARONE, M.A., GRUBER, K A., BURG, R G., "High Reactivity Metakaolin a New Generation Mineral Admixtures", Concrete International: Design and Construction, No 11, (1994), pp. 37-40.

9. DUBEY, A., BANTHIA, N., "Influence of High-Reactivity Metakaolin and Silica Fume on the Flexural Toughness of High-Performance Steel Fiber-reinforced Concrete ACI Materials Journal, V 95, No 3, (1998), pp. 284-292.

10. RAJA, G.L.V., "Mechanical and Fracture Characteristics of Steel Fibre Reinforced High Strength Concrete", Ph. D Thesis (1993), submitted to Indian Institute of Technology, Madras, India.

11. WILD, S., KHATIB, J.M., JONES, A., "Relative Strength Pozzolanic Activity and Cement Hydration in Superplasticised Metakaolin Concrete", Cement and Concrete Research, V 26, No 10, (1996), pp. 1537-1544.

BEHAVIOUR OF THE C_3S-$C_{12}A_7$-$CASO_4.2H_2O$-H_2O SYSTEM

S Yan

A Cai

J Zhou

Nanjing University of Technology

China

ABSTRACT. The effect of $C_{12}A_7$ and gypsum on the hydration of C_3S, hydration mechanism and hydration kinetics of C_3S in different systems were studied by various physical and chemical methods, such as quantitative XRD, and determination of combined water content and chemical shrinkage. The results show that $C_{12}A_7$ can accelerate the hydration of C_3S, especially its early hydration. When gypsum is present, the accelerating action of $C_{12}A_7$ is remarkable. The mechanism of faster hydration of C_3S is through the chemical reaction. Hydration kinetics equations of C_3S in different systems have also been brought forward.

Keywords: C_3S-$C_{12}A_7$-$CaSO_4$-H_2O system, Hydration, Kinetics equation

Yan Sheng was born in 1948 and graduated from Nanjing Institute of Chemical Technology, China in 1975. He is a professor at the Nanjing University of Technology and the director of the Institute of Chemical and Building Material Design & Research, He is mainly engaged in the research of advanced materials and new technologies. At one time or another, he presided and was involved in many key scientific research issues. He has published many monographs such as Quality Analysis and Control in Cement Production, Special Low-calcium Fly Ash Cement, and Study on Alkali Activated Slag Cement Immobilization of High Level Radioactive Waste, and Cement Plant Technology Design.

Cai Anlan was born in 1965. Now she works as adjunct professor at the Yancheng Institute of Technology, and reads for doctor's degree at Nanjing University of Technology, China. She is mainly engaged in the research of materials science of cement and concrete. A few papers written by her had been published in various periodical.

Zhou Jidong was born in 1963, and now he is a doctoral student. His area of research interest is in the materials science of cement and concrete.

INTRODUCTION

A study showed that if cement clinker minerals were in suitable proportion, properties of the system could be effectively improved [1]. The present paper studies the hydration process of C_3S in the presence of $C_{12}A_7$ and gypsum. Using quantitative XRD technique, the hydration kinetics and hydration mechanism of C_3S in the systems of C_3S-H_2O, C_3S-$C_{12}A_7$-H_2O and C_3S-$C_{12}A_7$-$CaSO_4.2H_2O$-H_2O were studied.

EXPERIMENTAL STUDIES

Materials

The preparation of individual minerals

C_3S: A mixture of chemically pure $CaCO_3$ and SiO_2 in a molar ratio of 2 was ground to pass through a 0.08 mm sieve, which yielded an intimately mixed fine powder. After a series of blending with water, shaping under a pressure of 200 kg/cm^2, drying for 1 h at (105±5) ℃, the fine powder was burned for 10 h in a siliconit stove at 1350℃ and cooled down to room temperature. The f-CaO content in the burned product was tested and the heat treatment repeated until the f-CaO content became 1.30%. The product was γ-C_2S, from which C_3S would be prepared. C_3S was prepared by adding 0.5 *wt*% CaF_2 and 1.50 *wt*% SO_3 to the mixture of γ-C_2S and $CaCO_3$ having a molar ratio of 1. The process of blending, shaping, burning was just the same as that of γ-C_2S except that the burning time was 6 h and the cooling rate was rapid. When the f-CaO content of the burned product was 1.40%, XRD technique was used to identify the product as C_3S.

$C_{12}A_7$: $C_{12}A_7$ was prepared by blending $CaCO_3$ and Al_2O_3. The molar proportions were $12CaCO_3 : 7Al_2O_3$. The process was just the same as that of C_3S made from γ-C_2S except that the burning temperature was 1280 °C and the f-CaO content was 0.13%.
C_3S and $C_{12}A_7$ were respectively grounded to pass through a 0.08 mm sieve, sealed in bottles to be used and tested to the standard consistence.

Table 1 Composition and symbol of the samples

SYMBOL	COMPOSITION, *wt*/%		
	C_3S	$C_{12}A_7$	SO_3
A_{01}	100	0	0
A_{02}	0	100	0
A_1	90	10	3.3
A_2	80	20	6.6
A_3	70	30	10
A_4	60	40	13.3

Composition and designation of the samples

A study of the C_3S-$C_{12}A_7$-$CaSO_4.2H_2O$-H_2O system showed that if the $C_{12}A_7/SO_3$ weight ratio is between 2.5 and 3.5, hydration rate of the system was the best [1]. We chose $C_{12}A_7/SO_3=3.0$, and blended $C_{12}A_7$, $CaSO_4.2H_2O$ and C_3S in different proportions. Table 1 shows the composition and designation of the samples.

Experimental Methods

(1) Measurement of compressive strength: The water to cement ratio of the pastes was adjusted to obtain standard consistency. 1.414 cm cubes were prepared and cured in standard conditions until testing. The test was only intended to compare the hydration behaviour for the samples in the study.
(2) Combined water content method for testing the hydration rates of samples at all ages: After a series of ceasing hydration (samples had w/c=0.5), grinding, filtrating, drying for 6 h at 60 ℃, burning to a constant weight at 950 ℃ in a muffle furnace, the samples' weight loss percentage were measured, of which the combined water content was calculated.
(3) The chemical shrinkage method was used to examine the hydration rates of samples at various ages.
(4) Quantitative XRD was used to determine the hydration rates of C_3S in different systems. Samples were scanned continuously at a rate of $5° \ 2\theta$/min using Cu K_α radiation.
(5) The working curve was made using the K value method with corundum (Al_2O_3) used as standard substance. The value of K=0.935 was used, so that the linear regression equation was :

$$y = -0.0095 + 0.935\,x$$

RESULTS AND DISCUSSION

Compressive Strength

The compressive strengths of samples listed in Table 1 at different ages are shown in Table 2.

Table 2 Compressive strength of samples at different ages, MPa

SYMBOL	AGE, days			
	1	3	7	28
A_{01}	2.35	6.86	13.72	19.80
A_{02}	6.86	2.54	9.80	11.76
A_1	5.49	13.72	20.97	26.66
A_2	5.68	17.64	25.09	29.79
A_3	5.68	15.29	19.80	25.68
A_4	5.29	11.17	15.88	21.56

There is evidence that compressive strengths of samples in the C_3S-$C_{12}A_7$-$CaSO_4.2H_2O$-H_2O system are higher than those of C_3S-H_2O system at all ages. With the change of the $C_{12}A_7$ content, the minerals appear to have an optimal mass proportion, *viz.* $C_3S : C_{12}A_7 : SO_3 = 80 : 20 : 6.6$ in the C_3S-$C_{12}A_7$-$CaSO_4.2H_2O$-H_2O system. Strength for this optimal proportions outclasses that of weighted sum of C_3S-H_2O system and $C_{12}A_7$-H_2O system. It can be shown that the system can achieve the best strength development by proper proportions of $C_{12}A_7$, $CaSO_4.2H_2O$ and C_3S.

Hydration Rate

The amount of combined water in the hydration products of samples A_{01}, A_{02} and A_2 at different hydration ages and their chemical shrinkage are shown in Table 3 and Table 4. The amount of combined water and the chemical shrinkage of samples in the C_3S-$C_{12}A_7$-$CaSO_4.2H_2O$-H_2O system outclasses that in weighted sum of C_3S-H_2O system and $C_{12}A_7$-H_2O system at all ages. It is obvious that the hydration rate of samples in the C_3S-$C_{12}A_7$-$CaSO_4.2H_2O$-H_2O system is accelerated in the presence of $C_{12}A_7$ and $CaSO_4.2H_2O$.

Table 3 The combined water content of samples at different ages, *wt%*

SYMBOL	AGE, days			
	1	3	7	28
A_{01}	4.55	6.35	8.78	11.66
A_{02}	32.85	34.91	35.47	43.01
A_2	14.17	20.48	25.73	28.75

Table 4 The chemical shrinkage of samples at different ages, ml/100g

SYMBOL	AGE, days			
	1	3	7	28
A_{01}	0.726	0.917	1.308	1.638
A_{02}	12.545	12.757	14.340	15.459
A_2	4.45	5.82	6.86	7.83

Hydration Products

DTA technique was used to distinguish the hydration products. The DTA curves for hydrated samples A_{01}, A_{02} and A_2 at 3 days are shown in Figure 1. It can be seen that the hydration products of C_3S (A_{01}) are CSH and Ca $(OH)_2$, and those of $C_{12}A_7$ (A_{02}) are C_2AH_8 and C_3AH_6. In the presence of $C_{12}A_7$ and $CaSO_4.2H_2O$, the hydration products of C_3S are CSH, AFt and $Ca(OH)_2$, with C_2AH_8 and C_3AH_6 not being observed.

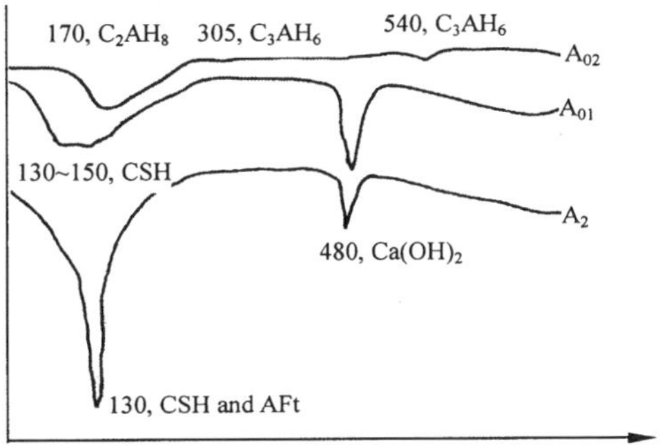

Figure 1 DTA curves for hydrated samples at 3 days

Hydration Kinetics of C_3S

Based on the results mentioned above, it is known that when C_3S, $C_{12}A_7$ and gypsum are in proper proportion, the hydration rate of the mix is faster than that of the individual mineral C_3S. It is inexplainable that if it is $C_{12}A_7$ that accelerates the hydration of C_3S, for the hydration rate of C_3S in the mixed system is unknown. The reasons why the mixed system has faster hydration rate may be:

(1) $C_{12}A_7$ accelerates the hydration of C_3S,
(2) C_3S accelerates the hydration of $C_{12}A_7$, and the hydration rate of C_3S is constant or slower,
(3) $C_{12}A_7$ and C_3S promote each other. If the hydration rate of C_3S in the mixed system is faster than that of the individual mineral C_3S (in the C_3S-H_2O system), it may be said that $C_{12}A_7$ accelerates the hydration of C_3S. So the quantitative XRD technique was used to analyse the hydration kinetics equations of C_3S in different systems.

Hydration kinetics of the individual mineral C_3S

Figure 2 shows the variation of quantitative XRD characteristic peaks of C_3S (A_{01}) paste at different hydration ages. The hydration degree of C_3S can be calculated from the quantitative XRD working curve of C_3S

$$y = -0.0095 + 0.935\,x$$

where $x = x_i$ and $y = x_\alpha I_i / I_\alpha$ (x is the content of phase in weight percent, I is the intensity of XRD peaks, and the subscript i represents the tested phase and αrepresents Al_2O_3 (contents of the tested phases are given in weight percent). The calculated results are shown in Table 5.

Table 5 Hydration degree of C_3S detected by XRD

AGE, days	1	3	7	28
HYDRATION DEGREE, %	20.78	26.80	35.78	49.70

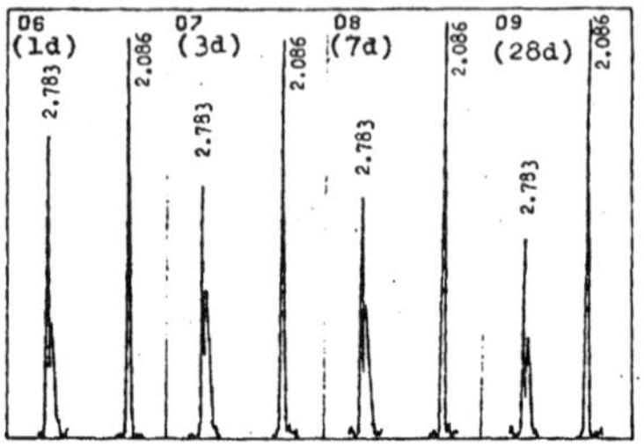

Figure 2 Variations of XRD peaks of C_3S paste at different ages

The experimental equation supposed by Budnikov[2] can be used to express the hydration kinetics of C_3S, *viz.* $L = K \lg t + B$
where L is the degree of hydration, t is the hydration age (days) and K and B are constants. By processing the $L \sim \lg t$ linear regression equation (data processed are shown in Table 6), the constants K and B in the equation are obtained as: K = 0.2002, and B=0.1926, and the hydration kinetics equation of individual C_3S is $L = 0.2002 \lg t + 0.1926$.

Table 6 Data processed

t	x (lg t)	y (L)	x^2	y^2	x y
1	0	0.2078	0	0.0432	0
3	0.4771	0.2680	0.2276	0.0718	0.1279
7	0.8451	0.3578	0.7142	0.1280	0.3024
28	1.4771	0.4970	2.1818	0.2470	0.7341
Σ39	2.7973	1.3306	3.1236	0.4900	1.1644

Hydration kinetics of C_3S in the C_3S-$C_{12}A_7$-H_2O system

The variations of quantitative XRD characteristic peaks of C_3S in the C_3S-$C_{12}A_7$-H_2O system (Samples of $C_3S/C_{12}A_7$ = 80:20) at different hydration ages are shown in Figure 3. Table 7 gives the calculated hydration degrees of C_3S in the system at different ages.

Table 7 Hydration degree of C_3S in C_3S-$C_{12}A_7$-H_2O system

AGE, days	1	3	7	28
HYDRATION DEGREE, %	64.90	66.63	77.03	83.79

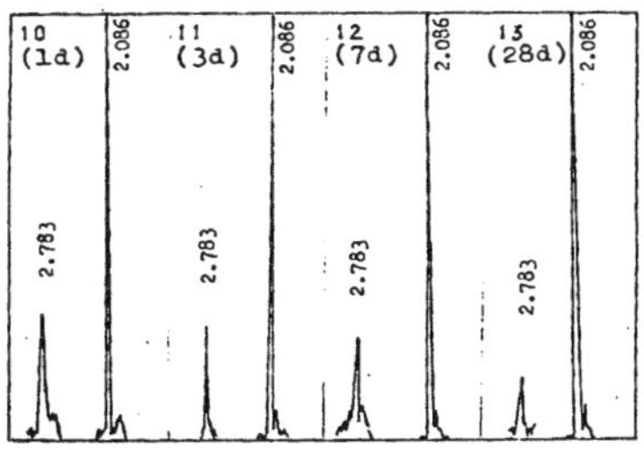

Figure 3 Variations of XRD peaks of C_3S in the C_3S-$C_{12}A_7$-H_2O system

By processing, we have K = 0.1380 and B = 0.6343. The hydration kinetics equation of C_3S in the C_3S-$C_{12}A_7$-$CaSO_4{\cdot}2H_2O$-H_2O system is $L = 0.1380 \lg t + 0.6343$.

2.4.3 The hydration kinetics of C_3S in the C_3S-$C_{12}A_7$-$CaSO_4.2H_2O$-H_2O system

The variations of quantitative XRD characteristic peaks of C_3S in the C_3S-$C_{12}A_7$-$CaSO_4.2H_2O$-H_2O system (A_2) at different hydration ages are shown in Figure 4. The calculated hydration degrees of C_3S in the C_3S-$C_{12}A_7$-$CaSO_4.2H_2O$-H_2O system at different ages are listed in Table 8.

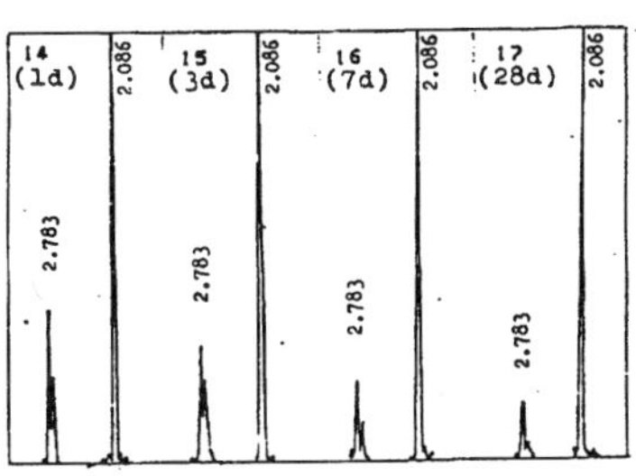

Figure 4 variations of XRD peaks of C_3S in C_3S-$C_{12}A_7$-$CaSO_4{\cdot}2H_2O$-H_2O system (A_2)

Table 8 Hydration degrees of C_3S in the C_3S-$C_{12}A_7$-$CaSO_4 \cdot 2H_2O$-H_2O system

AGE, days	1	3	7	28
HYDRATION DEGREE, %	50.42	66.30	85.3	96.00

By processing, we have K = 0.3174 and B = 0.5231. The hydration kinetics equation of C_3S in the C_3S-$C_{12}A_7$-$CaSO_4 \cdot 2H_2O$-H_2O system is L = 0.3174 lg t + 0.5231. The intercepts and slopes of the three hydration kinetics equations of C_3S in different system have shown that the hydration rates of C3S in different systems are in the order: C_3S-$C_{12}A_7$-$CaSO_4 \cdot 2H_2O$-H_2O system > C_3S-$C_{12}A_7$-H_2O system > C_3S-H_2O system. It is obvious that $C_{12}A_7$ can accelerate the hydration of C_3S, and with gypsum also present, the promoting action is remarkable.

Hydration Mechanism

It may be explained from cement chemistry standpoint why $C_{12}A_7$ can accelerate the hydration of C_3S in the presence of $CaSO_4.2H_2O$. With water, C_3S dissolves, which results in Ca^{2+} and OH^- coming into solution. In fact, C_3S reacts with water in $Ca(OH)_2$ (Ca^{2+} and OH^-) solution, which come from C_3S dissolution in water, and $C_{12}A_7$ in Ca^{2+} and $Al(OH)_4^-$ solution. When both C_3S and $C_{12}A_7$ are present, the equilibrium between C_3S and $Ca(OH)_2$ is destroyed because of the formation of C_3AH_6, which comes from the reaction between Ca^{2+} and $Al(OH)_4^-$, and the hydration of C_3S is promoted. As in the presence of gypsum, C_3AH_6 reacts with gypsum to form precipitation AFt which promotes the hydration of $C_{12}A_7$, thus the hydration of C_3S is accelerated.

CONCLUSIONS

- The compressive strengths of samples in the C_3S-$C_{12}A_7$-$CaSO_4.2H_2O$-H_2O system are much higher than those of the individual mineral C_3S at all stages especially at later ages.
- Hydration rates of samples in the C_3S-$C_{12}A_7$-$CaSO_4.2H_2O$-H_2O system are faster than that if the individual mineral C_3S at all stages especially at later ages.
- In the presence of gypsum, $C_{12}A_7$ can remarkably accelerate the hydration of C_3S.
- The hydration kinetic equation of C_3S in the C_3S-H_2O system is L = 0.2002 lg t + 0.1926, in the C_3S-$C_{12}A_7$-H_2O system is L = 0.1380 lg t + 0.6343 and in C_3S-$C_{12}A_7$-$CaSO_4.2H_2O$-H_2O system is L = 0.3174 lg t + 0.5231.
- The mechanism of $C_{12}A_7$ promoting the hydration of C_3S in the presence of gypsum is that chemical reactions quicken the hydration of C_3S.

REFERENCES

1. YAN S, ZHANG D C,. C_3S-$C_{12}A_7$-$CaSO_4 \cdot 2H_2O$-H_2O System, Nanjing Institute of Chemical Tech nology, Nanjing, China, 1984
2. BUDNIKOV P. P, et al, The Hydration of Tricalcium Silicate and β-Dicalcium Silicate, RILEM International Conference on Accelerated Hardening of Concrete, Moscow, 1964

DEVELOPING A NEW FIELD OF WASTE UTILIZATION IN CONCRETE WITH HIGH SOLID WASTE CONTENT

Y Bao

The Institute of Building Materials- Hubei Province

Y Zhang

HuaXing Cement Works

China

ABSTRACT. The treatment of solid wastes and toxic solid wastes is a very urgent and severe problem in the world today, because the quantity of solid waste is increasing year by year. This is particularly true for toxic solid wastes which come from the reduction of nuclear armaments and the treatment of unlawful nuclear materials. They must be treated in the place where they arise. Depending on the fly ash proportion, rate of carbonation, pore size distribution, density and impermeability, a series of tests on high fly ash content roller-compacted concrete showed its prospects of utilization in this field.

Keywords: Fly ash, Cement, Roller-compacted concrete, Strength, Impermeability, Porous structure distribution, Network, Toxic solid waste

Bao(Pao) Ying graduated from Nanjing Technological College and worked in various cement works. In the 1970s and 1980 she researched computer control and mathematic models before becoming a senior engineer in the Institute of Hubei Province Building Materials Research and Design. Her specialized field is now ancient history and archaeology.

Zhang Yunsheng is head of production and quality control at HuaXing Cement Works and specialized in new product development and research of cement. Prior to this he was a senior engineer and director of the Institute of Hubei Province Building Materials Research and Design.

INTRODUCTION

At present, the protection of the environment has become a severe problem which every country in the world must consider. One problem is properly treating solid waste including toxic and low radiation waste. It embarrasses many scientists and Governments. Based on China's forty years' practical experiences of construction, the authors of this paper provide a new field and a cheap method in trying to solve the above problem.

First, we must break through the traditional idea that concrete can only be used in industrial construction and not in other fields such as to purify the earth.

The authors considered blending high contents of fly ash and cement, which is then roller–compacted or pressed together with the solid waste. The solid waste is treated as artificial sand and aggregate to make into a mass concrete, which can be used as underground fillings, highway bases, and so on. It is not only a utilized base, but also a waste stack. If the bonded object is poisonous waste, the roller compacted concrete, as a bonding, stabilizing agent, can protect the soil and groundwater against pollution. This effect is due to the fact that the waste is isolated from its surroundings by the concrete. The roller–compacted concrete with high contents of fly ash has greater strength in the later age and exhibits good corrosive resistance.

PROPORTION AND TEST PROPERTIES OF HIGH FLY ASH AND CEMENT CONTENT ROLLER-COMPACTED CONCRETE

In China high fly ash and cement content roller-compacted concrete has been successfully applied in hydraulic engineering and highway bases. It can not only shorten the time of construction, but also save 15-30% of investment. The proportions of cementing materials used in one m^3 concrete were 65% fly ash, 16% slag, 19% Portland cement. The designed compressive strength of concrete at 90 days was 14.7 MPa. The quality of roller-compacted concrete has improved since 1985. See Table 1.

Table 1 Measured results of core drilled specimens

AGE (days)	COMPRESSIVE STRENGTH (MPa)	TENSILE STRENGTH (MPa)	SHEARING RESISTANCE (Mpa)	IMPERMEABILITY
2	2	/	/	/
7	6.5	0.35	/	/
28	11.5	0.99		≥S-6
90	18	1.96	1.33	≥S-6
180	23	/	1.76-1.96	/

The proportions of cementing materials used in another m^3 concrete were 57% fly ash and 43% Portland cement. The quality of roller-compacted concrete in engineering has improved since 1986. For test properties of roller-compacted concrete, see Table 2.

Table 2 Proportion and properties test of roller-compacted concrete

PROPORTION (kg/m^3)					PLASTICIZER	STIFFNESS
Cement	Fly ash	Water	Sand	Aggregate		
60	80	98	789	1370	0.2%	13 sec

COMPRESSIVE STRENGTH (MPa)						TENSILE STRENGTH (mpa)				
3d	7d	28d	90d	180d	365d	7d	28d	90d	180d	365d
4.1	6.4	9.8	20.5	23.5	24.4	0.33	0.67	1.93	2.58	2.77

ULTIMATE ELONGATION $\times 10^{-4}$m/m		STATIC MODULUS OF ELASTICITY (MPa)		ABSOLUTE TEMP RISE (℃)	
28d	90d	28d	90d	28d	final
0.66	0.98	14170	22653	13.2	14.2

IMPERMEABILITY 90d	DENSITY (kg/m^3)
$\geqslant$S-12=0.129×10^{-8}cm/s	2324

The sand content is 36.8%

Carbonation Test

Fly ash was substituted for cement at: 0, 10, 20, 30 and 40 %. Two specimens per sample of dimensions 10 × 10 × 10 cm^3were prepared with equal quantities of aggregate and sand. After curing for 28 days, one specimen was placed continuously in the atomizer room, the other in the air. The test curing in the atomizer room showed that the strength of specimens with or without additional fly ash grew continuously in the course of 15 years, but the speed of growth of fly ash concrete was larger than that of concrete without PFA added. In air the increase of strength on adding fly ash was smaller. All specimens cured in the atomizer room carbonated only at the surface layer after 20 years. Therefore we can infer that at water saturated condition and in the inner part of mass concrete, the fly ash contents are not influenced by carbonation.

Contrarily, for the specimens placed in air, the carbonation of concrete is much influenced by the fly ash contents, such as after 20 years the depth of carbonation of 20% fly ash concrete is 25 mm, while with 30% fly ash this increases to 33 mm; the more fly ash added, the deeper the carbonation. Therefore fly ash concrete must be used in underground engineering.

THE REACTION MECHANISM OF HIGH FLY ASH AND CEMENT CONTENT ROLLER-COMPACTED CONCRETE

Reducing Water and Enhancing Plasticity Effects

Fly ash consists of glassy particles of various dimensions and forms. The closed spherical particles in roller-compacted concrete act as a lubricant and look like a ball bearing, therefore it is referred to as solid plasticizer reducing the water demand by 10% (maximum) for a consistency of mortar is similar to that of no fly ash concrete. In China the proportion of coarse fly ash particles is generally 2-6% (the residue on a 4900 aperture/cm^2 sieve). The open or thin wall spherical particles are broken through roller compaction. They attract and combine each with other so that the tensile and compressive strengths are increased. This is a special property of fly ash rather than of any other pozzolanic material. Due to this reaction the density of roller-compacted fly ash concrete approaches 97% of the theoretic density.

Improvement of Pore Structure

The high fly ash content concrete possesses better pore structure than that of no fly ash concrete. If using fly ash instead of sand, due to its fine spherical particles, it is easier to completely fill the pores of concrete under vibration, therefore fly ash can improve the workability and denseness of concrete, resulting in advantageous changes in the pore structure. The proportions and measured properties of E1, E2, and E3 specimens after curing for 120 days are given in Table 3.

Table 3 The proportions and strength of concrete

No	F/C+F%	W/C %	Sand Ratio %	Plasticizer %	Comp. Strength MPa 7d	28d	120d	180d
E1	0	0.75	37	0.2	17.9	24.8	26.0	28.6
E2	50	0.75	33	0.2	9.1	13.7	23.5	29.2
E3	54	0.75	32	0.2	10.6	18.3	28.2	32.9

Table 4 The pore structure distribution of concrete

No	10000-55000Å (%)	1000-10000Å (%)	500-1000Å (%)	250-500Å (%)	100-250Å (%)	50-100Å (%)	25-50Å (%)
E1	4.03	29.40	20.81	18.12	15.43	9.39	2.82
E2	3.05	13.41	21.95	21.83	24.39	9.39	5.98
E3	1.28	9.33	21.18	22.51	26.75	11.55	7.40

From Table 4, we can see the following: In E1 the large pores >1000 Å represent 33.43% which is higher than that of E2 (16.46%) and E3 (10.61%). E1 is not advantageous for impermeable concrete. The pores with diameters <250 Å entail good impermeability, to concrete. Since the samples with <250 Å pores amount to 27.64% in E1, 39.76% in E2, and 45.70% in E3, the order is E3 > E2 > E1 regarding impermeability. Generally, as fly ash is added in concrete, it makes the pore structure distribution of concrete more rational to increase impermeability.

Dense, Active Network

The fine particles (<30μm) in fly ash separate the cement particles and completely fill the space between them. They become nuclei for crystal grains and then form an active connected network. During the initial period, the hydration products formed on the surface of cement particles precipitate and crystallize on the fly ash particles, then the new surface of cement particle is exposed so that the hydration proceeds continuously. This improves the latent energy of hydration and partly offsets the decreased cement content. Observing the specimens with SEM indicates that the above process needs about 7-14 days. The fly ash acts as micron scale aggregate at this period when the networks are weakly adhered by a gelatinous material and there are large pores contained, therefore the strength of concrete is low. The pozzolanicity of fly ash is seen after a great deal of C-S-H gel and $Ca(OH)_2$ is produced during the cement hydration. The $Ca(OH)_2$ corrodes the surface of fly ash particles and then stimulates the reaction among the $Ca(OH)_2$, SiO_2, and Al_2O_3 trapped in glass to form secondary C-S-H gels on the fly ash surface. After 14 days the specimens were observed with an X-ray diffractometer. Ettringite peaks were found in the pattern. Owing to the uninterrupted increase and crystallization of C-S-H gels we can observe the progressive growth of crystal size by a microscope with reflected light. At the corrosive surface of fly ash particles present, many fibre crystals, distributing into the space around these fibre crystals and the hydration crystals of cement in the near space connect with each other and overlap. All of them completely fill the pore space of networks and then the pore structure is changed from large to middle, middle to small. The variation of pore structure and strength of chemical solidification increases the strength and impermeability of fly ash concrete, that is higher than that with no fly ash or little fly ash added, especially the impermeable effect of high fly ash content roller-compacted concrete is better. This explains that in the later period the fly ash concrete activity supersedes that of ordinary concrete.

PROSPECTS FOR HIGH FLY ASH CONTENT CONCRETE

Based on the above research results, we called the mix with high fly ash and cement content a fly ash concrete adhesive-stabilizer; it can develop a new field of utilization of concrete for stabilizing solid wastes in the future. Many developing countries meet the same problem that the wastes pollute the soil and ground water. F.C. adhesive-stabilizer is submitted as a theoretical and practical example to solve the above question. Under wet conditions or in the inner part of a large body, the F.C. has an important role in stimulation activity each other for a long time to improve properties of concrete such as drying shrinkage, crack resistance, impermeability and low heat of hydration, other than an ordinary base engineering, it especially appropriates to mass volume concrete under ground and in harbour engineering.

The solid waste is treated as artificial sand and aggregate and then added to F.C. to make into a mass concrete or lump that is roller-compacted. This mass concrete can be used as fillings under ground, or in the seashore, highway base and soft base for marshland. It is not merely waste storage, but also a utilized base. If the bonded object is poisonous waste, the concrete admixed with F.C. can protect soil and ground water, because of its good impermeability. The authors consider that if the nuclear waste stockpile is not established in under ground concentrically in USA. They may try to use the F.C. to bond the nuclear waste and then roller-compact concrete in appropriate wasteland which should be coated by this concrete protective layer against radioactivity. After the toxicity of waste releases, the waste stockpile may be used again. The EC decided recently that poisonous wastes must be treated where it is produced. They may try to use the F.C. to treat the poisonous waste.

REFERENCES

1. LEA. F. M, The Chemistry of Cement and Concrete, Chemical Publication Co., New York, 1971.

2. LIU CONGXI, Some Theoretic Problems of Alkali-Aggregate Reaction. Changjiang Academy of Sciences Connects Reports, Wuhan, 1987.

3. BAO YING AND ZHANG YUNSHENG, The Research Reports of Mid-Heat and Grouted Cements. Hubei Research and Design Institute of Building Materials, 1988-1992.

4. ZHANG YUNSHENG, The Tests Reports of GB-175, GB-177 Revision. Changjiang Academy of Science and Huaxing Cement Works in Wuhan, Huangshi, 1978.

Notes: GB is the National Standard of China.

LABORATORY AND FIELD STUDIES ON PERFORMANCE OF PCC DRY BOTTOM ASH AND FLY ASH CONCRETE IN CAST-IN-PLACE DRILLED SHAFTS

N Ghafoori

Tennessee Technological University

A Hosin

Southern Illinois University

United States of America

ABSTRACT. The primary goal of this research investigation was to ascertain the performance of Pulverized Coal Combustion (PCC) dry bottom ash and fly ash concretes prepared by replacing a portion of Portland cement with fly ash and by partial/full substitution of natural fine aggregate with dry bottom ash when used in cast-in-place drilled shafts. Two PCC dry bottom ash and fly ash matrices and a control concrete were prepared in the laboratory and field at a constant slump of 19 mm. Laboratory-made sample were examined for fresh, hardened, and long-term durability characteristics. Field cores, tested for strength and deformation properties, were used to assess the influence of field curing conditions as compared to the concrete specimens cured under controlled conditions of the laboratory. Tests on the samples prepared and cured in the laboratory revealed that strength, stiffness, and durability characteristics of the dry bottom ash and fly ash concretes were similar to those of the reference mixture. The results obtained from the samples cored from the field drilled shafts showed that the properties of the selected concretes can be easily duplicated in field. For all practical purposes, the hardened characteristics of the field concretes were identical to those of the specimens prepared in the laboratory.

Keywords: Bottom ash, Fly ash, Coal, Combustion, Field studies

Nader Ghafoori holds a Ph.D from the University of Miami, Coral Gables, Florida, and is currently Professor and Chairperson of the Department of Civil and Environmental Engineering at Tennessee Technological University. His special interests are in the area of durability, strength, and behavior of concrete systems. He is member of various technical committees of ACI, ASCE, and TMS.

Alyass Hosin completed his MS in civil engineering at Southern Illinois University at Carbondale where he received his BS degree in civil engineering. His research interests are in concrete materials and structures.

INTRODUCTION

As the world population grows, so do the amount and type of by-products being generated by various industries. A significant portion of the waste produced today will remain in the environment for hundreds, perhaps thousands, of years. By-products originating from burning of coal constitute a significant amount of the total residues produced in the world. The creation of non-decaying coal solid by-products, combined with a growing consumer population, is on the verge of becoming a disposal crisis. One solution to this crisis lies in recycling of coal combustion residues into useful construction-related products [1].

In the United States, nearly 80% of the coal produced is used for electric power generation and about 15% of this amount is recovered as coal combustion by-products [2]. The most common coal combustion by-products recovered from burning of coal are fly ash and bottom ash. An estimated 16.20 million metric tons of fly ash and bottom ash is produced annually in the United States, and 123 million metric tons is already in stockpiles [5]. Typically, most of these ashes are disposed of by hauling to landfills. Because of the increasing costs associated with coal combustion ash disposal, limited available space, and the environmental regulations in place; the federal, state and local agencies, as well as the private sector have been taking an active role in developing alternative methods to make profitable and environmentally safe uses of the already abundant by-products [3].

Coal fly ash has long been recognized as a construction material and is in use since the 1940's in cement and concrete products, structural fills and embankments, and road bases/subbases [4]. However, to date, the use of bottom ash in construction related activities has been relatively limited. Several projects have progressed over the last few years for large volume use of fly ash and bottom ash to make value-added marketable products, e.g., ceramic tile, fiber-reinforced cement composites, bricks, and other building materials [7]. Use of fly ash and bottom ash in construction of deep foundations so far has been very limited, if any, mainly due to the lack of technical data to convince the engineering community that fly and bottom ashes could be used in deep foundations without jeopardizing long-term performance and structural integrity of foundations to resist the anticipated loads. Therefore, the proposed project was intended to develop viable mixtures utilizing pulverized coal combustion fly ash and dry bottom ash, and demonstrate their suitability for construction of cast-in-place drilled shafts.

To achieve the intended objective of this study, the execution of the project was divided into two distinct phases. Phase I of the investigation consisted of laboratory tests on three different mixtures using Portland cement, natural coarse and fine aggregates, and pulverized coal combustion fly ash and dry bottom ash to identify suitable mixtures that have potential of performing satisfactorily in actual field loading conditions. Dry bottom ash was used as a partial or full replacement of natural fine aggregate, and fly ash was used to partially replace a portion of Portland cement. A number of laboratory tests were performed to determine stiffness, strength, and durability characteristics of the trial matrices. The purpose of Phase II of the investigation was to determine the effect of the actual environmental conditions (curing) on the strength and stiffness of the selected Portland cement concretes. In this phase of the investigation, several short drilled shafts, 305mm in diameter and 1.5m deep, were constructed in the field and allowed to cure under field conditions for 90 days. Cylindrical and beam-shaped samples were then cored from these shafts and tested in the laboratory to determine the effect of curing under actual field conditions on strength, stiffness, and durability characteristics of the selected PCC matrices.

RAW MATERIALS AND MATRIX PROPORTIONS

The raw materials used in the investigation were: Portland cement Type I and Pulverized Coal Combustion (PCC) Class "C" fly ash as primary and secondary binders, respectively; PCC Class "F" dry bottom ash and natural sand as fine aggregates; crushed limestone as a coarse aggregate; and tap water.

ASTM C150 Type I Portland cement was used throughout the investigation. Its chemical properties are shown in Table 1. The PCC Class "C" fly ash, used as a secondary cementitious binder, had the chemical properties as presented in Table 1.

The chemical properties of the PCC dry bottom ash are also given in Table 1. The fineness modulus of the dry bottom ash was measured at 2.66. The PCC dry bottom ash had a porous internal structure resulting in an absorption value of 10%.

Table 1 Chemical Properties of PCC fly Ash and Dry Bottom Ash and Portland cement

CHEMICAL COMPOSITIONS	PORTLAND CEMENT (%)	PCC FLY ASH (%)	PCC DRY BOTTOM ASH (%)	STANDARD LIMITS FOR FLY ASH (ASTM C 618)
Silicon Dioxide (SiO_2)	21.88	31.73	46.84	---
Aluminum Oxide (Al_2O_3)	4.37	16.72	14.36	---
Iron Oxide (Fe_2O_3)	2.84	5.29	19.62	---
Total ($SiO_2 + Al_2O_3 + Fe_2O_3$)	29.09	53.74	79.85	50% Min. for Class C 70% Min. for Class F
Sulfur Trioxide (SO_3)	2.7	1.82	0.33	5.0% Max
Calcium Oxide (CaO)	62.62	30.39	7.24	>10% for Class C $\leq$10% for Class F
Magnesium Oxide (MgO)	4.00	5.47	1.04	---
Loss on Ignition	1.2	0.52	4.21	6% Max
Free Moisture	N/A	0.13	N/A	3% Max
Available Alkalies as Na_2O	0.49	1.40	1.73	1.5% Max
Potassium Oxide (K_2O)	N/A	N/A	1.20	---
Others (P_2O_5, TiO_2, and etc.)	8.64	4.64	2.10	---

The oven-dry and saturated surface dry specific gravities were 2.4 and 2.52, respectively. The silicious natural fine aggregate had an absorption value of 0.50%. Its fineness modulus was 2.8 and it had the oven-dry and saturated specific gravities of 2.65 and 2.66, respectively.

The crushed limestone coarse aggregate used in the experimental program had a nominal maximum size of 19 mm. The oven dry and saturated surface dry specific gravities were 2.64 and 2.66, respectively. The oven dry unit weight and void ratio were measured at 1508.9 kg/m^3 and 43.75%, respectively.

For the purpose of the investigation, two different Pulverized Coal Combustion dry bottom ash and fly ash concretes and an equivalent control mixture were prepared at a constant slump. The selected matrices had the following characteristics: (1) 100% by-volume replacement of natural fine aggregate with PCC dry bottom ash and 10% by-mass replacement of Portland cement with PCC fly ash (MA mixture), (2) 50% by-volume replacement of natural fine aggregate with PCC dry bottom ash and 20% by-mass replacement of Portland cement with PCC fly ash (MA mixture), and (3) control concrete containing no PCC dry bottom ash and fly ash (CM mixture). Table 2 details the constituents and quantities of the trial matrices.

Table 2 Constituents and Proportions of Trial Matrices

MIXTURE DESIGNATION	CEMENTITIOUS BINDERS (kg/m^3)		FINE AGGREGATES (kg/m^3)		COARSE AGGREGATES (kg/m^3)	WATER (kg/m^3)
	Portland Cement	PCC Fly Ash	PCC Dry Bottom Ash	Natural Sand		
CM	335	0	0	842	1044	177
ME	268	67	400	440	1041	198
MA	302	33	806	0	1043	232

PREPARATION AND TESTINGS

Laboratory mixtures were prepared using a counter-current electric mixer. Cylindrical and beam-shaped specimens were cured in a lime-saturated curing tank prior to the actual testings. The field concretes were made at a ready-mixed concrete plant, transported to the site by trucks, and placed and consolidated in the pre-drilled shafts using an external vibrator. After 90 days of curing under field conditions, cylindrical and beam-shaped specimens were cored from the drilled shaft and examined for various medium and long-term properties. The trial concretes were tested for workability, bleeding (ASTM C 232), time of setting (ASTM C 403), compression (ASTM C 39), splitting-tension (ASTM C 496), flexure (ASTM C 78), elastic modules and Poisson's ratio (ASTM C 469), and sealed shrinkage and swelling (ASTM C 157). The test specimens used to evaluate resistance to freezing and thawing were submerged in water and exposed to 50 cycles. Each freezing and thawing cycle consisted of 24 hours of freezing at -18°C followed by 24 hours of thawing in the room temperature 25±2 °C. Mass loss was evaluated as a percentage of the specimen's original air-dried mass.

RESULTS AND DISCUSSION

Short-term Properties

The selected mixtures were prepared at a uniform slump of 102 ± 6 mm, resulting in different water-to-cementitious materials ratios as shown in Table 3. The results of bleeding tests of the selected matrices are also documented in Table 3. The reference concrete produced the highest bleeding percentage which may be attributed to its higher water-to-cementitious materials ratio and less mount of available fines when compared to those of the PCC dry bottom ash and fly ash mixtures. The initial and final setting times of the trial matrices were also studied. As documented in Table 3, replacing a portion of Portland cement with PCC fly ash and partial or full substitution of natural fine with PCC dry bottom ash had no adverse influence on the setting times of the selected concretes.

Table 3 Fresh Characteristics of trial Matrices

MIXTURE DESIGNATION	SLUMP (mm)	WATER-TO-CEMENTITIOUS MATERIALS RATIO	BLEEDING (%)	SETTING TIME	
				Initial (min.)	Final (min.)
CM	102 ± 6	0.5	3.65	315	475
ME	102 ± 6	0.455	3.20	345	476
MA	102 ± 6	0.45	3.51	295	430

Medium-Term Properties

Compressive strength of the selected mixtures was measured at various curing ages and the results are shown in Table 4. After 7 days of curing, strength gain in the reference mixture was 82% of its 28-day strength as compared to 78 and 70% for the mixture ME and MA, respectively. The rate of strength gain beyond 28 days of curing was observed to be significantly higher in the PCC dry bottom ash and fly ash concretes (51% after 180 day curing) as compared to that displayed by the reference mixture (24% after curing age of 180 days). At the end of 180 days curing, the compressive strength of the PCC dry bottom ash and fly ash concretes surpassed that of the control matrix by an average of 7%. Late-age strength gain exhibited by the dry bottom ash and fly ash concretes may be attributed to the continued hydration of the Portland cement and fly ash, to a lesser extent, and the pozzolanic reactions provided by the PCC dry bottom ash to a larger extent.

Table 4 also presents the results of splitting-tension tests measured at various curing ages. Similar to the behavior observed for compressive strength, the splitting-tensile resistance of the trial composites increased with increases in curing age and additional gain in strength was higher in the PCC dry bottom ash and fly ash concretes when compared to that of the control mixture. At the end of 180 days curing, the concretes containing PCC dry bottom ash and fly ash had nearly 3.5% higher splitting-tensile strength than the equivalent reference mixture. The ratio of splitting-tensile resistance to compressive strength, measured at different curing ages, remained in the range that is typical for the cylinder compressive strengths between 7 to 62 MPa [6].

Also shown in Table 4 are the results of modulus of rupture tests conducted on the beam-shaped specimens for each matrix at different curing ages. The PCC dry bottom ash and fly ash concretes exhibited a slightly higher amount of flexural strength (3.5%) than that of the control matrix. The flexural to compressive strength ratios were observed to be 0.16 to 0.18 which are in the range typically seen in conventional concrete [6].

The unit weight of the trial composites were 2499.44, 2457.79, and 2394.83 kg/m^3 for the mixtures CM, ME, and MA, respectively. In line with the results of the unit weight, the PCC dry bottom ash and fly ash concretes produced elastic modulus slightly below that of the reference mixture. The 28-day modulus of elasticity of the selected mixtures are documented in Table 4. The 28-day modulus of elasticity of the concretes ME and MA was observed to be 6 and 14% lower than that of the control mixture, respectively. However, after 90 days of curing (see Table 7), the modulus of elasticity of the same mixtures was only 1 and 4% lower than of that obtained for the control concrete.

Table 4 Strength and Stiffness Properties of Trial Matrices

MIXTURE DESIGN	COMPRESSIVE (MPa)			SPLITTING-TENSILE (MPa)			FLEXURAL STRENGTH (MPa)			MOE (GPa)
	Curing Age (Days)			Curing Age (Days)			Curing Age (Days)			Curing Age (Days)
	7	28	180	7	28	180	7	28	180	28
CM	30.03	36.85	45.49	3.40	3.59	5.05	5.06	6.42	7.74	32.41
ME	26.60	33.91	49.33	3.08	3.62	5.34	4.76	6.14	7.98	30.41
MA	21.82	31.00	47.98	2.77	3.28	5.11	4.74	6.25	8.00	27.86

Long-term Properties

Table 5 presents sealed shrinkage and swelling strains at different concrete ages. The ultimate sealed shrinkage strain for the PCC dry bottom ash and fly ash concretes was slightly higher than that of the reference matrix. This finding indicates that the higher availability of cementitious materials in the concrete containing PCC dry bottom ash and fly ash creates a favourable environment to increase sealed shrinkage strains.

The ultimate swelling strains, as shown in Table 5, were also higher for the mixtures made with the PCC dry bottom ash and fly ash. The increase in the volume due to addition of PCC fly ash and dry bottom ash may be due to (1) slightly lower paste quality and (2) slightly higher permeability for the PCC dry bottom ash and fly ash concretes as evident by the absorption values shown in Table 6.

Table 5 Sealed Shrinkage and Swelling of Trial Matrices

MIXTURE DESIGNATION	SEALED SHRINKAGE STRAIN ($x10^{-6}$) Curing Age (Days)				SWELLING STRAIN ($x10^{-6}$) Curing Age (Days)			
	30	90	180	360	30	90	180	360
CM	-350	-600	-600	-600	100	100	100	100
ME	-400	-500	-570	-570	100	200	200	200
MA	-300	-550	-650	-650	50	100	150	150

The results of resistance to freezing and thawing in terms of cumulative mass loss, up to 50 cycles, are given in Table 6. While the matrix ME produced slightly higher resistance to freezing and thawing than that of the control concrete, the mixture MA exhibited a higher amount of mass loss. The decrease in resistance to freezing and thawing observed for the mixture MA may be attributed to the higher void structure and permeability as evident by the absorption results.

Table 6 Resistance to Freezing and Thawing of Trial Matrices

MIXTURE DESIGNATION	7 DAYS ABSORPTION %	CUMULATIVE MASS LOSS OF LAB.-MADE SAMPLES % Number of Freezing and thawing Cycles					CUMULATIVE MASS LOSS OF FIELD CORES % Number of Freezing and thawing Cycles				
		10	20	30	40	50	10	20	30	40	50
CM	3.8	0.00	0.01	0.02	0.06	0.18	0.00	0.02	0.29	0.48	0.77
ME	4.5	0.00	0.01	0.05	0.10	0.17	0.00	0.01	0.25	0.63	0.87
MA	5.6	0.00	0.01	0.07	0.53	0.65	0.00	0.02	0.35	1.43	2.62

Hardened Characteristics of Field Cores

The cores from drilled shafts were tested at 90 days of age for compressive strength, splitting-tensile resistance, flexural strength, modulus of elasticity and Poisson's ratio. The results are documented in Table 7.

The 90-day compressive strength of the field cores was found to be approximately 8% less than that of the equivalent laboratory-made specimens. The cores tested to obtain splitting-tensile resistance produced results that were 3 to 7% below those of the equivalent samples prepared and tested in laboratory. The flexural strength of the field cores of the matrices CM, ME, and MA after 90 days of field curing was 15, 13, and 7%, respectively, higher than that of the laboratory-made concretes. The results obtained for the modulus of elasticity and Poisson's ratio of the field cores were nearly the same as those obtained for the equivalent matrices prepared in the laboratory. Based on the above-mentioned results, it was concluded that the influence of different sample preparation and curing conditions (laboratory versus field) on the hardened properties of the trial concretes was rather insignificant and the laboratory results could be easily duplicated in field. Moreover, the moisture in the soil around drilled shafts was sufficient to prevent loss of moisture from the concrete placed in the shafts and helped to achieve a degree of hydration similar to that attained in the samples cured under controlled conditions of the laboratory.

Table 7 90-Day Hardened properties of Field Cores

MIXTURE DESIGNATION	COMPRESSIVE STRENGTH (MPa)		SPLITTING-TENSILE(MPa)		FLEXURAL STRENGTH(MPa)		ELASTIC MODULUS (GPa)		POISSON'S RATIO	
	Field	Field/ Lab	Field	Field/ Lab	Field	Field/ Lab	Field	Field/ Lab	Field	Field/ Lab
CM	45.98	0.93	4.78	0.97	6.28	1.15	34.49	0.97	0.143	0.89
ME	45.52	0.92	4.68	0.93	6.75	1.13	32.94	0.92	0.159	1.04
MA	38.63	0.92	4.47	0.96	6.03	1.07	29.04	0.86	0.168	0.98

CONCLUSIONS

Based on the results shown, the following conclusions are drawn:

1. Due to spherical particles of fly ash and bottom ash, the water-to-cementitious materials ratio to achieve a constant slump was higher for the reference concrete when compared to that exhibited by the PCC dry bottom ash and fly ash matrices.

2. The combined effect of higher water-to-cementitious materials ratio and less amount of fines available in the control concrete produced a higher bleeding value for the reference mixture than that shown by either of the PCC dry bottom ash and fly ash concretes.

3. Inclusion of the PCC dry bottom ash and fly ash had insignificant effect on the time of setting of the trial concretes.

4. The strength and elastic modulus increased with an increase in curing age. The rate of strength gain beyond 28 days of curing was observed to be higher in the PCC dry bottom ash and fly ash concretes when compared to that exhibited by the reference matrix. After 90 days of curing, the selected PCC dry bottom ash and fly ash mixtures offered higher strength characteristics.

5. Both PCC dry bottom ash and fly ash concretes exhibited higher volume instability than the reference concrete.

6. Replacement of a portion of Portland cement with PCC fly ash and inclusion of PCC dry bottom ash as a fine aggregate had an insignificant impact on the resistance to freezing and thawing of the trial concretes.

7. Tests conducted on the field cores revealed that the influence of sample preparation, and curing conditions (field versus laboratory) is insignificant, and the laboratory results can be easily reproduced in field.

ACKNOWLEDGEMENTS

This study was funded by the office of Coal Development and Marketing of the Illinois Department of Commerce and Community Affairs under Project number 99-1/2. 1B-6.

CONVERSION FACTORS

1 in. = 25.4 mm

1 lb (force) = 4.4482 N

1 psi = 6.895 kPa

1 kip = 4448.2 N

1 ksi = 6.895 MPa

1 ksi = 70.31 kgf/cm^2

1 ft^3 = 0.02832 m^3

1 lb/yd^3 = 0.594 kg/m^3

1 lb/ft^3 = 16.02 kg/m^3

1 oz/yd^3 = 22.61228 ml/m^3

F = 9/5 (C) + 32

REFERENCES

1. American Coal Ash Associates (ACAA). Coal Combustion Byproduct (CCB) Production & Use: 1966-1994. American Coal Association, Alexandria, VA, May 1966, 59 pp.

2. Energy Information Administration (EIA). Cost and Quality of Fuels for Electric Utility Plants. Office of Coal, Electric and Alternate Fuels, U.S. Department of Energy, Washington, D.C. August 1992, p. 27.

3. GHAFOORI, N., AND CHANG, W.F. Roller Compacted Concrete Utilizing Dehydrate Phophogypsum. Journal of Transportation Research Record, Vol. 1, No. 1301, 1990, pp. 139-148.

4. HELMUTH, R. Fly Ash in Cement and Concrete. Portland Cement Association, 1987, pp. 203-205.

5. MANZ, O.E. Worldwide Production and Utilization of Coal Ash in Concrete and Other Products. Proceedings of the 13th ACAA International Symposium on Use and Management of Coal Combustion Products (CCPs). Volume 67, 1997, no. 689, pp. 691-697.

6. NEVILLE, A.M. Properties of Concrete. John Wiley & Sons, Inc., New York, Fourth and Final Edition, 1996, p. 779.

7. SAYLAK, DONALD, ET AL. Applications for FGD Byproduct Gypsum. Symposium Proceedings Recovery and Effective Reuse of Discarded and Byproducts for Construction of Highway Facilities, 1993, pp. 63-71.

COMPOSITIONS & PROPERTIES OF SECONDARY MINERAL RESOURCES FOR DEVELOPING HEAT-INSULATING CEMENTLESS CONCRETE

A G Tuleyev
Governor of Kemerovo Region

N L Dobretsov
SBRAS

S I Pavlenko **K V Eryomkin**
Siberian State University of Industry
Russia

K Freidin
Ben-Gurion University
Israel

ABSTRACT The materials used for the investigation were fly ash and granulated slag from the Yurga thermal power plant, used molding sand from the Yurga machine building plant and the high alumina product (HAP) from the Yurga abrasion works. The bulk density, the absolute density and the surface area of the fly ash were 1150 to 1200, 2800 to 2850 kg/m^3 and 300 to 310 m^2/kg, respectively. According to its chemical analysis (31.73% free CaO, 1.87% SO_3, 1.06% LOI), the ash was a high-calcium type. It had good binding properties with the coefficient of quality more than 1. (C_q=CaO+MgO+Al_2O_3/SiO_2) Slag sand of 0.14 to 5 mm particle size from its chemical analysis met the State Standard requirements for use as an aggregate. The burnt sand incorporated 92% quartz sand, 6% bentonite clay and 2% water glass. After being used in foundry and removing of gritty scale, it had the form of black sand with the fineness modulus, bulk density and absolute density being 1.91, 1450 and 2380 kg/m$_3$, respectively. The burnt sand had a high content of SiO_2 in the amorphous state. No loss on ignition was observed, which increased the feasibility of chemically binding the free CaO into calcium silicates. The high alumina product was a powder containing above 88% Al_2O_3. The radionuclides contents of all the materials used was 3 to 5 times lower than allowed by the standards (370 Bq/kg). By grinding the three components (fly ash, burnt sand and HAP) in the planetary mill, a new cementless binder was obtained. In the process of grinding, the mechanochemical activation of the components as well as the interaction of the amorphous silica and HAP with free CaO took place. The study of the secondary resources permits the developing on their basis of a high-performance heat-insulating cementless concrete using the air-entraining admixture and the slag sand as an aggregate.

Keywords: Minerals, Cementless concrete, Heat insulating.

A.G Tuleyev is a Professor and Governor of Kemerovo Region in Russia.

S I Pavlenko is a Professor and Head of the Department of Civil Engineering at the Siberian State University of Industry, Russia. **K V Eryomkin** is a Postgraduate Student at the Siberian State University of Industry, Russia

N L Dobretsov is an academic and President of SBRAS, Russia

K Freidin is a Professor at the Ben-Gurion University, Israel

INTRODUCTION

To develop a new cementless binder [1] and heat–insulating cementless concrete on its basis, the Siberian State University of Industry (SSUI) in association with the Institute of Geology, Geophysics and Mineralogy, the Siberian Branch of the Russian Academy of Sciences (SBRAS), have carried out the study of the following materials: high-calcium fly ash and granulated slag from the Yurga thermal power plant, used molding sand ("burnt sand") from the foundry department of the Yurga machine building plant and high alumina product (HAP) from the Yurga abrasion works. The development of a new binder and heat-insulating concrete on its basis contributes much to protect the environment, saves energy and natural resources. The cost of these materials is lower by a factor of 1.5 to 2 compared with that for a cement concrete. The work has been performed in accordance with the "Higher School Research on Priority Trends of Science and Techniques" programme, Ministry of Education of Russia, as part of the project "Development of a new composition and energy-conservating technology for heat-insulating composite cementless concrete from secondary mineral resources" financed by the Kemerovo region Administration.

MATERIALS

The micromorphology and the qualitative composition of some phases of all the starting materials were studied using a JEOL JSM-35 electron microscope with the energy-dispersion KEVEX attachment (Figure. 1). Rock forming components were analyzed by X-ray fluorescent spectroscopy for total silicate content using a CPM-25 multichannel X-ray spectrometer controlled by the computing attachment. The phases were identified by a x-ray diffraction using a DRON-3 device (Figure. 2).

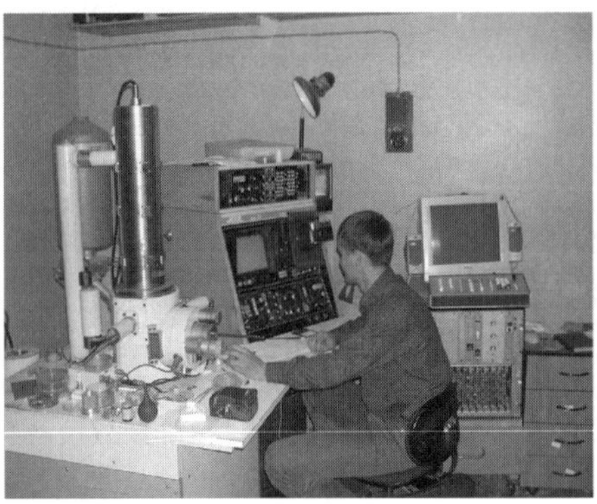

Figure 1 JSM-35 Electron Microscope

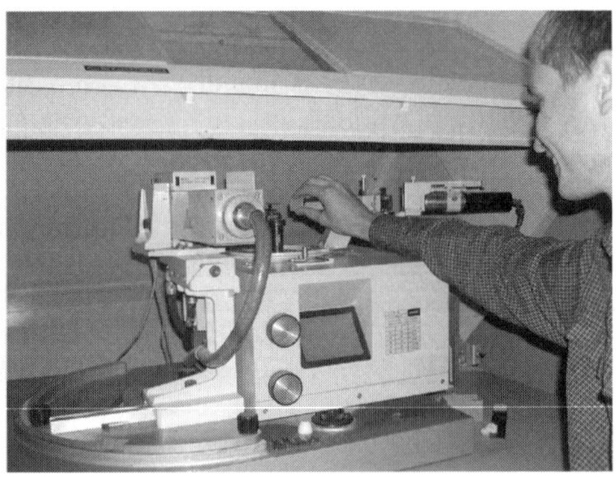

Figure 2 DRON-3 Device

The determination of toxic components contents (Zn, Pb, Cu, Cd) in a solid matter was made using a PyuE-Unicam SP-9 atomic-absorption spectrometer (Figure. 3) in air-acetylene and nitric oxide-acetylene flame, a Perkin-Eilmer 3030B device as will as a 3030Z spectrophotometer (Figure. 4) with electrothermal atomization and Zeiman background correction.

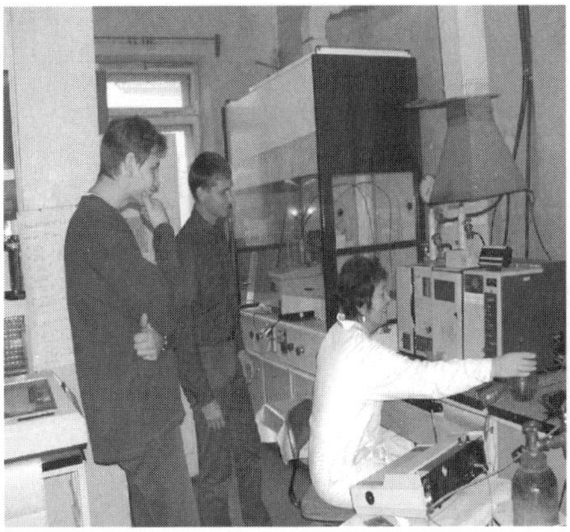

Figure. 3 SP-9 Atomic-Absorption Spectrometer

Figure 4 3030Z Spectrophotometer

Fly Ash from the Yurga Thermal Power Plant

Brown coals from the rich Kansk-Achinsk basin, KATPC [2, 3], containing little ash (up 10%) are used at the Yurga thermal power plant. The ash from these coals is of high-calcium containing 25 to 60% calcium oxide, small amounts of unburnt particles and alkaline components. These properties of the ash permit its use as a binder for low-cement concretes and to develop cementless concretes.

Physical properties and chemical analysis of fly ash and slag sand are given in Tables 1 and 2, respectively. The ash was mostly of a light grey colour due to a small amount of unburnt coal. It mainly consisted of spherical particles ranging in size from 0.5 to 20 μm with some of them having 30...40 μm in diameter.

The micrograph of the fly ash is given in Figure. 5.

Table 1 Physical Properties of Fly Ash and Slag Sand from the Yurga Thermal Power Plant

CHARACTERISTICS	VALUE	
	Fly Ash	Slag Sand
Bulk density, kg/m^3	1150–1200	1430
Grain density, kg/m^3	–	1870
Absolute density, kg/m^3	2800–2850	2350
Surface area, m^2/kg	300–310	–
Fineness modulus	–	3.1
Cylinder crushing strength, kg/m^2	–	13
Colour	light-grey	dark-green

Table 2 Chemical Analysis of Fly Ash and Slag Sand

COMPOUND	PERCENT	
	Fly Ash	Slag Sand
CaO	31.99	28.94
free CaO of total	17	–
SiO_2	25.32	35.68
Al_2O_3	7.24	7.96
Fe_2O_3 + FeO	8.85	19.82
MgO	16.56	2.16
Na_2O + K_2O	0.65	2.13
TiO_2	0.33	0.87
MnO	0.25	0.39
P_2O_5	–	0.38
SO_3	–	0.47
Loss on Ignition	8.55	0.62
TOTAL	99.74	99.42

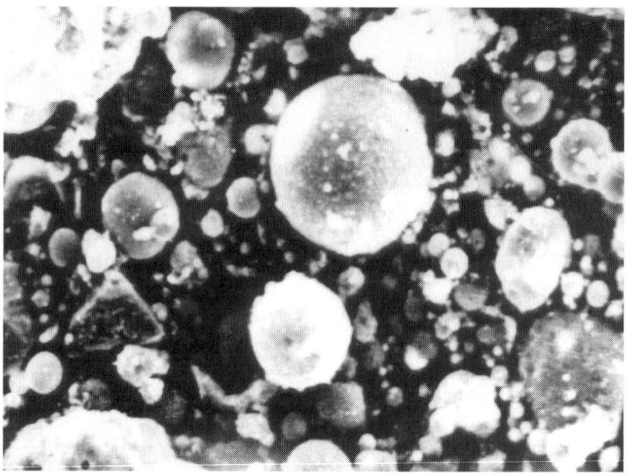

Figure. 5. Micrograph of the Ash from the Yurga Thermal Power Plant (x150)

The average specific surface area of the ash collected from all fields of electrostatic filters was 300 to 310 m^2/kg and according to the classification of Ivanov [2] it was a coarse-dispersed polymineral ash. The average values were taken to utilize 100 percent ash and not the selected fractions from different fields. The residue on the sieve No 008 was 10.3%. The ash was to be treated in combination with burnt sand and HAP. The ash had good binding properties with the coefficient of quality being more than 1: $C_q = (CaO + MgO + Al_2O_3)/SiO_2 = (31.99 + 16.56 + 7.24)/25.32 = 2.2$.

Mineralogy and Structure.

Fly ash from the Yurga thermal power plant contained 25% X-ray amorphous aluminate and silicate, 15-20% and 8-10% calcium and magnesium oxide, respectively, 8-10% quartz, 3% heamatite, 3-5% anhydrite and about 2% mullite. The contents of Portland cement clinker phases ($CaAl_2O4$, $Ca_3Al_2O_6$, $Ca_{12}Al_{14}O_{33}$) was about 15%. The X-ray diffractogram and the micrograph of the ash are given in Figures 6 and 7, respectively.

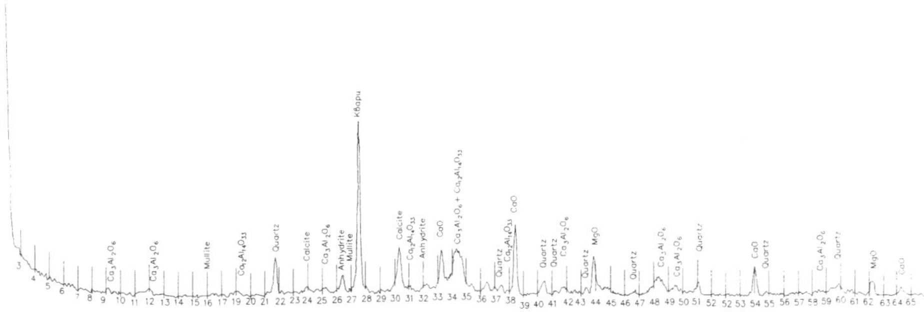

Figure 6 X-ray Diffractogram of Fly Ash from the Yurga Thermal Power Plant

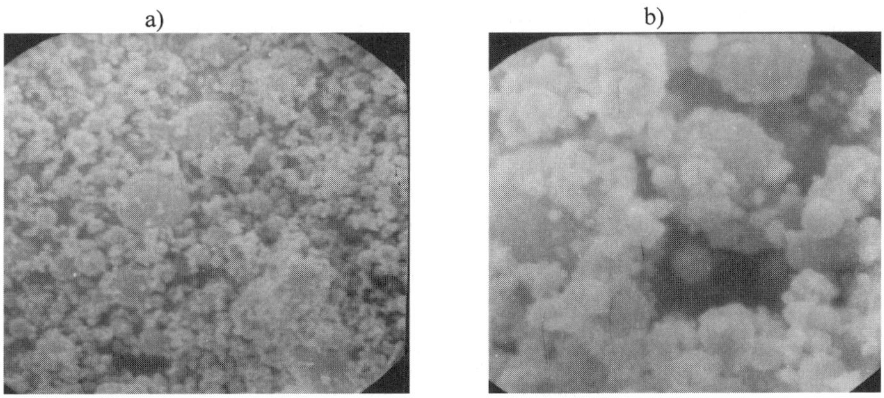

Figure 7 Electron Micrograph of the Ash from the Yurga Thermal Power Plant: a) x400, b) x2000

Granulated Slag from the Yurga Thermal Power Plant

Slag sand from the Yurga thermal power plant varied in colour from dark grey to dark green. It contained up to 96% grading fractions of 0 to 5 mm and 4% size fractions of above 5mm. It was a blend of a dense granulated slag removed in a molten state with water cooling and porous slag cooled by air. The porosity of the blend was 20 %. Figure. 8 shows the curve for slag sand screening.

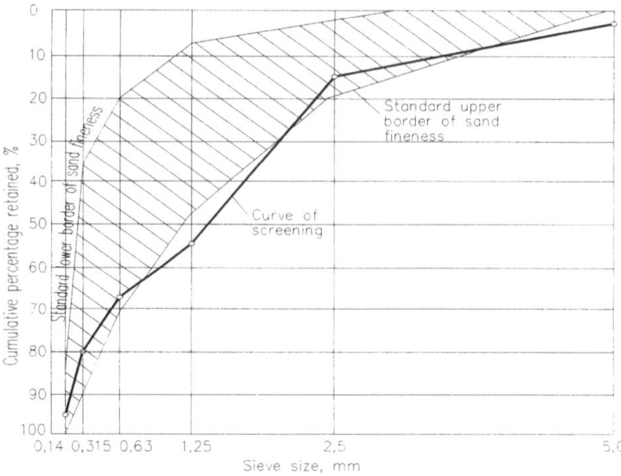

Figure 8 Curve of Screening of Slag Sand from the Yurga Thermal Power Plant Compared with a Standard Zone.

To comply with the fine concretes standard, the slag sand was ground in a roller crusher [3]. Weak grains of the slag sand were destroyed improving the concrete performance. Its physical properties and chemical analysis are given in Tables 1 and 2, respectively.

Used Molding Sand from Foundry

The molding blend for foundry incorporates 92% quartz sand, 6% bentonite clay and 2% water glass. After a primary use, the blend turns into "burnt sand". For repeated utilization, it is partly regenerated and then dumped. A ton of casting requires a ton of molding blend. So, its use for a new binder and concrete permits the solution of the ecological problem. Before its use in a new production, the blend was screened and disposed of gritty scale (mostly metallic). The fineness modulus of the burnt sand was 1.91 which corresponds to the average fineness of 2 to 2.5 mm. The burnt sand was monomineral quartz sand containing up to 2 % micaTraces of feldspars ($2\theta \approx 13\text{-}14^{o}$ and $27\text{-}28^{o}$) and clayey minerals ($2\theta \approx 3\text{-}13^{o}$), typical for natural sands were not observed (Figure 9).

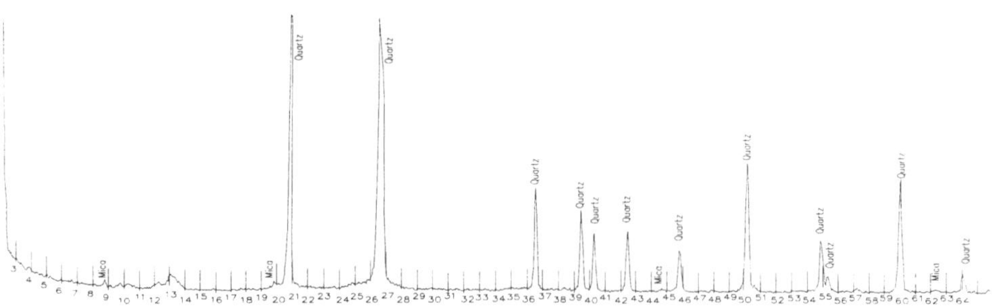

Figure 9 X-ray Diffractogram of Burnt Sand from the Yurga Machine Building Plant

It shows that during its service in foundry the burnt sand turns into an X-ray amorphous mass. The sand has a regular grain structure with the grain size of 0.5 – 0.25mm (Figure 10).

a) b)

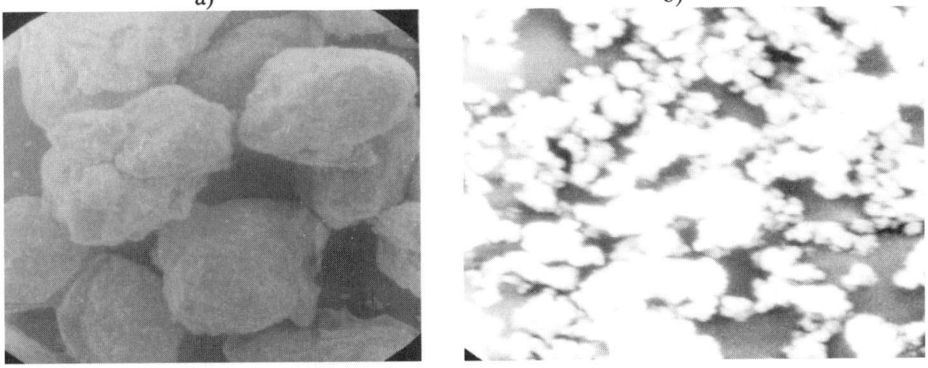

Figure 10 Electron Micrograph of the Burnt Sand from the Yurga Machine Building Plant: a) x48, b) structure after grinding in a planetary mill (x6000)

The chemical analysis of the burnt sand is given in Table 3. It had the high content (above 90%) of silicon dioxide and little amounts of aluminum oxide (4.4%) and ferric oxide (< 1). The high content of SiO^2, mostly in the amorphous state, permitted its interaction with free calcium, magnesium and aluminum oxides in the process of mechanochemical treatment in planetary mills [4].

Table 3 Chemical Analysis of Burnt Sand from the Yurga Machine Building Plant

COMPOUND	PERCENT
SiO_2	90.28
CaO	0.13
Al_2O_3	4.43
$Na_2O + K_2O$	1.11
MgO	< 0.1
Fe_2O_3 + FeO	0.96
TiO_2	0.26
MnO	< 0.03
P_2O_5	0.04
Loss on ignition	2.16
TOTAL	99.37

High Alumina Product (HAP)

The high alumina product was the waste from the abrasion works (fused electrocorundum) dumped as sludge. After drying it was a powder of grey colour. The initial material, which was not subjected to grinding, consisted mostly of isometric grains of about 1 mm in size (Figure 11).

a) b)

Figure 11 Structure and Morphology of the High Alumina Product:
a) x 130, b) x 400

The X-ray structural (Figure. 12) and electron microscopic analyses showed the presence of the following phases 40-45% $NaAl_7O_{11}$ compound, 30 to 35% corundum (Al_2O_3), 5 to 7% quartz and 5% $CaTi_2O_4$ compound.

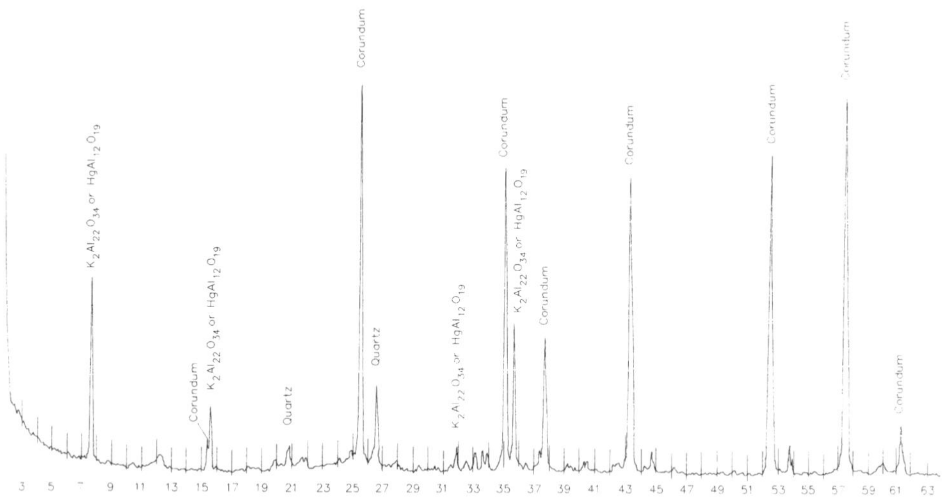

Figure 12 X-ray Diffractogram of the High Alumina Product

Besides, the HAP revealed the traces (about 5% of total) of potassic feldspar ($KAlSi_3O_8$) and kaolinite ($Al_2Si_2O_5(OH)_4$). Chemical analysis of the HAP is given in Table 4. This powder was used to enhance the strength and fire resistance characteristics of the future binder and concrete (5 to 10%).

Table 4 Chemical Analysis of the HAP

MATERIAL	OXIDES							
	Al_2O_3	SiO_2	Fe_2O_3+ FeO	TiO_2	MgO	CaO	Na_2O+ K_2O	LOI
HAP	84.0	4.51	5.17	2.36	1.64	1.14	0.34	0.12

Air-Entraining Admixture

The admixture used was the complex plasticizing air-entraining admixture developed and patented by us [5] which incorporated technical grade lignosulphonate (TGL) and secondary sodium alkyl sulfate (a detergent called "Progress"). The cost of the admixture is reduced by a factor of 2 compared with that for the aluminium powder. Providing the formation of closed pores in concrete instead of capillary pores, it eliminated the need of cutting off "the crust".

Study of the activated blend. Concrete is known to include the following compounds: 40 to 60% $3CaO{\bullet}SiO_2$, 15 to 35% $2CaO{\bullet}SiO_2$, 4 to 14% $3CaO{\bullet}Al_2O_3$, 10 to 18% $4CaO{\bullet}Al_2O_3{\bullet}Fe_2O_3$ [6]. These combinations, as well as the initial components for their synthesis, enter high-calcium ashes from the brown coals combustion, wastes of foundry and abrasion works, and therefore these by-products are used for producing cementless binders [7, 8, 9]. To obtain the new cementless binder, the blend consisting of 80% fly ash from the Yurga thermal power plant, 20% burnt sand and 5% HAP (above 100%) was ground in the AGO-3 planetary mill for 10 minutes to the optimum specific surface area of 750 m_2/kg.
The mineralogical composition and chemical analysis of the blend are given in Tables 5 and 6, respectively.

Table 5 Mineralogical Phase Composition of the Ground Blend

PHASE	PERCENT
X-ray phase	40-50
Quartz (Fe_2O_3)	10-15
Calcite ($CaCO_3$)	1-2
Anhydrite ($CaSO_4$)	2-3
Calcium oxide (CaO)	2-3
Magnesium oxide (MgO)	5-7
Corundum	5
Mullite ($Al_6Si_2O_{13}$)	not observed
$CaA_{12}O_4$	10-15
$Ca_3A_{12}O_6$	-
$Ca_{12}Al_{14}O_{33}$	-
Haematite (Fe_2O_3)	1-2
Magnetite ($FeFe_2O_4$)	not observed
Mica	–
Pyrrhoxain (augite, diopside)	–
α-iron (Fe)	–
Unidentified phase	5-7

Table 6 Chemical Analysis of the Ground Blend

COMPOUND	PERCENT
Na_2O	<0.3
MgO	11.86
Al_2O_3	9.99
SiO_2	35.18
P_2O_5	<0.03
K_2O	0.35
CaO	23.03
TiO_2	0.608
Ba	0.369
MnO	0.205
$Fe2O_3$	12.14
Loss on ignition	6.41
TOTAL	100.15

All the components were tested for the effective activity of radionuclides in accordance with the State Standard 30108-94 [10] (Table 7) and heavy metals contents (Table 8).

Table 7 Gamma-Spectrometric Analysis

MATERIAL	SPECIFIC ACTIVITY, Bq/kg			EFFECTIVE SPECIFIC ACTIVITY (A_{eff}), Bq/kg
	Th-232	Ra-226	K-40	
Fly ash	10.62	13.17	62.68	32.41
Burnt sand	13.2	12.3	123.6	40.1
HAP				40.3
Granulated slag	7.4	7.4	25.9	14.8
Ground Blend	11.43	12.9	84.5	36.26

Table 8 Heavy Metal Contents in Dumped Products, g/t.

SAMPLE	Cu	Zn	Cd	Pb	TOTAL
Fly ash	36	38	<0.5	14	88.0
Burnt sand	10	10	<0.5	2.4	22.5
HAP	19	120	0.005	0.68	139.7
Granulated slag	48	13	0.020	0.25	61.3
Ground Blend	62	34	<0.5	11	107.0

CONCLUSIONS

The investigation carried out on high-calcium ash from the Yurga thermal power station, used up molding sand from the Yurga machine building plant, high alumina product from the Yurga abrasion works and the complex air-entraining admixture proved the feasibility for the development of a new composite binder and lightweight concrete from secondary mineral resources which produced satisfactory performance and reduced the cost by a factor of 1.5 to 2 compared with that for cement concrete. The materials produced from the by-products contribute much to protect the environment.

REFERENCES

1. PATENT № 2196749. Cementless binder (authors: S.I. Pavlenko, V.I. Malyshkin, S.I. Merkulova, A.V. Aksyonov, K.V. Eryomkin, N.V. Zakharova). // M.: Bulletin, 2003, № 2.

2. IVANOV I.A. Lightweight concretes with ash from power plants. 2nd edition, revised and enlarged // Moscow: Stroyizdat, 1986, 136 p.

3. PAVLENKO S.I. Fine concretes from the industrial by-products // Moskow: Construction Higher Scholl Association publishers, 1997, 176 p.

4. ERYOMKIN K.V. The second chapter of the thesis (Advisor S.I. Pavlenko) “Study of the secondary mineral resources for developing heat-insulating concrete”. Post-graduates proceedings, Siberian State University of Industry (SSUI). Novokuznetsk, SSUI publishers, 2003, pp. 17-53.

5. PAVLENKO, S.I., Complex plasticizing air-entraining admixture in light-weight fine-grained ash slag concrete. Fourth CANMET/ACI International Conference on Superplasticizers and other Chemical Admixtures in Concrete. October 11-13, 1994. Centre Sheraton Hotel, Montreal, PQ, Canada // Edited by V.M. Malhotra, Montreal Canada, 1994, pp. 80-100.

6. CONCISE CHEMICAL ENCYCLOPEDIA. Edited by I.L. Knunyants // Moscow: Soviet encyclopedia, 1967, v. 5, p. 802.

7. PAVLENKO S.I., MALYSHKIN V.I., BAZHENOV Y.M. Fine composite cementless concrete from secondary mineral resources // Novosibirsk, SB RAS Publishing, 2000, 142 p.

8. AVVAKUMOV E.G., PAVLENKO S.I., KOSOVA N.W., LYAKHOV N.Z., MERKULOVA S.I., KUVSHINOV P.G., AKSYONOV A.V., ERYOMKIN K.V., MALYSHKIN V.I. Composite binder from mechanically activated industrial wastes // Novosibirsk: “Chemistry for stability” journal, SB RAS Publishing House, 2000, v. 8, № 5, pp. 657-660.

9. AVVAKUMOV, E.G., SENNA, M., KOSOVA, N.W., Soft Mechanochemical Synthesis a Basis for New Chemical Technologies // Kluwer Academic Publishers, Hardbound, ISBN 0-7923-7431-2, The Netherlands, 2001, 216 pp.

HEAT INSULATING CEMENTLESS CONCRETE MADE FROM SECONDARY MINERAL RESOURCES

A G Tuleyev
Kemerov Region
S I Pavlenko K V Eryomkin E G Avvakumov
Siberian State University of Industry
Russia
K Freidin
Ben Gurion University
Israel

ABSTRACT. A new composition of heat-insulating cementless concrete made from mineral by-products was developed. The concrete was developed on the basis of a new composite cementless binder incorporating the following materials: 80 % high-calcium ash, 10 to 15 % burnt sand and 5-10 % high alumina product. These three components were ground simultaneously in a planetary activator mill where the mechanochemical interaction between the amorphous silica of the burnt sand and free CaO of the ash occurred. The compressive strength and the optimum surface area of the binder were 40 to 60 MPa and 750 m^2/kg, respectively. The blend sets rapidly when mixed with water. The materials used for concrete were as follows: fly ash and furnace bottom ash sand from a thermal power plant, molding sands known as burnt sands from the foundry production and the high alumina product (HAP) from the abrasion works. The compressive strength and average density of the concrete were 0.25 to 5 MPa and 200 to 800 kg/m^3, respectively. The heat-insulating concrete was prepared in a turbulent mixer by introducing of a composite plasticizing and air-entraining admixture (technical grade lignosulfonate and the detergent called "Progress") into the mixture. The cost of the concrete is reduced by a factor of 1.5 to 2 compared with that for the conventional cement concrete.

Keywords: Insulation, Cementless concrete, Mechanical, Chemical

Dr A G Tuleyev is a professor and Governor of the Kemerovo Region, Russia

Dr S I Pavlenko is a professor and Head of the Department of Civil Engineering at the Siberian State University of Industry, Russia

K V Eryomkin is a Postgraduate Student at the Siberian State University of Industry, Russia

Dr E G Avvakumov is a professor and Head of the Laboratory of Mechanico-Chemical Reactions, Russia

Dr K Freidin is a professor at the Ben-Gurion University, Israel.

INTRODUCTION

Compositions and technologies for heat-insulating cementless concrete produced from industrial by-products have been developed by the Siberian State University of Industry (SSUI) in association with the Siberian branch of the Russian Academy of Sciences. Secondary resources were used for the concrete. They were as follows: high-calcium fly ash and granulated slag (also known as furnace bottom ash) from the combustion of brown Kansk-Achinsk Thermal Power Complex coals, molding sand from foundry, high alumina product from the abrasion works and the detergent called "Progress". The compressive strength and the average density of the concrete were 0.25 to 5 MPa and 200 to 800 kg/m^2, respectively. The cost is considerably reduced as compared with that of cement concrete. A new class 400-600 (40 to 60 MPa) binder was used. The work has been performed in accordance with the "Higher School Research on Priority Trends of Science and Techniques" programme, Ministry of Education of Russia, as part of the project "Development of a new composition and energy-conservating technology for heat-insulating composite cementless concrete from secondary mineral resources" financed by the Kemerovo region Administration.

MATERIALS

The concrete mixtures were made using the following materials: high-calcium fly ash from the Yurga thermal power plant, granulated furnace bottom ash (fba) of the same source, spent sand ("burnt sand") from the foundry department of the Yurga machine building plant, high alumina product from the Yurga abrasion works and the secondary sodium alkyl sulphate (the detergent called "Progress") used as an air-entraining admixture. As the properties of these materials have been described previously (Study of the Compositions and Properties of Secondary Mineral Resources for Developing the New Heat-Insulating Cementless Concrete), this paper presents the new cementless binder which was the main material for producing the heat-insulating concrete.

NEW COMPOSITE CEMENTLESS BINDER

The binder incorporated the following materials: 70 to 80 % high-calcium ash, 10 to 20 % burnt sand and 5 to 10 high alumina product. The three components were simultaneously ground in an AGO-9 planetary activator continuous mill (Figure 1) to the fineness of 750 m^2/kg. Due to the mechanochemical process, the free calcium oxide content of the blend was reduced from 12-15 % to 2-3 % forming mainly calcium silicates (interaction between the amorphous silica of the burnt sand and free CaO of the fly ash) and partly tricalcium aluminate [3, 4, 5]. The compressive strength of the cementless binder was 40 to 60 MPa. It may be used as a base for heat-insulating concrete with air- or gas-entraining admixtures.

COMPOSITIONS OF HEAT-INSULATING CONCRETE

Using the method of experiment planning and the three-dimensional integral scheme for proportioning cementless concrete, a heat-insulating concrete based on the above binder was obtained, the mixture proportions and properties of which are given in Tables 1 and 2, respectively.

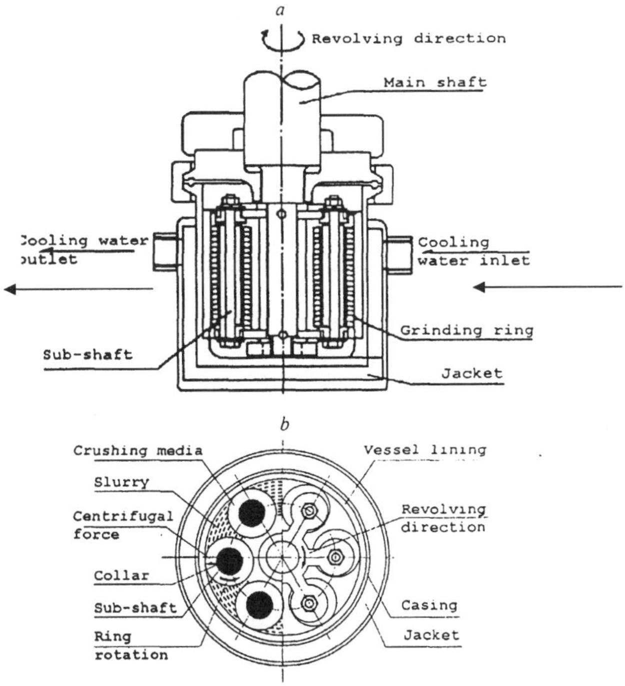

Figure 1 Schematic diagram of the AGO-9 activator mill

Table 1 Optimum Mixture Proportions for Heat-Insulating Concrete

MIXTURE №	CONCRETE COMPONENTS, kg/m^3				
	Cementless binder	Fba sand	Water	Technical grade lignosulfonate, % by weight	Air-entraining admixture, % by weight
1	140	–	75	0.3	2.0
2	350	120	184	0.3	1.0
3	400	168	212	0.3	0.6

Table 2 The Properties of Air-Entrained Ash-Slag Concrete

MIXTURE №	AVERAGE MIXTURE DENSITY, kg/m^3	DRY DENSITY, kg/m^3	COMPRESSIVE STRENGTH, MPa		COEFFICIENT OF HEAT CONDUCTIVITY, Wt/m°C
			at 28 days	at 90 days	
1	268	215	0.18	0.25	0.06
2	290	654	2.2	2.5	0.10
3	804	780	3.8	4.9	0.12

The heat-insulating concrete was prepared in a turbulent mixer (Figure 2) by introducing a composite plasticizing and air-entraining admixture (technical grade lignosulfonate and the detergent called "Progress" generally used in Kuzbass for washing the metallurgists and miners overall) into the mixture.

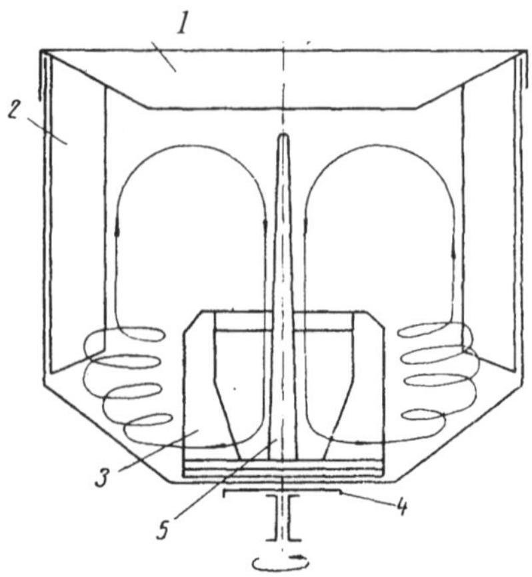

Figure 2 Technological scheme of the turbulent mixer:
1 – hopper, 2 – brake blades, 3 – rotor-mixer,
4 – protecting disk, 5 – cone dowel

The advantages of the concrete developed over the gas concrete based on cement and lime are in the exclusion of costly and energy-consuming above components, as well as aluminium powder, and the use of industrial by-products as replacement for them. The aluminium powder creates mostly capillary (and not closed) pores and also, forms the so-called "crusts" which are to be cut off.

Besides, the traditional cumbersome and energy-consuming ball mills of small efficiency were replaced with small, high-efficiency planetary activator mills (Figure 1) or vibrocentrifugal ones (Figure 3) designed by the Institute for Chemistry of Solids and Mechanochemistry, Siberian branch of the Russian Academy of Sciences (ICS&M SB RAS).

Their application permitted binding of the free CaO with silica and increased the compressive strength and durability of concrete.

The tests for heat conductivity show that the glass phase and the amorphous state of the slag (fba) component considerably decreased the heat conductivity of concrete even for the compressive strength of the cementless concrete equal to that of cement concrete.

CONCLUSIONS

The study carried out shows the feasibility of developing a new heat-insulating cementless concrete from secondary mineral resources thus contributing much to the environment protection and to the solution of economic and social problems.

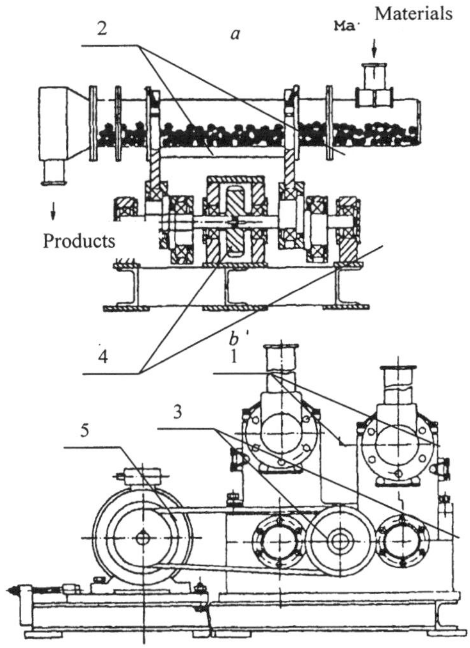

Figure 3 The VCM-10G vibrocentrifugal continuous mill:
a) side view, b) mill and electric motor arrangement
1 – grinding flasks, 2 – couriers, 3 – eccentric crankshafts, 4 – eccentrics, 5 – motor shaft

REFERENCES

1. Patent № 2196749. "Cementless binder" (authors: Pavlenko, S.I., Malyshkin, V.I., Merkulova, S.I., Aksyonov, A.V., Eryomkin, K.V., Zakharova, N.V.); Moscow, Bulletin, 2003, № 2.

2. TULEYEV, A.G., KULAGIN, N.M., PAVLENKO, S.I., TKACHENKO, V.V., The perspective of the Kuzbass construction industry revival in the organization of the production of efficient composite building materials from industrial by-products. Seminar proceedings: "Problems and the ways of creating composite materials and technologies from secondary mineral resources", January 29 – February 1, 2002, Novokuznetsk, Edited by professor S.I. Pavlenko // Novokuznetsk, SSUI, 2002, 122 pp.

3. PAVLENKO, S.I., MALYSHKIN, V.I., TKACHENKO, V.V., Aspects of Technology for Cement Mortar from Wastes of the Abakan TPP, Proceedings of the seventh CANMET/ACI International Conference "Fly Ash, Silica Fume, Slag and Natural Pozzolans in Concrete", 22-29 July, 2001, Chennai (Madras), INDIA. Editor V.M. Malhotra, Vol. 2, pp. 815-834, 2001 // Printed in USA, ACI, Farmington Hills, Michigan, SP 199-47.

4. TULEYEV, A.G., PAVLENKO, S.I., KULAGIN, N.M., AKSYONOV, A.V., BAZHENOV, YU.M., DOBRETSOV, N.L., LYAKHOV, N.Z., AVVAKUMOV, A.G., Mechanochemical Synthesis of the New Composite Binder from Secondary Mineral Resources. Journal "Izvestiya Vuzov, Chemistry and Chemical Technology" // Ivanovo: Publishing house IGCTU, Ministry of Education of Russia, 2002, № 6, pp. 126-129.

5. AVVAKUMOV, A.G., SENNA, M., KOSOVA, N.W., Soft Mechanochemical Synthesis a Basis for New Chemical Technologies // Kluwer Academic Publishers, Hardbound, ISBN 0-7923-7431-2. The Netherlands, 2001. 216 pp.

BRICK WASTE A SUPPLEMENTARY CEMENTING MATERIAL IN AUTOCLAVED BUILDING PRODUCTS

D S Klimesch

M Gutovic

A Ray

University of Technology

Australia

ABSTRACT. Fired-clay products such as bricks, tiles and pavers, are made in large volumes throughout the world and widely used as construction materials. A significant proportion of them end up being a waste material during either their production process or the demolition of buildings. High pressure steam curing or autoclaving is a proven and versatile method for utilising supplementary cementing materials such as fly-ash and blast furnace slag for the manufacture of cement-based building products. In this investigation, autoclaving has been used to treat blends of ordinary Portland cement with different amounts of ground brick waste. Our experimental data demonstrate the viability of this waste to be used by the construction industry as a supplementary cementing material. The paper presents results of salient engineering properties in relation to microstructural characteristics.

Keywords: Fired-clay products, Recycling, Tobermorite, Hydrogarnet, Mullite

D S Klimesch is a Research Fellow at the University of Technology, Australia.

M Gutovic is a PhD student at the University of Technology, Australia.

A Ray is an Associate Professor in the Department of Chemistry, Materials & Forensic Sciences at the University of Technology, Australia.

INTRODUCTION

The amount of Construction & Demolition (C&D) waste generated in the Sydney region alone exceeded 2 million tonnes in 2001. At present, an estimated 50% of this quantity is recycled for use in the construction of physical infrastructure, i.e. sub-base material for road works, and the rest is disposed of to landfills. A significant amount of C&D waste produced in Sydney represents clay-brick waste, generated during their production process and from demolition activities. The development of sustainable, cost-effective technical solutions for the use of a high volume of clay brick waste will have a significant impact on the State's solid waste stream as has been highlighted by several recent investigations [1, 2].

Previous work by the present authors has shown that the formation of 1.1nm tobermorite, the principal binder of most autoclaved calcium silicate based building materials, is enhanced by the addition of finely ground brick waste to blends of Portland cement (OPC) and quartz sand [3, 4]. To gain better insight into the behaviour of brick waste under autoclaving conditions, blends of OPC with different amounts of finely ground brick waste were prepared and evaluated for compressive strength and microstructural characteristics. The present article reports the findings of the effects of brick waste on phase formation and strength development.

EXPERIMENTAL

Raw Materials

The following raw materials were used:

(a) Goliath cement (OPC) containing SiO_2 20.0%, CaO 64.2%, Al_2O_3 4.5%, Fe_2O_3 3.7%, SO_3 3.5%, and having a fineness index (Blaine) of 350 m^2/kg was produced by Australian Cement, Auburn, NSW, Australia.

(b) Crushed brick waste less than 2.38 mm, was derived from a major clay-brick manufacturer in Sydney, Australia. The method of grinding used to produce ground brick fines having a surface area comparable to that of quartz sand, as used in commercial practice, was dry ball milling. Table 1 depicts the major oxides of the brick fines as determined by X-ray fluorescence. The mineralogical composition was determined qualitatively using X-ray powder diffractometry (XRD) using JCPDS powder patterns. The major crystalline phases, in decreasing order of abundance were: quartz, cristobalite, mullite and hematite.

Sample Preparation

OPC-brick fines mixtures were prepared using a water-to-total solids ratio of 0.36 as this yielded comparable workability for all the mixtures. Brick fines were added at 0, 10, 20, 30, 40, 50, 60, 70, 80 and 90 mass % as cement replacement. The chemical compositions of the pastes studied are presented in Table 2. Mechanical mixing was conducted in accordance with ASTM C 305-082. Pastes were cast into stainless steel moulds and consolidated on a vibratory table, followed by a 24 hour curing period in a moist cabinet. For each mix, three 50 mm cubes were made. Demoulded specimens were autoclaved for 8 hours, 6 hours of which were at 180°C under saturated steam.

After autoclaving, all cubes were allowed to air dry for 48 hours followed by compressive strength testing. After compressive strength testing, fractured pieces from each of the three cubes per mix were oven dried at 103° to 105°C overnight followed by disc milling for further analyses.

Table 1 Major oxides of brick fines used in this investigation (major oxides are in mass percent)

MAJOR OXIDES	
SiO_2	69.7
Al_2O_3	18.6
Fe_2O_3	7.5
K_2O	1.88
MgO	0.95
Na_2O	0.44
CaO	0.20

Table 2 Chemical compositions of pastes studied. Brick fines added are in mass percent

BRICK FINES ADDED	BULK MOLE Ca / Si	BULK MOLE Ca / (Al + Si)	BULK MOLE Al / (Al + Si)
0	3.44	2.72	0.21
10	2.48	1.94	0.22
20	1.84	1.43	0.22
30	1.38	1.07	0.23
40	1.04	0.80	0.23
50	0.77	0.59	0.23
60	0.55	0.42	0.24
70	0.38	0.29	0.24
80	0.23	0.18	0.24
90	0.11	0.08	0.24

Methods

Thermal analysis was carried out on powdered samples using a TA-instruments SDT 2960 simultaneous differential thermal and thermogravimetric analyser (DTA-TGA) at a heating rate of 10°C/min under flowing air (20mL/min) from 20° to 1000°C. Sample sizes were between 30 to 50 mg and were packed into a Pt-Rh crucible with 20 taps. All DTA-TGA curves were evaluated using the TA instruments data analysis software as detailed by Klimesch and Ray [5]. The second derivative differential thermal curves (DDT) were used for peak temperature determinations [6]. X-ray diffraction (XRD) analyses were conducted on powdered specimens using a Siemens D5000 diffractometer and copper $K_{\alpha 1}$ radiation from 3 to 60° 2θ at 0.02° 2θ per second. XRD patterns were $K_{\alpha 2}$ stripped and corrected for instrumental drift. Crystalline phases were identified using JCPDS powder patterns. Additionally, an X-ray line profile fitting program was used to determine integrated peak heights and areas as well as peak positions of major crystalline phases present.

The amount of acid-insoluble residue (AIR) was used as a measure of total unreacted material, as detailed by Klimesch and Ray [7]. In addition, the residues from triplicate AIR analyses were combined and the chemical composition determined by Inductively Coupled Plasma Optical Emission Spectrometry by the Genalysis Laboratory Services Pty. Ltd., Western Australia.

RESULTS AND DISCUSSION

Figure 1 shows that the highest strengths were obtained for brick fines additions between 40 to 50 mass %. Table 3 provides a summary of the phases identified by XRD and DTA.

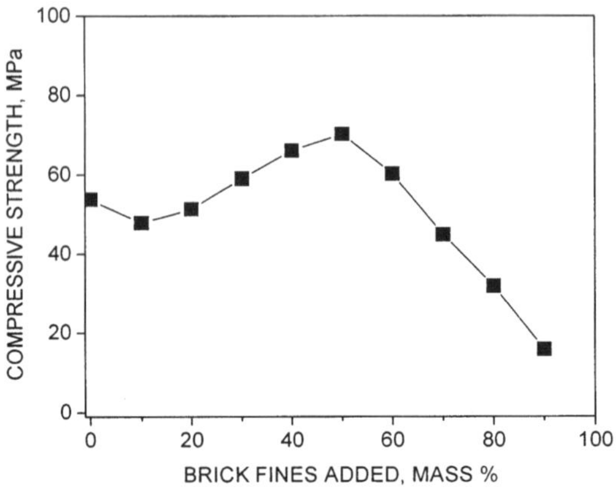

Figure 1 Average compressive strength of autoclaved OPC-brick fines mixtures.

Table 3 Phases identified by XRD and DTA.
Brick fines added are in mass percent

BRICK FINES ADDED	C-S-Hs	Tob*	Hyd	CH	α-C_2SH	$CaCO_3$	Q
0	✓	X	✓	✓	✓	✓	X
10	✓	X	✓	✓	✓	✓	✓
20	✓	✓	✓	✓	X	✓	✓
30	✓	✓	✓	✓	X	✓	✓
40	✓	✓	✓	✓	X	✓	✓
50	✓	✓	✓	X	X	✓	✓
60	✓	✓	✓	X	X	✓	✓
70	✓	✓	✓	X	X	✓	✓
80	✓	✓	X	X	X	✓	✓
90	✓	X	X	X	X	✓	✓

* see text for discussion

C-S-Hs = calcium silicate hydrates; Tob = tobermorite; Hyd = hydrogarnet; CH = $Ca(OH)_2$; α-C_2SH = alpha-dicalcium silicate hydrate; Q = quartz

i. In relation to Table 3, the following points are noteworthy: Tobermorite was considered absent when the diffraction line due to the (002) peak at around 1.1nm was not discernible. It is known that this peak is affected in height by the degree of order parallel to the c-direction, and that both absolute and relative intensities of peaks in the tobermorite pattern vary with the Ca/Si ratio and the degree of Al-substitution [8, 9].

ii. C-S-H with main d-spacings at around 0.304nm and 0.73nm, referred to as "fibrous C-S-H" by Kalousek [10], was present in all samples. The XRD peak in the 0.300nm region due to $CaCO_3$ interferes with reflections attributable to C-S-H.

The crystallinity of tobermorite was determined from XRD data by the peak intensity ratio of the 0.308nm and 1.1nm ($H_{0.308}$ /$H_{1.1}$) reflections [11]. In this context, the higher this ratio the less crystalline the material. The variations in the integrated area (arbitrary units) of the 1.1nm peak and the ratio of peak heights of the 0.308nm and 1.1nm peaks are depicted in Figure 2. Figures 3 and 4 show the variations in the integrated areas of the 0.304nm (Figure 3) and 0.505nm (Figure 4) peaks attributable to C-S-H and hydrogarnet, respectively.

1. From XRD data, the presence of Al-bearing tobermorite (Al-tobermorite) was manifested by the position of the (002) peak, the d-spacing ranging from 1.13nm to 1.14nm. Figure 2 shows that brick fines additions of 20 and 30 mass % promoted the formation of poorly crystalline Al-tobermorite. Additions of 40 to 60 mass % promoted the formation of highly crystalline Al-tobermorite, the amount reaching a maximum at 40 mass % followed by a steady decrease from 50 to 60 mass %. For additions greater than 60 mass %, Al-tobermorite amount and its crystallinity decreased significantly. The variations in Al-tobermorite amount can be explained, at least in part, in terms of the decreasing bulk Ca/Si and Ca/(Al+Si) molar ratios (Table 2). In other words, with decreasing OPC:brick fines ratio, the availability of CaO from cement decreases resulting in the formation of fewer hydration products.

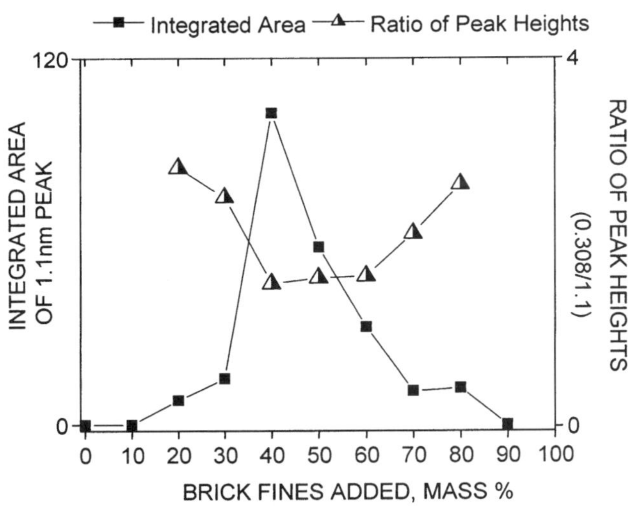

Figure 2 Showing the variations in the integrated area (arbitrary units) of the 1.1nm peak and the ratio of peak heights of the 0.308nm and 1.1nm peaks versus brick fines content.

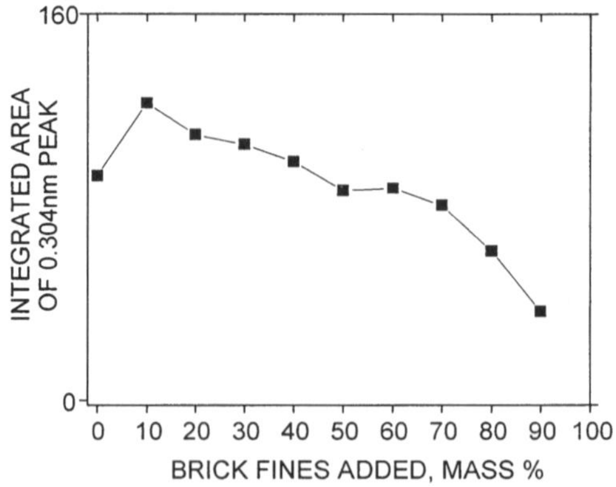

Figure 3 Variation in the integrated area of the 0.304nm peak attributable to C-S-H with increasing brick fines addition.

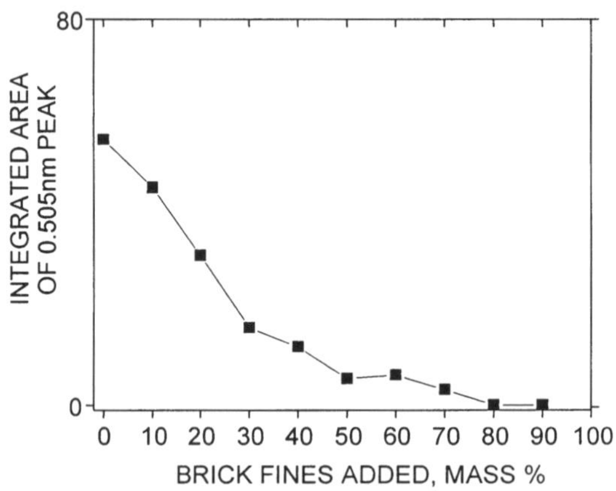

Figure 4 Variation in the integrated area of the 0.505nm peak attributable to hydrogarnet (where present) with increasing brick fines addition.

2. The formation of "fibrous C-S-H" diminished slightly and then appreciably for brick fines additions of 20 to 50 mass % and greater than 60 mass %, respectively (Figure 3). Concurrently, the amount of hydrogarnet decreased monotonically for brick fines additions of 10 to 70 mass %. Again, these results can be explained, at least in part, in terms of the decreasing bulk Ca/Si and Ca/(Al+Si) molar ratios (Table 2).

From the standpoint of strength development, the present data reinforce previous findings in that: first, there exists a good correlation between the total sum of tobermorite and C-S-H present and the compressive strength; and second, some optimum degree of binder crystallinity exists [12]. In this context, the present data show that the highest strengths obtained (Figure 1) correspond to the presence of highly crystalline tobermorite (Figure 2) and "fibrous C-S-H" (Figure 3).

3. It may be noted that the total bulk composition, i.e. total bulk Ca/(Al+Si) molar ratio, does not equal the total reacted composition, since unreacted brick fines remained in all the specimens. More specifically the amount of unreacted material, as determined by the acid-insoluble residue (AIR), ranged from 2.7% to 75.1% for 10% to 90% brick fines additions, respectively. From analyses of the oxide composition of the AIR, the actual reacted Ca/(Al+Si) molar ratios have been determined and are presented in Figure 5. From this figure it is evident that for reacted Ca/(Al+Si) molar ratios between 1.05 and 0.92, maximum strength development occurred with highly crystalline tobermorite being the main binder (Figure 2) along with "fibrous C-S-H" (Figure 3).

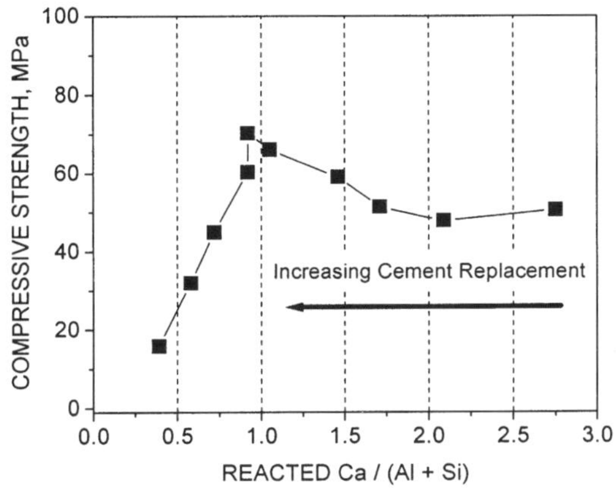

Figure 5 Showing the variations of reacted Ca / (Al + Si) molar ratio versus compressive strength for increasing cement replacement with brick fines additions.

CONCLUSIONS

1. Brick fines additions of 20 and 30 mass % promoted the formation of poorly crystalline Al-tobermorite while the compressive strength increased concurrently. Additions of 40 to 60 mass % promoted the formation of highly crystalline Al-tobermorite, the amount reaching a maximum at 40 mass % followed by a steady decrease from 50 to 60 mass %. The compressive strength reached a maximum between 40 to 50 mass % brick fines additions. For additions greater than 60 mass %, Al-tobermorite amount and its crystallinity decreased significantly. Concurrently, the formation of "fibrous C-S-H" diminished slightly and then appreciably for brick fines additions of 20 to 50 mass % and greater than 60 mass %, respectively.

2. For reacted Ca/(Al+Si) molar ratios between 1.05 and 0.92, maximum strength development occurred with highly crystalline tobermorite being the main binder along with "fibrous C-S-H".

3. The use of waste brick fines in combination with OPC for the production of hydrothermally cured calcium silicate based materials is a viable option for the future.

REFERENCES

1. BAKOSS, S.L., SRI RAVINDRARAJAH, R., Recycled C&D Materials for Use in Roadworks and other Local Government Activities: Scoping Report, Centre for Built Infrastructure Research, University of Technology, Sydney, 1999, pp 1-68.

2. KLIMESCH, D.S., SRI RAVINDRARAJAH, R., BAKOSS, S.L., New Technologies for use of Recycled Brick as Construction Material: Final Report to NSW Waste Boards, Centre for Built Infrastructure Research, University of Technology, Sydney, 2001, pp 1-66.

3. KLIMESCH, D.S., RAY, A., Incorporating brick fines in cement-based building materials: Proceedings of the International Conference on Composites in Construction, Figueiras et al. (Eds.), Porto, 2001, pp 47-49.

4. KLIMESCH, D.S., RAY, A., GUERBOIS, J.P., Differential scanning calorimetry evaluation of autoclaved cement based building materials made with construction and demolition waste: Thermochimica Acta, Vol. 389, 2002, pp 195-198.

5. KLIMESCH, D.S., RAY, A., The use of DTA/TGA to study the effects of ground quartz with different surface areas in autoclaved cement:quartz pastes. Part 1: A method for evaluating DTA/TGA results: Thermochimica Acta, Vol. 289, 1996, pp 41-54.

6. KLIMESCH, D.S., RAY, A., Use of the second-derivative differential thermal curve in the evaluation of cement-quartz pastes with metakaolin addition autoclaved at 180°C: Thermochimica Acta, Vol. 307, 1997, pp167-176.

7. KLIMESCH, D.S., RAY, A., Metakaolin additions to autoclaved cement-quartz pastes: Evaluation of the acid-insoluble residue: Advances in Cement Research, Vol. 9, 1997, pp 157-165.

8. DYCZEK, J.R.L., TAYLOR, H.F.W., X-ray determination of tobermorite, quartz, and alpha-dicalcium silicate hydrate in autoclaved calcium silicate materials: Cement and Concrete Research, Vol. 1, 1971, pp 589-605.

9. KALOUSEK, G.L., Crystal chemistry of hydrous calcium silicates: I, Substitution of aluminum in lattice of tobermorite: Journal of the American Ceramic Society, Vol. 40, 1957, pp 74-80.

10. KALOUSEK, G.L., Tobermorite and related phases in the system CaO-SiO_2-H_2O: Journal of the American Ceramic Society, Vol. 51, 1955, pp 989-1011.

11. SUN, B., SU, E., The crystallinity and rate of carbonation of hydrated calcium silicates: Guisuanyan Xuebao, Vol. 12, 1984, pp 281-285.

12. TAYLOR, H.F.W., Cement Chemistry, 2nd ed, Thomas Telford, London, 1997, p 342.

INDUSTRIAL BY-PRODUCTS AS POSSIBLE ALTERNATIVES TO GGBS AND PFA IN THE PRECAST CONCRETE MARKET

L J Davies G Hunt

K P Williams B I G Barr

Cardiff University

United Kingdom

ABSTRACT. It is becoming increasingly important to minimise the amount of waste going to landfill and find alternative uses for these materials. This paper presents data to show that several industrial by-products which are currently being landfilled are capable, after a minimum of pre-treatment and blending, of replacing up to 50% of OPC in producing precast concrete. It has been found that modified cements can be formulated which exhibit better high early strengths (7-day) than comparable materials formed from GGBS and PFA. These modified cements are shown to be acceptable in terms of commonly used characterisation parameters quoted in the BS EN 196 series. Extensive leaching test results demonstrate that leachate concentrations are within acceptable levels. In an attempt to understand the mechanisms responsible for these early strengths, measurements of pozzolanicity of the raw materials have been recorded together with a systematic study of the crystal formations.

Keywords: Industrial by-product, Modified cement, Pozzolanicity, XRD

L Davies, is a graduate of Environmental Engineering at Cardiff University, Wales and is currently studying for a MPhil concerned with the environmental assessment of cement substitutes from industrial by-products. Louise is a member of the Chartered Institute for Water and Environmental Management (CIWEM).

G Hunt, is the Managing Director of Environmental and Industrial Evaluations (E&IE) Limited. He has 20 years experience in the construction industry and is currently a visiting honorary research fellow at Cardiff University.

K P Williams, is the Head of the Materials and Minerals Division in the School of Engineering, Cardiff University. He has been involved with recovery and reprocessing of mineral waste materials for over 25 years and has published extensively in this field. Currently a member of the Welsh Assembly Waste Policy Advisory Group (sub-committee on Composting).

B I G Barr, is the Carillion Professor of the Built Environment, School of Engineering, Cardiff University. He is Vice Chairman of the Structures and Buildings Board of the ICE, a former member of council of the ICE and a Fellow of the Transport Research Foundation.

INTRODUCTION

Since landfill sites in general are becoming overcrowded and expensive for waste producers, then wherever possible, material going to disposal should be minimised. If the production of the waste can not be prevented, then it is attractive to create an alternative use in another process before considering disposal. The benefits of this recycling can be economical, due to lower costs of removing the waste, and environmental, such as reduction in pollution and contamination.

The successful use of industrial by-products in concrete extenders or secondary aggregates depends on a number of factors. For example, readymixed concrete and precast (non-structural) concrete often require very different strength performance properties. Also, the economic factors would be of great importance and are generally influenced by the cost of by-product disposal, by the cost of transportation of the by-product to a processing site and by current environmental legislation.

It is not only the mineralogical and chemical composition of the industrial by-products, but also their particle size distribution and particle shape, which will govern their potential for inclusion as a cement substitute. After this potential has been assessed it is then necessary to consider the compatibility of the physical, chemical and mineralogical properties of the substitute, with the corresponding properties of the Ordinary Portland Cement (OPC) and aggregates that will be used in the concrete mix [1].

This paper presents data to show that several industrial by-products which are currently being landfilled are capable, after a minimum of pre-treatment and blending, of replacing up to 50% of OPC in producing precast concrete. The tests carried out include chemical analyses for composition using ICP AES and pozzolanicity measurements, mineralogical analysis by X-ray diffraction, and physical analyses generating particle size distributions and scanning electron microscopy images, as well as measurements of compressive strength.

MATERIALS

The cement replacements presented in this paper have been composed of raw materials currently being landfilled or stockpiled and could be the subject of Waste Management regulations in the future. The materials used are from the Iron, Steel and Quarrying Industries. For this paper, the materials have been given 'RM' references due to confidentiality requirements. Table 1 presents the chemical and physical properties of the OPC, the Ground Granulated Blast-furnace Slag (GGBS), the Pulverised Fuel Ash (PFA) and all the RMs used to create the cement replacements used in these trials. Prior to the tests reported in Table 1 being carried out, all the RMs were dried and ground to nominally <40μm using a Sceibechik Tema Mill.

To achieve compliance with the EN 196 limits for Loss On Ignition (LOI), Free Lime (FL) and Insoluble Residue (IR) of 5%, 1% (2.5% if the soundness test has also been passed) and 5% respectively, it was necessary to thermally treat RM8, RM9, RM16 and RM18 so that their inclusion in the blend could be maximised.

Table 1 Chemical and physical properties of cement and the cement replacements

Test procedure/Analysis method		Comp %	OPC	GGBS	PFA	RM3	RM8	RM9	RM16	RM18
	In house ICP analysis	CaO	62.33	37.40	6.16	42.52	21.22	36.78	55.36	59.65
		SiO_2	22.95	34.73	43.84	34.31	41.15	3.99	2.44	1.24
		Al_2O_3	4.66	12.35	25.18	11.79	16.21	0.47	0.66	0.23
		MgO	1.17	7.94	2.06	8.26	11.61	0.16	0.54	2.49
		Fe_2O_3	2.92	0.31	9.25	0.80	3.51	32.51	0.42	0.17
		Na_2O	0.18	0.17	0.64	0.37	0.38	0.95	0.13	0.09
		K_2O	0.67	0.53	1.68	0.95	0.18	0.90	0.00	0.00
		P_2O_5	0.05	0.01	0.59	0.00	0.00	0.00	0.11	0.00
		TiO_2	0.43	0.61	1.13	0.48	0.44	0.01	0.04	0.01
		MnO	0.04	0.54	0.10	0.51	2.29	0.11	0.02	0.01
		Cl^-	0.04	0.01	0.01	0.01	0.01	0.01	0.02	0.02
	EN196	LOI	1.00	-0.50	6.10	1.80	-0.30	12.7	41.7	41.3
		FL	0.93	0.50	2.80	0.30	0.03	0.10	0.90	0.10
		IR	0.44	0.09	57.79	1.06	16.11	0.80	2.43	0.70
		ρ	3150	2910	2330	2950	2920	2990	2920	2730
		SSA	370	375	320	210	134	981	124	411

Keys LOI = Loss On Ignition; FL = Free Lime; IR = Insoluble Residue; ρ = Density (kg/m^3)
SSA = Specific Surface Area (m^2/kg).

The particle shape of the 5 RMs, GGBS and PFA can be seen in Figure 1 and these images were taken using Scanning Electron Microscopy (SEM).

EXPERIMENTAL PROCEDURES AND RESULTS

The compressive strength results for concretes with 25% OPC being replaced by either a raw material after their heat treatment, GGBS or PFA can be seen in Figure 2. The concrete cubes were made using a mix of 7.2kg coarse and 4.8kg fine aggregate, 2.5kg of cementitious material and 1.250 litres of water which gave a water to cementitious material ratio (w/cm) of 0.50. At 1, 2 and 7 days, compressive strengths were measured. These results show that even on their own, the RM replacements each have desirable effects on the strength performance of concrete, and each could be used as potential cement replacements in the pre-cast concrete industry.

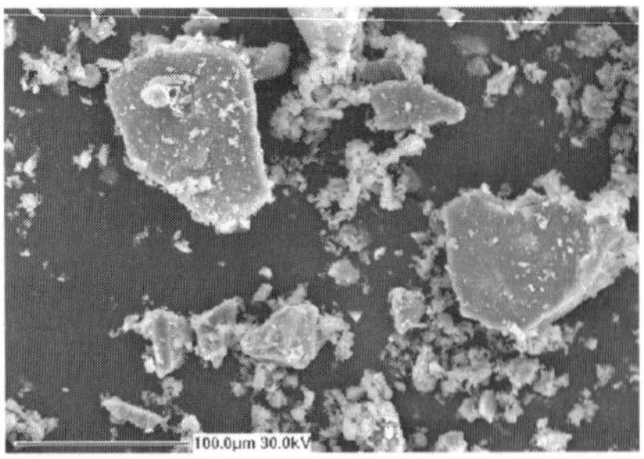

Figure 1a RM3

Figure 1e RM18

Figure 1b RM8

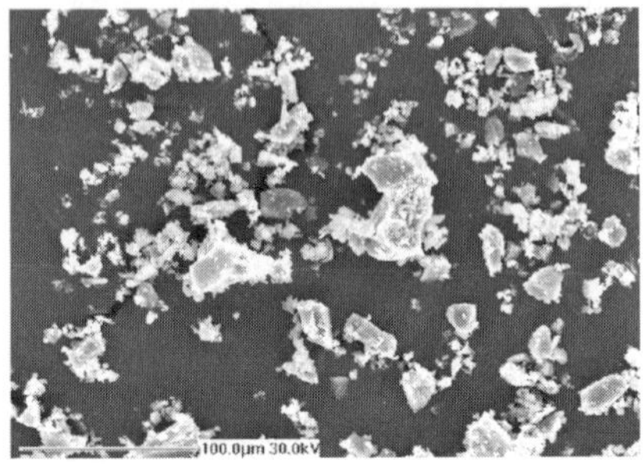

Figure 1f GGBS

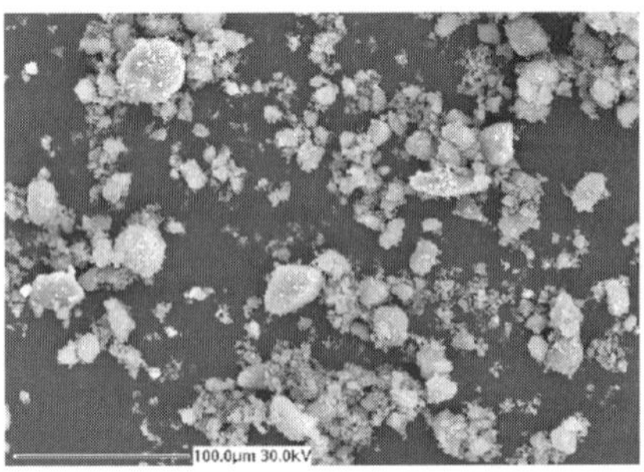

Figure 1c RM9

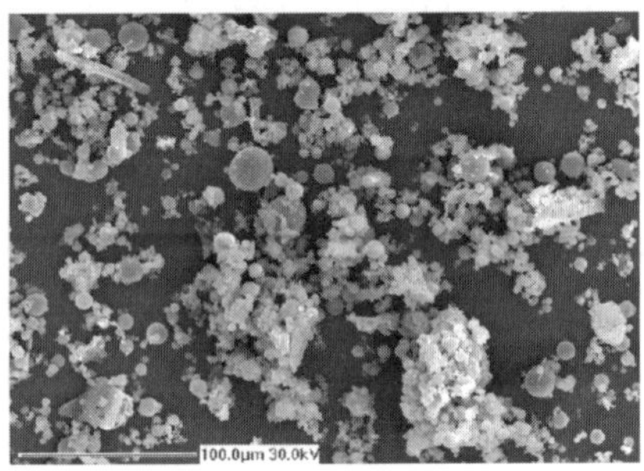

Figure 1g PFA

Figure 1d RM16

Figure 1a-g Images taken using a Scanning Electron Microscope (SEM) of the Raw Materials, GGBS and PFA all at x300 magnification

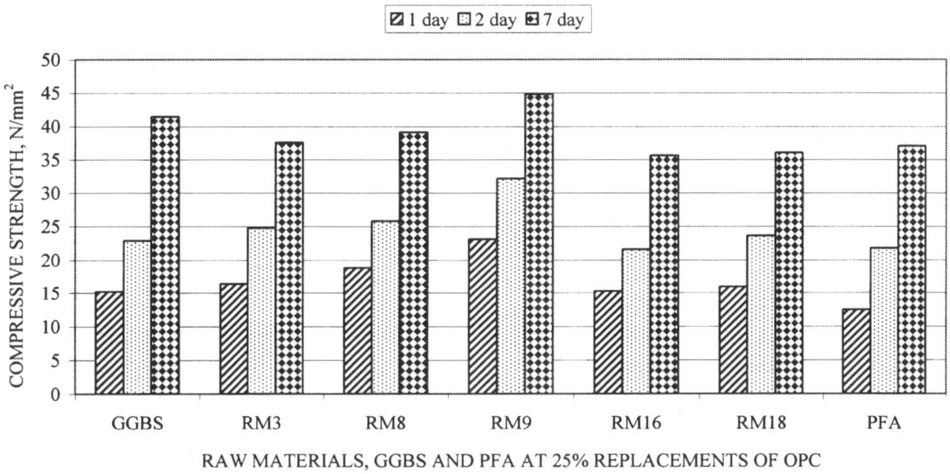

Figure 2 Compressive strength gain over 7 days of concrete composed of 25% replacements of OPC with GGBS, PFA and the Raw Materials

A blend containing RM3, RM8, RM9 and RM18 in specific proportions was prepared. This blend was given the reference B18. Replacement levels of 25% and 50% of B18 were combined with OPC for comparison with a currently widely used cement replacement, GGBS. The GGBS was obtained from Llanwern Works in South Wales, UK, with a chemical modulus ((CaO% + MgO% + Al_2O_3%) / SiO_2%) of 1.7. A second blend containing specific proportions of RM3, RM8, RM9 and RM16 was prepared. This blend was given the reference B19. Replacement levels of 25% and 50% of B19 were also combined with OPC for comparison with another widely accepted cement replacement, PFA. The PFA was obtained from a processing plant in Rugeley, UK with a chemical modulus of 0.76. All replacements were made by mass. Table 2 shows the combinations of the blended cements reported in this paper. The chemical and physical properties of the blended cements can be seen in Table 3.

Table 2 Composition and Compressive Strength of OPC and the cement replacements

MIX REF.	OPC	BLEND 18	BLEND 19	GGBS	PFA	COMPRESSIVE STRENGTH, N/mm²		
						1 day	2 day	7 day
B18(25%)	75%	25%	-	-	-	14.5	20.5	30.5
GGBS(25%)	75%	-	-	25%	-	13.5	20.5	32.5
B19(25%)	75%	-	25%	-	-	13.5	21.0	31.5
PFA(25%)	75%	-	-	-	25%	13.0	21.5	32.5
B18(50%)	50%	50%	-	-	-	7.5	11.0	17.3
GGBS(50%)	50%	-	-	50%	-	7.5	12.5	24.0
B19(50%)	50%	-	50%	-	-	7.5	11.0	16.9
PFA(50%)	50%	-	-	-	50%	6.5	12.0	19.0

Table 3 Chemical and physical properties of the cement substitutes at 25% replacement

Test procedure/Analysis method		Comp %	B18(25%)	GGBS(25%)	B19(25%)	PFA(25%)
	In house ICP analysis	CaO	57.77	56.10	56.69	48.29
		SiO_2	23.14	25.90	24.11	28.18
		Al_2O_3	5.55	6.58	5.90	9.79
		MgO	2.46	2.86	2.57	1.39
		Fe_2O_3	3.21	2.27	3.67	4.50
		Na_2O	0.22	0.18	0.24	0.29
		K_2O	0.66	0.64	0.69	0.93
		Na_2O eq	0.66	0.60	0.70	0.93
		P_2O_5	0.04	0.04	0.04	0.19
		TiO_2	0.40	0.47	0.41	0.60
		MnO	0.16	0.16	0.20	0.06
		Cl^-	0.03	0.03	0.03	0.03
	EN196	LOI	1.25	0.63	1.30	2.28
		FL	0.81	0.82	0.75	1.40
		IR	0.91	0.35	1.09	14.78
		ρ	3090	3090	3100	2950
		SSA	360	371	372	358

Keys LOI = Loss On Ignition; FL = Free Lime; IR = Insoluble Residue; ρ = Density (kg/m^3)
SSA = Specific Surface Area (m^2/kg).

For each blended cement containing either B18 or B19 at 25% or 50% replacement levels, mortar prisms were made using a standard sand, in compliance with EN 196-1, with a sand to cementitious material ratio (s/cm) of 3. The w/cm was 0.50 and the prisms were made and cast according to the procedures set out in EN 196-1.

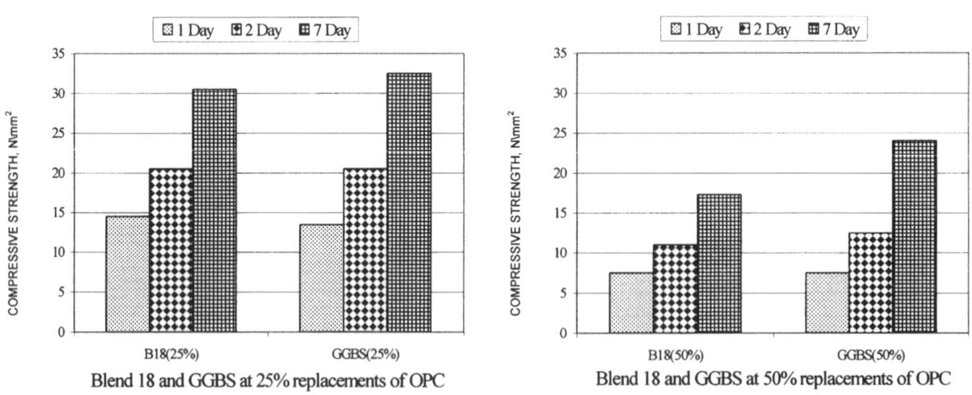

Figure 3 Comparison of strength development of Blend 18 and GGBS at 25% and 50% replacement levels in mortar prisms

At 1, 2 and 7 days, compressive strengths were determined on the two pieces of prism resulting from the flexural strength test. The compressive strength results reported in Table 2 represents the average of these two values and the results can be seen in graphical form in Figures 3 and 4.

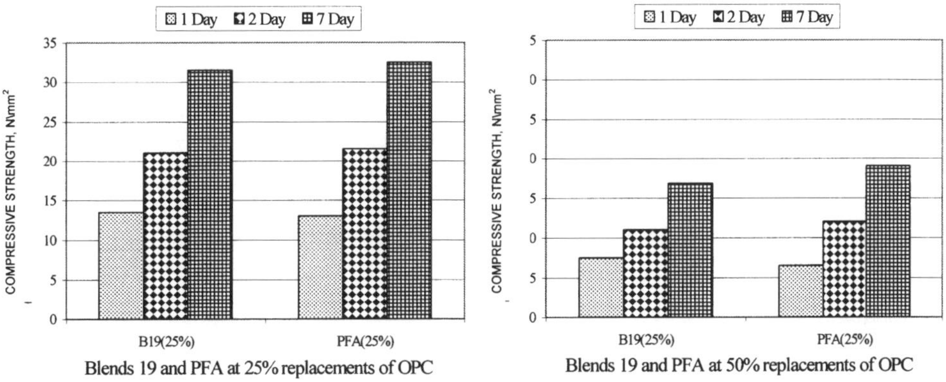

Figure 4 Comparison of strength development of Blend 19 and PFA at 25% and 50% replacement levels in mortar prisms

The particle size distributions presented in Figure 5 were generated from a HELOS Particle Size Analyser and were determined at the laboratories of Castle Cement Ltd in Ketton.

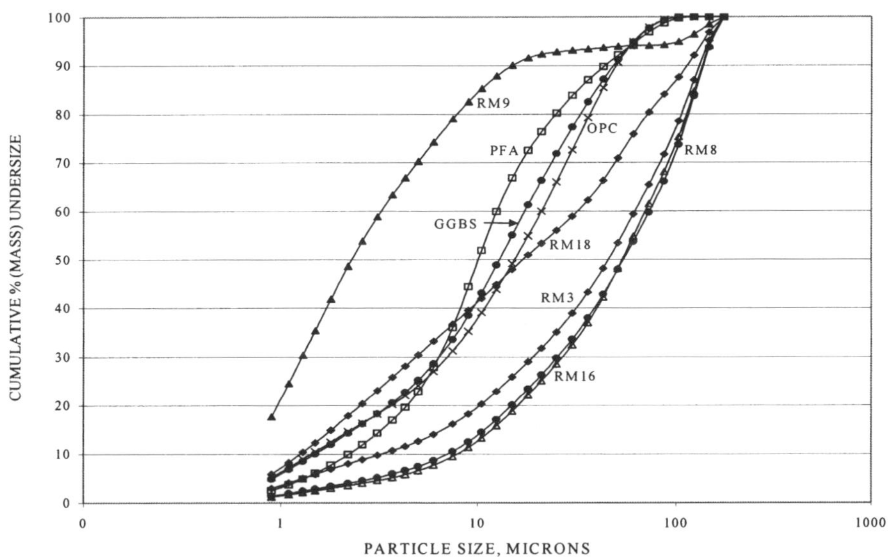

Figure 5 Particle size distributions of OPC, GGBS, PFA and the Raw Materials.

The leachability results presented in Table 4 were obtained by carrying out the modified EPA 1312 Leach Extraction Test on the concrete cubes produced from OPC and 25% replacement of OPC with GGBS, PFA, RM3, RM8 and RM9 at 2 and 7 days after casting. It is worth noting that the solids to liquid ratio in the leach test was 1:2 and maximum particle size was 4mm. In all other respects the method was the same as EPA 1312. Because of the lower amount of water used, (solids:liquid ratio in 1312 is 1:20) the modified test will exaggerate any soluble contaminant concentration in the leachate.

Table 4 Leachate analysis of crushed concrete cubes produced from OPC, and 25% replacement of OPC with GGBS, PFA, RM3, RM8 and RM9 at 2 and 7 days after casting

Element	OPC (ppm)		GGBS (ppm)		PFA (ppm)		RM3 (ppm)		RM8 (ppm)		RM9 (ppm)	
	2d	7d	2d	7d	2d	7d	2d	7d	2d	7d	2d	7d
Ca	607.5	635.4	798.0	600.0	840.0	573.8	635.8	n.a.	653.5	514.7	660.2	510.4
Al	2.2	1.9	2.4	1.9	2.3	1.9	1.5	n.d.	1.6	0.6	1.6	0.8
Mg	n.d.	n.d.	n.d.	n.d.	n.d.	n.d.	n.d.	n.d.	n.d.	n.d.	n.d.	n.d.
Fe	n.d.	n.d.	n.d.	n.d.	n.d.	n.d.	n.d.	n.d.	n.d.	n.d.	n.d.	n.d.
Na	69.0	72.8	68.2	56.0	71.1	57.8	154.9	120.7	148.3	127.0	142.4	120.2
P	n.d.	n.d.	n.d.	n.d.	n.d.	n.d.	n.d.	n.d.	n.d.	n.d.	n.d.	n.d.
Zn	n.d.	n.d.	<0.1	n.d.	n.d.	n.d.	0.2	<0.1	0.2	n.d.	0.7	n.d.
V	n.d.	n.d.	n.d.	n.d.	n.d.	n.d.	n.d.	n.d.	n.d.	n.d.	n.d.	n.d.
Ti	n.d.	n.d.	n.d.	n.d.	n.d.	n.d.	n.d.	n.d.	n.d.	n.d.	n.d.	n.d.
Mn	n.d.	n.d.	n.d.	n.d.	n.d.	n.d.	n.d.	n.d.	n.d.	n.d.	n.d.	n.d.
Cr	n.d.	n.d.	n.d.	n.d.	n.d.	n.d.	<0.10	n.d.	<0.10	n.d.	1.0	0.8
Cd	n.d.	n.d.	n.d.	n.d.	n.d.	n.d.	n.d.	n.d.	n.d.	n.d.	n.d.	n.d.
Pb	<0.2	<0.2	<0.2	<0.2	<0.2	<0.2	<0.2	<0.2	<0.2	n.d.	<0.2	n.d.
Ni	<0.1	0.2	<0.1	0.1	<0.1	0.1	<0.1	n.d.	<0.1	n.d.	<0.1	n.d.
Cu	n.d.	n.d.	n.d.	n.d.	n.d.	n.d.	n.d.	n.d.	n.d.	n.d.	n.d.	n.d.
Sr	10.8	10.7	13.1	8.6	13.0	10.1	11.4	4.8	11.4	5.4	8.4	3.4
Ba	6.2	3.6	7.3	4.2	8.1	4.4	3.8	1.4	3.2	1.5	1.1	0.4
As	<0.5	<0.5	n.d.	n.d.	<0.5	<0.5	n.d.	n.d.	n.d.	n.d.	n.d.	n.d.

Key n.d. = not detected, n.a. = not available

The pozzolanicity test taken from EN 196-5:1987, compares the quantity of calcium hydroxide in an aqueous solution in contact with the hydrated pozzolanic cement with the quantity of calcium hydroxide which saturates a solution of the same alkalinity. If the former concentration is lower than the latter, then the pozzolanicity of the cement is considered to be satisfactory. The underlying principle is that the pozzolanic activity consists of fixing of calcium hydroxide by the pozzolana so that the lower the resulting quantity of calcium hydroxide the higher the pozzolanicity [2]. The results from this test can be seen in Table 5.

Table 5 Pozzolanicity test results for the Raw Materials

	RM3	RM8	RM9	RM16	RM18
Pozzolanicity	Positive	Positive	Positive	Positive	Positive

The XRD data for the raw materials used to make B19 are presented in Figure 6. The major peaks have been identified either manually or by using software kindly supplied by Bruker AXS Inc. It is interesting to note that the crystalline phases in RM16 are mainly Calcite, Lime and Portlandite. RM9 and RM3 on the other hand only contain one crystalline phase each, Calcium Fluorite and Melilite respectively. This Melilite phase was expected to be seen in this material and has been previously well documented. The peak pattern of RM8 was more difficult to assign to well known phases. The best estimate of the major phases present are Calcium Aluminium Oxide and Calcium Iron Oxide.

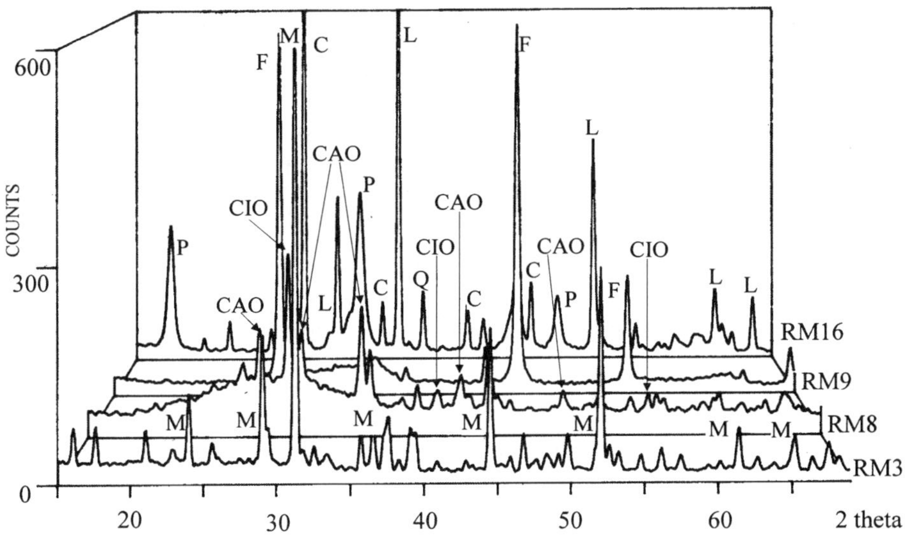

Key C = Calcite, L = Lime, P = Portlandite, Q = Quartz, M = Melilite, F = Fluorite, CAO = Calcium Aluminium Oxide, CIO = Calcium Iron Oxide.

Figure 6 X-ray diffractogram of RM3, RM8, RM9 and RM16 (B19) after pre-treatment.

DISCUSSION AND CONCLUSION

It has been demonstrated that suitable replacements for OPC are found amongst common waste materials that are currently either sent to landfill or stockpiled at the premises of the producers. Early strength properties for concretes made by replacing 25% of OPC with these materials are similar to those obtained for commonly used substitutes such as GGBS and PFA. However, it may be that cements containing these wastes as single replacements do not comply with limits set out in European standards, such as EN 196. It is possible to avoid this problem by using blends of raw materials that are also capable of mimicking early strength characteristics of GGBS and PFA cements.

Chemical compositions of leachates generated from concretes containing the individual substitute materials showed negligible release of any potential environmentally harmful elements. RM9 for example contains 32.51% iron oxide (Table 1), but no iron was detected in the corresponding leachate (Table 4), even for concrete that was only 2 days old.

The question of why these non-traditional substitutes are capable of exhibiting acceptable strength properties has recently received attention in the literature [3, 4, 5]. The understanding of the mechanisms involved is still not complete, although the consensus seems to be that the microfiller effect is dominant over pozzolanic contributions. The enhanced strength effects are the results of synergy between the hydration of the OPC and the presence of pozzolans and microfiller materials. The effect of the latter is greatest when the particle size is at a minimum. All of the raw materials tested during the present study exhibited positive pozzolanicity so it would be expected that the material exhibiting the finest particle size should give rise to the greatest early strength. Examination of the strength data of Figure 2 clearly shows that concretes formed from 75 % OPC and 25 % RM9 have the highest early strengths. In the same manner it is also clear that concretes containing RM16 exhibit the lowest early strengths. However, the pattern is not uniform for the other materials, highlighting more complex considerations such as morphology and differing crystalinity. For example, it has been demonstrated in Figure 6 that the silica in RM8 (41.15% from Table 1) occurs as an amorphous phase, a feature which is commonly accepted as being desirable in a cement replacement. For this reason, this material performs better than would be expected from its relatively coarse size.

ACKNOWLEDGEMENTS

We would like to acknowledge the help from Andy Croft of Castle Cement Ltd for producing the Particle Size Distribution Data.

REFERENCES

1. SAMARIN, A. Wastes in Concrete: Converting Liabilities into Assets, Proceedings of the Intern. Seminar on Exploiting Wastes in Concrete, Dundee, Eds. Dhir and Jappy, Thomas Telford Publishing, London, 1999, pp 1-21.

2. NEVILLE, A.M. Properties of Concrete, Longman, Harlow, 1995, pp 84.

3. GOLDMAN, A and BENTUR, A. The influence of microfillers on enhancement of concrete strength, Cement and Concrete Research, Vol 23, 1993, pp 962-972.

4. GOLDMAN, A and BENTUR, A. Properties of cementitious system containing silica fume or non-reactive microfillers, Advn. Cem. Bas. Mat., Vol 1, 1994, pp 209-215.

5. ISAIA, G.C, GASTALDINI, A. L. G and MORAES, R. Physical and pozzolanic action of mineral additions on the mechanical strength of high-performance concrete, Cement and Concrete Composites, Vol 25, 2003, pp 69-76.

FGD GYPSUM - A PROMISING INDUSTRIAL BY-PRODUCT FOR THE GREEK CEMENT INDUSTRY

G Tzouvalas

S Tsimas

National Technical University of Athens

A J Papageorgiou

Titan Cement Co

Greece

Abstract. The majority of cement plants in Greece mainly use natural gypsum to prevent rapid cement setting. The partial or total replacement of gypsum by materials, which contain calcium sulphate, has been instigated by two facts. Firstly, the increasing availability of low-cost by-products containing calcium sulphate and secondly, the prospect of the gypsum quarries to mine, in the near future, a rock that is a mixture of gypsum and anhydrite. FGD gypsum, a waste material of the desulphurisation process in coal burning power plants, to convert SO_3 content of flue gases into gypsum, is an important alternative source of chemical gypsum. The aim of this paper is to discuss results from laboratory and industrial scale trials, where calcium sulphate bearing materials (CSBM) such as natural gypsum, anhydrite and FGD gypsum have been used in various proportions in the cement production. At the same time, a hydration study of cements with FGD is presented. The addition of FGD gypsum increases setting time without affecting the compressive strength profile. The degree of dehydration of the dihydrated calcium sulphate regulates setting and strength performance of the cement partially replaced with either anhydrite or FGD gypsum.

Keywords: Portland cement, Gypsum, Anhydrite, FGD gypsum, Setting, Strength

George Tzouvalas is a Chemical Engineer and post-graduate student in the National Technical University (NTU) of Athens. His major interests are in aluminosilicate technology and he is currently undertaking research into the evaluation of alternative setting retarders in the production of cement.

Stamatis Tsimas is a Chemical Engineer and Associate Professor of Applied Inorganic Chemistry at the National Technical University (NTU) of Athens. His major interests are in chemistry and technology of aluminosilicates, specialising in cement technology and upgrade techniques of industrial minerals and byproducts. Prof. Tsimas has published over 50 papers in related Journals, Congresses and Conferences.

Aristotelis J Papageorgiou is a research Chemist in TITAN Cement Co. His major interests are in the cycle of heavy metals in the cement circuit, the use of alternative fuels in cement kilns, the cement components and cement chemistry, technology and microscopy.

INTRODUCTION

In the cement industry, gypsum is added during grinding of Portland cement clinker in order to delay the rapid reaction between C_3A and water and to regulate cement setting properties. The initial product of the C_3A-sulphate reaction is ettringite. It is assumed that ettringite forms initially on the reacting C_3A surface, a more-or-less impermeable coating that impedes diffusion of the ions needed to form the hydrates that cause setting. Retardation has also been attributed to an underlying layer of hydrous alumina or AFm phase [1, 2, 3]. In order to achieve the best setting behaviour, the availability of sulphate should – by the addition of an appropriate type and quantity of sulphate-bearing substance - be just so controlled that the hydration portion of C_3A is combined exclusively as ettringite [3, 4]. Nevertheless, if the gypsum content goes beyond a certain limit, it can give rise to expansion in the concrete. For this reason, cement standards fix a maximum for the gypsum content (in the form of SO_3) of cement. The need for gypsum in cement increases in accordance with the increasing amount of C_3A and alkalis in the clinker and the fineness of the cement [1, 4, 5].

For many years now, the Greek cement industry has been using mineral/natural gypsum, which is primarily a dehydrated product of calcium sulphate ($CaSO_4 \cdot 2H_2O$). Continuous mining of existing resources leads to their gradual depletion creating the need to find alternative raw materials to be used for the control of setting time [6]. At the same time, due to environmental concerns, it is essential to examine the possible use of by-product gypsum, such as phospho-, fluoro-, citro-, titano-, boro- and FGD- gypsum as setting retarders [5, 6, 7, 8]. Phosphogypsum and FGD gypsum are the best candidates for the cement industry in Greece. Phosphogypsum (PG), a by-product from phosphoric acid production and the fertilizer industry, contains some impurities (P_2O_5, F, organic matter), which retard the hydration of OPC to a large extent, but does not change the strength development [6]. In this case, the effect of the impurities from these materials on the properties of cement remains to be studied. Desulphogypsum or FGD gypsum is an apparent waste material of the desulphurisation process of residual gases with limestone in coal burning power plants [6, 7, 8]. Strong efforts have been made to utilize FGD gypsum from the technical and economic point of view. At the same time FGD gypsum has gained recognition in the European Union and in the OECD as being a product [9]. The chemical reactions that can take place in order to convert flue gases into gypsum are the following:

$$SO_2 + H_2O \rightarrow H^+ + HSO_3^- \quad (a) \qquad H^+ + HSO_3^- + 1/2O_2 \rightarrow 2H^+ + SO_4 \quad (b)$$

$$2H^+ + SO_4 + Ca(OH)_2 \rightarrow CaSO_4 \cdot 2H_2O \quad (c)$$

There are two major methods of producing FGD gypsum: the wet-type scrubbing process, which is the most commonly used (90%) and the dry or semidry scrubbing process. The wet-type scrubbing process contains two alternative aspects: i) the slurry scrubbing method, where a mixture of $CaO/CaCO_3$ with water (slurry) is used, ii) the Clear Solution Scrubbing process (CSS) where either a solution of Na and $CaO/CaCO_3$ is used (Dual Alkali Process, DAP) or a solution of sulphate aluminium at pH 3-4 is used (Dowa Process, DP). The dry or semidry process also contains two alternative aspects: I) the Spray Dry Scrubbing (SDS) or the Spray Semidry Scrubbing (SSS) where the residual gases react with the dry absorbent ($Ca(OH)_2$ or $CaCO_3$) which is sprayed into the absorber and ii) the Limestone Injection Process (LIP) where the absorbent is straightly inserted into the boiler [10].

EXPERIMENTAL

Three calcium sulphate-bearing materials (CSBM), as are natural gypsum (as reference material), natural anhydrite and FGD gypsum, have been interground in different proportions with clinker for an hour. The chemical analysis of these materials, according to ASTM C 471M - 95, the hydrated forms of calcium sulphate, as well as their solubility at 25°C are shown in Table 1, while the chemical analysis of clinker is shown in Table 2. The % addition of each material in mixtures with clinker and the percentage of SO_3, measured by XRF, are shown in Table 3. The specific surface obtained was in the range of 3700 to 3900 cm^2/g. In order to achieve a SO_3 value about to 3.5%, which is as it is derived from Fig. 1, and discussed in the next chapter, the optimum percentage for the addition of the calcium sulphate-bearing materials (CSBM), clinker was ground with suitable mixtures of gypsum and FGD gypsum (CGF) and anhydrite and FGD gypsum (CAF). The addition of the mixtures of the CSBM was in the range of 6.5 – 6.8%, while clinker's admixing ratios varied from 93.5 to 93.2% respectively. The mixtures of calcium sulphate-bearing materials expressed as natural gypsum or anhydrite/FGD gypsum were: 0/100, 20/80, 40/60, 50/50, 60/40, 70/30, 100/0. The specific surface obtained was in the range of 3650 to 3850 cm^2/g.

Table 1 Chemical analysis and solubility of calcium sulphate-bearing materials

	COMBINED WATER	SO_3	SiO_2	CO_2	CaO	R_2O_3*	MgO	SOLUBILITY (g/100g H_2O)
Gypsum	19.30	43.41	0.65	2.51	32.40	0.03	0.92	0.260
Anhydrite	1.81	51.31	2.77	3.44	37.63	0.05	2.14	0.198
FGD	18.05	42.19	0.30	1.58	33.40	0.10	0.10	0.273

- R_2O_3: $Fe_2O_3 + Al_2O_3$

Table 2 Chemical analysis of clinker

	SO_3	SiO_2	CaO	R_2O_3*	MgO	C_3S	C_2S	C_3A	C_4AF
Clinker	0.71	22.00	66.15	8.52	1.92	63.87	14.89	6.65	11.23

* R_2O_3: $Fe_2O_3 + Al_2O_3$

Table 3 SO_3 content of cement mixtures * with different materials vs. their % addition.

SAMPLE	%	% SO_3	SAMPLE	%	% SO_3
CG1	4.0	2.37	CF1	4.0	2.46
CG2	4.5	2.58	CF2	4.5	2.69
CG3	5.0	2.93	CF3	5.5	3.12
CG4	5.5	3.12	CF4	6.5	3.47
CG5	6.5	3.54	CF5	7.0	4.06
CG6	7.0	3.93	CF6	7.5	4.36
CG7	7.5	4.31			

* CG: clinker-gypsum, CF: clinker-FGDgypsum

During industrial production of cement, in the cement mill, gypsum undergoes partial dehydration at 110-130^{0}C forming hemihydrate ($CaSO_4 \cdot 0.5H_2O$) and in some cases total dehydration, at 170-190^{0}C, forming soluble anhydrite (γ-$CaSO_4$) [11, 12]. In order to investigate the influence of gypsum and FGD gypsum dehydration to the compressive strength and initial setting time of cement, clinker was ground at a wide range of temperature (130-250^{0}C) with mixtures of CSBM [100% gypsum (CG), 75% gypsum/25% FGD (CGF)] and filler (limestone) to produce cement (CEM I) in industrial scale.

It is understandable though that these ratios are nominal since gypsum was not 100% pure in its composition. The temperature at the exit of the mill was always maintained below 118^{0}C by spraying the necessary quantities of water at the inlet and/or the middle of the mill. The quantity of sprayed water and the mill dew point for each trial were calculated through the construction of mill heat balances. All these with the clinker temperature, the SO_3 values obtained for cements and the admixing ratios of dihydrate and anhydrite to the CSBM are shown in Table 4. The specific surface obtained was in the range of 3500 to 3650 cm^2/g. Once anhydrite and hemihydrate have been formed in the cement, the degree of gypsum dehydration may be calculated as follows:

$$\%\ \text{Dehydration} = 100 * \frac{\%CaSO_4 \cdot 0.5H_2O * \dfrac{MWCaSO_4 \cdot 0.5H_2O}{MWCaSO_4 \cdot 2H_2O}}{\%CaSO_4 \cdot 2H_2O + \%CaSO_4 \cdot 0.5H_2O * \dfrac{MWCaSO_4 \cdot 0.5H_2O}{MWCaSO_4 \cdot 2H_2O}} \quad [5]$$

The calcium sulphates hydrates content (dihydrate, hemihydrate, anhydrite) in cement were determined by Differential Thermal Analysis (Netzsch STA 409 – DSC/TG). The results and the degree of gypsum dehydration are summarized in Table 5. At the same time, physico-mechanical properties such as setting time and compressive strength of the produced cement were determined based on EN standards.

Table 4 Clinker temperature, SO_3 content of industrially produced cements, dew point, sprayed water and admixing ratios of dihydrate and anhydrite to CSBM

	CG1	CG2	CGF1	CGF2	CGF3
Clinker temperature (^{0}C)	250	192	250	189	133
% SO_3	2.63	2.67	2.78	2.92	2.95
% dihydrate	3.70	3.70	4.73	4.54	4.71
% anhydrite	0.64	0.64	0.36	0.43	0.35
H_2O (kg/h)	5894	4146	6252	3966	2142
Dew point (^{0}C)	73	66	76	65	53

Table 5 Calcium sulfate hydrates and % dehydration in industrially produced cements

	T CLINKER, °C	DIHYDRATE, %	HEMI-HYDRATE, %	ANHYDRITE, %	DEHYDRATION, %
CG1	250	1.70	1.69	0.64	54.13
CG2	192	1.17	2.13	0.64	68.31
CGF1	250	2.04	1.95	0.34	53.14
CGF2	189	0.94	3.02	0.43	79.19
CGF3	133	1.84	2.40	0.35	60.70

RESULTS AND DISCUSSION

The compressive strengths at 1, 2, 7, 28 days, according to EN 196-2, for FGD gypsum are compared with those for gypsum in Figure 1. It is derived from this figure that the optimum percentage of SO_3 relating to the compressive strength, is 3.5% for both cements. It is also shown that the results of compressive strength of cements with FGD and natural gypsum are in the same order. This is clearer for SO_3 percentages around the optimum addition of 3.5%.

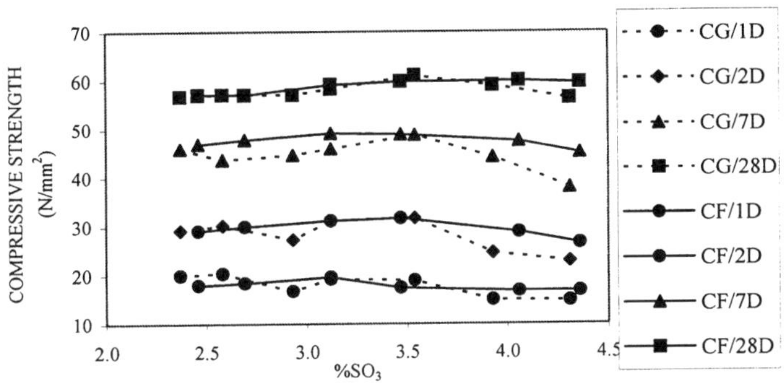

Figure 1 Comparison of compressive strengths of CG and CF cements vs. % SO_3.

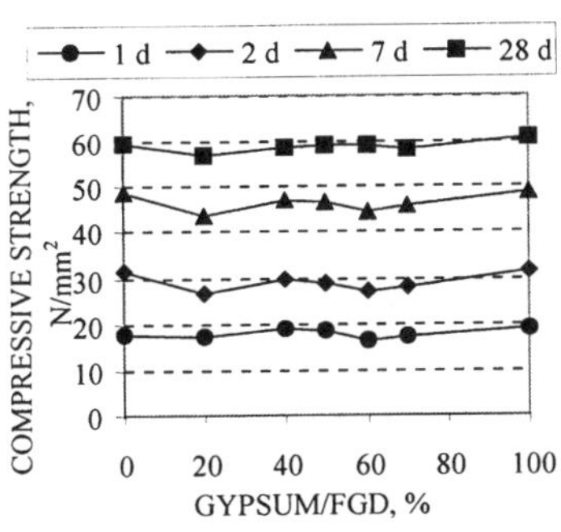

Figure 2 Compressive strengths of CGF cements.

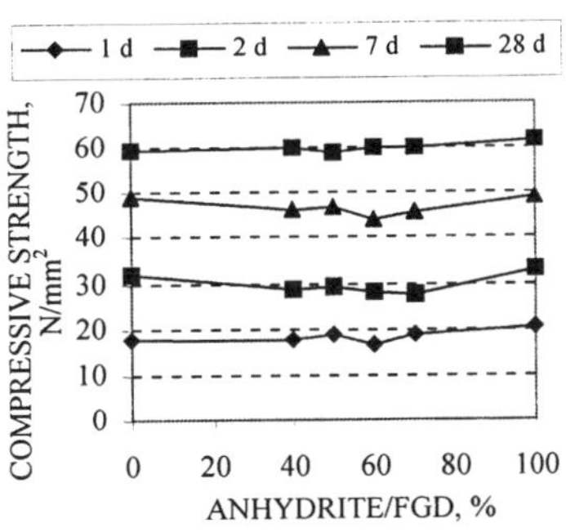

Figure 3 Compressive strengths of CAF cements

Compressive strength results (EN 196-1) for cements with mixtures of sulphate-bearing materials (CEM I) up to 28 days are shown in Figures 2 and 3. We must point out that no significant differences, except the 20/80 mixtures, are observed concerning the compressive strength of mixtures. The compressive strengths of CGF and CAF cements showed the same values for all examined ratios. It is also evident from Figure 4 that in the admixture gypsum/FGD gypsum, dihydrate of both origins exhibits a similar behaviour in cement strength performance. Optimum compressive strength is accomplished when the ratio between dihydrate content of gypsum and of FGD gypsum ranges from 0.7 to 1.6.

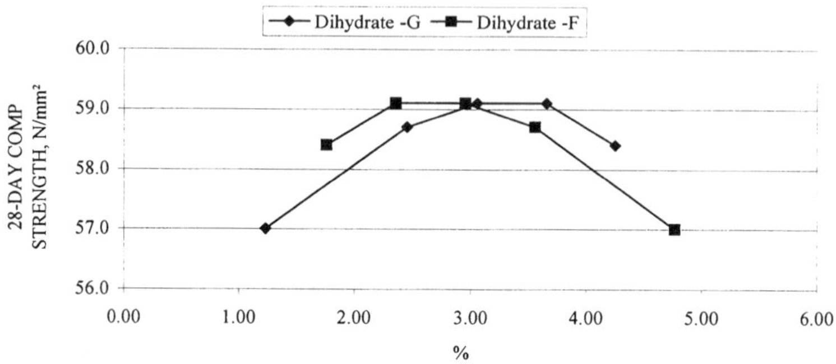

Figure 4 Cement with admixtures of gypsum and FGD gypsum: 28-Day Compressive Strength vs. Dihydrate content of a different origin.

As it is also shown in Figure 5, cements with FGD gypsum showed relatively higher setting times. The initial setting times of all mixtures versus the FGD content gypsum in the cement sample are plotted in Figure 6. It is evident that the setting time increases as more dihydrate of FGD gypsum replaces dihydrate of gypsum and anhydrite in the admixture. It is also extracted that FGD is a very good controller of setting, as it can give, in mixtures with gypsum and anhydrite, a large range of setting time (80 - 170 min). FGD gypsum is a better retarder than gypsum, although they both contain the same chemical compound, i.e. $CaSO_4 \cdot 2H_2O$.

The difference in reactivity of the two CSBMs is attributed to the difference in the morphology and the crystal lattice of the dihydrate. Gypsum is found in the shape of compressed sheets as a result of geological effects caused by heat and pressure changes. The theoretical crystalline system of gypsum is monoclinic. FGD gypsum is crystallized in rounded hexagons that is an indication that soluble anhydrite of hexagonal crystal lattice is formed. Moreover, higher solubility of FGD gypsum can be a fact (Table 1), which prolongs the setting process. It must be also pointed out that FGD gypsum contains a small percentage of $CaSO_3 \cdot 0.5H_2O$, which may retard the setting furthermore. In order to study the hydration mechanism of FGD gypsum through scanning optical microscope, suitable quantities of C_3A/C_4AF, which were separated from clinker, and FGD gypsum were mixed and hydrated. In the first ten minutes of hydration, there are a lot of pores and FGD crystals of about 50-μm sizes, which are approaching those of C_3A. The reaction starts and surface tensions are developed around the grains of FGD and C_3A, while voids are reduced (45 minutes).

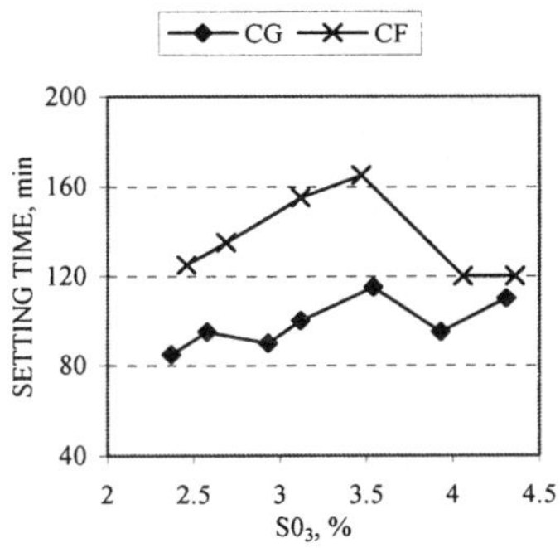

Figure 5 Setting times of CG and CF cements.

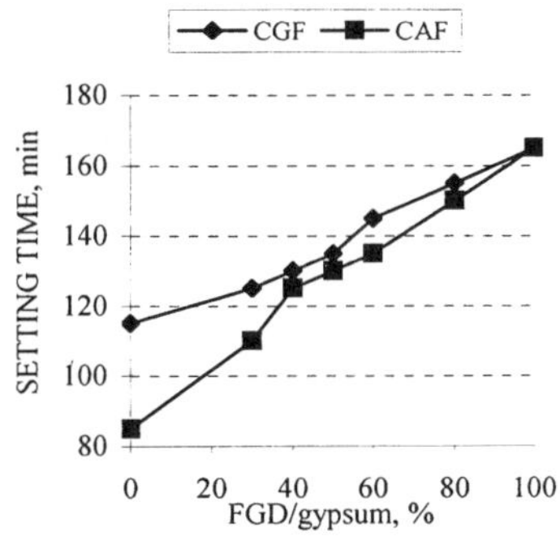

Figure 6 Setting times of CGF and CAF cements.

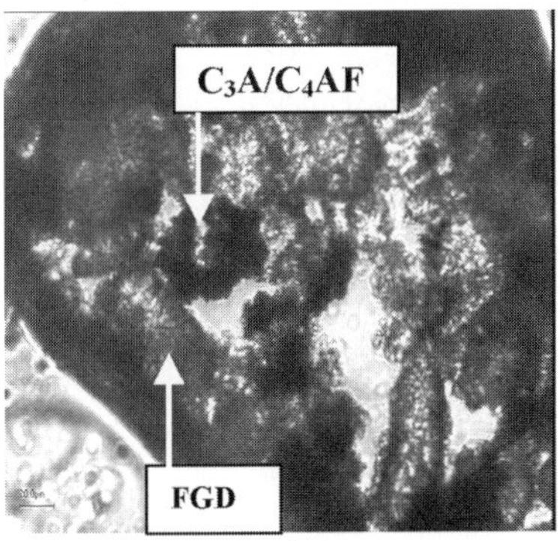

Figure 7 C_3A/C_4AF and FGD crystals after 10 min of hydration

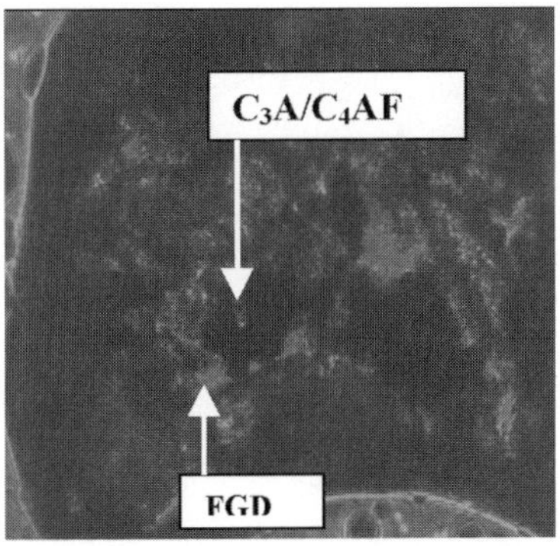

Figure 8 C_3A/C_4AF and FGD crystals after 45 min of hydration

From 80 to 100 minutes, surface tensions are increased and, at the same time, FGD and C_3A are combined forming bridges of hydration products, which fill voids of the paste and retard C_3A hydration. After 2 hours of hydration, C_3A seems to react with water and a compact mass of hydration products is formed. Setting has obviously initiated, there is no free water and early strength development is in progress. We must point out that the initial setting time in laboratory-produced cements ranged from 125 to 165 minutes that is confirmed by the hydration study through optical microscopy.

As far as the industrial production is concerned, temperatures in the interior of the mill did not exceed 130^{0}C and gypsum dehydration was limited to the formation of hemihydrate and not γ-$CaSO_4$. The behaviour of each CSBM in cement setting was confirmed using the industrial samples. The cement samples with the highest three setting times contained FGD gypsum. In cement setting, the presence of hemihydrate and anhydrite has a shortening effect (Figure 11).

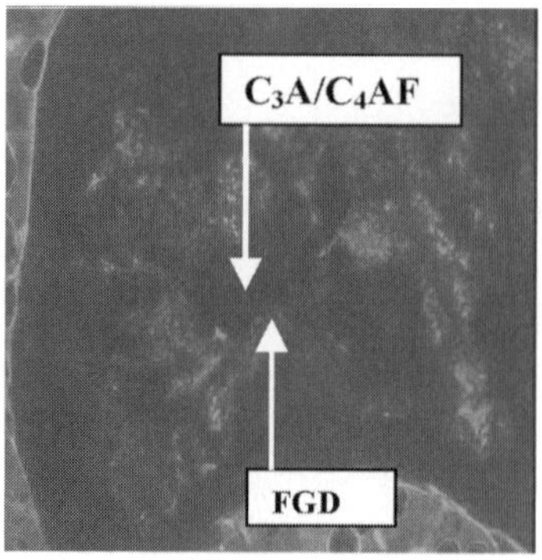

Figure 9 C_3A/C_4AF and FGD crystals after 90 min of hydration

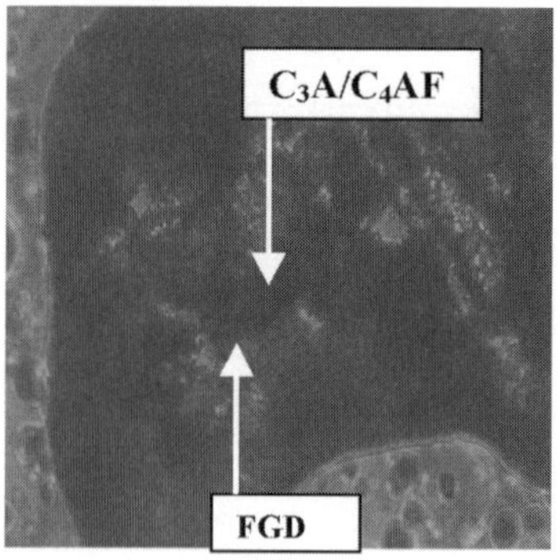

Figure 10 C_3A/C_4AF and FGD crystals after 130 min of hydration

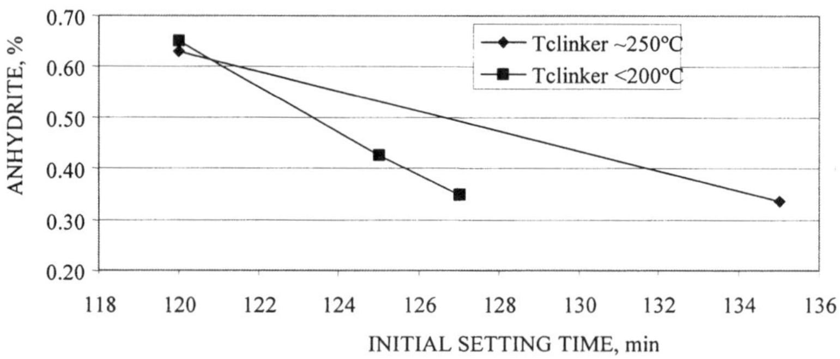

Figure 11 Cement with various admixtures of gypsum and FGD gypsum produced in industrial scale: anhydrite content vs. initial setting time

As illustrated in Figure 12, the differences in reactivity are also reflected in the respective dehydration profile (dynamic step 1°C/min, isothermal step 30 min). In gypsum, water is slowly removed, starting at temperatures below 90°C; whereas in FGD gypsum, water is rapidly lost forming a sharper peak, starting at temperatures above 95°C. This means that at temperatures below 100-110^{0}C, less hemihydrate and anhydrite is formed when FGD gypsum is used as setting retarder and so setting time is increased comparing with natural gypsum.

On the other hand, the forms of calcium sulphate have a little positive effect on the differentiation of the 28-day compressive strength. Still, hemihydrate is considerably more effective than the rest of the $CaSO_4$ forms, since higher compressive strengths were achieved in samples with high hemihydrate content (Figures 13 and 14). In this case, the temperature of clinker was between 130-190^{0}C (Table 5) and the dehydration exhibited a maximum value for dew point in the range of 55 to 65^{0}C, as it is derived from Figure 15.

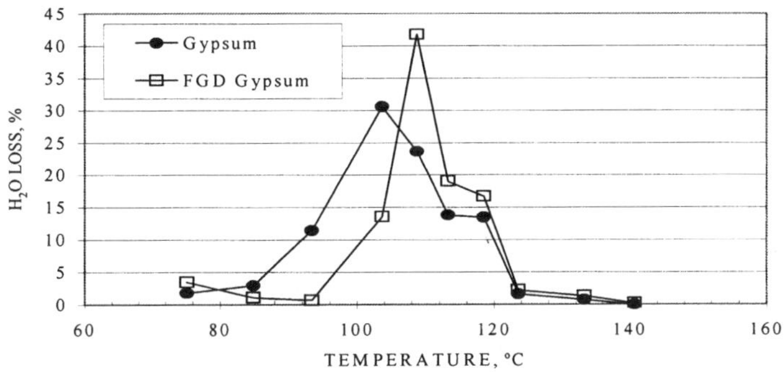

Figure 12 Gypsum and FGD gypsum dehydration profile.

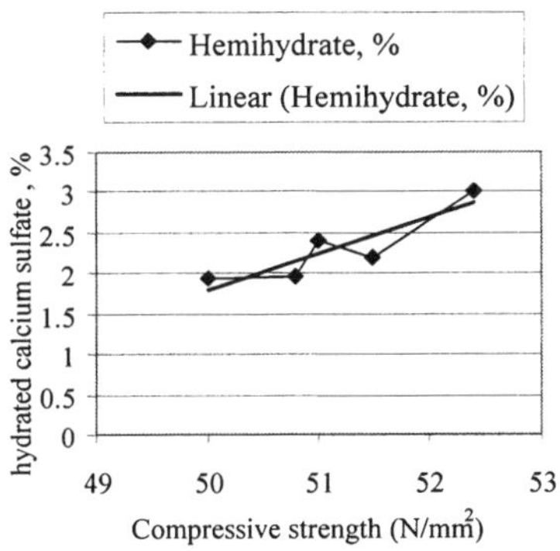

Figure 13 28-day compressive strength vs. the ratio of hemihydrate in industrially produced cement.

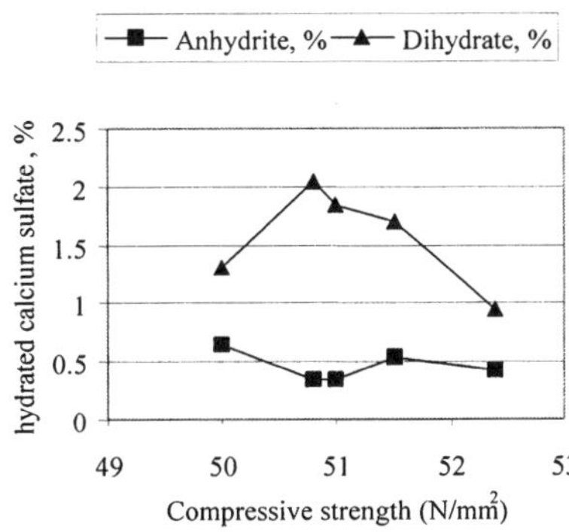

Figure 14 28-day compressive strength vs. the ratio of dihydrate and anhydrite in industrially produced cement.

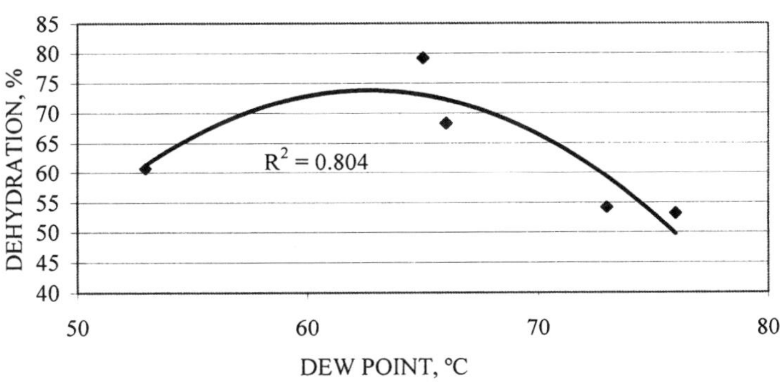

Figure 15 Cement with various admixtures of gypsum and FGD gypsum produced in industrial scale: Gypsum Dehydration vs. Cement mill Dew point.

CONCLUSIONS

The study of laboratory and industrially produced cements with a diversified setting retarder composition showed that FGD gypsum is a suitable alternative to gypsum. FGD gypsum showed about 1 h delay in its initial setting time compared with natural gypsum. This must be attributed to the different solubility of each CSBM and consequently to the different quantities of soluble SO_3, which exist in cement after its mixing with water and react with C_3A, retarding its hydration. FGD gypsum showed similar behaviour to gypsum in compressive strength development, especially in the range of the optimum SO_3 addition (about to 3.5%). The use of mixtures of sulphate bearing materials with gypsum seems to be advantageous for the precise control of setting time. However, the high moisture content of FGD gypsum (~10%) could cause handling and feeding problems, when it is used in large percentages. Gypsum and FGD gypsum dehydration and formation of hemihydrate occurs in industrial cement mills, that has a diverse effect on setting and compressive strength depending on the composition of the setting retarder. By controlling the temperature of the clinker and the relative humidity within the mill through the quantity of sprayed water in the mill, optimum degree of dehydration is observed at a dew point about 55 – 65^0C.

REFERENCES

1. TAYLOR H. F. W., Cement Chemistry, Thomas Telford, 2nd edition, 1997, pp. 182-183, 218-241.
2. Lea's Chemistry of Cement and Concrete, Arnold, 4th edition, 1998, pp. 260-262.
3. SKALNY J., JAWED J., TAYLOR H. F. W., Studies on hydration of cement-recent developments, World Cement Technology, September 1978, pp. 183-193.
4. LOCHER F. G., RICHARTZ W., SPRUNG S., Setting of cement – Effect of adding calcium sulphate, ZKG, vol. 6, 1980, pp. 271-277.
5. THEISEN K., Relationship between gypsum dehydration and strength development in Portland Cement, ZKG, vol. 10, 1983, pp. 571-577.
6. TSIMAS S., DELAGRAMMATIKAS G., PAPAGEORGIOU A., TZOUVALAS G., Evaluation of industrial byproducts for the control of setting time of cements, Cement and Concrete World, vol. 6, 2001, pp. 54-62.
7. OZKUL HULUSI M., Utilization of citro- and desulphogypsum as set retarders in Portland cement, Cement and Concrete Research, vol. 30, 2000, pp.1755-1758.
8. MANJIT SINGH, Influence of blended gypsum on the properties of Portland cement and Portland slag cement, Cement and Concrete Research, vol. 30, 2000, pp.1185-1188.
9. HAMM H., Coping with FGD gypsum, ZKG, vol. 8, 1994, pp.443-449.
10. ROULIA M., Flue gas desulphurisation technologies with calcium compounds, Chimika Chronika, vol. 3, 1999, pp.73-77.
11. ROY S., GHOSH S. N., Case study of rising cement mill temperatures – incidence of false set, ZKG, vol.54 (No 4/2001), pp. 206 – 208
12. STRYDOM C. A., POTGIETER J. H., Dehydration behavior of a natural gypsum and a phosphogypsum during milling, Thermochimica Acta, vol. 332, 1999, pp. 89-96.

EFFECT OF HEAT TREATMENT ON THE BEHAVIOUR OF MONTMORILLONITE CLAY IN THE PRESENCE OF LIME

A Taher
Al-Azhar University
Egypt

Abstract The activation of clays by heat treatment will produce artificial pozzolana which reacts with Ca^{+2} or $Ca(OH)_2$ in presence of water to form new building materials. In this study the montmorillonite clay was activated by burning at different temperatures (550,700,800,900 & 1000°C) then mixed with lime at different compositions. The dry mixtures were mixed with water at w/c ratios of 0.5,0.6 &0.7 by weight .The resulting masses was then molded and the produced specimens were cured in an autoclave at 10 atmosphere pressure for different periods. The change occurring in the physico-chemical properties were investigated by compressive strength and chemically-combined water measurements, also the change in morphology and microstructure of hardened pastes were investigated using SEM tests. The results of this study indicate that the optimum activation temperature for montmorillonite clay to produce artificial pozzolana is 1000°C and the optimum mix composition with lime to produce binding compounds is 80% burnt mintmorillonite clay : 20% lime $Ca(OH)_2$.

Keywords: Heat treatment; Montmorillonite clay; Lime.

A Taher is a Assistant Professor in the Chemistry Department at Al-Azhar University, Egypt.

INTRODUCTION

Burnt clays are very interest for the production of building materials; this is due to the formation of a mixture of amorphous silica and alumina in the stoichiomertric proportion of the original materials [1] The addition of thermally or chemically activated clay to mortar increases the early hydration due to adsorption and ion-exchange process occurring on the surface [2]. The clay minerals dissociate between 500 and 700°C into the amorphous alumina-silicate mixture, whereas $CaCO_3$ decomposes into CaO between 600-800°C, above 800°C, free CaO content decreases and disappears almost completely at 950°C due to formation of Wollastonite($CaO.Al_2O_3$),Anorthite($CaO.Al_2O_3.2SiO_2$) and Gehlenite ($2CaO.Al_2O_3.SiO_3$)[3]. Mixing of $Ca(OH)_2$ and water with calcined clay will produce Gehlenite hydrate ($2CaO.Al_2O_3.SiO_3.8H_2O$) and CSH (I) with some of $4CaO.Al_2O_3.13H_2O$ [4]. The amount of C_2ASH_8 increases with curing time and hydrogarnet is also formed at early stages of hydration in the mixes of lower lime contents and increases on the prolonged hydration due to the lower pH of reaction medium[5].

The thermal activation of clay doesn't produce completely amorphous materials, but it shows pozzolanic activity when mixed with lime[6]. An increase of the treatment temperature of clay can have a positive effect (increase of vitreous, zeolitic and clayey phase activity) or a negative effect (specific surface area decreases, and insoluble residue increases)[7]. The reactivity of natural pozzolana with lime are changed by thermal treatments[8],however, it is not possible to compare the changes from one natural pozzolana to another, due to slow rearrangement caused by intense thermal heating of the pozzolana. The SEM studies of three kaolinite materials from different sources before and after firing at 300,500,700,900,1100 and 1250°C showed that the raw material of clay consists of hexagonal particles and finer grains , while after heating at 700-1100°C they converted to poorly crystalline structure[9]

EXPERIMENTAL WORK

Three types of autoclaved specimens were prepared by mixing lime with thermally treated montmorillonite clay which were heated at 550,700,800,900 and 1000°C for 4 hours. Four dry mixtures were prepared from various proportions of burnt montmorillonite clay and lime with mass ratios,90:10(I),80:20(II),70:30(III) and 60:40(IV) using ethanol to attain complete homogeneity of the mixtures, then dried overnight at 105°C. Each dry mixture was mixed with lime for three minutes at water/solid ratios of 0.5,0.6 and 0.7 by weight. The resulting mass was then pressed by hand molding pressure into stainless steel cylindrical moulds of 3.14 cm^2 cross section and 2 cm height.

The moulds were vibrated for one minute to remove any air bubble and voids. Immediately after molding, the cylindrical specimens were cured in humidity cabinet at 100% relative humidity at room temperature (23 $\pm$2°C) for 24 hours in order to attain the setting of specimens. The specimens of each mixture were autoclaved in saturated steam at 10 atmospheres for 2,6,12 and 24 hours. Compressive strength tests were carried out on the dried samples [10]. The chemically-combined water contents were determined after stopping the hydration of hardened pastes and by using the ground dried specimens[11]. The morphology and microstructure of some dried samples were investigated using scanning electron microscopy (SEM).

RESULTS AND DISCUSSION

The compressive strength values of hardened specimens made from the autoclaved lime-artificial pozzolana (montmorillonite clay burnt at different temperatures) pastes are graphically shown in Figure 1. The results of compressive strength show very low values at low firing temperatures of clay (550-800°C); this effect is mainly attributed to the low hydraulic activity of artificial pozzolana made at low firing temperatures of clay. However, the strength values are relatively high values at high firing temperature of clay (900&1000°C) for all of the autoclave lime- burnt montmorillonite clay mixes under investigation. This is attributed to the hydraulic activation of the artificial pozzolanas made at high temperatures. Therefore, the hydration products obtained, mainly as calcium silicate hydrates and/or calcium aluminosilicates, are responsible for the later strength development during the final stages of hydration. Mix II at burnt clay temperature 1000°C has the highest strength values at all stages of the hydrothermal solidification process; hence, it is concluded that the desired temperature of thermal treatment of montmorillonite clay which yield a reasonable hydraulic reactivity is 1000°C and the reasonable mix composition of burnt montmorillonite clay:lime is 80:20 weight%, respectively.

The results of chemically-combined water contents,(W_n,%) of the hydrothermally hardened burnt montmorillonite clay-lime pastes are graphically represented in Figure 2. For all the hardened burnt montmorillonite clay-lime pastes the combined water content decreases with increasing autoclaving time up to 6 hours, this is mainly associated with the transformation of the initially formed high-lime calcium silicate hydrate into low-lime form as a result of the hydrothermal interaction between the initial hydration coatings and the anhydrate parts of pozzolana grains. Some mixes show an increase in the combined water content up to 24 hours autoclaving (II and IV) and others show a decrease in the W_n-values. The decrease in combined water content occurred for the samples at later autoclaving time (24hours) my be attributed to the adsorption of part of water on the surface of CSH as a results of its higher surface area (higher pozzolanas content). This adsorption makes the determination of reacted water inadequate [12]. Thus, it can be concluded that the strength of hardened product for a certain mixture is related to the chemically-combined water content, i.e., it increases as the related water content increases and visa versa.

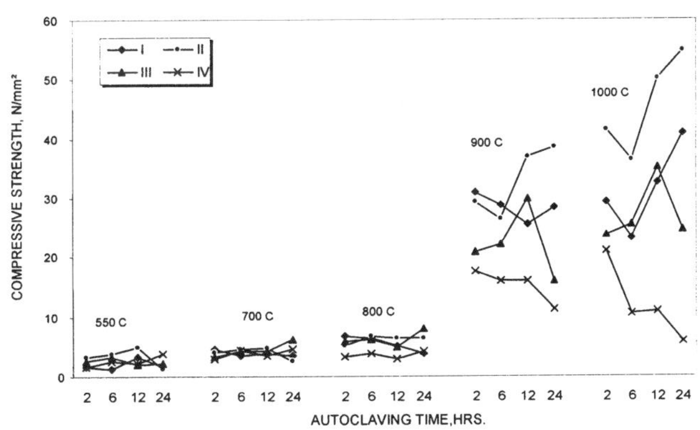

Figure 1 Effect of mix composition of burnt montmorillonite clay-lime pastes at different firing clay temperatures on the compressive strength as a function of autoclaving time.

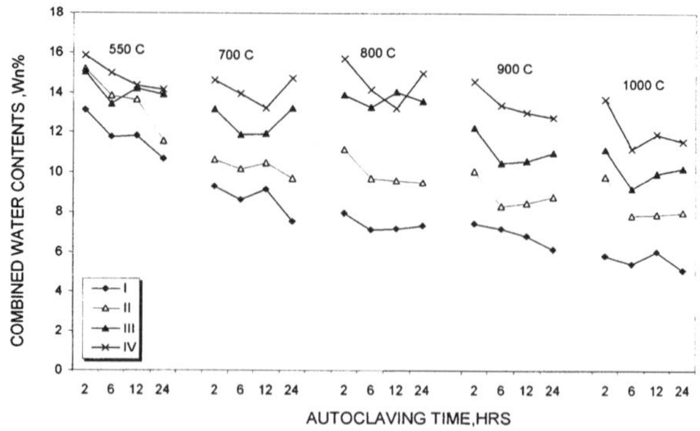

Figure 2 Effect of mix composition of burnt montmorillonite clay-lime pastes at different firing clay temperatures on the chemically combined water contents as a function of autoclaving time

The results of the physico-chemical characteristics of the hardened pastes made from lime and artificial pozzolana (burnt montmorillonite clay at different temperatures) mixes I-IV can be summarized from the view of the effect of firing temperature used for activation of clay on the hydraulic reactivity of the autoclaved pastes made from these mixes. The compressive strength results obtained for various lime-burnt montmorillonite clay are shown in Figure 3.

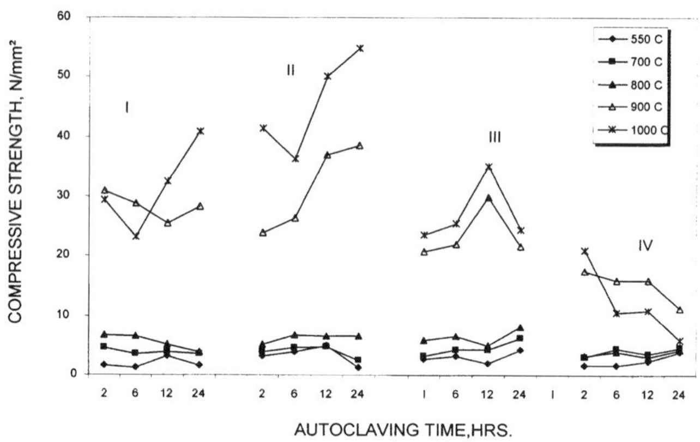

Figure 3 Compressive strength of different mixes(I-IV) at different firing clay temperatures as a function of autoclaving time.

The results indicate that the increase in firing temperature of montmorillonite clay causes a marked increase in the compressive strength values especially for the mixes having high pozzolana contents. Mix II (80% montmorillonite clay burned at 1000°C: 20% lime) has the highest compressive strength values. The strength values of the autoclave specimens made from mixes (III and IV) containing low pozzolana contents are very low in comparison with those containing high pozzolana contents (mixes I and IV) at all autoclaving age and for all firing temperature of the clay. This is mainly attributed to the formation of larger amounts of calcium silicate hydrate(CSH) in case of the autoclaved mixes richer in pozzolana contents, and formation of α-dicalcium silicate hydrate(α-C_2SH) in case of the autoclaved mixes having low pozzolana contents(high lime contents). The CSH possess very strong binding characteristics as compared with weaker binding character of the α-C_2SH[13].

The results of chemically-combined water contents ;(W_n%), are shown in Figure 4 for the hydrothermaly hardened lime- burnt montmorillonite clay (mixes I, II, III and IV) pastes treated at various autoclaving ages. The results show that combined water contents of the autoclaved lime pozzolana pastes decreases with increasing autoclaving age up to 6 hours for all mixes (I-IV) and at all firing temperatures of montmorillonite clay. This may be attributed to the formation of low-lime CSH from the initially formed high-lime CSH. Later, the combined water contents increases gradually up 12 hours autoclaving, and finally decreases or increases up to 24 hours autoclaving; this result may be attributed to the formation of hydrated phases with a high specific area. Generally, the autoclaving mixes containing burnt montmorillonite clay, burned at 550°C, have the highest combined water contents while those containing montmorillonite clay burned at 1000°C have the lowest combined water contents at all stages of the hydrothermal process. This result may be due to the formation of lime-rich calcium silicate hydrate mainly as (CSH-I) with a high water content in case of autoclave mixes containing burnt montmorillonite clay at 550°C as well as the presence of uncalcined parts of clay minerals which have a very high specific surface area.

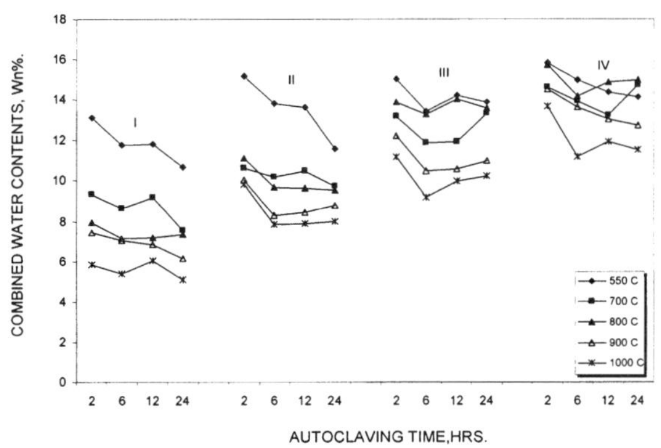

Figure 4 Chemically-combined water contents (Wn%), of different mixes(I-IV) at different firing temperatures as a function of autoclaving time.

Eight samples were investigated using scanning electron microscopy (SEM) as representative for the hardened lime- burnt montmorillonite clay pastes. Five of these samples were of mix II at the firing temperatures of clay 550,700,800,900 and 1000°C and autoclaved for 24 hours; their SEM micrographs are shown in Figures 5(a&b), 6(a&b),7(a&b),8(a&b) and 9(a&b&c),respectively. The other three samples were of the same mix II at the firing temperature of clay 1000°C and autoclaved for 2,6,12 hours; their SEM micrographs are shown in Figures 10, 11(a&b) and 12, respectively.

Figure 5a

Figure 5b

Figure 6a

Figure 6a

Figure 7a

Figure 7b

Figure 8a

Figure 8b

Figure 9a

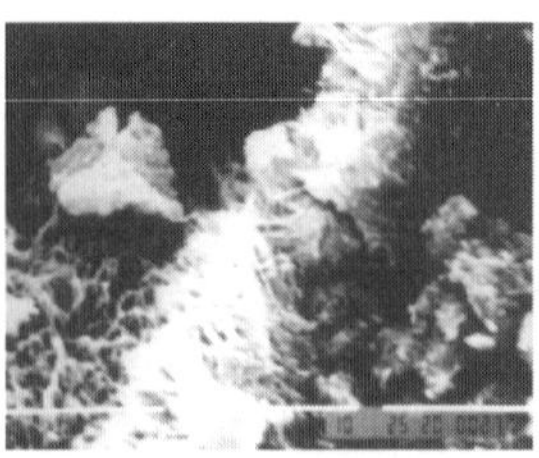

Figure 9b

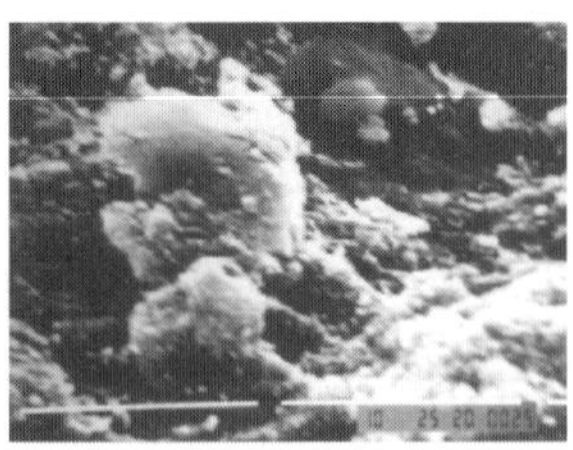

Figure 9c

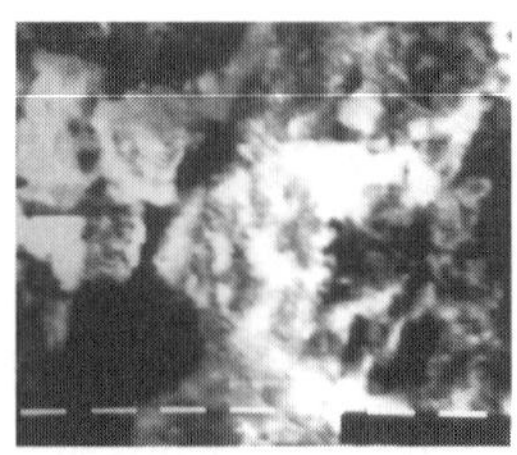

Figure 10

Figure 11a

Figure 11b

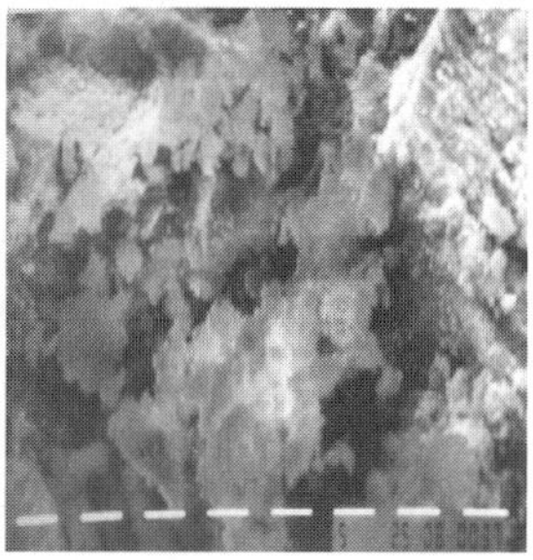

Figure 12

In the final stage of hydrothermal reaction(24 hours) of mix II, containing burnt montmorillonite clay burned at 550°C , the hydration products appeared as semi-crystalline foils of tobermorite-like calcium silicate hydrates(CSH); the cubic hydrogarnet crystals were also distinguished in the structure, Figure5 (a &b). After 24 hours autoclaving of mix II, containing burnt montmorillonite clay burned at 700°C, the SEM micrographs displayed a mixture of various hydration products mainly as nearly amorphous and short fibers of CSH Figure 6 (a&b). The SEM micrographs obtained for the autoclaved specimens of mix II, containing burnt montmorillonite clay burned at 800°C, after 24 hours autoclaving displayed ill-crystallized and nearly amorphous CSH as well as the cubic hydrogarnet crystals Figures 7 (a&b). The microstructure obtained after 24 hours autoclaving of mix II, containing burnt montmorillonite clay at 900°C, showed a dense structure of nearly amorphous CSH as representing the main hydration products Figures 8 (a&b). In the final stage of hydrothermal solidification process of mix II, containing burnt montmorillonite clay at1000°C, the SEM micrographs displayed close-textured, dense and massive structure composed of interlocking fibers and crumpled foils as well as particles with irregular appearance Figure 9(a&b). The micrographs obtained for mix II containing burnt montmorillonite clay at 1000°C after 2,6 and 12 hours autoclaving showed the formation of nearly amorphous hydrates after 2hours autoclaving Figure 10 and a massive structure of ill-crystallized and fibrous particles of CSH after 6 hours autoclaving Figure 11(a&b) as well as a dense structure of microcrystalline and particles of CSH after 12 hours autoclaving Figure 12.

CONCLUSIONS

From this study we can concluded that:

- The suitable firing temperature for artificial pozzolana prepared from montmorillonite clay which yield a high hydraulic reactivity in presence of lime is 1000°C and the reasonable mix composition is 80% burnt montmorillonite clay : 20% lime.
- The chemically-combined water contents increase by increasing lime contents and the highest values obtained at low firing temperature of montmorillonite clay.
- The closed-textured, dense and massive structure of pastes prepared from 80% burnt montmorillonite clay at 1000°C and 20% lime at final stage show that the hydration products were mainly calcium silicate hydrate and calcium alumino silicate hydrate.

REFERENCES

1. MACHENZIE,R.C;GREENE,K.I, AND SCHEELTZ,N.,Econ.Geol.,46,p.311(1951).
2. A.F.GALAL,E.E.HECKAL,K.KH.AL.NOAIMI AND S.A.ABO-EL-ENEIN,Twelfth international on Cement Microscopy,April 2-6(1990).
3. GOOSH,S.N.,Advances in Technology,Wheaton Co.Ltd.Exter(1981)
4. MURRA,M.,"Hydration reaction and hardening of calcined clays and related minerals".“I-Preliminary investigation on metakaolinite",Cem.Concr.Res.,13(2),259-66 (1983).
5. SERRY,N.A.;TAHA,A.S.;EL-HEMALY,S.A.S.& EL-DIDAMONY,H.,"Metakaolin lime hydration products", Thermochim.Acta,79,103-110,(1984).
6. AMBROISE,J.;JOSEPH,D.J. and Pera,J.,Amer.Ceram.Soc.Bull.,66,p.1731(1987)
7. COSTA,U AND MASSAZZA,F.,"Influence of the thermal treatment on the activity of some natural pozzolanas",il Cemento,74,105-122(1977).
8. ROSSI,G. and Foechelli,L.,"Porous structure and reactivity with lime of some Itallian pozzolans" , il Cemento,73,215-221(1976).
9. TORILLO,A.R.;MARTINEZ,E.E.;CABILLON,S.L.; BENDIA,E.L.,MANALAC,G.C. and Shumosaka, K., "Kaolinite-rich clays of the Philipines",J.Geol,Soc,Phlipp.,36(2),32-46(1982).
10. BS 1881,Part 4,’Methods of testing Concrete for Strength “ (1970).
11. S.BRUNAUER, L.E.CAPELAND, the chemistry of concrete, Scient.Am.210(1964)83.
12. KALOUSEK,G.L.,J.Am.Concr.Inst.,51,989,(1955).
13. RAMACHANDRAN,V.S.,”Application of differential thermal analysis in cement chemistry”, Chem.Publ.Co.INC.,New York ,(1969).

ALKALI ACTIVATION OF PFA: EFFECT ON STRENGTH DEVELOPMENT

M R Jones

L J Csetenyi

E Csetenyi **R K Dhir**

University of Dundee

United Kingdom

ABSTRACT The compressive strength development of mortars with 30%, 50% and 100% by mass replacement of Portland cement by pulverised fuel ash, activated by alkalis was studied. The considered activators were based on NaOH, $Ca(OH)_2$, and waterglass (2:1 SiO_2:Na_2O) in various concentrations. Two different temperature curing regimes, *i.e.* 20°C and 60°C water storage after moulding at 24 hours, were tested. 40mm cube strength was measured at 1, 3, 7, 28, and 91 days. The results show lower strength development rates than PC control and indeed literature references report paste strength development cured at temperatures typically between 65 and 85°C. The results underline the difficulty of alkali activating PFA at practical temperatures that could be used in precast concrete production.

Keywords: Alkali activated, PFA, Mortar, Strength development, Heat curing

M R Jones is a Reader in the Concrete Technology Unit and head of the Division of Civil Engineering, University of Dundee. He is a chartered civil engineer with extensive work experience both in industry and academia. His research focuses mainly on binder technology, concreting and concrete durability.

L J Csetenyi is a Research/Teaching Fellow in the Concrete Technology Unit, Division of Civil Engineering, University of Dundee. His research interests include interactions and utilisation of waste materials and by-products in cement and concrete.

E Csetenyi is a Research Assistant in the Concrete Technology Unit, Division of Civil Engineering, University of Dundee. Her work focuses on the chemistry and technology of application of waste alkalis and other by-products to enhance performance of Portland cement blends containing high levels of PFA.

R K Dhir OBE is Director and Professor of the Concrete Technology Unit at the University of Dundee. His is a member of numerous national and international technical committees and has published extensively on many aspects of concrete technology, cement science, durability and construction methods..

INTRODUCTION AND BACKGROUND

Pulverized-fuel ash (PFA) has been used successfully for over 50 years in concrete construction. The use of this material in concrete offers a number of economic, environmental and technical benefits, including improved fresh concrete properties, long-term strength development and durability performance. However, low emission systems at power stations have changed the nature of PFA, in turn, have affected its contribution to concrete performance. In addition, if the relatively slow initial reaction rate of PFA could be accelerated, could lead to wider applications. A number of techniques exist to enhance the contribution of PFA to concrete, including (i) physical processing to remove coarser particles, (ii) mechanical grinding to increase fineness, (iii) thermal processing, and (iv) increased curing temperature and/or duration [1-3]. However, there are economic implications associated with many of these. One further method for enhancing the contribution that PFA can make to strength may be to an additional react it with an admixed alkali to stimulate break down of the glass phase of PFA, thereby accelerating pozzolanic reactions. Alkali activation can offer a practical, economic and environmental sustainable way to enhance the early reaction rate of PFA, if suitable waste materials arising from other chemical processes can be obtained. Previous research in this area has covered activation of PFA and other industrial by-products increasingly used as cement components in concrete construction, and has demonstrated the feasibility for alkali-activation of PFA [4-6], with reaction products similar those obtained with metakaolin (MK) identified. However, whilst research covering GGBS and MK [7-9] indicated that activation is most effective (both economically and technically) with addition of alkali to the mix water, work with PFA has tended to focus on storage of alkali/PFA mixtures at elevated temperatures (55 to 85°C), followed by drying and grinding, to produce a processed PFA [5, 6]. The separate technological steps associated with pre-treatment and successive application in concrete, however, undermine the cost effectiveness of the overall process. Present work, therefore, has concentrated on in-situ activation of PFA during the mixing process and identifying suitable alkali systems.

EXPERIMENTAL PROGRAMME

Starting materials and sample preparation

The test cements were a CEM I 42.5 Portland cement (PC) and a low CaO content PFA conforming to BS 3892: Part 1 (LoI: 10.8%, 45μm sieve residue: 5.5%), both of UK origin, were used to make the control mixes at 30% and 50% cement replacement of PC. The PC and PFA compositions are given in Table 1. Analytical grade reagents of NaOH, $Ca(OH)_2$ and waterglass (2:1 SiO_2:Na_2O) identified as being the most effective in the literature were selected as the alkali admixtures, based on these. A siliceous sand to zone M (BS 822) of saturated surface dry condition was used in the mortars. The control and test mixes are given in Table 2. Flow of fresh mortar was also measured, as given in Table 3.

Mortars were mixed at 0.5 w/c ratio for 10 minutes to make 20kg batches using a laboratory forced action concrete mixer set to 20rpm. Mortar bars of 40×40×160 mm were cast in three gang moulds conforming to BS EN 196-1 and stored sealed for 24h either at 20°C or 60°C followed by curing under water at 20°C or 60°C. Specimens of 100% PFA made with 8M NaOH as the mixing water were too weak to demould until 7d and, therefore, had a longer period of sealed curing.

Table 1 Composition of cement and PFA used

MAJOR OXIDE, % by mass	CEM I 42.5 cement	PFA (BS 3892: Part1)
CaO	64.78	4.06
SiO_2	21.16	48.09
Al_2O_3	5.07	30.70
Fe_2O_3	2.41	4.10
MgO	2.05	1.08
MnO	0.05	0.04
TiO_2	0.24	1.70
K_2O	0.85	0.97
Na_2O	0.71	0.28
P_2O_5	0.07	0.84
Cl	0.06	0.01
SO_3	2.92	0.76
	Total: 100.37	Total: 92.63

Table 2 Mix proportions and curing temperatures

	DESIGNATION OF TEST SAMPLES	CEMENT		ALKALI ACTIVATOR			CURING TEMP.
		PC	PFA	NaOH	$Ca(OH)_2$	Waterglass*	
Control	P2	100%	-	-	-	-	20°C
Control	3F2 3F6	70%	30%	-	-	-	20°C 60°C
Control	5F2 5F6	50%	50%	-	-	-	20°C 60°C
	3F5C2 3F5C6	70%	30%	-	1.5M	-	20°C 60°C
	5F5C2 5F5C6	50%	50%	-	2.5M	-	20°C 60°C
	3F3W2 3F3W6	70%	30%	-	0.9M	2.5M**	20°C 60°C
	5F3W2 5F3W6	50%	50%	-	0.9M	2.5M**	20°C 60°C
	3F5N2 3F5N6	70%	30%	1.5M	-	-	20°C 60°C
	3F8N2 3F8N6	70%	30%	2.5M	-	-	20°C 60°C
	5F5N2 5F5N6	50%	50%	2.5M	-	-	20°C 60°C
	5F8N2 5F8N6	50%	50%	4M	-	-	20°C 60°C
	F8N2 F8N6	-	100%	8M***	-	-	20°C 60°C

NB: w/(PC+PFA)=0.5, sand/(PC+PFA)=3.0, activator contents expressed as mol/l in mixing water (even if not dissolving completely and thus forming a slurry).
The last digit of designation of samples relates to curing temperature.
* SiO_2:Na_2O by weight: M_s=2:1, fine grain solid, apparent molar weight taken as 180g
** Equals to 5M Na content. Waterglass/lime ratio taken from [11]
*** Similarly high Na levels are used in [6, 10]

RESULTS AND DISCUSSION

Flow of mortars

The effect of alkali additions on the flow characteristics of the test mortars is shown in (Table 3). In some cases (5F5N and 5F5C) mortars were earth dry and difficult to compact in the moulds. However, in nearly all cases the flow was reduced compared to control mixes. This suggested that some form of viscosity modification was occurring with the presence of the alkali activator. This could be either a change in the viscosity of the mix water of agglomeration of particles into flocs. In the present study it was not possible to identify which of these processes or indeed other effect was dominant. The observation is, however, significant for the practical application of alkali-activation. Again, it was outside the scope of the study to look at whether this problem could be offset by the use of an additional plasticising chemical admixture.

Table 3 Flow of mortars (cm) measured according to BS EN 1015-3:1999

MIX	P	3F	5F	3F5N	3F8N	5F5N	5F8N	F8N	3F5C	5F5C	3F3W	5F3W
FLOW	14.0	14.2	13.0	12.0	13.2	10.7	12.0	14.0	13.7	10.7	12.5	14.0

NB Mixing 100% PC mortar and standard sand using a BS EN 196-1 mixer gives a flow value of 22.5cm and 17.5cm with the test sand.

Strength development

Cube compressive (40mm) and tensile (prism flexure) strengths of mortar bars were measured at 1, 3, 7, 28, and, in selected cases, 91 days using a standard test machine conforming to BS EN 196-1 (Tables 4 and 5).

Table 4 Tensile and compressive strength of mortars cured at 20°C

SAMPLE		TENSILE STRENGTH, MPa					COMPRESSIVE STRENGTH, MPa				
		1d	3d	7d	28d	91d	1d	3d	7d	28d	91d
Control	P2	3.5	5.4	6.3	7.7	9.3	15.0	15.7	32.3	38.3	50.4
Control	3F2	2.6	5.1	5.8	8.6	9.3	10.1	18.7	26.7	34.7	39.5
Control	5F2	1.4	2.9	7.2	6.1	7.7	5.0	10.4	14.7	22.9	29.0
3F5C2		2.4	4.0	4.8	9.0	8.8	*10.5*	*19.0*	22.4	30.8	34.8
5F5C2		1.4	2.4	3.1	6.2	7.2	*5.7*	10.3	13.3	15.6	26.8
3F3W2		1.6	2.4	3.7	4.0	4.7	3.6	5.1	5.7	10.4	15.1
5F3W2		*2.2*	*3.5*	4.6	4.2	5.2	*5.3*	7.3	6.5	14.7	17.1
3F5N2		2.4	2.8	3.4	5.3	7.0	7.3	11.0	14.7	10.6	21.0
3F8N2		n.m.	*5.3*	*5.9*	7.0	7.3	n.m.	7.9	10.0	20.9	20.8
5F5N2		*1.8*	2.4	2.9	5.0	6.6	3.2	5.2	8.0	10.6	18.3
5F8N2		n.m.	2.8	4.2	5.6	7.0	n.m.	5.0	5.5	17.4	19.5
F8N2		-	-	-	-	-	-	-	-	-	-

NB n.m. = not measured; - = not measurable; *italics* = indicates strength greater than control

Table 5 Tensile and compressive strength of mortars cured at 60°C

	SAMPLE	TENSILE STRENGTH, MPa					COMPRESSIVE STRENGTH, MPa				
		1d	3d	7d	28d	91d	1d	3d	7d	28d	91d
Control	3F6	5.2	7.4	8.5	10.1	n.m.	24.9	37.4	51.5	51.5	n.m.
Control	*5F6*	3.1	6.0	7.8	8.3	n.m.	11.9	31.2	37.7	39.5	n.m.
	3F5C6	4.8	7.0	6.7	*12.9*	9.6	*27.0*	*38.0*	38.7	45.3	44.4
	5F5C6	*3.7*	5.6	*8.7*	*11.2*	*10.7*	*16.1*	30.7	30.5	39.6	43.7
	3F3W6	3.3	3.6	3.8	3.0	3.1	8.0	8.5	8.5	13.6	14.5
	5F3W6	*3.8*	4.7	5.7	5.1	4.8	9.5	12.2	10.5	20.3	22.6
	3F5N6	3.8	4.0	n.m.	5.5	5.4	15.0	15.0	n.m.	10.6	19.7
	3F8N6	n.m.	6.0	5.8	5.4	5.4	n.m.	14.4	13.1	17.6	20.7
	5F5N6	*3.4*	3.3	n.m.	4.5	4.2	11.5	11.1	n.m.	9.8	14.5
	5F8N6	n.m.	*6.9*	6.8	6.4	5.2	n.m.	21.4	15.7	25.2	26.5
	F8N6	-	-	-	0.8	n.m.	-	0.6	0.6	2.8	1.9

NB n.m. = not measured; - = not measurable; *italics* = indicates strength greater than control

Mortar prisms with 100% PFA content and 8M NaOH as activator in the mixing water did not show any recordable strength development. The first attempt to demould hot cured prisms after 3d resulted in the specimen braking in half, hence the 3d compressive strength value measured as an average of the two halves, but with no tensile strength result obtained. The remaining F8N specimens were then demoulded after 7days; unfortunately F8N2 specimens (kept at 20°C) were still too weak and disintegrated when immersed in the water bath.

Results were recorded as average of three prisms for tensile strength and average of six split prisms for compressive strength. Although the tests were carried out without delay when removing the prisms from the storage tanks, there was some slight drop of strength or lessening of rate of increase detected for certain hot cured mixes after 7d (e.g. 5F3W6). This was attributed to thermal shock of bars.

As expected, curing at elevated temperature results in higher strengths and more rapid strength gain particularly at early periods of hydration. As a general trend, however, the test mixes did gain as much strength as the control samples with perhaps lime admixed samples being slightly better at early ages and on a par thereafter. It is likely that this was due to leaching of the activator, especially Na^+, leaving behind a more porous, less dense and , thereby, weaker matrix. Another factor could be the lack of sufficient supply of Ca^{2+} for C-S-H formation in mixes high in Na^+ activator (common ion effect). Similarly, the pH may not be high enough in the lime and waterglass admixed samples for significant breakdown of the glassy phase of PFA.

CONCLUSIONS AND PRACTICAL IMPLICATIONS

There does seem to be a discrepancy between the positive outcomes reported in the literature for the alkali activation of PFA and what is likely to be possible on a practical level in the concrete industry. It appears that to be able to gain any noticeable advantage to strength properties either excessively high curing temperatures of highly concentrated activator solutions or both would have to be used. Neither route would seem to offer any real benefits to industry. Theoretical activation of PFA using admixed alkalis is clearly not straightforward. It should also be recognised that the artificial activation of PFA without the presence of PC did not work in these studies. This underlines the significance of selecting the appropriate curing temperature and suggests that the hydration of PC provides an almost perfect environment for the promotion of the pozzolanic reactions. Thus it is difficult to improve upon the chemical / maturity environment offered by PC at least for the activation of a fine PFA. Sealed curing of the product against water curing is noted to avoid leaching of activators. Finding the right balance to promote both dissolution of PFA and formation of C-S-H and C-A-S-H gels from its components still remains to be addressed in any practical manner that could be adopted by industry.

ACKNOWLEDGEMENT

The project has been supported by the Department of Trade and Industry of the UK under the grant Ref. 39/3/679 (CC 2256) and by Ash Resources (Rugby Cement), Blue Circle Cement, British Board of Agrément, Castle Cement, National Ash and TXU Europe – Ash Sales, which is gratefully acknowledged.

REFERENCES

1. DHIR, R. K., JONES, M. R. and BOOTH, A. P. G., Feasibility of processing PFA to enhance its material characteristics for use in concrete. Internal Report, CTU, University of Dundee, 1998, 108 pp.
2. DHIR, R. K., BYARS, E. A. and McCARTHY, M. J., Strength development of Portland/fly ash cement binders. Second International Symposium: A valuable resource. Capetown, South Africa, 1994, pp 209-224.
3. SHI, C. and Day, R. L., Comparison of different methods for enhancing reactivity of pozzolans, Cement and Concrete Research, Vol. 31, 2001, pp 813-818.
4. WANG, S-D. Review of recent research on alkali-activated concrete in China. Magazine of Concrete Research, Vol. 43, No. 154, March, 1991. pp 29-35.
5. FAN, Y., YIN, S., WEN, Z. and ZHONG, J., Activation of fly ash and its effects on cement properties. Cement and Concrete Research, Vol. 29, 1999, pp 467-472.
6. PALOMO, A., GRUTZECK, M. W. and BLANCO, M. T., Alkali-activated fly ashes: A cement for the future. Cement and Concrete Research, Vol. 29, 1999, pp 1323-1329.
7. TALLING, B. and BRANDSTETR, J., Present state and future of alkali-activated slag concretes. *In:* Malhotra, V. M. (Editor). Fly ash, silica fume, slag and natural pozzolans. Proceedings of the Third International Conference, Trondheim, Norway. ACI SP-114, Vol. 2, 1989. pp 1519-1546.
8. PALOMO, A., BLANCO-VARELA, M. T., GRANIZO, M. L., PUERTAS, F., VAZQUEZ, T. and GRUTZECK, M. W., Chemical stability of cementitious materials based on metakaolin. Cement and Concrete Research, Vol. 29, 1999, pp 997-1004.
9. GRANIZO, M. L. and BLANCO, M. T., Alkaline activation of metakaolin: An isothermal conduction calorimetry study. Journal of Thermal Analysis, Vol. 52, 1998, pp 957-965.
10. ROY, A., SCHILLING, P.J. and EATON, H.C., Alkali Activated Class F Fly Ash Cement, US Patent 5,435,843, 25 July 1995
11. COLLINS, F.G. and SANJAYAN, J.G., Early age strength and workability of slag pastes activated by sodium silicates, Magazine of Concrete Research Vol. 53, (5), pp. 321-326, 2001

ACTIVATING FLY-ASHES: ENLARGING THE CONCEPT OF CEMENTITIOUS MATERIAL

A Fernández-Jiménez

A Palomo

Eduardo Torroja Institute (CSIC)

Spain

ABSTRACT. The alkali activation of waste materials (especially those coming from industrial and mining activities) has become an important area of research since it is possible to use these materials to synthesise inexpensive and ecologically sound cement-like construction materials. In the present paper, the mechanism of activation of a fly ash with highly alkaline solution is described: The main product of reaction is an amorphous aluminosilicate gel having a structure similar to that of zeolite precursors. XRD and SEM/EDX have been used in the characterisation process of the material. The study of technological properties has shown that the AAFA mortars and concrete have: rapid gaining of mechanical strength, low shrinkage and excellent adherence between the steel reinforcement and the matrix.

Keywords: Fly ash, Alkaline activation, Zeolite precursors, Technological properties

A Fernández-Jiménez has a Ph.D in Chemistry and fellowship in the Eduardo Torroja Institute (CSIC), Madrid, Spain. She is specialised in alkaline cements.

Angel Palomo has a Ph.D. in Chemistry and is a senior researcher of CSIC, at the Eduardo Torroja Institute (CSIC), Madrid, Spain. He is specialised in alkaline cements, but he has experience in investigation of different topics related to Portland cement (clinkering, hydration,durability,etc.).

INTRODUCTION

Every year, 1500 millions of tons of Portland cement are produced, contributing to the global pollution of the atmosphere (the manufacture of one ton of cement produces one ton of CO_2). Moreover, this amount increases every year, making it difficult to adjust growth policies with the reduction of the pollution levels. Alkaline cements are an interesting alternative to the Portland cement as they present similar or even better binding properties, low cost and high ecological benefits.

The alkaline activation of materials [1-7] is a chemical process that provides a rapid transformation of some specific structures, partial or totally amorphous and/or meta-stables in compact cemented frameworks. These materials display not only cementing characteristics: high mechanical strengths [6-8], but also ceramic properties (smoothness, glassy state and shiny appearance). Moreover, workability of the fresh mixture is good, allowing low liquid/solid ratios and short setting times (2-5 hours).

In this paper the rudiments of the alkaline activation process are described as well as the process of production of some concrete specimens.

FUNDAMENTAL ASPECTS OF ALKALINE ACTIVATION

In general, two types of cementitious materials can be produced by alkaline activation: 1st those based on Si and Ca, and 2nd those based on Si and Al. The alkali activation of metakaolin or fly ashes (FA) is a typical example of the second situation [1-7].

The alkali activation of fly ashes (AAFA) is a singular procedure by which the grey powder proceeding from the coal combustion (FA) is mixed with certain alkaline activators (alkaline dissolutions) and then the mixture is cured under a certain temperature to make solid materials. From the microstructual point of view, the appearance of the products of reaction is a consequence of the dissolution of the starting fly ash and the subsequent condensation of species formed in the reactional medium. In previous works [6,8,9], it was found that the main reaction product formed is an amorphous aluminosilicate gel (X-ray amorphous) difficult to characterise, but quite similar to some "*zeolite precursor*". According to the investigations of Davidovits [4] and Palomo [9] a total zeolite crystallisation could occur at long term.

There are relevant elements establishing a rational parallelism between the process of zeolite's synthesis [10-14] and the one of the obtention of the cement formed from alkaline activation of metakaolin or fly ashes [8,15,16]. In general, it is admitted that several stages are produced during the synthesis of zeolites from alkaline leached aluminosilicates. In a first stage, the aluminosilicate mineral is dissolved in the alkali solution, which favours the formation of different zeolite precursors from dissolved species (nucleation), and in a second stage, different nuclei reach the critical size and the crystal growth starts. During the nucleation period, both thermodynamic and kinetic parameters have a big influence. In the particular case of the activation of fly ashes (taken into account the experimental conditions used) the crystal growth from the zeolitic nuclei formed is extremely slow. Therefore, an amorphous cementitious matrix (alumino-silicate gel or "*zeolite precursor*") is preferentially formed and stabilised [9].

THE PROCESS OF FABRICATION OF THE NEW CEMENTITIOUS MATERIAL

In order to establish an elemental basis of comparison between the technological performance of a traditional Portland cement and that of an activated fly ash, some data of interest are presented:

Main variables involved in the alkali activation process

The chemical composition of a Portland cement and two Spanish fly ashes (all of them used in this work) is presented in Table 1. From this Table 1 is clearly inferred that the fly ashes are basically formed by silica and alumina while the Portland cement is essentially composed by lime and silica.

Table 1 Chemical composition of cement and fly ash

	[1]L.o.I	[2]IR	SiO_2	Al_2O_3	Fe_2O_3	CaO	MgO	SO_3	K_2O	Na_2O	SiO_2 Reac
CEM[4] I 42.5	3.48	1.65	18.4	5.96	2.17	62.83	1.43	3.50	0.54	0.2	
Fly Ash 1	1.80	0.40	51.51	27.47	7.23	4.39	1.86	0.15	3.46	0.70	
Fly Ash 2	3.59	0.32	53.09	4.80	8.01	2.44	1.94	0.23	3.78	0.73	50.44

[1]L.o.I. = Loss on ignition ; [2]IR. = Insoluble Residue;
[3] Reactive SiO_2 calculated according to the Spanish standard UNE 80-225-93
[4] Type I Cement according to the Spanish standard UNE 80300:2000 IN

The main variables involved in the process of activation of fly ashes are similar to those involved in the hydration of a conventional mortar or concrete, although some differences are also evident (see Table 2). One of the most important factors affecting the process of alkali activation is that related with the curing conditions (time and temperature of the thermal treatment). It is very well known that thermal curing of Portland cement mortars and concrete is the origin of some problems affecting the durability of the material. Thus, the interest of having a cementitious material alternative to Portland cement, is not only because of the maintenance of the performance of the traditional cement but to dispose of a material not generating additional problems when it is thermally cured. In this sense, it must be taken into account that a lot of precast elements are produced in thermal curing conditions.

Table 2 Some factors affecting the quality of the hardened material.

MORTARS AND CONCRETE OF OPC	MORTARS AND CONCRETE OF AAFA, WITHOUT OPC
water/ cement ratio curing temperature < 50°C	Alkaline dissolution/ FA ratio Curing temperature (preferably > 65°C) pH of alkali solution >11.5 Curing time .

Preparation of AAFA mortars and concrete

From the physical aspect point of view, OPC mortars and concrete and AAFA mortars and concrete don't present estimable differences as it can be observed in the Figure. 1. Nevertheless certain basic variations exist in the dosage concept for mixtures. In the case of the new material one can work with mortars or concretes with a high content of fly ash. It is partly due to the low density of the FA and also because the AAFA mortars and concrete present low shrinkage.

Figure 1 Physical aspect of a fresh and hardened concrete of (a) OPC (b) AAFA,without OPC

Rate of heat release

In Figure. 2 the rate of heat evolution developed during the AAFA activation and the OPC hydration is sowed. The calorimetric profile of Portland cement has been studied by numerous authors [6], so the interpretation of curves from Figure. 2(a) is well known. However the process of the activation of fly ashes is not so well known. In any case in AAFA systems the activation reaction of the ashes is very slow at ambient temperature.

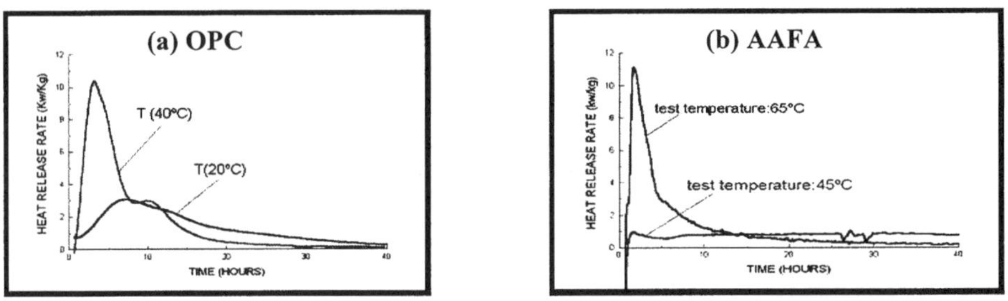

Figure 2 Rate of heat evolution of (a) OPC at 20°C and 40°C and (b) AAFA at 45°C and 65°C.

REACTION DEGREE AND CHARACTERISATION OF THE REACTION PRODUCTS

One of the most interesting items described in this paper is that related to the calculation of the reaction degree and the study of the microstructure of the main phases constituting the reaction products.

In Table 3, the reaction degree attained in activated ashes at 85°C, is showed as a function of the curing time. Data were obtained from samples attacked with HCl 1:20 [9]. These results show that for a reaction time of 5 hours, the degree of reaction is about 38.1%-41%; however, this degree increases till to 50 % when the time of reaction is extended to 7 days.

Table 3 Degree of reaction (%)

	5H	8H	20H	7 DAYS
Fly ash 1	41	44	48.1	50.0
Fly ash 2	38.1	41.84	51.7	52.51.

X-Ray Diffraction

Results obtained by XRD (see Figure 3) show that the original fly ash is basically constituted by a major vitreous phase (halo registered between $2\theta = 20°$ and $2\theta = 35°$) and for some minor crystalline phases (quartz, mullite, hematite, magnetite and some CaO). When the fly ash is put in contact with the alkaline solution, the main reaction product formed is the already mentioned "alkaline aluminosilicate gel" with a low-order crystalline structure [4,9], which does not produce a pattern but appears with low and scattered bands. Additionally, some minor crystalline phases like hydroxi-sodalite ($Na_4Al_3Si_3O_{12}OH$) and especially herschelite ($NaAlSi_2O_6 \cdot 3H_2O$), have also been detected. Finally, it appears that crystalline phases originally existing in the fly ash (quartz, mullite, etc.) are not altered during the activation process.

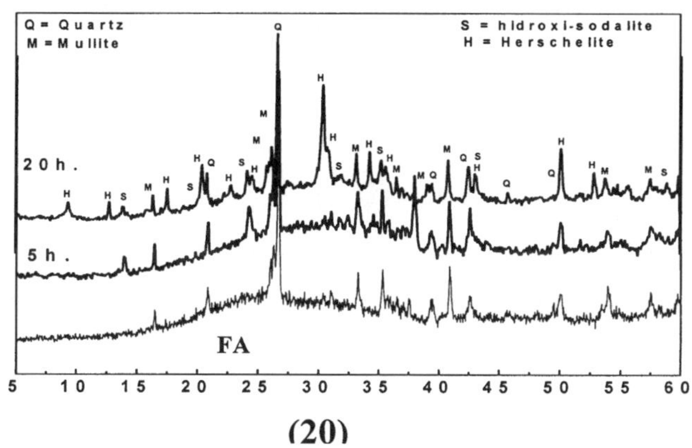

Figure 3 XRD spectra, of initial fly ash and AAFA with NaOH 8M at 85°C for 5h and 20h.

Characterisation by SEM/EDX

The micrograph in Figure.4(a) shows the general aspect of the original fly ash before the activation takes place. As it can be seen in the Figure, the fly ash is mainly constituted by compact or hollowed spheres of different size, but with a regular smooth texture.

Also in Figure. 4.(b) the microstructure of AAFA pastes, is presented. In said Figure. 4(b) together with some unreacted spheres there exists the amorphous aluminosilicate gel (points 3 and 4). From the EDX analysis an averaged compositional ratios of "Si/Al" = 1.6-1.8 and "Na/Al" =0.46-0.68 has been deduced for that gel. These ratios are similar to those obtained by chemical analysis and NMR spectroscopy [9]. Also it can be seen that some ash spheres are partially covered with reaction product (point 5). This might be suggesting that the precipitation of the reaction products forms, in a short period of time, a layer on still unreacted spheres which would make its activation difficult. This fact could justify the moderate degrees of reaction already observed in this type of systems. Finally the presence of a group of small crystalline particles (probably zeolite crystals) was also detected (see point 6). These particles are richer in Na and Si than the rest of the matrix (Si/Al = 2.16 and Na/Al= 1.3).

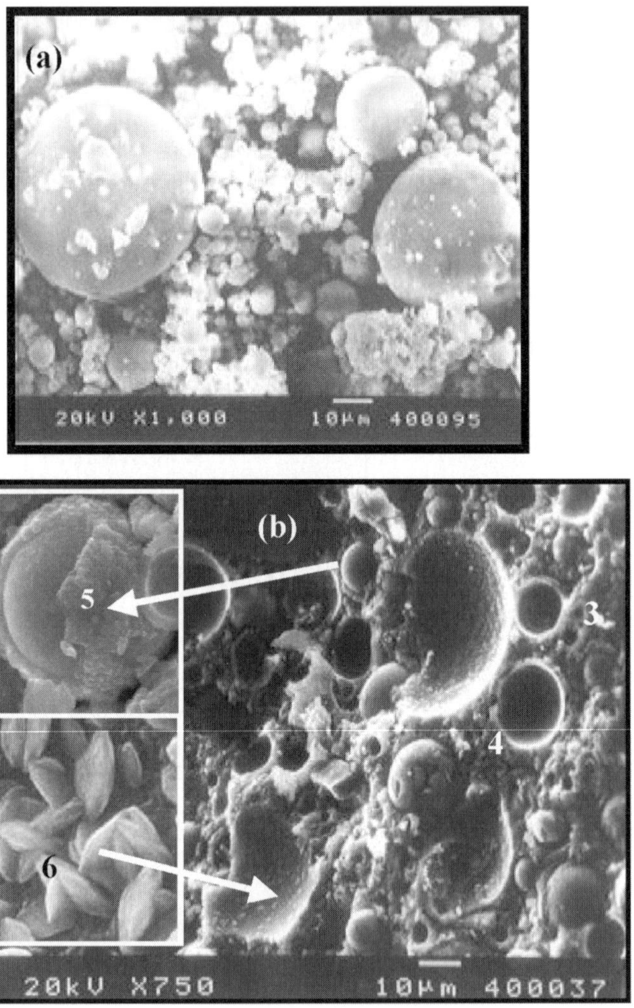

Figure 4 (a) Starting fly ash; (b) paste of AAFA.

TECHNOLOGICAL PROPERTIES OF AAFA MORTARS AND CONCRETE

The main technological properties of AAFA mortars and concrete are: rapid gaining of mechanical strength, low shrinkage and excellent adherence between the matrix, the aggregate and also the steel reinforcement.

Mechanical strength

The compressive mechanical strength as evaluated using prismatic specimens (4x4x16cm mortars and 15x15x15cm concrete) according to the Spanish standard UNE-80-101-88. In Figure. 5(a) the evolution of compressive strength of two OPC mortars respectively cured at 25°C and 40°C, has been plotted versus time and compared with the strength evolution of two AAFA mortars thermally cured for 20h at 85°C. It is observed that the ash mortars present a very advanced development in the course of some few hours (higher than the OPC mortar in the same short period of time). The later evolution of mechanical strength in all the cases shows a tendency of progressive increment although with relative slowness. In Figure. 5(b) the evolution of compressive strength for AAFA concrete is shown. Again, the mechanical strength development of the new material seems to be quicker than in an OPC concrete.

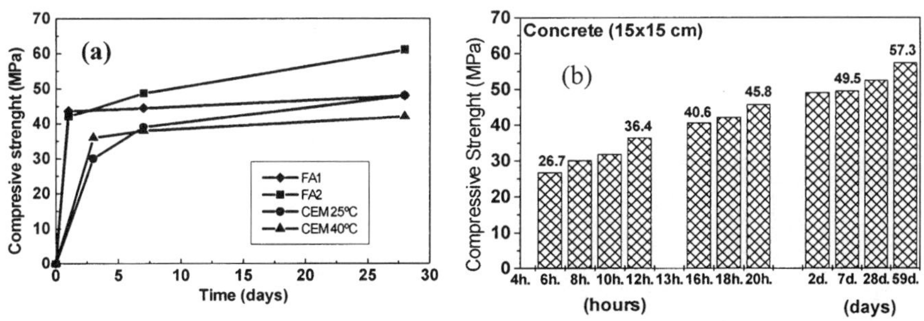

Figure 5 Compressive mechanical strengths of: (a) OPC and AAFA mortars and (b) AAFA concrete. The AAFA mortars and concrete were produced by activating the fly ashes with NaOH 8M, and curing for 20h at 85°C.

Shrinkage behaviour

Shrinkage tests were carried out according to the ASTM C 806-87 standard. In Figure. 6 the shrinkage of OPC and AAFA mortars is shown. Under laboratory environment, OPC mortars show a substantial shrinkage, due basically to the drying and elimination of the free water in the mortar. The alkaline activated fly ash mortars show, however, a very slight shrinkage reaching at the end of the test a value around 0.02%. The conclusion is that alkali activated fly ash mortars are largely stable with very few dimensional variations under the studied conditions.

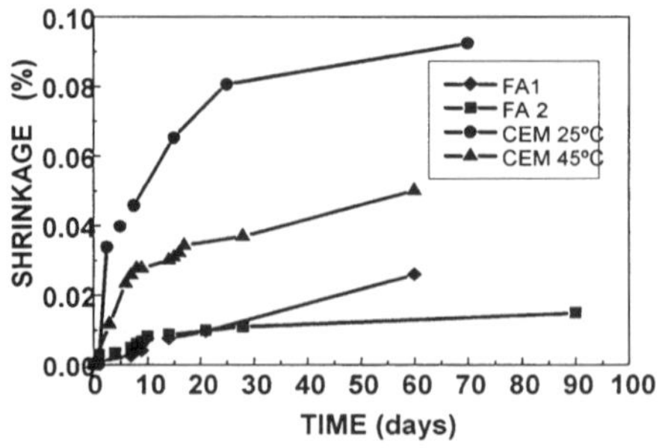

Figure 6 Shrinkage of OPC and AAFA mortars.

Test of adherence (pull-out)

The pull-out test carried out was that standarised by the RILEM/CEB/FIP (see Figure. 7), in which cubic specimens (20x20x20 cm) of AAFA concrete and steel bars of 70cm length and 16 mm of diameter were used. According to the Spanish Standards the minimum τ that should be reached with bars of 16 mm of diameter is 9.70 N/mm² (being τ = tension of adherence). In our particular case a τ value about 17.5 N/mm^2 was obtained (see Figure. 7(b)). For this tension (τ) value the bars embedded in two of the tree specimens tested, slipped (Figure 7(b)), the third one (Figure.7(c)) was broken down.

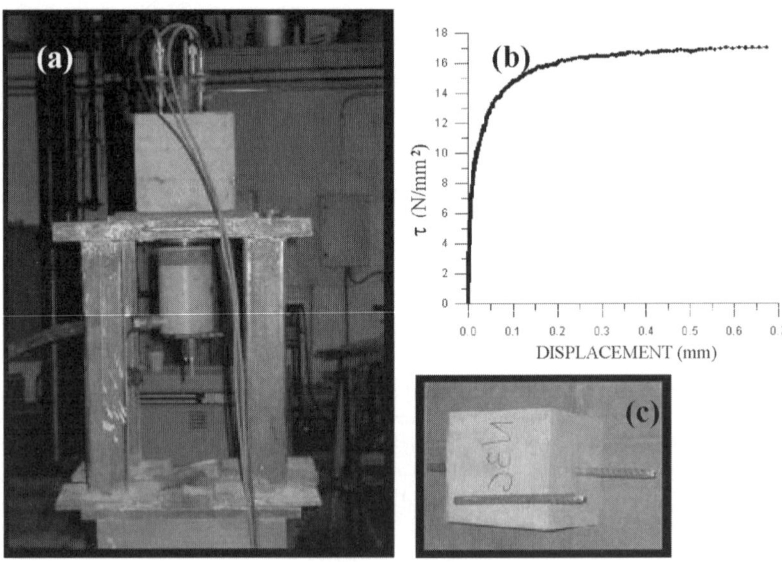

Figure 7 Pull-out tests

CONCLUSIONS

The main reaction product of alkali-activated fly ashes is an alkaline silicoaluminate gel: an amorphous compound displaying the same characteristics than zeolites precursors with a Si/Al ratio close to 1.7. Additionally the presence of a small content of some zeolite type Hydroxi-sodalite, Herschelite etc. has also been detected.

The study of the technological properties shows that the AAFA mortars and concrete have: rapid gaining of mechanical strength, low shrinkage and high cohesive bond between the concrete matrix and the steel bars embedded in said matrix. These properties have a considerable significance in the modern precast construction industry.

ACKNOWLEDGEMENTS

The authors thank the General Direction of Scientific Investigation for financial support (project COO-1999-AX-038), and to J.L. García and A. Gil for their co-operation in preparing the mechanical tests. A. Fernandez-Jimenez thanks the Government of Madrid for the Postdoctoral Scholarship.

REFERENCES

1. MALEK RIA. AND ROY DM. Structure and properties of alkaline activated cementitious materials. 97th annual meeting of the Am. Ceramic Soc. Cincinati, OH, (1995).

2. BARBOSA FFV. MACKENZIE JDK. AND THAUMATURGO C. Synthesis and characterisation of materials based on inorganic polymers of alumina and silica: sodium polysialate polymers" Int. J. of Inorganic Materials, 2, p.309-317, (2000).

3. .KRIVENKO P.V, Alkali cements. First International Conference of Alkaline Cements and Concretes, Ukraine, Kiev, pp.12-129 (1994).

4. DAVIDOVITS J. "Properties of geopolymer cements" First International Conference of Alkaline Cements and Concretes, Ukraine, Kiev, pp.131-149 (1994).

5. GLUKHOVSKY VD. ROSTOVSKAJA GS. AND RUMYNA GV. High strength slag-alkaline cements. 7th Int. Cong. on the Chemistry of Cem,. Paris. 3, p.164-168, (1980).

6. PALOMO A. GRUTZECK MW. AND. BLANCO MT. Alkali activated fly ashes: A cement for the future. Cement and Concrete Research, Vol 29, n° 8, p.323-1329, (1999).

7. XIE Z H. and X I YP Hardening mechanisms of an alkaline-activated class F fly ash. Cem. Concr. Res. Vol. 31, p.1245-1249, (2001).

8. FERNÁNDEZ-JIMÉNEZ A. AND PALOMO. Microstructure of alkali activated fly ash mortars: effect of the activator. Cem. Concr. Res. (Submitted for publication).

9. PALOMO A., ALONSO S., FERNÁNDEZ-JIMÉNEZ A., SOBRADOS I. AND SANZ J. Alkaline activation of fly ashes. A ^{29}Si NMR study of the reaction products. J. Am. Ceramic. Soc. (Submitted for publication).

10. MONDRAGON F., RINCON F.,. SIERRA L, ESCOBAR J., RAMÍREZ J. AND FERNÁNDEZ J., New perspectives for coal ash utilization: synthesis of zeolitic materials. Fuel Vol. 69, p.263-266, (1990)

11. J.L. LAROSA, S. KWAN AND M.W. GRUTZECK, Zeolite formation in class F fly ash blended cement pastes. J. AM. Ceram. Soc. Vol. 75, nº 6, p.1574-1580, (1992)

12. N. MURAYAMA, H. YAMAMOTO, J. SHIBATA, Mechanism of zeolite synthesis from coal fly ash by alkali hydrothermal reaction. Int. J. Miner. Process. 64, p.1-17, (2002).

13. X. QUERL, N. MORENO, J.C. UMAÑA, A. ALASTUEY, E. HERNÁNDEZ, A. LÓPEZ-SOLER, F. PLANA, "Síntesis of zeolites from coal fly ash: on overview" Inter. National J. of Coal Geology, nº 50, pp413-423, (2002).

14. STEENBRUGGEN G., HOLLMAN G.G. "The synthesis of zeolites from fly ash and the properties of the zeolite products" J. Geochem. Explor. 62, pp.305-309, (1998)

15. VAN JAARSVELD J.G.S. AND VAN DEVENTER J.S.J. Effect of the Alkali Metal activator on the properties of fly ash-based geopolymers. Ind. Eng. Chem. Res. Vol 38, nº 10, p.3932-3941 (1999).

16. GRANIZO ML. ALONSO S. BLANCO-VARELA M.T AND PALOMO A. Alkaline activation of metakaolin. Effect of calcium hydroxide in the products of reaction. J. Am. Ceramic Society, Vol.85, nº1 p.225-231, (2002).

CAPACITY OF CHROME IONS FIXATION BY BLAST FURNACE SLAG

G Laforest

J Duchesne

Laval University

Canada

ABSTRACT. Storage of industrial wastes in landfills often generates leachates with high heavy metal concentrations. Conventional solidification/stabilisation process using ordinary Portland cement shows a lack of mineralogical control on some elements, especially on Cr ions. Addition of blast furnace slag seems to allow fixation of Cr ions, in particular Cr^{6+}. The aim of this study was to compare and especially quantify the effect of blast furnace slag and Type 1 ordinary Portland cement in the presence of solutions rich in Cr ions. Results show that the blast furnace slag exerts a solubility control on Cr and Mo concentrations and that there is no ionic competition between these two ions. Blast furnace slag completely fixes Cr^{6+} present at a concentration of 1000 ppm in a sample with a slag to solution ratio of 1:5. X-ray diffraction results on this sample reveal the presence of calcite, merwinite and two calcic chromium oxides ($Ca(CrO_2)_2$ and $CaCrO_3$). These oxides were not detected by scanning electron microscope analysis, which confirmed the presence of calcite and highlighted the presence of calcium silicate hydrates and hydrated calcium aluminate phases.

Keywords: Blast furnace slag, Binding isotherm, Waste solidification/stabilisation, Leaching, Chromium, Molybdenum, Hydraulic binders.

G Laforest is a Ph. D. candidate at Laval University, Québec City, Canada. Her research interests are in the areas of industrial waste stabilization and reuse.

J Duchesne is Professor at the Department of Geological Engineering at Laval University, Québec City, Canada. Her research field includes concrete durability, supplementary cementing materials characterization and utilization and industrial wastes stabilization.

INTRODUCTION

Everywhere in the industrialized world, the production of industrial wastes is in constant increase. Storage in landfills is the simplest, fast and economical solution for the industries. Nevertheless, this solution is not very acceptable since the storage of industrial wastes can generate leachates containing high heavy metal concentrations. To avoid the propagation of contamination, heavy metals must be fixed in an insoluble phase, because they cannot be destroyed or deteriorated by chemical or thermal processes.

The concept of solubility control includes two aspects about the solubility of the elements. The concentration in solution of an element can depend either on the solubility of secondary precipitates or of their availability to the leachate solutions and by their diffusive flux into solution from the leaching of the primary phases over time. If a mineral solubility control exists for an element, its concentration in water flowing through the waste can be readily predicted with chemical equilibria models. To differentiate between these two classes of elemental behaviour, Reardon et al. [1] recommended conducting leaching tests on a particular waste material at least at two different water/solid ratios. They point out that if an element's concentration does not double when the water/solid ratio is halved, there must be a solid phase control on its concentration in solution. The processes of solidification/stabilisation (s/s) aim at carrying out a mineralogical control on metals to be stabilized. Recent researches showed that there was lack of chemical fixation on the concentrations of Cr in cement-based s/s technology [2, 3]. Thus Cr, in particular Cr^{6+}, remains free in the leachate. According to Duchesne [2] and Allan et al. [4], some phases of the blast furnace slag seem to be able to exert a solubility control on chromium. This is particularly interesting considering that blast furnace slag is itself a by product of the iron industry.

Chromium can be found under 3 states, whose valences are +2, +3 and +6. Cr^{3+} and Cr^{6+} are the more frequent states. Cr^{6+} is an acidic form, which forms chromates $(CrO_4)^{2-}$ and dichromates $(Cr_2O_7)^{2-}$, the other valences are basic. The traditional technique of stabilization of Cr^{6+} is initially to reduce it to Cr^{3+} in acid medium using a reducing agent. The most used reducing agents are ferrous sulphate ($FeSO_4 \bullet 7H_2O$), sodium metabisulfite ($Na_2S_2O_5$) and the sodium hydrosulphite ($Na_2S_2O_4$). Cr^3+ is then precipitated, in basic medium, as an almost unsoluble hydroxide. This technique is difficult and expensive since it involves two stages and that a reducing agent must be used. There is a solution, which comprises only one stage since it consists of reducing Cr^{6+} to Cr^{3+} all while solidifying it, and this, without the stage of acidification. The treatment is carried out in a mixture of the residue and several reagents of which one is a reducing agent. That is faster, but rather expensive since several reagents intervene the process.

The blast furnace slag seems to have the capacity to reduce Cr^{6+} to Cr^{3+} and to solidify the residue. The reducing agent and the acidification are thus not necessary. Indeed, the activity field (conditions of particular Eh and pH) of the slag is found in the zone of stability of chromium, i.e. when chromium is found in the $Cr(OH)_3$ form. With regard to Portland cement, its field of activity is located in the zone of chromate $(CrO_4)^{2-}$, which is a very soluble form of Cr^{6+} in solution.

According to several researchers [3, 4], the addition of blast furnace slag can fix Cr, but also Cr^{6+}. Kindness et al. [3] noticed that in the presence of blast furnace slag, Cr^{6+} is reduced to Cr^{3+} that can precipitate as the stable and insoluble form $Cr(OH)_3(s)$. The Cr^{3+} can also substitute for Al^{3+} in most of the calcium aluminate hydrates formed by the hydration of slag. Blast furnace slag has a reducing potential that would create an environment where the Eh values will be lower than in a system composed mainly of Portland cement [3]. There are also economical and environmental advantages to use blast-furnace slag in replacement of Portland cement. In fact, the utilization of slag permits to save energy considering that the production of Portland cement needs 4000 MJ by metric ton, while the production of ground granulated blast-furnace slag requests only 25 to 33% of this amount. Thus, the saving of energy coupled with the re-use and the valorization of the slag are advantageous for the environment.

The purpose of this study was to evaluate the effect of various binders (Type I ordinary Portland cement, ternary cement and blast furnace slag) on the concentration of Cr in leachates. The influence of the presence of another ion (Mo) in solution on the fixation of Cr was evaluated. Then, the research project aims to give quantitative information on the interaction mechanisms of chromium by isotherm adsorption analysis.

MATERIALS

Ground granulated blast furnace slag (GGBS) is a waste product of iron manufacture, and is mainly composed of calcium silicate and aluminosilicate. If allowed to cool slowly, it crystallizes to give a material having no cementing properties. If cooled rapidly, it forms a glass, which is a latent hydraulic cement. The crystalline part of slag is mainly distributed between akermanite ($Ca_2MgSi_2O_7$), merwinite ($Ca_3Mg(SiO_4)$) and gehlenite ($Ca_2Al_2SiO_7$) minerals. The chemical composition of blast furnace slag shows little variation in a plant and is homogenous from particle to particle. North American blast furnace slags are composed mainly of CaO (38%), SiO_2 (36%), MgO (12%) and Al_2O_3 (11%). Fe_2O_3, SO_3, MnO and K_2O remain in the major constituents, but in minor proportions (1%) [5]. Its MgO content can vary in different areas. In fact, in North America, the typical MgO content is approximately 12%, while the average concentration is only 6% in European countries [6]. The surface area of most GGBS varies from 340 to 550 m^2/kg. This variation is mainly due to the different methods of crushing. The density is less variable with a value close to 2900 kg/m^3. The density of the slag studied was 2915 kg/m^3 (ASTM C188-78).

The hydration of the slag cements showed that these products give the same hydrates as Type 1 ordinary Portland cement (OPC). The hydration of slag cements forms less portlandite ($Ca(OH)_2$) and, in parallel, releases less heat. The C-S-H (calcium silicate hydrate), produced by the slag, has a Ca/Si ratio smaller than that of the C-S-H produced by Portland cement [7]. These C-S-H gels have a greater capacity of fixing of alkalis and other metals. The slag reacts slowly with water (approximately 28 days), but its hydration can be activated chemically, thermically or mechanically.

The blast furnace slag studied here comes from Algoma Steel (Sault Ste-Marie, Canada). This slag, cooled quickly, is largely of a glassy form with a small proportion of a melilite solid solution that is formed, more precisely, of merwinite ($Ca_3Mg(SiO_4)$) and gehlenite ($Ca_2Al_2SiO_7$). Two types of cements were used in this study. The first one was a Type I ordinary Portland cement. The second cement was a ternary blend composed mainly of clinker, ground granulated blast furnace slag and silica fume. Chemical compositions of the cements are given in Table 1.

Table 1 Chemical composition of materials.

OXIDES (% mass)	GGBS	OPC	TERNARY CEMENT
CaO	37.31	62.49	56.27
SiO_2	36.77	19.75	25.90
MgO	13.91	2.62	3.83
Al_2O_3	7.77	4.41	5.13
MnO	1.02	0.05	0.23
Fe_2O_3	0.85	2.92	2.61
K_2O	0.43	0.89	0.76
TiO_2	0.36	0.17	0.37
Na_2O	0.31	0.28	0.33

METHODS

Static Leaching Test

A series of batch leaching experiments were performed by reacting samples of cementitious material at 3/1 and 6/1 water/solid ratios in 60 ml high density polyethylene bottles. The bottles were suspended using carousels maintained horizontally in a water batch controlled at 25±1°C. The carousel was rotated between 10 and 20 rpm for 90 days.

The cementitious materials tested were GGBS, ternary blend, and mixtures of OPC with GGBS at a mass ratio of 50:50 (GGBFS 50) and 25:75 (GGBFS 75). The cementitious materials were put in contact with synthetic solutions containing 20 or 40 ppm Cr. A solution was also prepared with both 20 ppm Cr and 20 ppm Mo.

Samples from the solution were obtained at time intervals (7, 28, 56 and 84 days) and filtered through 0.22 μm membrane filters. The samples were acidified by addition of HCl 1:1 for cation analyses (Cr and Mo) on a Perkin Elmer Aanalyst 100 Atomic Absorption Spectrophotometer (AAS). All solutions were kept at 4°C until analysis.

Chromium Binding Isotherm

The chromium binding isotherm method was adapted according to the procedure of Delagrave et al. [8] to quantify the capacity for Cr absorption by different binders. The test was divided into two series. The first series was prepared with GGBS or OPC as pulverulent materials. Approximately 10g of GGBS or OPC was introduced into 50ml solutions of various Cr^{6+} concentrations (10, 20, 40, 50, 100, 250, 500, 750 and 1000 ppm). The immersion solution was alkaline and made up of lg/l of NaOH and 4.65 g/l of KOH. The solution was alkaline in order to accurately reproduce the conditions within hydrated cementing materials. These mixtures were placed in polyethylene bottles suspended in a water bath maintained 25°C for 3 weeks. After this period, the solutions were filtered on membranes of 0.22μm and acidified by addition of HCl 1:1. The Cr concentration was then analyzed by AAS.

The second series of tests was made on hydrated cement or hydrated slag at a water to binder ratio of 0.4. This series was set up in order to simulate the contact between already hydrated materials and solution rich in Cr^{6+} ions. Cylindrical samples (diameter = 2.3 cm and height = 5cm) were moulded in hermetic containers and stored at 100% relative humidity for 80 days. The pastes were crushed in order to pass the 2-mm sieve. A mass of approximately 10g was introduced into 50ml of solution of various Cr^{6+} concentrations as for the first series. After 3 weeks, the solutions were filtered, acidified and their Cr concentration was determined by AAS. Only the mixtures with hydrated GGBS sample had a reaction period of 150 days.

The amount of bound chromium was calculated according to $C_b = [(C_0-C_1)V] / W$ where C_b was the bound chromium content (mg Cr/g sample), C_0 and C_1 were initial and equilibrium chromium concentration (mg/l), V was the volume of the chromium solution and W was the sample weight.

X-ray Diffraction (XRD) and Scanning Electron Microscopic Analyses (SEM)

GGBS was put in contact with 1000ppm Cr^{6+} solution in order to identify Cr-bearing phases. A sample of the reacted solid was analysed by Siemems D5000 X-ray diffractometer using Cu Kα radiation generated at 20mA and 40kV. Specimens were step-scanned as random powder mounts from 5-65° 2θ at 0.02° 2θ steps integrated at 1.2 s step-1.

The same sample was also observed under a JEOL LSM-840A Scanning Electron Microscope equipped with an energy dispersive X-ray analysis system (EDXA).

RESULTS AND DISCUSSION

Static Leaching Test Results

Figure 1 presents the effect of various binders on the Cr concentrations of the leachates. First, one can note the very drastic fall of the Cr concentrations, which were 20 ppm at the beginning of the test. The results show that the Cr content decreases very quickly to obtain values near 0.03 ppm after only 7 days of contact between the contaminated solution and the solids. Later on, Cr concentrations fall under the detection limits. The sample containing only GGBS is the most effective. It suppresses the Cr content below the detection limit after 56 days of reaction. The quantity of GGBS in the mixtures has an effect on the removal of chromium ion from the solution. In fact, the ternary blend, which contains less slag according to its MgO content, offers the weakest performance but by being only slightly inferior. It should be noted that all the binders studied were powerful to reduce the Cr concentrations of the contaminated solution. After only 7 days of reaction, the initial Cr concentration of 20 ppm was lower or equal to 0.03ppm. A very small quantity of GGBS makes it possible to obtain chemical fixation of Cr ions.

Figures 2 and 3 present the final Cr or Mo concentrations of the leaching solutions in contact with the samples containing only slag. This test was carried out with two liquid/solid ratios (3/1 and 6/1) in order to make it possible to conclude on the presence or not of mineralogical control.

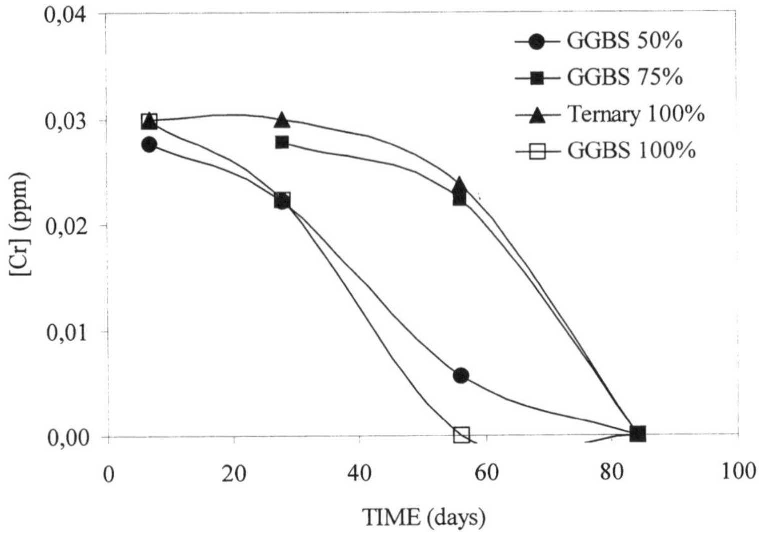

Figure 1 Effect of the type of binder on the Cr concentrations in leachates; water/solid = 3/1 and 20 ppm Cr initial solution.

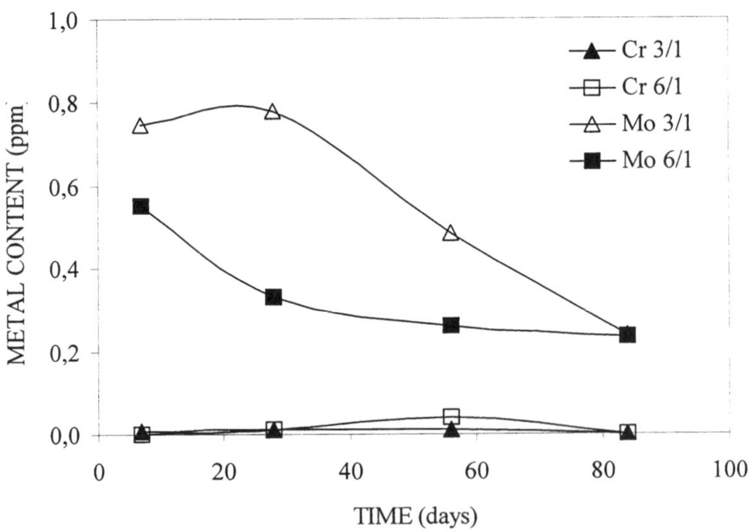

Figure 2 Concentration of metals in solution for samples of GGBS in contact with the solution containing 20 ppm Cr and 20 ppm Mo.

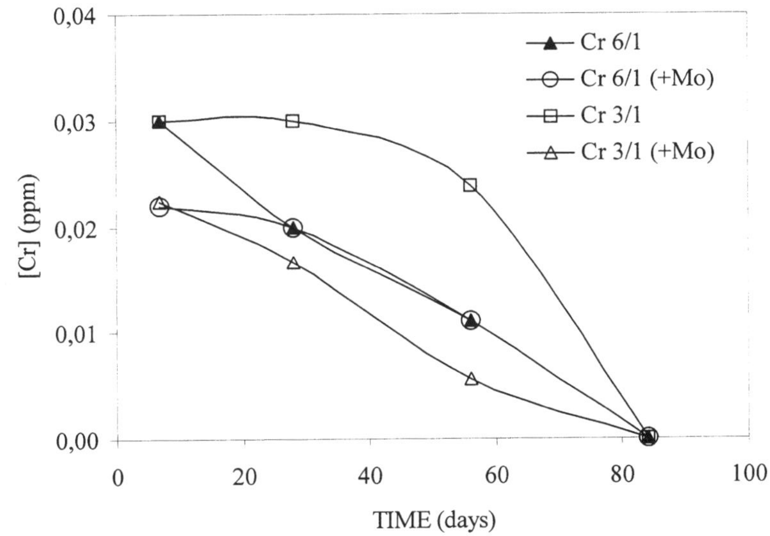

Figure 3 Effect of Mo ion on the Cr concentrations of leachates in contact with Ternary blend; (20 ppm Cr + 20ppm Mo or only 20 ppm Cr initial solutions).

Figure 2 shows that Cr ions are almost completely confined in the solid phases, and this, irrespective of the initial water/solid ratio. Molybdenum ions are not completely stabilized by the GGBS, even after 84 days of reaction. However, as it is revealed in each case, concentration values measured at the water/solid ratio of 3/1 are always lower than the double of those observed at the 6/1 ratio. We can conclude that a solid phase being present in the slag or in the slag hydration products, has a solubility control on both Cr and Mo concentrations.

Figure 3 shows that the Ternary blend exerts a mineralogical control on Cr ions in the presence or not of Mo ions. It is also noticed that chromium fixation is better or equal in presence of Mo ions.

Figures 2 and 3 present results with regard to the influence of the presence of another ion in solution. In fact, molybdenum ions were introduced to verify whether another ion of similar behaviour would have a negative influence on the fixation of Cr ions. According to data obtained, there is no ionic competition between molybdenum and chromium ions, at least for the concentrations studied.

Chromium Binding Isotherm Preliminary Results

The Cr fixation capacity ions of slag were studied using isotherms. Four stabilizing materials were tested: OPC, GGBS, both being either hydrated or not. Table 2 compares the initial Cr^{6+} concentration in the different solutions with the Cr concentrations obtained after the period of contact with the four stabilizing materials studied.

Table 2 Chromium binding isotherms data.

INITIAL Cr^{6+} (ppm)	FREE CHROMIUM (ppm)				BOUND CHROMIUM (mg/g sample)			
	Hydrated OPC	Hydrated GGFBS	OPC	GGBS	Hydrated OPC	Hydrated GGFBS	OPC	GGBS
10	0.06	0.52	0.30	0.00	0.05	0.05	0.05	0.05
20	0.14	0.52	0.46	0.00	0.10	0.10	0.10	0.10
40	0.46	0.52	0.83	0.00	0.20	0.20	0.20	0.10
50	0.71	0.52	1.12	0.00	0.24	0.24	0.25	0.25
100	1.98	0.69	2.48	0.00	0.49	0.50	0.48	0.49
250	9.42	4.86	8.42	0.00	1.19	1.20	1.21	1.23
500	29.51	11.82	14.08	0.00	2.33	2.43	2.41	2.49
750	34.20	66.98	20.02	0.00	3.54	3.38	3.63	3.71
1000	46.94	111.23	32.34	0.00	4.72	4.42	4.78	4.93

The results show that the concentrations of free chromium ions in solution depend on the initial Cr^{6+} concentrations. The type of equation obtained for a typical Freudlich isotherm is $\text{Log } C_b = a \log C_f + b$ where a and b are the adsorption constants, C_b is bound chromium and C_f is free chromium.

The Freudlich equation can graphically represent the behaviour of the hydrated slag with an accuracy of 97% according to $C_b = 0.7255 \log C_f + 0.5868$. The OPC and hydrated OPC samples adopt linear relations with coefficients of 99% and 97%, respectively. The chromium binding isotherms data, presented in Table 2, show that it is impossible to appreciate a relation with the GGBS sample for the initial solution below 1000 ppm Cr^{6+} concentration. In fact, all the chromium ions were bound by the slag.

The Freudlich equation seems to be adequate for a possible modelling of the adsorption of Cr ions by the hydrated GGBS sample. In order to check the possibility of observing this type of relation for other materials, it would be important to widen the range of concentrations being studied and/or the reaction time. To adequately quantify the capacity for GGBS adsorption, the test will have to be carried out again with concentration between 1000 and 5000ppm to refine the first portion of the curve.

X-ray Diffraction (XRD) and Scanning Electron Microscopic Analyses (SEM) Results

Figure 4 and 5 present the XRD and SEM results obtained on a sample of GGBS in contact with a 1000 ppm Cr^{6+} solution.

The XRD presented in Figure 4 made it possible to identify merwinite, primary phase of unhydrated slag, with calcite and two types of calcic chromium oxides [$Ca(CrO_2)_2$ and $CaCrO_3$]. The presence of these oxides was not validated at the time of the study by electron microscopy. Moreover, the presence of calcite was retained and confirmed, thereafter, with the SEM. C-S-H without chromium ions was detected with the SEM. Aluminate phases were not detected.

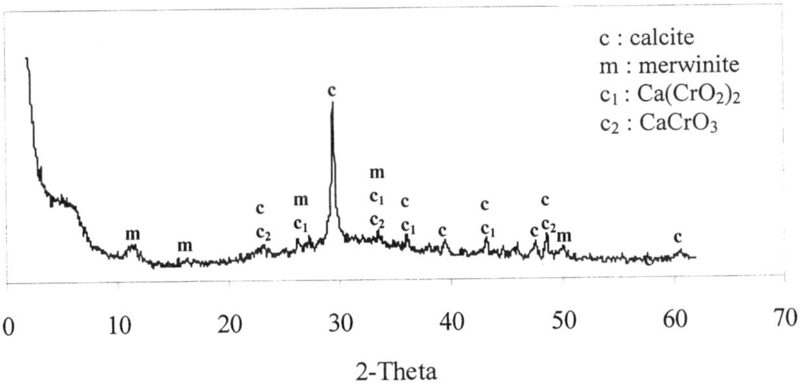

Figure 4 XRD results of GGBFS in contact with a 1000 ppm Cr^{6+} solution for 3 weeks, (λ = 1.54184 Angström CuKα).

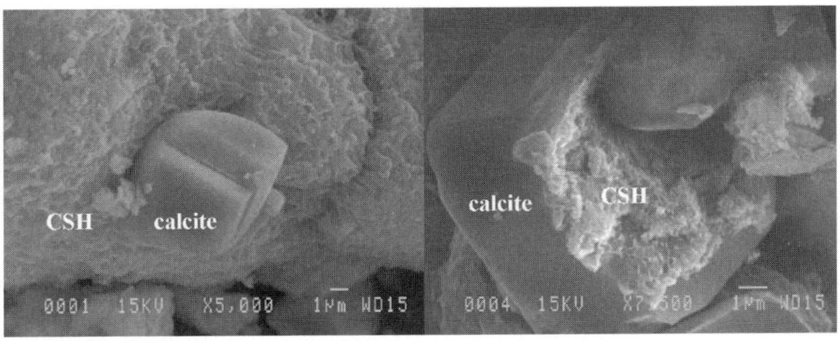

Figure 5 SEM micrographs of GGBFS in contact with a 1000 ppm Cr^{6+} solution.

CONCLUSION

The static leaching results show that all the stabilizing materials are effective in reducing Cr ions from solution. The sample containing only GGBS is the most effective. It suppresses Cr content below the detection limits. Cr ions are almost completely confined in the solid phases, irrespective of the initial water/solid ratio. Molybdenum ions are not completely stabilized by GGBS,. but this phase has a solubility control on Mo ions. Also, no ionic competition between molybdenum and chromium ions was observed.

The Chromium binding isotherm made it possible to quantify the capacity for absorption of Cr ions by the stabilizing materials. The Freudlich equations represent the absorption of chromium ions by the hydrated slag, while a linear equation represents the unhydated and hydrated cement samples. It was not possible to establish a relation for the GGBS sample, because all the Cr ions were fixed in the solids for concentrations less than 1000 ppm in Cr^{6+} ions

ACKNOWLEDGEMENTS

This study has been supported by the National Science and Engineering Research Council of Canada (NSERC) and by the Fonds pour la Formation de Chercheurs et l'Aide à la Recherche of the Province of Québec (FCAR).

REFERENCES

1. REARDON, E.J., CZANK, C.A., WARREN, C.J., DAYAL, R., JOHNSTON, H.M., Determining controls on element concentrations in fly ash leachate. Waste Management & Research, Vol. 13, No. 2, October 1995, pp 435-450.

2. DUCHESNE, J., Réduction des concentrations en Cr et en Mo dans des lixiviats générées par des poussières de four de cimenterie. Stabilisation des Déchets & Environnement 1999, Lyon Villeurbanne, pp 21-25.

3. KINDNESS, A., MACIAS, A., GLASSER, F.P., Immobilization of chromium in cement matrices. Waste Management, Vol. 14, No. 1, 1994, pp 3-11.

4. ALLAN, M.L., KUKACKA, L.E., Blast furnace slag-modified grouts for in situ stabilization of chromium-contaminated soil. Waste Management, Vol. 15, No. 3, 1995, pp 193-202.

5. AHMED, Y.H., BUENFELD, N.R., An investigation of ground granulated blastfurnace slag as a toxic waste solidification/stabilization reagent. Environmental Engineering, Vol. 14, No. 2, 1997, pp 113-131.

6. TAYLOR, H.F.W., Cement Chemistry. Academic Press, London, 1990, 475pp.

7. DUCHESNE, J., BÉRUBÉ, M.A., Effect of supplementary cementing materials on the composition of cement hydratation products. Advanced Cement Based Materials, Vol. 2, 1995, pp 43-52.

8. DELAGRAVE, A., Chloride binding capacity of various hydrated cement paste systems. Advanced Cement Based Materials, Vol. 6, No. 1, June 1997, pp 28-35.

ALAKALI-ACTIVATED CEMENTS AS IMMOBILISATION MATRICES FOR ARSENIC WASTES

A Fernandez-Jiminez **A Palomo**

Eduardo Torroja Institute (CSIC)

Spain

E E Lachowski **D E Macphee**

University of Aberdeen

United Kingdom

ABSTRACT. A variety of natural and industrial waste product materials rich in SiO_2 and Al_2O_3 may be activated with certain alkaline materials to produce cementitious systems which when cured under mild temperature conditions, set and harden, in the presence of appropriate amounts of water, to give a very compact paste. In some cases, e.g. where fly ash is the activated material, fast setting and rapid strength development have been reported. In addition to their application in construction, their role as host matrices for waste materials is beginning to attract interest and it is this application that the study reported here addresses. Arsenic immobilisation is not achieved satisfactorily in Portland cement matrices because calcium arsenate phases, which form readily, decompose when the matrix carbonates. This paper describes the results of some of the leaching and microstructural studies on alkali-activated fly ash matrices containing arsenic with additional iron and manganese. Arsenic leaching from fly ash matrices is high, and considerably increased by the presence of manganese. The role of iron is more complicated and depends on its speciation.

Keywords: Arsenic, Immobilisation, Fly ash, Geopolymer, Leaching, Electron Microscopy

A Fernandez-Jimenez is a postdoctoral research fellow at the fellowship in the Eduardo Torroja Institute (CSIC), Madrid, Spain. She specialises in alkaline cements.

D E Macphee is a senior lecturer in the Department of Chemistry, University of Aberdeen. His interests include novel cement and concrete materials and glasses and their applications.

E E Lachowski is a research officer in the Department of Chemistry, University of Aberedeen who specialises in the electron microscopy of materials, in particular cement and concrete.

A Palomo is a senior researcher of CSIC, at the Eduardo Torroja Institute (CSIC), Madrid, Spain. He is specialised in alkaline cements, but he has experience in investigation of different topics related to Portland cement (clinkering, hydration, durability, etc.).

INTRODUCTION

Despite its toxicity, arsenic is still in use in the semi-conductor industry, the production of alloys and of some herbicides and pesticides [1-3]. Arising as a by-product of metal extraction processes, its production greatly exceeds world demand and consequently, environmentally acceptable disposal options must be identified to deal with existing and future accumulation of arsenical wastes. Arsenic chemistry is complex however, involving a variety of valence states in anionic and cationic species. As inorganic arsenic, the As^{3+} form is more toxic than the +5 form and is also more difficult to immobilise. Arsenic combines with many metals to form arsenates [4]. The solubility of As_2O_3 in water is high (2.05 g/l at 25°C) and as the oxide is amphoteric, its is soluble in acids and bases.

A number of CFS (Chemical Fixation and Solidification) processes specifically for arsenic fixation have been described in the literature [1-10]. These convert the waste to a solid that is safer to deal with than an aqueous waste stream. Amongst the most frequently studied matrices for fixing heavy metals and other hazardous elements are those based on Portland cements. However, it has been found that these matrices are not effective in fixing arsenic (Kyle and Lunt [11], Shively et al [12]) due to the reaction of the cement with the environment and subsequent release of arsenic into the aqueous phase by leaching.

This paper focuses on an alternative binder system. Geopolymers are representative of a class of cementitious materials derived from the alkaline activation of readily available natural or industrial by-product aluminosilicate materials. These alkali-activated cements are very attractive environmentally for a number of reasons. Their performance as construction materials is in many ways comparable with Portland cement but they differ from Portland cements because they do not require the thermal decomposition of limestone in their manufacture. This alone represents a considerable benefit in terms of reducing global CO_2 emissions. Further, the use of industrial by-products is consistent with the growing trend towards waste re-utilisation.

Here, we describe the properties of geopolymers derived from the coal combustion by-product, fly ash. The solidification of alkali-activated fly ashes is thought to arise from a series of solution and polycondensation processes in which a randomly arranged, three dimensional aluminosilicate network is established [13]. These systems, in which the framework aluminosilicate is negatively charged due to the incorporation of the AlO_4 tetrahedral units, require corresponding numbers of positive charges, normally provided by cations held tightly within the geopolymer matrix. It is the compact 3-D aluminosilicate matrix coupled with the potential for ion exchange which make geopolymers attractive matrices for the binding of metal wastes.

However, a previous study [14] on arsenic immobilisation in alkali-activated fly ash and alkali-activated metakaolin matrices concluded that the influence of other components of the system may have a positive effect. It was shown that the fly ash matrices gave slightly better performance than the metakaolin ones and a tentative correlation was drawn between degree of As immobilisation and Fe content of the fly ash. The affinity of arsenic for ferric hydroxide flocs is well known and it was considered reasonable that Fe released from the fly ash was adsorbing As within the geopolymer pore structure and limiting its mobility during leaching. The objective of this study was to further investigate this hypothesis.

EXPERIMENTAL

Synthesis of Geopolymers

The compositions of the matrices studied are presented in Table 1. The arsenic content was constant (1 wt% added as $NaAsO_2$) and Fe_2O_3 (5 wt%) and MnO_2 (1 wt%) were included in selected systems. The bulk chemical composition of fly ash used is presented in Table 2. The fly ash is extremely heterogeneous and consists mainly of a glassy phase but also has significant amounts of mullite, spinel and quartz, so that its reactivity is not uniform.

Table 1 Matrices studied

SAMPLE	FLY ASH (%)	Fe_2O_3 (%)	As (%)	As (ppm)	MnO_2 (%)
FA1	100	0%	0	0	0
FA3	95	5%	0	0	0
FA5As	99	0%	1%	10.000	0
FA7As	94	5%	1%	10.000	0
FA9AsMn	93	5%	1%	10.000	1%

Table 2 Chemical composition of fly ash

	LoI	IR	SiO_2	Al_2O_3	Fe_2O_3	CaO	MgO	SO_3	K_2O	Na_2O
Fly ash	1.80	0.40	51.51	27.47	7.23	4.39	1.86	0.15	3.46	0.70

LoI= Loss on ignition IR= Insoluble residue

The fly ash was activated using an NaOH (8M) solution (containing the appropriate amount of arsenate solution) using a liquid/solid ratio of 0.35. Pastes were mixed to give complete homogenisation, then poured and compacted into metallic moulds. Samples were then cured in an oven at 85°C (20 hours). After curing, the samples were removed from the oven and de-moulded.

Leaching

The capacity of the different matrices for arsenic fixation was evaluated by means of the TCLP and ANSI/ANS methods [4,15,16]. TCLP is a dynamic leaching test using acidified water as leachant. Specimens were finely granulated and sieved to pass a 9.5 mm but retained in a 4.5 mm mesh. Fluid #2 [4] was used as leachant, prepared by diluting 5.7 ml of acetic acid in 1000 ml of deionised water. The pH of the resulting fluid is 2.88 ± 0.05. The leaching test was carried out with 100g of sample and 3 litres of *Fluid #2*. Solid and leachant were put into a tightly closed vessel with constant agitation for 18 hours, at 22 ± 3°C. Liquid and solid phases were then separated. Aqueous As, Al, Fe and Mn concentrations were determined by Flame Atomic Absorption Spectroscopy (Spectra AA-10 atomic absorption spectrophotometer). The sodium content of the leachate was analysed by Flame Emission Spectroscopy (FES) using a Corning 400 Flame Photometer. The silicon analysis was carried out by spectrophotometry using a sodium molybdate reagent on a DIONEX DX120 ion chromatograph with an AD20 absorbance detector.

The quasi-dynamic ANSI/ANS method involved leaching from cylindrical test samples (13 x 6 mm), with and without arsenic. After the curing, the specimens were demoulded and stored in plastic containers containing 30 ml of water such that the volume of leachant / matrix surface area ratio was 10mm. All leach data were referenced against the data obtained 30 s after immersion. Leachant replacements and analyses (by ICP) were executed at various times up to 2160 hours

Electron microscopy

Fresh fractures were coated with gold and examined in a JEOL 2000 EX Temscan operating at 20kV in the SEM mode. Thin sections for TEM examination were prepared from petrographic thin sections by argon milling with an Atom Tech mill operating at 5-6kV. The sections were coated with carbon and observed with a JEOL 2000 EX Temscan operating at 200kV. EDX analysis was done with a Link AN10/85S EDX analyser. Quantification was done by the Cliff-Lorimer method [18], which gives results in terms of the ratios of the constituents. Only sample FA9AsMn was studied by electron microscopy.

RESULTS AND DISCUSSION

Leaching

Table 1. Leaching results obtained by TCLP method.

MATRICES	LEACHING ELEMENTS *									
	Na		Si		Al		Fe		As	
	ppm	%	ppm	%	ppm	%	ppm	%	ppm	%
FA1	449	26.11	58.85	1.0	49.37	1.2	9.23	0.7	-	-
FA3	423	24.7	38.04	0.7	22.77	0.6	26.44	1.3	-	-
FA5As	662	37.36	70.13	1.2	9.65	0.2	7.74	0.6	119.8	49.2
FA7As	668	37.94	74.69	1.4	2.46	0.1	4.08	0.2	134	55.1
**FA9AsMn	603	34.02	65.9	1.2	17.19	0.4	12.52	0.6	167.4	68.35

*Final pH of TCLP test ≅ 4-4.5. **FA9AsMn [Mn] = 1.52 ppm (0.98%)

Table 1 shows the result of leaching tests. Overall the leachability of As is high. Comparison of the specimens with As (FA5As), As + Fe (FA7As) and As +Fe + Mn (Fa9AsMn) shows that addition of Fe increases the leachability of As and that this is further increased by the presence of Mn. The leachability of Fe is low and fairly insensitive to the composition of the matrix. The leachability of Si, which is also low, is relatively insensitive to these additions, but Al becomes even less soluble. Na leachability is high throughout, but does show some variation. The variation with time of leaching is shown for Si, Al, Fe and As in Figures 1 and 2, expressed as per 1. Although the values for Si, Al, and Fe continue to rise, they are very low, and although the As values appear to be levelling off, they are very high. The presence of Mn has a severe deleterious effect on As leaching, although the low leach rates of Si and Al indicate that the matrix itself is not severely affected.

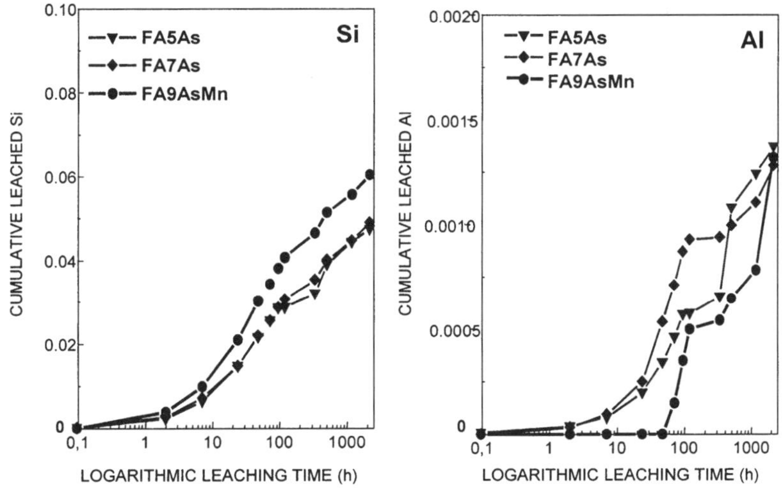

Figure 1 Cumulative leaching of Si and Al expressed as per 1, against time. Data obtained through the ANSI/ANS method

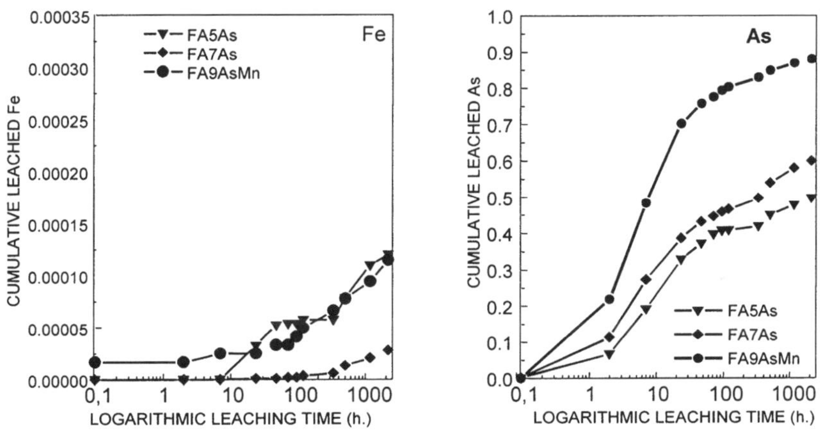

Figure 2 Cumulative leaching of Fe and As expressed as per 1, against time.

Porosity measurements [19] suggest that the addition of Fe and Mn causes a small increase in total porosity, as shown in Figure 3. There is, however no significant difference between the porosities of FA5As and FA9AsMn, suggesting that porosity is not a significant factor in determining leaching.

Electron Microscopy

SEM shows the geopolymer to consist of a number of distinctive morphologies dispersed in a relatively featureless aluminosilicate matrix, shown in Figures 4 and 5. There is considerable variation in the reaction of different fly ash particles, some being almost completely consumed, their relicts surrounded by hollow shells of hydration product, while others are largely intact. This variability has also been reported previously [20].

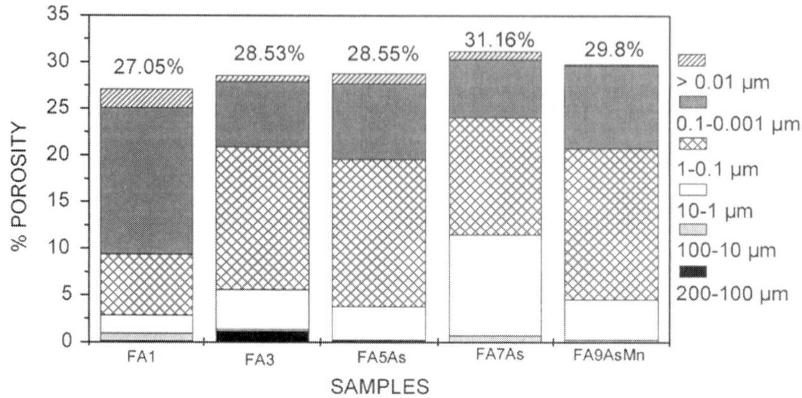

Figure 3 Porosity distribution of fly ash pastes.

The EDX resolution in the SEM is too coarse to extract much meaningful compositional

information, so thin sections were prepared and studied by TEM/EDX, where the analytical resolution is close to the diameter of the beam. Figures 6-9 are transmission electron micrographs of the geopolymer matrices featured in the scanning images (Figures 1-4). The familiar spherical morphologies of the fly ash particles are again evident as discs amongst other features including hydration rims and the protruberances associated with them but, this time, more detail can be derived both from morphological and compositional information, the latter being available from energy dispersive X-ray analysis (EDX). Transmission microscopy is capable of providing the resolution necessary to focus on small microstructural features enabling microanalysis within hydration rims. Such information is useful in correlating compositions with specific morphologies. Figure 10 summarises the raw analytical data obtained from EDX acquired from the points numbered 1 to 24 in Figures 6-9. The values are expressed as atomic percentages normalised from all the elements analysed (Na, Mg, Al, Si, S, K, Ca, Ti, Fe and As) and take no account of the presence of oxygen or water. Mn levels were too small to be resolved. Although they are not absolute percentages they are, nevertheless, a useful indicator of relative amounts. Note that the penetration of beam electrons into the specimen at these points may lead to interaction of the electrons with underlying phases although analytical points were chosen to minimise such interferences. Electron diffraction showed that the hydration products were amorphous.

Although the fly ash is predominantly an amorphous aluminosilicate, it also contains other phases such as quartz, mullite and spinels. Indeed the glassy phase contains other chemical constituents, see Figure 10 and points 8 (Figure 6), 13 (Figure 8) and 15 and 22 (Figure 9).

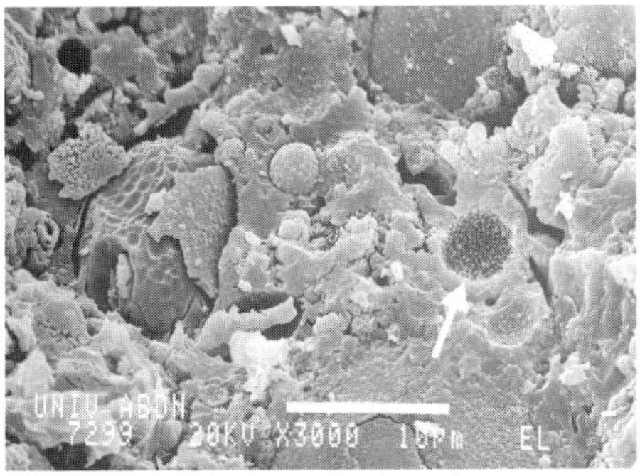

Figure 4 SEM micrograph of fly ash geopolymer showing fly ash particles and Fe_2O_3 (arrowed) in an aluminosilicate matrix

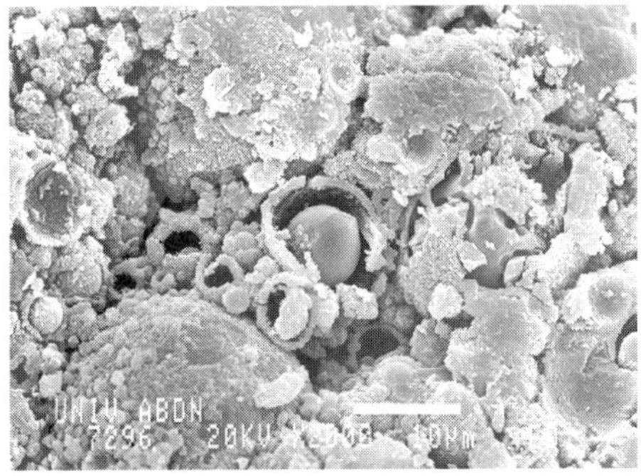

Figure 5 SEM micrograph of fly ash geopolymer showing fly ash particles with various degrees of reaction

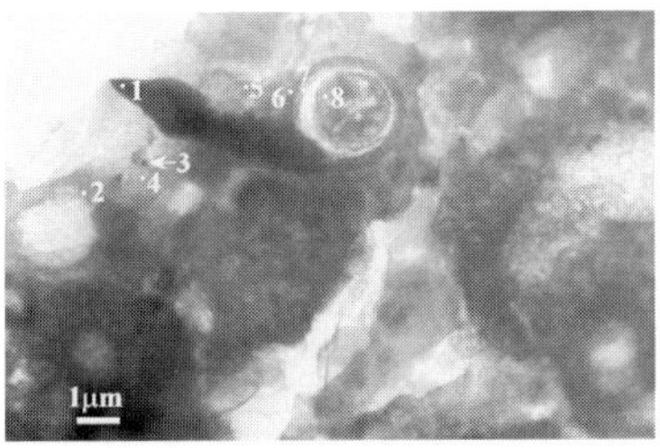

Figure 6 TEM micrograph of argon-milled section. Numbers correspond to analytical points referred to in text.

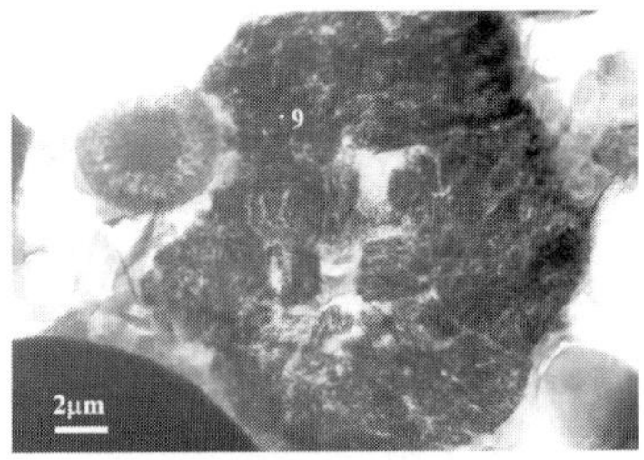

Figure 7 TEM micrograph of argon-milled thin section showing large, unreacted cluster of Fe_2O_3 crystals. Number corresponds to analytical point referred to in text

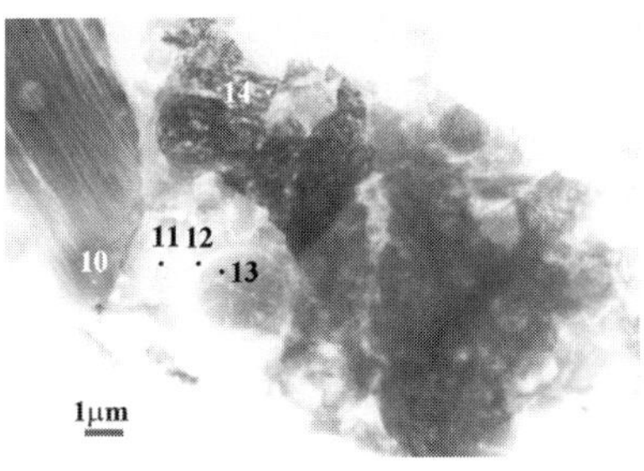

Figure 8 TEM micrograph of argon-milled thin section showing quartz, PFA, Fe_2O_3 and hydration product. Numbers correspond to analytical points referred to in text

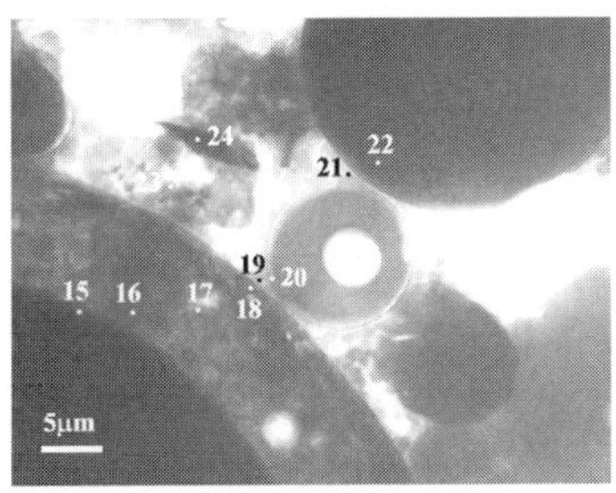

Figure 9 TEM micrograph of argon-milled thin section showing PFA with hydration rims. The dark particle bottom left is an iron-rich PFA grain, readily distinguishable from the added Fe_2O_3. Numbers correspond to analytical points referred to in text.

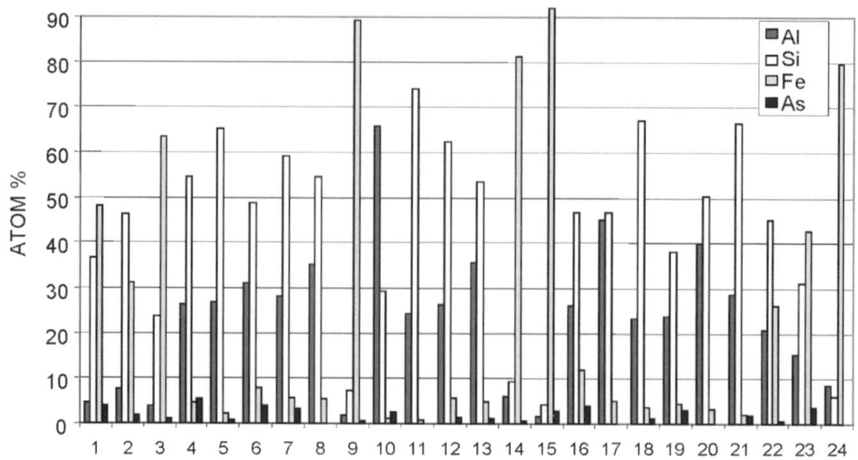

Figure 10 EDX analyses of points numbered in Figures 6-9

Figure 6, point 1 indicates a crystalline (by electron diffraction) iron silicate and Figure 8, point 10 is well crystallised quartz (with underlying hydration product according to the analysis - Figure 10). Those PFA grains which give featureless sections are probably wholly glassy, as in Figure 9, while the textures seen in Figures 6-8 suggest the presence of refractory crystals such as mullite. Arsenic is not associated in high levels with hydration product. This is consistent with the high leach rates observed. The highest relative As concentration in any of the analyses (Figure 10) is associated with points 1, 4, 6, 7, 10, 15, 16, 19 and 23. With the exception of point 1 (the crystalline iron silicate), the others are in (or close to) the hydration rims of fly ash particles. It is clear from the analysis at point 9 that the arsenic is *not* concentrated in the iron oxide cluster but the association of arsenic with iron originating from the fly ash (which contains 7.23% iron expressed as Fe_2O_3) cannot be ruled out. A reasonable correlation is observed between the As:Fe (As/Si:Fe/Si) ratios at the points associated with hydration products only, i.e. points 4, 6, 7, 10, 15, 16, 19 and 23, (see Figure 11).

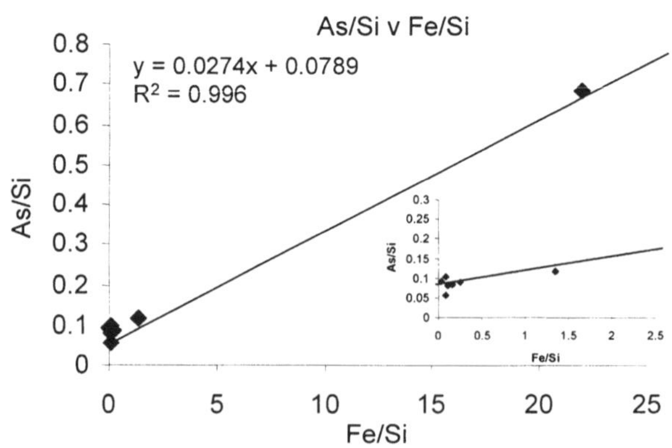

Figure 11 Effective As/Fe ratio derived from EDX of 'iron-rich' hydration product

This implies that it is the speciation of iron which is important in binding arsenic. The alkaline conditions in which the fly ash reacts is quite likely to provide an environment in which *aqueous* iron would exist as a hydroxy complex or precipitate as an oxyhydroxide (e.g. goethite). The behaviour of iron as Fe_2O_3 would be quite different. The lower reactivity of Fe_2O_3 (active in the protection of steel in concrete) would limit the conversion of this iron source to iron compounds with adsorption characteristics for As. Consequently, the Fe_2O_3 additions have not clarified the position regarding the role of Fe in As immobilisation within fly ash-derived geopolymers.

CONCLUSIONS

The leaching characteristics of alkali-activated PFA geopolymers containing As and morphological and compositional development have been studied. EDX analysis shows that only small amounts of As are incorporated in the hydration product. There is an apparent association of As with Fe arising from the hydration of the PFA, but iron added to the paste as Fe_2O_3 does not react with the As. We conclude therefore that it is the chemical form of Fe available to the aqueous phase that defines the immobilisation performance of fly ash geopolymer matrices and that the hydration conditions at the reacting fly ash surface provides the necessary chemistry for arsenic uptake by iron-containing hydration products. The absence of distinct arsenic compounds within the hydration products implies that the small amount of As that is held within the matrix is therefore chemically bound, rather than physically encapsulated. No explanation for the effect of Mn on As leachability was found

ACKNOWLEDGEMENTS

To the Royal Society/CSIC for awarding a grant associated with this research, also to the Regional Government of Madrid for awarding a post-doctoral grant. The Spanish Directorate General of Scientific Research for funding the project COO-1999-AX-038 associated with this research.

REFERENCES

1. MAMBOTE, R.C.M-, REUTER, M.A., VAN SANDWIJK, A and KRIJGSMAN, P., Immobilization of arsenic in crystalline form from aqueous solution by hydrothermal processing above 483.15°K, Minerals Engineering, Vol. 14, 2001, pp 391-403

2. WELHAM, N.J., MALATT, K.A. and VUKCEVIC, S., The stability of Iron phases presently used for disposal from metallurgical systems-a review, Minerals Engineering, Vol. 13, 2000, pp 991-931

3. BREED, A.W.,. HARRISON, S.T.L AND HANSFORD, G.S., Technical Note, A preliminary investigation of the ferric leaching of a pyrite/arsenopyrite flotation concentrate, Minerals Engineering, Vol. 10, 1997, pp1023-1030

4. CONNER, J.R. Chemical fixation and solidification of hazardous wastes Ed. VNR, New York, 1990

5. TURNER, R.R., Oxidation state of arsenic in coal ash leachate, Environ. Sci. Technol., Vol 25 , 1981, pp1062-1064

6. ENRIQUE, E., MONFORT, E., CANTAVELLA, V., NEGRE, F., Leaching kinetics of toxic elements immobilized in cement matrices (I). Selecting a test method, Materials Engineering, Vol. 8, 1997, pp151-169

7. ENRIQUE, E., MONFORT, E., CANTAVELLA, V., NEGRE, F., Leaching kinetics of toxic elements immobilized in cement matrices (II). Proposal of a kinetic model, Materials Engineering, Vol. 8, 1997, pp 227-248

8. MACPHEE, D.E. and GLASSER, F.P., Immobilization science of cement systems, Mat. Res. Soc. Bull., Vol. 18, 1993, pp 66-71

9. BONEN, D. and SARKAR, L., The present state-of-the-art of immobilization of hazardous heavy metals in cement-based materials, Engineering Foundation Conference on Advances in Cement and Concrete. (ASCE, 1994) pp 481-497

10. BATCHELOR, B., Leach models: theory and application, Journal of Hazardous Materials, Vol. 24, 1990, pp 225-266

11. KYLE, J.H. and LUNT, D., An investigation of disposal options for As_2O_3 produced from roasting operations, Proc. 5th Extractive Metallurgy Conference, 1991, pp 347-353.

12. SHIVELY, W., BISHOP, P., GRESSS, D. and BROWN, T., Leaching test of heavy metals stabilized with Portland cement J. W.P.C.F., Vol. 58, 1986, pp 234-241

13. PALOMO, A., GRUTZECK, M.W. and BLANCO, M.T., Alkali-activated fly ashes a cement for the future Cement. and Concrete Research, Vol 29, 1999, pp.1323-1329

14. JAMIESON, L., The leaching of arsenic from geopolymer matrix, University of Aberdeen Undergraduate Project Report, 2000 (unpublished)

15. PALOMO, A. and LÓPEZ DE LA FUENTE, J.I., Alkali-activated cementitous materials: alternative matrices for the immobilisation of hazardous wastes. Part I. Stabilisation of boron, Cem. Concr. Res. Cem. Concr. Res. 33, 2003, pp 281-288.

16. PALOMO,A. and PALACIOS, M., Alkali-activated cementitous materials: alternative matrices for the immobilisation of hazardous wastes. Part II. Stabilisation of Chromium and lead, Cem. Concr. Res. 33, 2003, pp 289-295.

17. PALOMO, A., ALONSO, S., SOBRADOS, I. and SANZ, J., Alkaline activation of fly ashes. A ^{29}Si NMR study of the reaction products, J. Am. Ceram. Soc. (Submitted).

18. CLIFF, G. and LORIMER, G.W., The quantitative analysis of thin specimens, J. Microsc., Vol 103, 1975, pp 203-207

19. FERNANDEZ-JIMENEZ, A., MACPHEE, D.E.,. LACHOWSKI, E.E and. PALOMO, A.,Fixing arsenic in alkali activated cementitious matrices, J.Am.Ceram.Soc. (Submitted).

20. COOK, D.J. and. CAO, H.T, An Investigation of the Pore Structure in FlyAsh/OPC Blends, Proc 1st Int Congr. Pore Structure and Materials Properties, Versailles, 1987, pp 69-76.

THEME FOUR:

PERFORMANCE OF PORTLAND CEMENT CONCRETE IN THE ENVIRONMENT

Keynote Paper

DURABILITY OF PORTLAND CEMENT SYSTEMS IN AGGRESSIVE ENVIRONMENTS

Y Maltais
SIMCO Technologies Inc
J Marchand
University of Laval
Canada

ABSTRACT. Portland cement systems are now commonly used for the construction of structures and buildings for residential, commercial and industrial applications. In many instances, these elements are exposed to severe environments, and their long-term performance is of concern. The main results of a comprehensive investigation of deterioration processes that may affect the behavior of Portland cement systems exposed to chemically-aggressive environments is presented. As part of this investigation, well-cured cement paste discs were exposed to deionized water, sodium sulfate and sodium chloride solutions. Degradation experiments were conducted under saturated and unsaturated conditions. At the end of the exposure period, microstructural alterations were investigated by microprobe analyses, scanning electron microscope observations and energy-dispersive X-ray analyses. Test results provide information on the basic aspects of various degradation phenomena such as chloride penetration, decalcification and external sulfate attack. For each phenomenon, the influence of factors such as water/cement ratio, type of cement and water transport is reviewed. The implication of these results on the prediction of the long-term performance of hydrated cement systems in chemically aggressive environments is discussed.

Keyword: Durability, Portland cement, Water/cement Ratio, Type of cement, Chloride penetration, Sulfate attack, Decalcification

Yannick Maltais is project engineer at SIMCO Technologies Inc. (Quebec City, Canada). He holds a B.A. in multidisciplinary engineering from Université du Québec at Chicoutimi (Canada) and a M.Sc. degree in civil engineering from Laval University (Canada) where he is currently completing Ph.D. studies. His research interests are in the area of concrete durability and service-life prediction.

Jacques Marchand is professor of civil engineering at Laval University (Canada) and director of the research center on concrete infrastructures (CRIB), which includes 26 principal researchers from six different universities in Canada. He is also President of SIMCO Technologies Inc. (Quebec City, Canada). His research interests are in the area of concrete durability and numerical modeling.

INTRODUCTION

Over the past decades, laboratory investigations and field experience have clearly emphasized the fact that concrete in contact with aggressive solutions may suffer from premature degradation. For instance, in North America, chloride ions (originating from de-icing salts and seawater) are at the origin of most reinforcing steel corrosion problems, which remain by far the primary cause of the rapid and on-going deterioration of highway, bridges and marine structures [1]. The penetration of chloride, magnesium and sulfate ions in hydrated cement systems is also known to significantly modify the microstructure of these materials [2-4]. In too many cases, these microstructural alterations have a detrimental influence on the engineering properties of concrete and contribute to reduce the service-life of the structure. Despite the large number of papers published on the durability of hydrated cement systems exposed to aggressive solutions, it is still rather difficult to find studies specifically devoted to the influence of chloride and sulfate ions penetration on all the different aspects of the microstructure of these materials. Most studies tend to concentrate on a single aspect of the problem. There is also very little systematic data on the influence of cement type and water-cement ratio on the kinetics of degradation. In addition, most published investigations concern experiments carried out in saturated conditions. This is most unfortunate considering that, in practice, most concrete structures are usually not maintained in fully saturated conditions.

This paper summarizes the main results of a comprehensive investigation on the behavior of hydrated cement systems exposed to aggressive ionic solutions. Test program was designed to generate quantitative information on the influence of water/cement ratio, type of cement and type of solution on the kinetics of degradation. One of the key objectives of this investigation was to generate reliable data for the validation of a numerical model developed to predict the service-life of concrete structures in aggressive environments [5].

MATERIAL AND MIXTURE CHARACTERISTICS

Neat cement pastes (W/C = 0.40 and 0.60) were prepared using an ordinary Portland cement (Canadian CSA Type 10) and a sulfate-resisting Portland cement (Canadian CSA Type 50). The chemical and mineralogical compositions of the two cements are given in Table 1. Both cements are commercially available on the Canadian market. All mixtures were prepared using deionized water and without any chemical admixture. Mixtures were batched under vacuum (at 10 mbar) to prevent the formation of air void during mixing. Mixtures were cast in plastic cylinders (diameter = 7.0 cm; height = 20 cm). Molds were sealed and rotated for the first 24 hours to prevent bleeding. Cylinders were then demolded and sealed with an adhesive aluminum foil for 18 months at room temperature.

EXPERIMENTAL PROCEDURES

At the end of the curing period, cylinders were sawn in 15-mm thick discs. Samples were then vacuum saturated in a 300 mmol/l sodium hydroxide solution for a 24-hour period. Experiments were performed during three months, in saturated and unsaturated conditions, using the following solutions: deionized water, sodium sulfate solution (prepared at 10, 50 and 100 mmol/l) and sodium chloride solution (prepared at 250 mmol/l). During the experiments, the aggressive solutions were systematically renewed on a weekly basis in order to maintain the exposure conditions constant and the pH around 10.0-10.5. The detailed test program is given Table 2.

Table 1 Chemical and mineralogical composition of cements

OXIDE	CEMENT (%)		COMPOUND	CEMENT (%)	
	CSA Type 10	CSA Type 50		CSA Type 10	CSA Type 50
SiO_2	19.78	21.45	C_3S	59	62
Al_2O_3	4.39	3.58	C_2S	12	16
TiO_2	0.22	0.21	C_3A	7	2
Fe_2O_3	3.00	4.38	C_4AF	9	13
CaO	62.04	63.93			
SrO	0.26	0.07			
MgO	2.84	1.81			
Mn_2O_3	0.04	0.05			
Na_2O	0.32	0.24			
K_2O	0.91	0.70			
SO_3	3.20	2.28			
LOI	2.41	0.86			

Table 2 List of the degradation tests performed on neat cement pastes

N°	CODE	TEST PERIOD	MOISTURE CONDITION	CONC.
1	T10-0.40-3M-S-H_2O	3 months	Saturated	0 mmol/l
2	T10-0.40-3M-US-H_2O	3 months	Unsaturated	0 mmol/l
3	T10-0.60-3M-S-H_2O	3 months	Saturated	0 mmol/l
4	T10-0.60-3M-US-H_2O	3 months	Unsaturated	0 mmol/l
5	T50-0.40-3M-S-H_2O	3 months	Saturated	0 mmol/l
6	T50-0.40-3M-US-H_2O	3 months	Unsaturated	0 mmol/l
7	T10-0.60-3M-S-10 mmol/l Na_2SO_4	3 months	Saturated	10 mmol/l
8	T10-0.40-3M-S-50 mmol/l Na_2SO_4	3 months	Saturated	50 mmol/l
9	T10-0.40-3M-US-50 mmol/l Na_2SO_4	3 months	Unsaturated	50 mmol/l
10	T50-0.40-3M-S-50 mmol/l Na_2SO_4	3 months	Saturated	50 mmol/l
11	T10-0.60-3M-S-50 mmol/l Na_2SO_4	3 months	Saturated	50 mmol/l
12	T10-0.60-3M-US-50 mmol/l Na_2SO_4	3 months	Unsaturated	50 mmol/l
13	T10-0.60-3M-US-100 mmol/l Na_2SO_4	3 months	Unsaturated	100 mmol/l
14	T10-0.60-7d-S-250 mmol/l NaCl	7 days	Saturated	250 mmol/l
15	T10-0.60-14d-S-250 mmol/l NaCl	14 days	Saturated	250 mmol/l
16	T10-0.60-28d-S-250 mmol/l NaCl	28 days	Saturated	250 mmol/l
17	T10-0.60-3M-S-250 mmol/l NaCl	3 months	Saturated	250 mmol/l

Before their immersion in the aggressive solutions, samples of series 1 (saturated conditions) were first coated (on all their faces except one) and then immersed in the test solutions. The sides of the samples of series 2 (unsaturated conditions) were coated. A relative humidity gradient was created between the two faces of the discs of series 2. One face was directly placed in contact with the solution and the other was placed in contact with a CO_2 free environment maintained at a relative humidity of 65%. In order to avoid carbonation, nitrogen was added on a regular basis in the compartment. All experiments were conducted at 23° C.

At the end of the degradation experiments, samples were broken in small pieces and then immersed in isopropylic alcohol for a minimum period of 14 days. After this period, samples were dried under vacuum for 7 days. Once the drying process over, samples were impregnated with an epoxy resin, polished, and coated with carbon.

Microstructural alterations were investigated by means of microprobe analyses using a Cameca SX-100 operating at 15 kV and 20 nA. For each sample, measurements were performed at intervals varying from 4 μm to 16 μm on four distinct imaginary lines extending from the external surface in contact with the aggressive solution toward the internal part of the material. A 2-mm x 2-mm (or 4-mm^2) elemental mapping was also performed for a few selected samples. At each point, the total content of calcium, sulfur, sodium, silicon, aluminum, and chloride (if required) was determined. The degradation of selected samples was also investigated by scanning electron microscope (SEM) observations carried out on polished sections. X-ray energy dispersion analyses were performed to establish the chemical composition of the phases observed. All measurements were carried out at 15 kV.

TEST RESULTS AND DISCUSSION

Chemical Degradation Using Deionized Water

Figures 1 presents typical calcium and sulfur profiles (determined by the microprobe analyses and expressed in counts per second – cps) for samples exposed for three months to the deionized water. For all samples, the location of the calcium-depleted zone(s) was determined on the basis of the calcium, aluminum, silicon and sulfur microprobe profiles measured from the exposed surface (in contact with the aggressive solution) up to the opposite face of the discs.

Test results indicate that all samples were altered (to various degrees) by the exposure to deionized water. The thickness of calcium depleted zone after three months of exposure is given in Table 3 for all samples. Test results vary from 960 μm up to 1950 μm. Data emphasize the significant influence of water/cement ratio and type of cement. As can be seen in Figure 1 and in Table 3, the use of a sulfate-resisting cement (Type 50) had a detrimental influence on the kinetics of degradation, increasing the thickness of the calcium-depleted zone by nearly 60% for both saturated and unsaturated test conditions.

These observations on the influence of the sulfate-resisting cement on the diffusion properties of the cement paste mixtures are in good agreement with previous data obtained on similar systems [6]. As part of another phase of this project, chloride migration experiments were conducted on all the neat paste samples cured for 12 and 24 months in sealed conditions. Details on the migration procedure can be found elsewhere [7]. All migration test results were analyzed using the method described in reference 8. Diffusion coefficients derived from these calculations are given in Table 4. As can be seen, the use of a sulfate-resisting cement was systematically found to have a detrimental influence on the ionic diffusion properties of the hydrated cement pastes. Similar results were also observed on concrete mixtures prepared with the same cements [9].

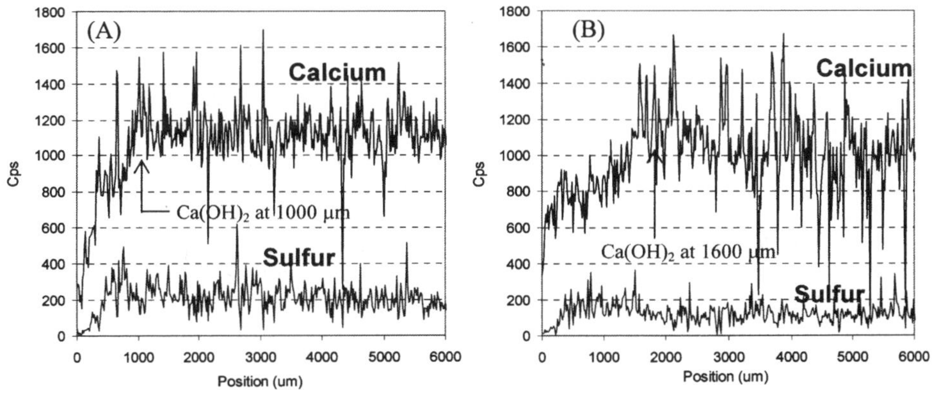

Figure 1 Calcium and sulfur content analysis after immersion in pure water for 3 months (A: W/C = 0.40 – CSA Type 10 – Saturated; B: W/C = 0.40 – CSA Type 50 - Saturated)

Table 3 Depth of the calcium-depleted zones for the experiments in deionized water (measured from the exposed surface)

N°	CODE	DEPTH (μm)
1	T10-0.40-3M-S-H_2O	1000
2	T10-0.40-3M-US-H_2O	960
3	T10-0.60-3M-S-H_2O	1950
4	T10-0.60-3M-US-H_2O	1710
5	T50-0.40-3M-S-H_2O	1600
6	T50-0.40-3M-US-H_2O	1530

Table 4 Diffusion coefficients of OH^- ions in the different mixtures

MIXTURE (n°)	CEMENT TYPE	WATER/ CEMENT	OH^- DIFFUSION COEFFICIENT (m^2/s) 1 year of curing	2 years of curing
A	CSA type 10	0.40	2.8E-11	2.2E-11
B		0.50	5.2E-11	4.8E-11
C		0.60	9.2E-11	8.7E-11
D		0.80	23.3E-11	--
E	CSA type 50	0.40	4.3E-11	3.9E-11
F		0.60	23.0E-11	16.1E-11

Water/cement ratio has also a major influence on the kinetics of calcium leaching. Test results indicate that a reduction of the water/cement ratio from 0.60 to 0.40 tends to reduce the kinetics of leaching by nearly a factor of two. These results emphasize, once more, the importance of using low water/cement ratios to increase the durability of cement-based materials to chemical attack [10, 11].

Test results summarized in Table 3 also indicate that moisture transport has only a limited influence on the kinetics of calcium leaching. Indeed, the thickness of the calcium-depleted zone appears to be slightly reduced for the samples tested in unsaturated conditions.

Globally, the microstructural analyses confirm that calcium hydroxide is the first calcium-bearing phase to be influenced by affected by the exposure to deionized water. This observation is in good agreement with those published by Delagrave et al. [10] and Vernet [12]. The microprobe analyses also demonstrate that C-S-H are not stable in presence of deionized water. In fact, as pointed out by Revertégat et al. [13], Berner [14] and Faucon [15], C-S-H decalcification begins when the amount of available calcium hydroxide is no longer sufficient to restore the pore solution equilibrium. When all the calcium hydroxide has been dissolved, the equilibrium of the system is controlled by the C-S-H that undergo partial decalcification. The decalcification of C-S-H leaves a residual skeleton of "silica gel" with almost no residual mechanical strength [16, 17].

Finally, microprobe analyses emphasize the intricate nature of the degradation mechanisms. A close examination of the profiles shows a slight increase of the sulfur content in the calcium-depleted zones even if the sulfate concentration of the test solution was equal to zero. This phenomenon was observed for all tested samples, and can be easily seen in Figure 1(b) where the sulfur content slightly increases between 350 μm and 1500 μm.

Chemical Degradation in Presence of Sodium Sulfate Solutions

Figure 2 presents the calcium and sulfur profiles for the 0.60 water/cement ratio (CSA Type 10) samples immersed for three months in the sodium sulfate solutions prepared at 0, 10 and 50 mmol/l. These profiles are typical of the degradation observed for the other samples exposed to the sodium sulfate solutions.

Globally, the microprobe analyses emphasize the marked influence of cement type on the kinetics of calcium leaching. After 3 months, the depth of calcium leaching for the 0.40 water/cement ratio samples exposed to the 50 mmol/l sodium sulfate solution ranges from 1015 μm (for ordinary Portland cement - CSA Type 10) to 1475 μm (for the sulfate-resistant Portland cement - CSA Type 50). These results are in good agreement with those obtained for the samples exposed to deionized water (see previous section) and confirm the sensitivity of the CSA Type 50 mixtures to calcium leaching.

Test results also confirm the marked influence of water/cement ratio on the kinetics of degradation. Indeed, for the experiments carried out at 50 mmol/l under saturated condition, the average thickness of the calcium-depleted zone ranges from 1030 μm for the 0.40 water/cement ratio (CSA Type 10) samples to 1630 μm for the 0.60 water/cement ratio (CSA Type 10) samples. Similar results were also obtained for degradation experiments performed in unsaturated conditions.

Microprobe analyses indicate that concentration in sodium sulfate of the aggressive solution seems to have a mixed influence on the kinetics of calcium leaching. The point is illustrated in Figure 2 for the samples tested in saturated conditions. For instance, at low sulfate concentration, the thickness of calcium-depleted zone is increased from 1950 μm (at 0 mmol/l) to 2230 μm (10 mmol/l). However, at high sulfate concentration (i.e. 50 mmol/l), the depth of the calcium-depleted zone appears to be limited to 1630 μm. Similar results were obtained for the experiments carried out in unsaturated conditions, for which the depth of calcium leaching was reduced from 1710 μm (at 0 mmol/l) to 1535 μm (at 50 mmol/l) and 1500 μm (at 100 mmol/l).

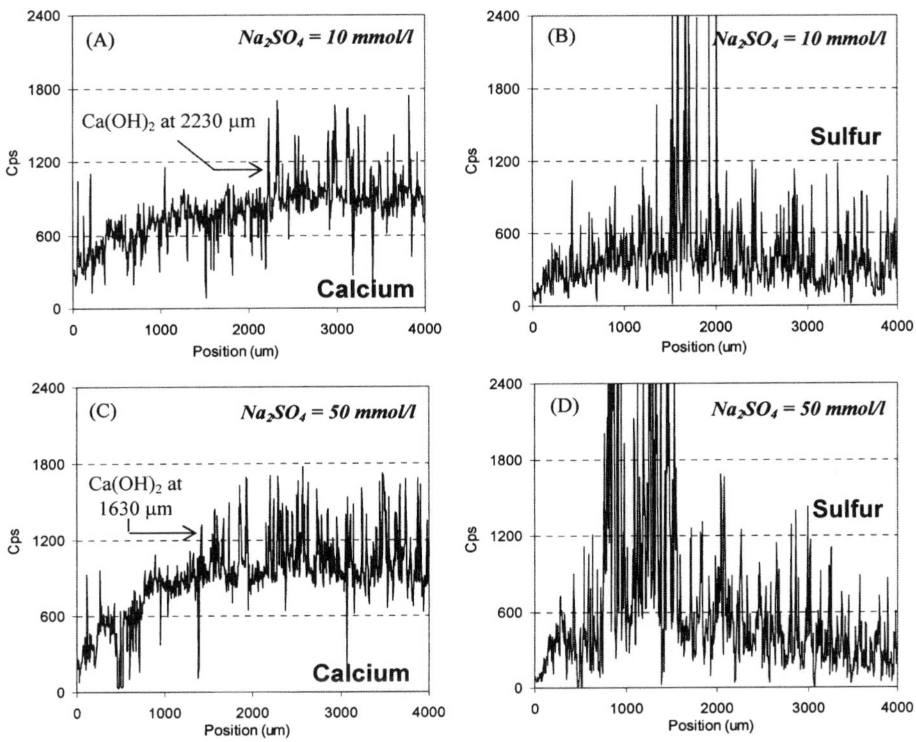

Figure 2 Calcium and sulfur profiles for the 0.60 water/cement ratio samples immersed for three months in different sodium sulfate solutions

Test results reveal the presence of a narrow band of high sulfur content near the external surface of all samples exposed to the sodium sulfate solutions, whatever the concentration (10, 50 or 100 mmol/l). The microprobe analyses presented in Figure 2 indicate that the sulfur content in this band increases with the sulfate concentration of the external solution. EDS analyses showed that the sulfur band is made of a mixture of gypsum and ettringite crystals. It could be observed during the SEM observations that gypsum was mainly located in long veins parallel to the exposed surface.

The presence of a band of gypsum in the samples exposed to the 10 mmol/l sodium sulfate solution is noteworthy. It is often erroneously assumed that gypsum formation is only possible for exposure to very high sulfate concentrations. This result is also in good agreement with petrographic examinations of field samples that often reveal the presence of gypsum near the surface of concrete in contact with the soil [18-20].

The presence of the high-sulfur content band can probably explain the ambiguous influence of the sodium sulfate concentration on the kinetics of calcium leaching. The massive formation of calcium sulfate crystals has probably filled the pores of the material, which has contributed to locally reduce the transport of ions. The influence of gypsum formation on the transport properties of the material should, however, be considered with caution. In many instances, SEM observations revealed the presence of large cracks running parallel to the exposed surface of the material and located within or right next to the gypsum band. The presence of these cracks certainly contributes to increase the transport of moisture and ions within the material. Furthermore, these observations tend to contradict the assumption of Hime and Mather [21] that gypsum formation is innocuous to concrete.

Table 5 gives for each sample the position and mean thickness of the high sulfur zone (identified as *Zone A*). On the one hand, the type of cement and the water/cement ratio have a strong influence on the results. On the other hand, sodium sulfate concentration and moisture flux only marginally affect the position and thickness of the sulfur band.

Table 5 Position and thickness of zones A and B

N°	TEST CONDITIONS	ZONE A (μm)			ZONE B (μm)		
		Begin	End	Thickness	Begin	End	Thickness
7	T10-0.60-S-10 mmol/l	1355	2005	650	2005	4785	2780
8	T10-0.40-S-50 mmol/l	810	985	175	985	2570	1585
9	T10-0.40-US-50 mmol/l	790	970	180	970	2618	1645
10	T50-0.40-S-50 mmol/l	950	1305	360	1305	4060	2755
11	T10-0.60-S-50 mmol/l	760	1560	800	1560	3735	2145
12	T10-0.60-US-50 mmol/l	985	1630	645	1630	3900	2270
13	T10-0.60-US-100 mmol/l	980	1435	455	1435	3700	2265

In addition to the high sulfur band, a second zone (named *Zone B*) of lower sulfur content has also been identified in samples exposed to the sodium sulfate solutions. This second zone can be seen right next to *Zone A*. Typical microprobe results are given in Figure 3. Analyses indicate that ettringite is the only sulfate-bearing compound in this second zone. No gypsum crystal could be detected outside *Zone A*. It should finally be pointed out that the parameters affecting the position and thickness of *Zone B* are the same as those affecting the characteristics of *Zone A*.

Globally, these results are in good agreement with the conclusions of Brown et al. [18, 19] who observed zonal formation of various compounds in field concrete samples exposed to sulfate-contaminated soils. The presence of layers of gypsum and ettringite was found to be indicative of the transport of aggressive ions in the microstructure of concrete.

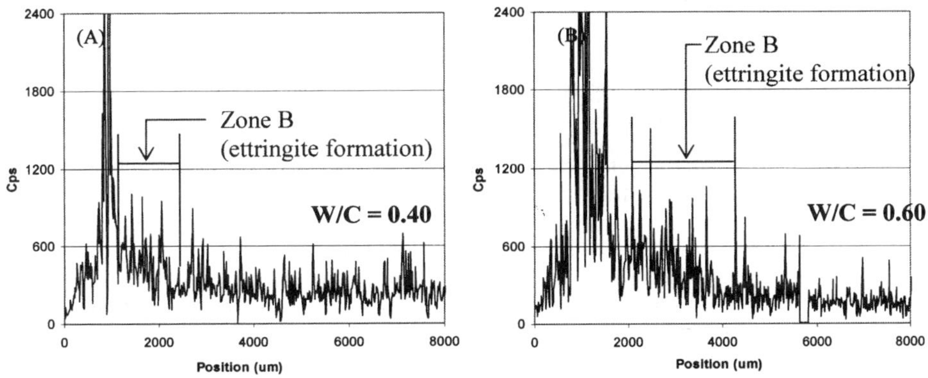

Figure 3 Sulfur analyses for the 0.40 and 0.60 water/cement ratio samples exposed to the 50 mmol/l sodium sulfate solution in saturated conditions

Chemical Degradation in Presence of a Sodium Chloride Solution

Figures 4 and 5 show the main results obtained for the 0.60-water/cement ratio (CSA Type 10) samples immersed in a 250-mmol/l sodium chloride solution. Microprobe analyses show calcium (Figure 4) and chloride (Figure 5) profiles after 7, 14, 28 and 91 days of testing.

Test results presented in Figure 4 indicate that the presence of chloride ions in solution has a significant influence on the kinetics of calcium leaching. Indeed, after 3 months of immersion, the calcium-depleted zone extends over 3000 μm. The depth of degradation for the companion sample exposed to deionized water was limited to 1950 μm According to these results, the presence of chloride ions in the aggressive solution increases the kinetics of calcium leaching by a factor of 1.5. Similar results were reported by Delagrave et al. [10]. Test results are also in good agreement with those published by Gégout et al. [11].

The accelerating effect of the presence of chloride ions on the kinetics of decalcification can be explained by their influence on the transport of OH^- ions in solution. In the material, the penetration of chloride ions is compensated by the counter-diffusion of hydroxide ions in order to maintain the electroneutrality of the pore solution. The counter-diffusion process affects the local equilibrium of the system and initiates the dissolution of calcium hydroxide to restore the OH^- concentration in solution. Previous studies have shown that chloride ions at low concentrations have only a limited effect on the solubility of calcium hydroxide crystals [22, 23].

Figure 7 graphically illustrates the evolution of the depth of decalcification after 7, 14, 28 and 91 days of immersion. The thickness of the calcium-leached zone increases rapidly with time. Results also indicate that decalcification progresses as a non-linear function of time. Microprobe analyses have also been used to study the formation of chloride-bearing phases. Test results presented in Figure 5 confirm the rapid penetration of chloride ions into the samples. For instance, after only 3 months of immersion, the chloride content was fairly constant over the entire thickness of the sample. It is also interesting to point out that in the highly decalcified zone, almost no chloride could be detected.

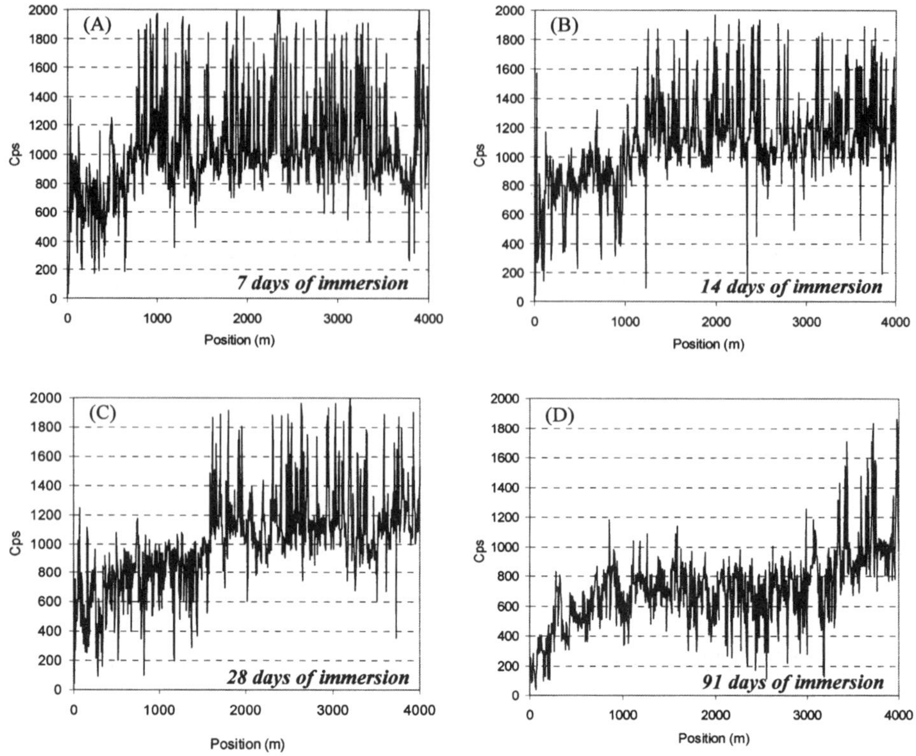

Figure 4 Decalcification of the 0.60-water/cement ratio (CSA Type 10) samples after 7, 14, 28 and 91 days of immersion in a sodium chloride solution (at 250 mmol/l)

SEM observations were performed on the sample immersed for 14 days in the chloride solution. These observations (mainly carried out at the vicinity of the calcium hydroxide dissolution front) revealed the presence of numerous chloroaluminate crystals scattered across the observed area. Chloroaluminate formation was confirmed by EDS analyses.

The crystals were rather small (typically 5 μm x 10 μm) as compared to the gypsum crystals (typically 20 μm x 40 μm or more) observed in other samples. Chloroaluminate crystals were found to be limited to a band extending from the surface of the sample up to 6500 μm after 7-day, 7500 μm after 14-day and 13750 μm after 28-day. After 91 days of testing, chloroaluminates were found over the entire thickness of the sample.

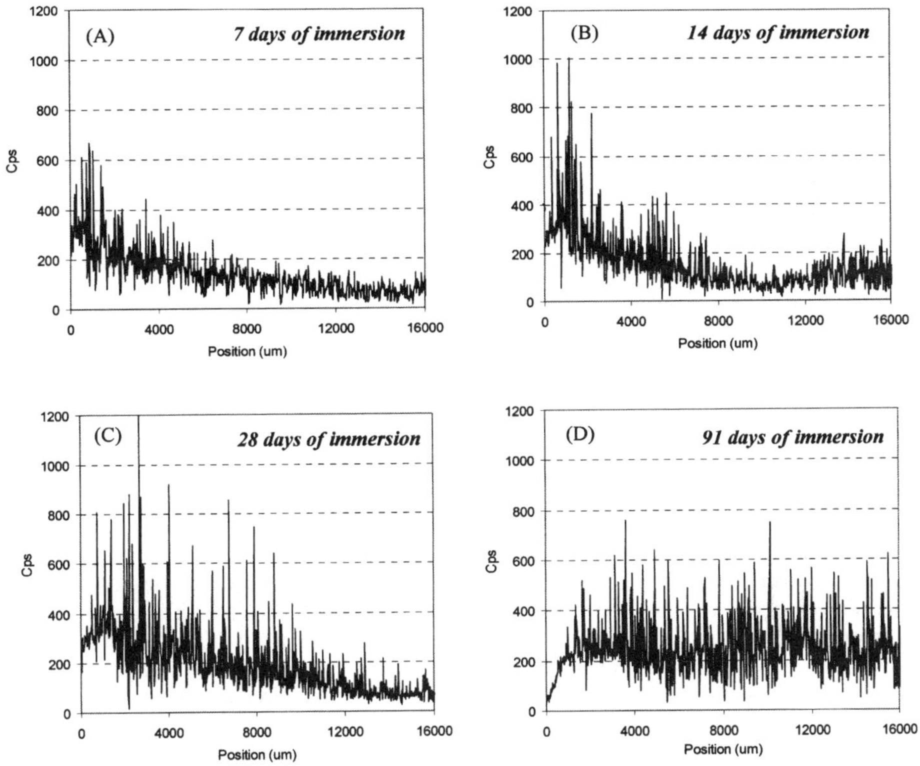

Figure 5 Chloride contents of the 0.60 water/cement ratio (CSA Type 10) samples after 7, 14, 28 and 91 days of immersion in a sodium chloride solution (at 250 mmol/l)

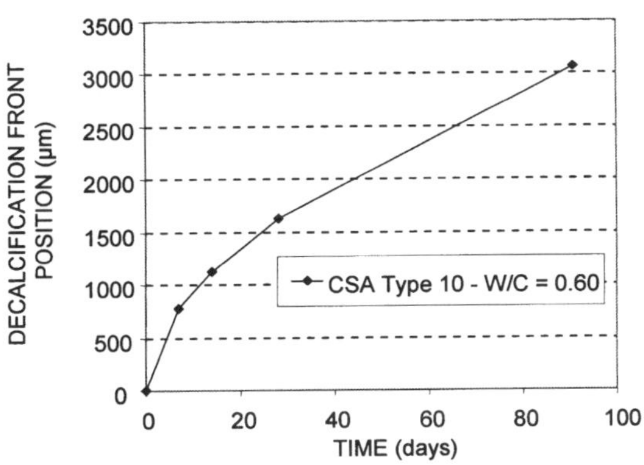

Figure 6 Position of the decalcification front after the immersion of the 0.60-water/cement ratio (CSA Type 10) samples in a sodium chloride solution (at 250 mmol/l)

CONSEQUENCES ON THE DEVELOPMENT OF NUMERICAL MODELS

The results of this investigation have numerous implications on the development of numerical models for the prediction of the microstructural degradation of hydrated cement systems exposed to chemically-aggressive environments. For instance, results obtained for the samples in contact with deionized water and low sodium sulfate solutions clearly show the complexity of most deterioration problems. The penetration of external ions and the leaching of species already present in the pore solution lead to a complete reorganization of the microstructure of the paste. Unfortunately, microstructural alterations are not limited to one single phase (or a "family" of phases) [2].

Accordingly, numerical models should be developed in such a way to be able to address all aspects of the problem. In that respect, the development of transport models coupled to chemical equilibrium codes appears to be the most promising way to model the deterioration of hydrated cement systems. Chemical equilibrium codes allow taking into account the contribution of all solid phases and ionic species on the reorganization of the material. Over the past decade, the chemical equilibrium problems in hydrated cement systems have been extensively investigated [23, 24]. Results of these studies are available to those working on the development of numerical models.

The comparison of the kinetics of decalcification between the pastes exposed to deionized water and those in contact with the chloride solutions also emphasizes the complexity of ionic transport mechanisms in porous media. It is now well established that the diffusion of a given ionic species in an electrolyte is coupled to that of other ions in solution [25]. In many instances, the coupling effect cannot be neglected [26]. The development of single-ionic-species models presents very few advantages and may even yield grossly erroneous results.

Finally, microprobe analyses and SEM observations strongly suggest that microstructural alterations may have a significant influence, at least locally, on the transport properties of the material. The treatment of this aspect of the degradation of hydrated cement systems offers an area of collaboration between researchers working on the elaboration of durability models and those that focus on the development of microstructure/hydration models.

CONCLUSIONS

Microprobe analyses and SEM observations carried out in this investigation confirm that the type of solution has a dominant influence on the nature and the kinetics of alteration of hydrated cement systems. Samples exposed to deionized water were all affected, to various degrees, by $Ca(OH)_2$ dissolution and C-S-H decalcification. In addition to the precipitation of Friedel's salt, exposure to chloride solutions was found to accelerate the kinetics of decalcification of the pastes. Ettringite and gypsum formation was observed in all samples exposed to sodium sulfate solutions.

Water/cement ratio was found to have a major influence on the kinetics of degradation. These results emphasize, once more, the importance of using low water/cement ratios to increase the durability of cement-based materials to chemical attack. Results also emphasize the significant effect of the type of cement and the relatively weak influence of moisture transport on the extent of degradation.

REFERENCES

1. FEDERAL HIGHWAY ADMINISTRATION, Corrosion cost and preventive strategies in the United States, FHWA-RD-01-156, Appendix D – Highway Bridges, 2002, p. 74

2. GLASSER, F.P., Role of chemical binding on diffusion and mass transport, in Materials Science of Concrete – Ion and Mass Transport in Cement-Based Materials, American Ceramic Society (USA), 2001, p. 97-112.

3. TAYLOR, H.F.W., Cement chemistry, Academic Press (UK), 1997, p. 475.

4. SKALNY, J., MARCHAND, J., ODLER, I., Sulfate attack of concrete, E & FN Spon (UK), 2001.

5. SAMSON, E., Numerical modeling of degradation phenomena of hydrated cement systems, Ph.D. Thesis, Dept. of Civil Engineering, Laval University (Canada), 2003 (in preparation).

6. MALTAIS, Y., Investigation of transport and degradation mechanisms of partially saturated cement systems, Ph.D. Thesis, Dept. of Civil Engineering, Laval University (Canada), 2003 (in preparation).

7. JACOBSEN, S., MARCHAND, J., BOISVERT, L., Effect of cracking and healing on chloride transport in OPC concrete, Cement Concrete Research, Vol. 26, N° 6, 1996, p. 869-882.

8. SAMSON, E., MARCHAND, J., SNYDER, K.A., Calculation of ionic diffusion coefficients on the basis migration test results, Concrete Science & Engineering/Materials and Structures, Vol. 36, N° 257, 2003, p. 156-165.

9. MALTAIS, Y., MARCHAND, J., SAMSON, E., Influence of cement type and water/cement ratio on the ionic diffusion properties of concrete, 2003 (in preparation).

10. DELAGRAVE, A., GÉRARD, G., MARCHAND, J., Modelling calcium leaching mechanisms in hydrated cement pastes, in Mechanisms of Chemical Degradation of Cement-Based Materials, E & FN Spon, London (UK), 1997, p. 38-49.

11. GÉGOUT, P. REVERTÉGAT, E., MOINE, G., Action of chloride ions on hydrated cement pastes : Influence of the cement type and long time effect of the concentration of chloride, Cement and Concrete Research, Vol. 22, 1992, p. 451-457.

12. VERNET, C., Chemical stability of hydrates – Defense mechanisms of concrete against chemical aggressions, La durabilité des bétons, Presses de l'école nationale des Ponts et chaussées, 1992, p. 129-170, (in French).

13. REVERTÉGAT, E., GÉGOUT, P. MOINE, G., Effect of pH on the durability of cement pastes, Cement and Concrete Research, Vol. 22, N° 2/3, 1992, p. 259-272.

14. BERNER, U.R., Modeling the incongruent dissolution of hydrated cement minerals, Radiochinica Acta, Vol. 44/45, 1988, p. 387-393.

15. MENETRIER, D., Contribution to kinetic study of the initial hydration period of $3CaO.SiO_2$, Ph.D. Thesis, Université de Bourgogne, Dijon, France, 1977, 261 p.

16. GÉRARD, B., Influence of the couplings chemistry-mechanics-mass transfer on the long-term durability of concrete nuclear waste storage facilities, Ph. D. Thesis, Dept. of Civil Engineering, ENS-Cachan/Université Laval (Canada), 1996, 289 p.

17. CARDE, C., FRANÇOIS, R., Effect of the leaching of calcium hydroxide from cement paste on the mechanical and physical properties, Cement and Concrete Research, Vol. 27, 1997, p. 539-550.

18. BROWN, P.W., BADGER, S., The distributions of bound sulfates and chlorides in concrete subjected to mixed NaCl, $MgSO_4$, Na_2SO_4 attack, Cement and Concrete Research, Vol. 30, N° 10, 2000, p.1535-1542.

19. BROWN, P.W., DOERR, A., Chemical changes in concrete due to ingress of aggressive species, Cement and Concrete Research, Vol. 30, N° 3, 2000, p. 411-418.

20. DIAMOND, S., LEE, R.J., Microstructural alterations associated with sulfate attack in permeable concretes, in Materials Science of Concrete Special Volume: Sulfate Attack Mechanisms, American Ceramic Society (USA), 1999, p. 123-173.

21. HIME, W.G., MATHER, B., Sulfate attack or is it? Cement and Concrete Research, Vol. 29, N° 5, 1999, p. 789-792.

22. DUCHESNE, J., REARDON, E.J., Measurements and prediction of portlandite solubility in alkali solutions, Cement and Concrete Research, Vol. 25., N° 5, 1995, p. 1043-1053.

23. DAMIDOT, D., GLASSER, F.P., Thermodynamic investigation of the CaO-Al_2O_3-$CaSO_4$-$CaCl_2$-H_2O system and the influence of Na_2O, 10th Congress on the Chemistry of Cement, Göthenburg, Sweden, 1997.

24. DAMIDOT, D., GLASSER, F.P., Thermodynamic investigation of the CaO-Al_2O_3-$CaSO_4$-H_2O system at 25° C and the influence of Na_2O, Cement and Concrete Research, Vol. 23, 1993, p. 221-238.

25. MARCHAND, J., Modeling the behavior of unsaturated cement systems exposed to aggressive chemical environments, Materials and Structures, Vol. 34, 2001, p. 195-200.

26. SAMSON, E., MARCHAND, J., ROBERT, J.L., BOURNAZEL, J.P., Modeling the mechanisms of ion diffusion transport in porous media, International Journal of Numerical Methods in Engineering, Vol. 46, 2000, p. 2043-2060.

DETERIORATION MODES OF NORMAL CEMENT MORTARS SUBJECTED TO SULFATE SOLUTIONS

H Y Moon

S T Lee

Hanyang University

South Korea

ABSTRACT. The objective of this work is to understand the deterioration modes of normal cement pastes and mortars immersed in three types of sulfate solutions for 510 days. In order to achieve the goal, x-ray diffraction (XRD) and differential scanning calorimetry (DSC) are presented in this experimental work. Reduction in compressive strength and expansion of normal cement mortars were also measured to evaluate resistances to sulfate solutions with different attacking sources. The mortar and paste were prepared by using water-binder ratio (w/b) of 0.45. Conclusively, the deterioration by sodium sulfate attack was primarily due to the formation of ettringite and thaumasite, and the main deterioration mode in magnesium sulfate solution was the gypsum formation. In the concomitant presence of chloride and sulfate ions, the increase of gypsum solubility in pore solution of cement matrix exhibited the mitigating effect on the sulfate attack.

Keywords: Sulfate attack, XRD, DSC, Expansion, Reduction in compressive strength, Deterioration mode, Chloride ions.

H Y Moon is a Professor in the Department of Civil Engineering at Hanyang University in South Korea. His main research is on binder technology, concrete durability and high strength concrete.

S T Lee is a Researcher of Department of Civil Engineering at Hanyang University in Korea. His main interest is durability and chemical attack of reinforced concrete structures.

INTRODUCTION

Sulfate deterioration of concrete is commonly observed in structures exposed to soil or groundwater containing a high concentration of sulfate ions. Sulfate attack mechanism is not obvious because of its complexity of chemical reaction. However, it is generally accepted that sodium sulfate attack of hydrated cement matrix takes place because of the reaction of sulfate ions with calcium hydroxide and calcium aluminate hydrate forming gypsum and ettringite. Further, in case of magnesium sulfate attack, brucite ($Mg(OH)_2$), which has low solubility and is relatively impermeable, is produced and a relatively great amount of gypsum leads to drastic reduction on strength. Investigations on the effects of cations accompanying sulfate ions were carried out by Cohen and Bentur [1] and Park et al. [2].

Since concrete is increasingly used in marine environment, the performance of this concrete in the chloride-sulfate environments also needs to be investigated. A review of literature [3] on the role of chloride ions on sulfate attack indicated that the presence of the ions in sulfate environments mitigated the sulfate attack in both normal and blended cement matrix.

Therefore, a better understanding of sulfate attack mechanisms through this paper will enhance the ability of design for concrete structures exposed to sulfate attack.

EXPERIMENTAL PROCEDURE

Materials and Test Solutions

In this investigation, ASTM C 150 Type I cement was used. The chemical composition, mineralogical components of the cement are shown in Table 1. River sand was used as fine aggregate. The specific gravity, absorption and fineness modulus of the fine aggregate are 2.60, 0.80% and 2.80, respectively. The weight ratio of fine aggregate to cement was 2.0.

Superplasticizer used in manufacturing mortar specimens was a polycarbonic acid-based chemical admixture.

5% sodium sulfate (NS), 5% magnesium sulfate (MS) and 3.5% sodium chloride plus 5% sodium sulfate (CS) were used as test solutions for immersion. These solutions were refreshed every 4 weeks. For comparison, tap water was also used as control solution.

Table 1 Chemical composition and mineralogical components of cement

CHEMICAL COMPOSITION (%)							MINERALOGICAL COMPONENTS (%)			
SiO_2	Al_2O_3	Fe_2O_3	CaO	MgO	SO_3	I.O.L	C_3S	C_2S	C_3A	C_4AF
20.2	5.8	3.0	63.3	3.4	2.1	2.1	54.9	16.6	10.3	9.1

Test Techniques

Expansion measurement of prism mortar specimens with the size of 2.5 × 2.5 × 28.5 mm was performed according to ASTM C 1012, and tested on three mortar specimens withdrawn from test solutions. The expansion values were averaged.

The reduction in compressive strength of cube mortar specimens was calculated as follows.

$$\textit{Reduction in compressive strength (\%)} = [(A - B) / A] \times 100$$

where, A is the average compressive strength of mortar specimens cured in tap water (MPa) and B is the average compressive strength of mortar specimens immersed in test solutions (MPa). Compressive strength tests were performed on each mixture at 0, 28, 91, 180, 270, 360 and 510 days of immersion. At each interval, the compressive strengths of five mortar specimens drawn from tap water and test solutions were measured and then their values were averaged.

For XRD analysis, the surface and inner part of mortar specimen, stored in sulfate test solutions, were used and finely grounded by hand, respectively. XRD was conducted using the RINT D/max 2500 (Rigaku) X-ray diffractometer. For the detailed items of XRD, CuKα radiation with a wavelength of 1.54λ at a voltage of 30 kV, scanning speed of 3 °/min. and current of 30 mA were used.

DSC (DSC 2010 Differential Scanning Calorimeter, TA instruments, U.S.A) analysis was conducted to evaluate the products by hydration and chemical reaction. The amount of paste sample powder used was approximately 10 mg. The powder was heated from ambient temperature to 560 °C at a rate of 10 °C / min in an N_2 atmosphere.

DETERIORATION OF NORMAL CEMENT MORTAR

Expansion

Expansion data of normal cement mortars stored in NS, MS and CS solutions are plotted in Figure 1. The expansion was observed to be higher in mortar specimen placed in NS solution compared to those placed in MS and CS solutions. In particular, the expansion of the mortar specimen placed in NS solution was about 0.51% at 210 days of exposure and then the mortar was disintegrated due to the excessive expansion after the age. However, the expansion of mortar specimen placed in MS solution was observed to be lower than that placed in NS solution. For mortar specimen in NS solution, the critical time, which was previously stated by Santhanam et al. [4], was much shorter compared to that of mortar specimen in MS solution, as shown in Figure 1. The role of cations accompanying sulfate ions on sulfate attack was studied Al-Amoudi et al. [5]. In the concomitant presence of chloride and sulfate ions in attacking source, the migration effect of sulfate attack of mortar specimen was clearly observed. For example, after 180 days of exposure, the expansion of mortar specimen placed in NS solution was observed to be higher about 0.25%, whereas for mortar specimen placed in CS solution, the expansion observed was not of high magnitude as the value of about 0.03% at the age. The beneficial effect of chloride ions, therefore, on sulfate attack was significant in the present study.

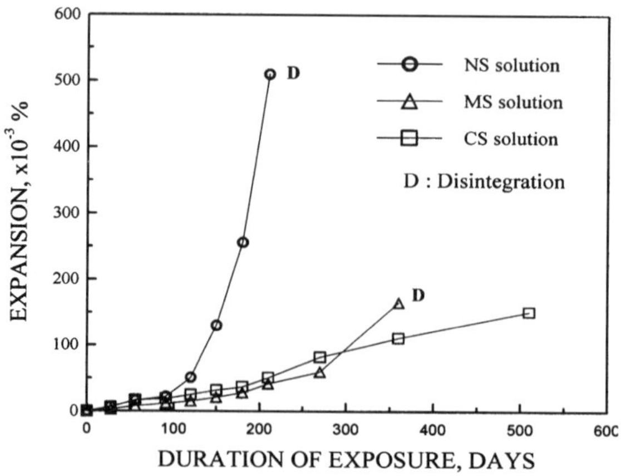

Figure 1 Expansion of mortar specimens immersed in sulfate solutions

Reduction in Compressive Strength

The data on reduction in compressive strength of mortar specimens placed in NS, MS and CS solutions are shown in Figure 2. The reduction in compressive strength of mortar specimens was almost similar in NS and MS solutions up to about 270 days of exposure. However, after that time the reduction in compressive strength was observed to be higher in mortar specimens in NS solution than those in MS solution.

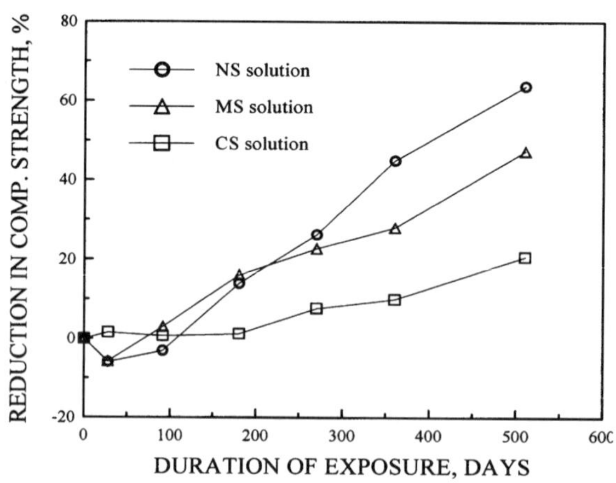

Figure 2 Expansion of mortar specimens immersed in sulfate solutions

After about 510 days of sulfate exposure, the reduction in compressive strength placed in NS solution was about 1.4 times that in MS solution. These data indicated that the reduction in compressive strength in NS solution was more pronounced in mortar specimens compared to that in MS solution. The reduction in compressive strength of mortar specimens placed in NS and MS solutions except CS solution, was initially negative, and beyond 180 days of exposure the strength reduction was evident.

The negative values signify as increase of compressive strength in NS and MS solutions, which is attributable to the filling up of the pores by reaction products, densifying the mortar specimens in early period of exposure [6, 7].

Comparing with reduction in comparing strength of mortar specimens placed in NS and CS solution, the remarkable difference, in particular at later period of exposure, was displayed. After 510 days of exposure, for example, the values of reduction in compressive strength in NS and CS were about 63.8% and 20.7%, respectively. This trend elucidates the predominant role of chloride ions in mitigating sodium sulfate attack

MICROSTRUCTURAL OBSERVATIONS

X-ray Diffraction

Deteriorated normal cement paste samples subjected to sulfate solutions for 510 days were examined by XRD. The XRD patterns of the paste samples are presented in Figure 3. Figure 3 (a) indicates the XRD pattern of paste sample subjected to NS solution. Form the XRD trace, it is observed that the main thaumasite peak (at 9.2°) was very intense compared with the ettringite peak (at 9.1°). In particular, it is difficult to distinguish peak for ettringite from that for thaumasite, especially if both may be present in a paste sample. However, unlike ettringite peak patterns, the thaumasite peaks are present at around 9.5, 23.4, 26.0 and 28.0°. The corresponding XRD pattern also indicated the strong peaks for ettringite and gypsum at around 32.2° and 11.7°.

Figure 3 (b) presents the XRD pattern of paste samples subjected to MS solution for 510 days. The XRD trace shows maximum intensity peak for gypsum at 11.7°. Comparing to the peak for gypsum under NS solution, it is observed to be much stronger intensity peak for gypsum in paste sample under MS solution. This XRD analysis suggests that the main deterioration caused under MS solution is the formation of gypsum. Another important observation of the XRD pattern is the absence of peak for portlandite. Additionally, the existence of brucite peaks at 18.6 and 37.9° in the sample supports the possibility that magnesium ions oriented from the MS solution react with calcium hydroxide at the surface part of the sample. On the other hand, there was no proof for the presence of magnesium silicate hydrates, which causes the strength loss in mortar specimens, from XRD analysis because of its poor crystallization.

XRD trace of paste sample subjected to CS solution, as shown in Figure 3 (c), demonstrated the presence of Friedel's salt peaks at around 11.1 and 27.5°, indicating the reaction with Cl^- ions and hydrated products in the normal cement system. Additionally, the gypsum was entirely eliminated or reduced to negligible levels. From the XRD results, it was found that intensive peaks for portlandite and ettringite were also examined.. However, the structures of ettringite formed in this environment are different from those formed in NS solution, as stated by Al-Amoudi et el. [8]. Ettringite crystallizes from the pore solution when chloride is present, because of its higher solubility in chloride-bearing environments, and so does not cause a considerable expansion.

These XRD results well support the deterioration degree (Figure 1 and 2) mechanisms by sulfate attack.

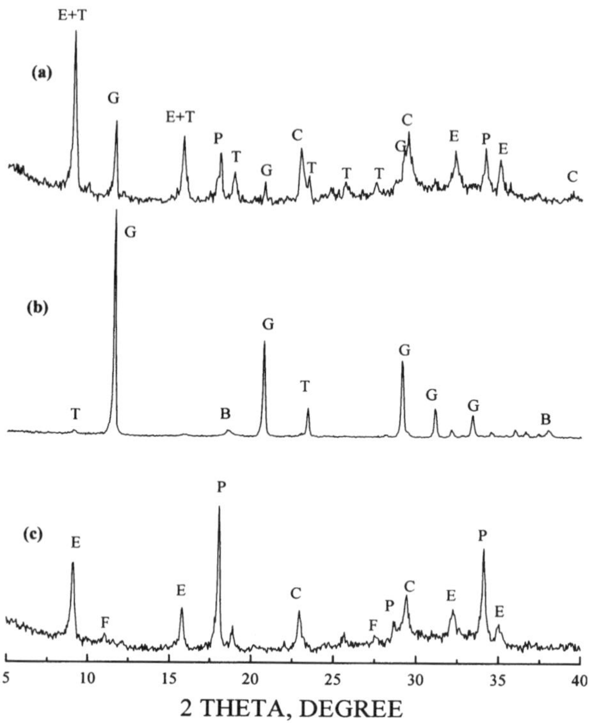

Figure 3 XRD patterns of pastes immersed in sulfate solutions for 510 days ((a) : NS solution, (b) MS solution, (c) : CS solution)

Note : E = ettringite, G = gypsum, T = thaumasite, P = portlandite, C = calcite, B = brucite, F = Friedel's salt.

Differential Scanning Calorimetry

The DSC curves of deteriorated materials obtained from normal cement paste samples exposed to sulfate solutions for 510 days are exhibited in Figure 4.

For the curve of paste sample under NS solution (Figure 4 (a)), there are four remarkable endothermic peaks at around 85, 124, 145 and 460 °C. The endotherm at 85 °C was attributed to dehydration of water in paste sample powder. The endotherm for ettringite and thaumasite was observed at around 125 °C, showing the enthalpy of 157.9 J/g. Furthermore, the endothermic peak at around 145 °C indicated the formation of secondary gypsum in the paste system by sulfate attack. As expected, a small amount of portlandite remained in the cement paste, indicating the endothermic peak at about 462 °C. At that temperature, the enthalpy was about 30.87 J/g.

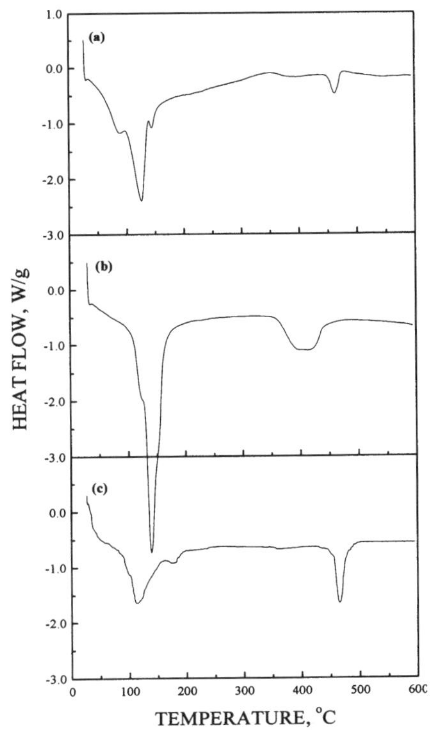

Figure 4 DSC curves of pastes immersed in sulfate solutions for 510 days ((a) : NS solution, (b) MS solution, (c) : CS solution)

Figure 4 (b) displays DSC curve of powder sample drawn from normal cement paste exposed to MS solution. The first major peak between 110 and 139 °C corresponded to the formation of secondary gypsum and thaumasite. XRD result presented in Figure 3 (a) showed the absence of ettringite due to the decomposition of it, as could be confirmed in DSC curve. Another important observation in DSC curve was the absence of portlandite at around 460 °C, which also well agree with XRD result conducted on the corresponding paste sample. Additionally, although the intensity peak for brucite from XRD trace was somewhat weak, the secondary major peak for brucite, which had an enthalpy of 209.1 J/g, was observed to be higher at around 399 °C. However, there was no confirmation for the presence of M-S-H by DSC curve.

DCS data on the paste sample conducted by thermal analysis system are plotted in Figure 4 (c). At about 460 °C, the first main peak indicated that a relatively great amount of portlandite is present.

CONCLUSIONS

This paper describes a detailed study on the different sulfate attack mechanisms and deterioration modes accompanying attacking sources.

In NS solution, the expansion and reduction in compressive strength of mortar specimens were more pronounced compared to those in MS and CS solution. Microstructural observation such as XRD and DSC confirmed the main deterioration mode in NS solution was the formation of ettringite and thaumasite. On the other hand, in paste sample subjected to MS solution, no ettringite was detected by XRD trace. In the environment, gypsum formation can be a deterioration mode of the mortar specimen. The data on reduction in compressive strength emphasized that the deterioration of mortar specimen in MS solution is attributable to the gypsum formation. In the concomitant presence of chloride and sulfate ions, the increase of gypsum solubility in pore solution of cement matrix exhibited the mitigating effect on the sulfate attack.

REFERENCES

1. COHEN, M. D., BENTUR, A., Durability of Portland Cement-Silica Fume Pastes in Magnesium Sulfate and Sodium Sulfate Solutions, ACI Material Journal, Vol. 85, 1988, pp 148-157.

2. PARK, Y. S., SUH, J. K., LEE, J. H., SHIN, Y. S., Strength Deterioration of High Strength Concrete in Sulfate Environment, Cement and Concrete Research, Vol. 29, 1999, pp 1397-1402.

3. AL-AMOUDI, O. S. B., MASLEHUDDIN, M., ABDUL-Al, Y. A. B., Role of Chloride Ions on Expansion and Strength Reduction in Plain and Blended Cements in Sulfate Environments, Construction and Building Materials, Vol. 9, No. 1, 1995, pp. 25-33.

4. SANTHANAM, D., COHEN, M. D., OLEK, J., Modeling the Effects of Solution Temperature and Concentration during Sulfate Attack on Cement Mortars, Cement and Concrete Research, Vol. 32, 2002, pp 585-592.

5. AL-AMOUDI, O. S. B., RASHEEDUZZAFAR, MASLHUDDIN, M., ABDULJAUWAD, S. N., Influence of Chloride Ions on Sulphate Deterioration in Plain and Blended Cements, Magazine of Concrete Research, Vol. 46, No. 167, 1994, pp 113-123.

6. BROWN, P. W., An Evaluation of Sulfate Resistance of Cements in a Controlled Environment, Cement and Concrete Research, Vol. 11, 1981, pp 719-727.

7. IRASSAR, W. G., BATIC, F., Effect of Low Calcium Fly Ash on Sulfate Resistance of OPC Cement, Cement and Concrete Research, Vol. 19, 1989, pp 194-202.

8. AL-AMOUDI, O. S. B., RASHEEDUZZAFAR, MASLEHUDDIN, M., ABDULJAUWAD, S. N., Influence of Chloride Ions on Sulfate Deterioration in Plain and Blended Cements, Magazine of Concrete Research, Vol. 46, No. 167, 1994, pp 113-123.

PERFORMANCE OF CONCRETE STRUCTURES IN THE MARINE ENVIRONMENT OF KARNATAKA, INDIA

S Devadas Bhat

B R Samaga

National Institute of Technology, Karnataka

India

ABSTRACT. Well-made concrete often survives better than might otherwise be expected, simply because of its impermeability and its ability to withstand the ingress of corrosive materials. The less the surplus voids the greater is the resistance to deterioration. Corrosion of steel is a multibillion-dollar problem worldwide. However, there are many structures, which show early deterioration, namely those exposed to aggressive environments. In the past and even at present times, too much emphasis is placed on concrete compressive strength rather than on environmental factors, which are known to affect concrete durability. This is one of the main reasons for serious deterioration of concrete structures that is prevalent today. The long-term behaviour of concrete structures has shown that their main cause of distress is reinforcement corrosion. One of the most aggressive exposure conditions for concrete is the marine environment. In these condition chloride penetration and chloride induced reinforcement corrosion rates can be very high, often leading to a reduced service life. This paper describes a series of case studies of different types of concrete structures, subjected to marine environment in the West-Coast of Karnataka (INDIA), that have suffered extensive deterioration due to corrosion.

Keywords: Corrosion, Marine environment, Chlorides, Durability.

Mr S Devadas Bhat is a teaching fellow in the Department of Civil Engineering at Manipal Institute of Technology, Manipal India. Presently pursuing his PhD from the National Institute of Technology, Karnataka, India. His research interests include long-term durability of concrete in coastal environment, corrosion of reinforcing steel in concrete.

Dr B R Samaga is a Professor in the Department of Civil Engineering, National Institute of Technology, Karnataka, India. He received his M.Tech from the Karnataka Regional Engineering College, Surathkal, and his PhD from University of Roorkee, India. His research interests includes Durability of Concrete in Coastal-environment, Use of supplementary cementitious materials in concrete, low cost housing.

INTRODUCTION

Concrete is an important construction material. Its importance continues to increase as its application area continues to grow. However, in marine environment it faces severe threat from corrosion. Corrosion of steel in reinforced concrete structures is a multibillion-dollar problem worldwide. While corrosion of steel reinforcing bars is rarely responsible for major structural failures, serviceability of the structure is significantly affected by reinforcement corrosion. The main reasons for the poor performance of concrete structures are poor workmanship and lack of knowledge of the deterioration mechanisms, which result in insufficient planning and wrong estimation of environmental effects. Nowadays, it is well know that the durability of materials and structures depends both on the environmental conditions at the exposed surfaces of the structures and on the material resistance to the action of aggressive substances. The understanding of these two main aspects is the key for the design and execution of durable structures and for the rational approach to the repair of damaged structures.

Significance of the Study

The paper presents two case studies where structures exposed to the marine environment showed early deterioration due to reinforcement corrosion. In these case studies the causes of the poor performance of the structures referred to above are well illustrated. Several degree of deterioration associated with different exposure conditions are described and steel corrosion in reinforced concrete is illustrated. Repair strategies for extending the service life of those structures are also presented in this paper [1].

Case Studies

Bridge:

This example refers to the deterioration of a 40 year old reinforced bridge constructed as part of the National Highway 17. This passes the coastal regions of Karnataka parallel to the West Coast. The bridge is located at a distance of 2 km from the seaboard. The total length of the bridge is 154 meters, which consists of 5 spans of 27.5 meters of R.C hollow box girders.

The bridge is exposed to an atmospheric marine environment and the concrete is never in direct contact with seawater. The central part of the bridge is more exposed to the wind, and therefore to the salt from blown spray and salt-laden mist. A very good quality concrete was achieved in the construction. Not much maintenance was done in 40 years. The main findings of the investigation were significant delaminating and spalling of concrete occurring in the deck slabs and beam junctions. Figure 1 shows the recorded information of the typical concrete deterioration observed during the inspection of the deck slab. Higher incidence of deterioration was observed in the downward faces and lower corners of the beams. The main cause of the concrete deterioration of the bridge was poor execution quality since the concrete quality was very good. Poorly made construction joints, honeycombing was frequently observed and thickness of concrete cover presented a high scatter. The inspection showed that there was a significant area of the beams with a concrete cover thickness less than 2 cm. From the observation it is clear that the rate of corrosion is very high as the 32-mm diameter bar is reduced to 12 mm in the duration of 40 years.

Figure 1 Typically corroded deck slab

Carbonation measurements were carried out on randomly selected RCC members to assess the effectiveness of cover concrete for protecting the reinforcement against corrosion. Diluted alcohol mixed with phenolphthalein indicator was used. If the concrete surface remains colourless after spraying the solution, it indicates that the concrete has carbonated as has lost its alkalinity. If it turns pink, this indicates that the concrete has not carbonated and is alkaline. In this bridge, the carbonation front is within the cover concrete zone (about 5-10mm) depth even in the areas of deficient concrete placing. Chloride penetration test was carried out to assess the effectiveness of cover concrete for protecting the reinforcement against corrosion. By conducting chemical analysis test, the average chloride level was found to be 3 times higher than the normally accepted limits (IS-456-2000) [2]. Half cell potential difference measurement test was carried out using a copper-copper sulphate half cell and multimeter; the measured potential difference was between –350 to –400 mV, which indicated that the corrosion of rebar was severely affected in regions of the deck slab [3].

The deterioration of a cross-section at the junction of deck slab with bearings is shown in Figure 2, which is observed as the critical junction. The reasons may be poor quality of materials (River bed sand, which contains salt content) and poor quality of concrete mix responsible for chloride penetration and subsequent rapid steel corrosion.

Figure 2 Deterioration at slab and bearing junction

The repair strategy adopted for this bridge was cleaning the deteriorated concrete portion with sand blasting. Local patch repair with cement mortar was used followed by an application of protective coating. The basic principles behind this repair methodology are the following: stopping the anodic process where the reinforcement is corroding, application of a surface barrier to control the ingress of further aggressive substances and also to limit the moisture content of the concrete. This last aim of the surface protection is very important since it is impossible in practice to assure that all the contaminated concrete is removed in a patch repair methodology. A surface protection impermeable to water and permeable to water vapour allows concrete to dry out raising its resistivity and controlling in this way the electrolytic conduction process of the corrosion mechanism [4].

Building:

An RCC structure constructed 20 years back, located at Malpe on the West Coast of Karnataka, suffered major distress mainly due to corrosion of steel leading to cracking and subsequent spalling of concrete. The building is located just 300 meters from seaboard. The structure is constructed for the purpose of a guesthouse, remained unoccupied till now. The building is a RCC frame, single storey, having an area of 2325 sq. meters, as shown in Figure 1 and has a proposed swimming pool and the structure is facing west for sea view. It is observed that corrosion induced cracking has affected the RCC columns located mainly on the periphery of the building and also the interior column, to some extent. The physical characteristics (eg: width, length) of cracks, spalling, deflection and other forms of evidence for damages were the measurable parameters employed to provide information to establish the structure deterioration.

The extent of corrosion of the reinforcing steel ranges from negligible to very severe. The diameter of the reinforcement has been reduced to nearly 50% in a span of 20 years. In one or two places, which are located on the outer periphery, there is no continuity of the reinforcement as shown in Figure 3. Half-cell potential measurements indicated active corrosion of the embedded reinforcement in most external structural components.

Figure 3 A typical corroded column

A number of concrete samples collected from different locations have been analyzed for alkalinity (pH), water-soluble chloride and sulphates in concrete. The pH of the sample lies between 7.8 to 8.9, which shows that all the samples are almost carbonated. The pH of good concrete should be between 12 to 13. The carbonation has proceeded to the full depth of concrete cover. The value of the total sulphates is within the specified limit. Low pH values coupled with high chloride content are the main cause of the corrosion.

The building has been built with a characteristic strength of 20 MPa. No durability criterion was considered. Documentary information about the structure indicated that the concrete mix contained a cement content of about 300 kg/m^3 and the water cement ratio was 0.7. The placing, consolidation and curing of concrete were inadequate, showing some segregation zones and poor concrete joints.

An extensive number of tests were performed using rebound hammer with a view to assess the concrete strength of the building. Regarding the compressive strength, a characteristic value of 17.5 MPa was obtained for columns and 20 MPa for bottom slabs. This means that the specified strength has not been achieved. The deterioration process developed along the years and extended to several parts of the building. The deterioration rate depends mainly on the exposure conditions.

Observations

The investigation shows that the reinforced concrete at Beach guesthouse at Malpe has suffered unusually severe and rapidly progressing deterioration, with the most extensive damage occurring in structures exposed to the outer portions shown in Figure 4. It is common experience that buildings in coastal region require extra attention in maintenance after construction. However, these buildings have neither been completed for the purpose of their use, nor have been put to any other use on periodic basis for maintaining their structural integrity [5]

Field and laboratory studies indicate that the primary cause of deterioration was corrosion of the reinforcing steel due to the presence of chloride ions. Chlorides at the steel surface have originated from the sea water (used to mix concrete), salt bearing soils and ground water, marine salts carried by the wind.

Tests show that in some structural components, the corrosion may have been intensified by inadequate concrete cover over the reinforcement, high water/cement ratio, and the presence of hairline cracks with subsequent accelerated ingress of moisture, oxygen, and chlorides to the steel surface. In general, it appears that the corrosion process in reinforced steel in this region is a result of complex synergistic effects of chlorides and unfavorable environmental conditions prevailing in this region as shown in Figure 4.

Rehabilitation Measures

For the purpose of formulating the extent of rehabilitation, concrete structures are classified into various damage classes and the following proposal is given for their rehabilitation in the West coast of Karnataka [6].

Figure 4 Typical distressed column, beam and slab

Class A Damage

All cracked concrete is removed so as to expose the reinforcement. The removal of concrete is extended to a level till no corrosion is found. Clean the surface with wire brush or grit blasting. Apply the epoxy or any other suitable coating to the cleaned steel and concrete surface. Apply the layer of concrete mortar on the treated surface to bring the cross-section of the member to its original shape.

Class B Damage

Removal of loose concrete to a depth of 30 mm and replacing by cement mortar in layers of 20 mm over a layer of chicken mesh 22G, 12 mm hexagonal opening in each layer. Proper curing is carried out till the newly laid concrete gains strength.

Class C Damage

Removal of concrete to a depth of 50 mm and replacing by cement concrete with 10 mm aggregates after applying epoxy based bonding coat on the surface. Welded mesh of 10G x 10G - 100mm x 100mm opening was recommended for use in each layer of concrete. Superplasticizers were also suggested to be added in the concrete to improve workability. Proper curing is carried out till the newly laid concrete gains the strength.

Class D Damage

Removal of loose concrete till no corrosion is found. Replacement of steel reinforcement if required, additional reinforcement including stirrups is required to be provided and the same to be welded to the existing reinforcement. Application of epoxy based bonding coat on the concrete surface. Replacement of concrete after wrapping a layer of welded mesh of 10G x10G - 100mm x 100mm opening. Proper curing is carried out till the newly laid concrete gains the strength.

Class E Damage

All such members, which fall under this vulnerability level are to be removed and replaced by properly designed new members. All exposed concrete surfaces on the periphery of the building must be provided with suitable coating.

CONCLUSIONS

The assessment of the structures referred to in this paper shows that, when defective structures are exposed to aggressive environments, such as marine environment, very high deterioration rates can develop leading to serious damage conditions in very short time periods.

The inspection carried out showed that the principal mechanism responsible for the extensive deterioration of the structures studied was chloride induced reinforcement corrosion. This mechanism led to extensive delaminating and spalling of the reinforced concrete cover. The repair costs, both financial and environmental of the presented structures are very high. To avoid these unacceptable situations, planning and execution must be based on sound design and good quality workmanship.

REFERENCES

1. BHAT DEVADAS S., SAMAGA B.R., " condition assessment and rehabilitation of distressed RC building in west-coast of Karnataka" accepted for publication in international seminar on Industrial structures. Jan 8-10, 2003

2. IS-456-2000 " Plain and Reinforced concrete" code of practice Bureau of Indian Standards.

3. HARI K. , SRINIVASAN T., Nov " case study on rehabilitation of berthing structure" symposium on Rehabilitation of Berthing structures Nov 22, 2002. Dept. of Ocean Engineering, IIT Madras, 2002 pp99-110.

4. BHAT DEVADAS S., B.R. SAMAGA. " Deterioration and life expectancy of concrete structures in the environment of west coast of Karnataka" on proceedings of World Conference on concrete material and structures, University of MARA , Malaysia May 14-16, 2002 pp 449-461

5. GUPTA V.K., ACHAL Kr. MITTAL. K.C.MAITHANI " Rehabilitation of Distressed RCC Frame structure at coastal region- A case study" proceedings of National conference on "Advances in concrete Technology ICI-Patiala Sep 21-22 .2000 pp170-178.

6. MANI K., SREENATH H.G., RAJAMANE N.P. proceedings of " Advanced course on High performance materials and methodology for Construction and Rehabilitation of concrete structures " SERC Chennai. 2000 pp 202-224.

INVESTIGATION OF SOME FACTORS AFFECTING THE DURABILITY OF UNDERGROUND SEWERS

I A Kudryavtsev

A A Vasiljev

Belarussian University of Transport

Belarus

ABSTRACT. The kinetics of carbonization of solutions containing calcium hydroxide in different concentrations in contact with air was been studied. The obtained results have enabled to identify the features of the kinetics and the mechanism of reactions of $Ca(OH)_2$ with CO_2 in static and hydrodynamic conditions as a function of the phase interface area, the thickness of the electrolyte and the initial concentration of $Ca(OH)_2$. The features of the mechanism have changed some ideas about the interaction of CO_2 with the moisture in the pores of condensed calcium silicates.

Keywords: Concrete, Durability, Sewers, Interfaces

I A Kudryavtsev is a Professor at the Belarussian University of Transport, Belarus

A A Vasiljev is an Engineer at the Belarussian University of Transport, Belarus

INTRODUCTION

Failures of gravity sewers have multiplied in recent years in a number of urban communities in the NIS. The cause is damage to concrete in the vaulted roof of the sewer followed by its caving-in under the effect of external loading. Such failures stretch to a significant distance imparting considerable material loss and grave harm to the environment.

It has been established that the main cause of corrosion of the roof is the damage of concrete by hydrogen sulfide H_2S, its concentration depending on the exposure to sewage, its temperature, the biological oxygen demand of the water, the pH of the sewage, the concentration of sulfides, the sewage flow, the area of the roof and the turbulence [1]. Yet, the statistics of failures point at one more factor, vis. 60–70 % of the failures occur at crossings with motor- or railways or in sites exposed to intensive vibration. The physical wear of the sewers related to the factors mentioned corresponds to problematic locations. [2]. It should be added that the actual data about the rate of emergencies obtained by various researchers proves that the components of the sewage system have a non-stationary pattern of dependable service.

Samples from the jacket of the sewer at the failed part have revealed that the roof was corroded 5 times quicker than the gutter. The pH has been selected as a criterion characterizing concrete corrosion (the reinforcement begins to depassivate at pH < 11.5). The study of the causes of failures has shown that the same indicators exist practically over a sufficiently long stretch of the sewer characterizing its performance. Water-saturated clay sand makes up the bed, the initial designed slope and flooding are identical, the depth slope is 7 m. The process of construction was impeccable. Yet, the settlement of the sewer turned out to be different (Figure 1).

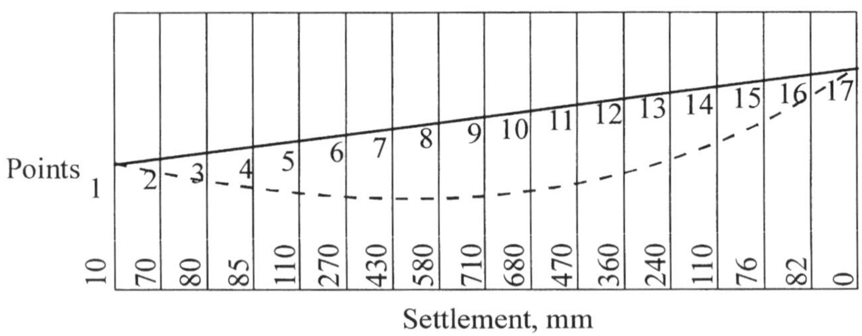

Figure 1 Settlements of sewer (points, settlement, mm)

Our observations have revealed that a crater up to 1.5 m deep appears directly above the sewer. Yet, main railway tracks run across the sites of the maximum settlement (at a distance 6-7 m from the axis). Table 1 lists some characteristics of the sewer portions in question.

Table 1 Some characteristics of sewer portions

PARAMETERS	NUMBER OF POINTS						
	2	6	9	10	12	15	16
Level of vibroacceleration transmitted from rail traffic to lining, m/s^2	10	15	25	25	15	10	5–8
Soil type in which sewer runs	Water saturated clay sand						
Soil corrosive attack	$\frac{\text{no}}{\text{yes}}$	$\frac{\text{no}}{\text{yes}}$	$\frac{\text{no}}{\text{yes}}$	$\frac{\text{no}}{\text{yes}}$	$\frac{\text{no}}{\text{yes}}$	$\frac{\text{no}}{\text{yes}}$	$\frac{\text{no}}{\text{yes}}$
Depth of soil from surface, m	8–9	8–9	8–9	8–9	8–9	8–9	8–9
Area of chips and through cracks per roof run m	0.10–0.18	0.17–0.23	0.40–1.55	0.50–1.60	.40–1.20	0.17–0.30	.08–0.15
Roof flooding	0.4–0.6	0.4–0.6	0.4–0.6	0.4–0.6	0.4–0.6	0.4–0.6	0.4–0.6

Note. The numerator – the corrosive attack of the soil during erection, the denominator – the attack 8 years later

It is noteworthy that the maximum corrosion of the roof occurs in areas where the vibration affects the sewer most intensively. This conclusion is corroborated both at the failure area spot and other locations directly adjacent to the sources of vibration.

When examining the condition of the sewers, their roofs, in the first place, it is noteworthy that the moment of deflection, stress and strain change at the points of crossing with motor- and railways. These factors depend both on the type of traffic, the depth of the structure and the flooding of the sewer. In particular, the smaller the flooding of the sewer H/D (H – the level of flooding, D – the diameter), the higher is the moment of deflection in the roof portion (Table 2).

Table 2 Moments of deflection as function of roof width

CHARACTERISTICS	H/D						
	0.95	0.90	0.85	0.80	0.75	0.70	0.65
Roof width l, m	1.25	1.70	2.00	2.20	2.45	2.55	2.70
Moment of deflection M_{max}, kNm	31.25	57.84	80.02	96.81	120.05	130.05	145.08

When the sewer is compressed by the soil, the roof thrust reduces and the moment of deflection grow. Also, it is known from our earlier studies that vibration boosts water accumulation in the pores and naturally reduces the soil strength that finally leads to the subsidence of the sewer [2, 3].

The level of vibration affecting the sewer depends on its flooding. Analysis of the vibration on the outer wall of the lining has revealed that the latter is exposed when water flows to areas of lower vibration produced by the passing train when the sewer is more flooded and vice versa. It means that the flow increases the tube's rigidity, yet, in case there is no vibration hum, the acceleration over the lining is directly proportional to the flooding (Figure 2).

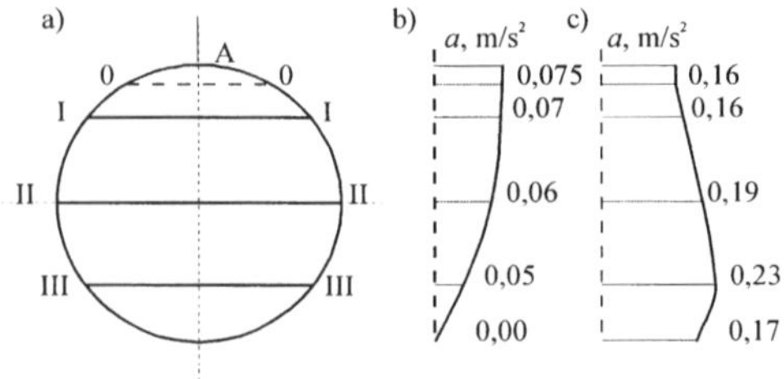

Figure 2 Diagram of vibration tests: (a) diagram of sewer flooding (A – point of installation of vibrosensors); (b) epure of vibration of sewer during sewage flow; (c) epure of total vibration in sewer caused by vibration due train traffic

Since the effect of vibration on the rate of corrosion of the lining and the shell of the sewer has been proven experimentally, the condition of vaulted roofs of sewers should be monitored thoroughly using TV monitors to assess the condition.

We have also established that water permeability of the reinforced shell exposed to vibration increases so that the underground water seeps more freely through the roof. As the corrosion intensifies and correspondingly the pores and cracks in the lining are filled up, the volume of water from the soil builds up, and more miniscule particles from the soil above the sewer are washed out (the erosive process evolves). In this case, the porosity coefficient increases and the soil strength in the road bed declines [4]. As soon as the corrosion of the reinforcement exceeds 10-15 %, the protective layer detaches from the reinforcement boosting the water permeability of the lining even more and impairing its resistance to the point when the roof collapses. According to our tentative assessments, the roof may cave in after 80 % corrosion of the concrete and 30 % corrosion of the reinforcement. In this case the situation approaches the stage of failure.

The basic requirement for the design of building structures according to the existing norms is formulated as the following inequality:

$$R_d - S_d > 0 \text{ or } \Theta_d > 0,$$

where R_d and S_d – calculated values of the load bearing capacity and the loading; Θ_d – the reliability limit determined with the calculated values of the load bearing capacity and the loading.

It is noted above that the strength characteristics of structures and sometimes the loading change the service life of the structure. In particular, the shell loses its strength characteristics during the flow of sewage and becomes affected by more vibration hum meaning that the stress-strain state changes. The ultimate safety under different service conditions depends on the designed service life (t_d).

Figure 3 shows how the safety limit declines in time.

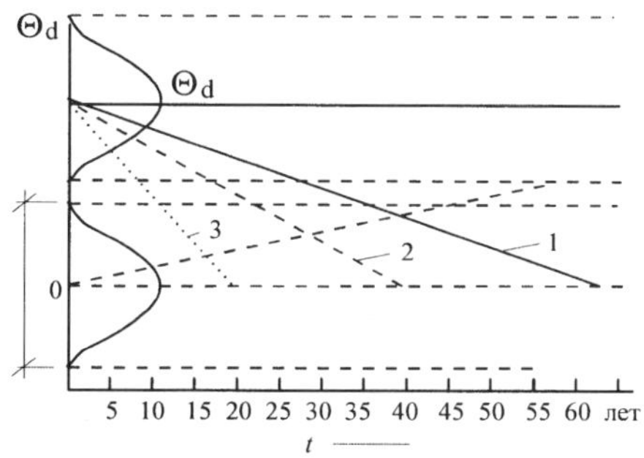

Figure 3 Service diagram of sewers: 1 – normal condition service; 2 – service without aeration; 3 – service without aeration at crossings with main railway lines

The zero line shows the minimal value of Θ_d that corresponds to the required safety level. The initial value should be above this level. The values of the ratio R_d/S_d obtained by calculations are listed in Table 3.

Table 3 Indicators of ratio R_d/S_d

SERVICE CONDITIONS	TERM OF SERVICE, YEARS				
	0	5	10	20	30
With vibration	1.60	1.35	1.15	0.94	0.75
Without aeration	1.60	1.48	1.20	0.98	0.81
With aeration	1.60	1.55	1.51	1.36	1.11

The results apparently show the case when R_d is smaller than S_d.

CONCLUSIONS

1. The sewers passing under various types of transport lines or adjacent to them need continuous monitoring to ensure safety of the traffic.

2. Factors have been identified causing premature failure of sewers. It has been established that, in addition to the bioprocesses, the rate of corrosion of the components of reinforced concrete lining and shells is strongly influenced by the level of vibration affecting the sewer.

3. It is shown that the stress-strain state of the sewer is a variable value that, together with the loading of the train on the shell, is affected, to a certain extent, by the vibration hum created by the traffic.

REFERENCES

1. G.YA. DROZD, N.V. SYTNICHENKO, I.V. SATIN, AND YA.A. GUSENTSEVA, Biological Factor as a Cause of Failure of Sewerage Systems (in Russian), Water Supply and Sanitary Engineering, no 1, pp. 22–24, 2002.

2. I.A. KUDRYAVTSEV, Analysis of Causes of Failure of Collectors Erected with Mechanized Shield (in Russian), Constriction and Architecture, no 1, pp. 32–35, 2000.

3. I.A. KUDRYAVTSEV, Effect of Vibration on Foundations (in Russian), 274 p., Belorussian State Univ. of Transport, Gomel, 1999.

4. A.S. ALESHIN, I.A. KUDRYAVTSEV, AND N.V. KUZNETSOV, The Effect of Vibration on the Processes of Migration of Fluids in Rock (in Russian), Transaction of the USSR Acad. of Sci., vol. 315, no 4, pp. 182–184, 1990.

EXPERIMENTAL INVESTIGATION OF CONCRETE PERFORMANCE UNDER MULTI-AGGRESSIVE CYCLE EXPOSURE CONDITIONS

L Zheng

M R Jones

University of Dundee

United Kingdom

ABSTRACT. Currently, concrete durability performance is generally examined by a series of individual durability tests, such as carbonation test, freeze/thaw test or chloride ingress test etc. However, in practice, a lot of the concrete structures are exposed in a multi aggressive environment. A highway bridge may suffer from chloride ingress, carbonation and freeze/thaw attack coincidently. Even a normal exposure structure generally undertakes carbonation freeze/thaw attack and wet/dry weathering at the same time. For the structure service life design or estimation, it seems not enough to know the concrete performance just under every individual durability test because those individual factors will affect each other in a multi-aggressive exposure condition. Indeed, survey from the real structure can supply some useful information, it is also scattering and difficult to be used in modelling the structure deterioration process. To address the problem, this paper reports the experimental investigation results of concrete performance under several multi-aggressive cycle exposure conditions. These cycling exposures are specially designed to simulate the most common multi-aggressive exposure conditions occurred in normal and highway constructions. The investigation will bridge the laboratory individual tests and survey on real structure and will be beneficial to improve the methodology of structure service life design and estimation.

Keywords: Carbonation, Freeze/thaw, Chloride, Multi-aggressive, Durability

Li Zheng is a research fellow in the Concrete Technology Unit in the Division of Civil Engineering, School of Engineering, at the University of Dundee, UK. He was an associate professor in the Department of Civil Engineering at Zhejiang University, PR China, prior to the current post. His research focuses mainly on concrete durability, mix optimisation and mathematical modelling.

M R Jones is a chartered civil engineer and reader in the Concrete Technology Unit in the Division of Civil Engineering, School of Engineering, at the University of Dundee, UK. His research focuses mainly on binder technology, concrete durability and repair and maintenance.

INTRODUCTION

Sustainable construction requires the best use of resources to accurately meet the performance requirements for our infrastructure. To take this forward, it was thought that both durability design for a new structure and service life estimation for an exist one are fundamental [1, 2]. This design/estimation process will unavoidably involve numerical calculations comparable with those for structural importance and safety [3]. During this numerical calculation, exposure condition is unsuspected as a major factor, which needs to be considered. With the introduction of new national/European standards, BS EN 206/BS 8500 [4, 5], a new exposure classification system has been proposed for use in the UK, and is significantly different from BS 5328 [6] system to merit investigation.

In quantifying concrete durability performance, specifications for different individual durability tests, such as carbonation test, freeze/thaw test or chloride ingress test etc, are established. However, in practice, concrete structures are generally exposed in a multi aggressive environment. A highway bridge may suffer from chloride ingress, carbonation and freeze/thaw attack coincidently. A normal exposure structure may undertake carbonation, freeze/thaw attack and wet/dry weathering at the same time. It has been noted deterioration of concrete is accelerated when it is exposed to some kind of multiple damaging processes [7, 8]. Therefore, for the structure service life design or estimation, it seems not enough to know the concrete performance just under every individual durability test because those individual factors are interactive in a multi-aggressive exposure condition. Survey from the real structure do supply some useful information, however, it is scattering and difficult to be used in modelling the structure deterioration process. To address the problem, this paper reports the experimental investigation results of the concrete performance under several multi-aggressive cycle exposure conditions. These cycling exposures are specially designed to simulate the most common multi-aggressive exposure conditions occurred in normal and highway constructions. Some different results were found between the traditional individual durability tests and multi aggressive exposure tests carried out in this research work. This work was supposed to establish a bridge between the laboratory test and construction site investigation and the results will be beneficial to the structure service life design and estimation.

EXPERIMENTAL PROGRAMME

Materials

A single source of cement strength class 42.5N Portland cement (PC) conforming to BS12 and BS EN 197 -1 was obtained for use in the tests. Additions used in the project included Pulverised Fuel Ash (PFA) conforming to BS 3892-1, Ground Granulated Blastfurnace Slag (GGBS) conforming to BS 6699, Metakaolin (MK) to Agrément certified (98/3540), Silica Fume (SF) to Agrément certified (85/1568) and limestone (LS) filler to BS7979.

The coarse aggregates used were natural local gravels of 20mm maximum size, in two single size fractions, 20-10mm and 10-5mm conforming to pr EN 12620. Fine aggregate used was natural sand conforming to zone M of BS882 and pr EN 12620 with 56.4% by mass passing the 600μm sieve. All aggregates were used in a laboratory dry condition.

Multi-aggressive Exposure Conditions

Multi-aggressive exposure cycle tests were designed for simulating the 'Normal' and 'Highway' natural exposures. The intention of this was to simulate the most common natural exposure conditions, in which different primary, secondary and tertiary durability requirements are needed. These tests, which can be considered as a first approach, were part of the efforts to set up a series of "standard" environments, which were easy to reproduce in the laboratory and to treat theoretically, but which were as similar to the real environment as possible. Detailed description of the simulated multi-aggressive exposure cycle tests for 'Normal' and 'Highway' exposures are given in Figure 1.

In 'Normal' multi-aggressive exposure condition, carbonation resistance is considered as the primary durability requirement and the freeze/thaw resistance is the secondary requirement. It was considered as a combined environment of XC3-4 and XF3 according to the BS EN 206/BS 8500 exposure classification. The cycle designed for the estimation of durability under this condition (See Figure 1a) was 3 weeks carbonation plus 1 week of freeze/thaw attack. The carbonation environment was the accelerated carbonation (4% CO_2, 20°C, and 55% RH), the freeze/thaw environment was the modified cycle (from 20°C to -10°C over 24 hours).

In replicating the highway environment in the laboratory by means of accelerated testing, the cycle was devised as illustrated in Figure 1b. In this exposure condition, chlorides are the primary aggressive mechanism, carbonation the secondary and freeze/thaw the tertiary. Elements of simulated rainfall and drying were also included. The sequence in which the test specimens were subjected to the various aggressive environments was sought to mirror the highway environment: the specimens were first subjected to a salt mist (3 Molar chlorides) for 14 days, which simulated spray from passing vehicles which contains de-icing salts prior to a period of freezing; the concrete was then subjected to freeze/thaw cycles (1 day); a period of drying was then simulated under accelerated conditions (4 days) which was followed by 7 days accelerated carbonation. Finally, rainfall (2 days shower) is predicted before the specimens were subjected to next chloride spray. It was considered as a combined environment of XD1 and 3, XC3-4, and XF2/4 according to the BS EN 206/BS 8500 exposure classification.

The cycle duration was four weeks for the designed multi-aggressive exposure conditions. Inevitably there are a number of combinations and periods of attack which could be performed. Although the length was arbitrary, it included weighted elements of the multi-aggressive environment in order of relative importance.

Concrete Mix Proportions

According to best practice in construction and previous study, optimal cement combinations vary with the exposure conditions [9]. Therefore, different cement combinations were used for different exposure conditions.

A binary cement of 70%PC+30%PFA is used for normal exposure test. Two different strength classes, C35 and C45, were designed to meet the different exposure classes. For each strength class, three different mix design methods were applied. One is the BRE method [10], and the other two are the particle packing optimisation methods developed by Jones, et al (DUN mixes) [11, 12] and Dewar (SIM mixes) [13]. In these particle packing optimised mixes, optimised LS filler content was determined by particle packing models [12].

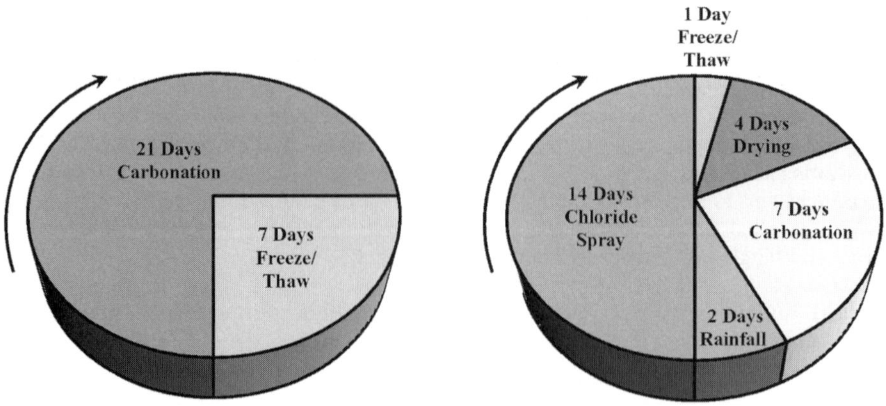

a. 'Normal' Multi-Aggressive Cycle

b. 'Highway' Multi-Aggressive Cycle

Carbonation: 4% CO_2, 20°C, 55%RH

Drying: 20°C, 40%RH

Freeze/thaw: +20/-10°C over 24 hours

Rainfall: 15 minutes shower every 6 hours

Chloride spray: 12 hours 3M chloride mist spray followed by 12 hours drying with a fan

Figure 1 Simulated Multi-Aggressive Cycle Exposure Conditions

For highway exposure, two series of mixes were tested. Mix proportions in Series 1 were designed using the same 3 methods as for normal exposure mixes, however, the cement combination was 60%PC+40%GGBS. In Series 2, ternary cement combinations, 50%PC+40%GGBS+10%SF (TSF mixes) and 45%PC+40%GGBS+15%MK (TMK mixes), were further tested because these additions, SF and MK are widely used for preparing high chloride resistant concretes [14-17]. Further improvement in chloride resistant can be obtained using ternary through its optimised chemical combinations [18, 19].

Mix constituent proportions for different strength and exposure classes are given in Table 1.

EXPERIMENTAL RESULTS

Normal Exposure

Properties of carbonation depth, surface scaling, weight changes and relative pulse velocity were tested after each single exposure phase of the 'Normal' exposure cycle. Comparing tests under single exposure were also carried out.

Figure 2 shows the carbonation development at equivalent exposure duration under cycling test in comparison with that in the individual accelerated carbonation test. It can be seen that the carbonation development was delayed with the periodic freeze/thaw exposure. This is due to the sample absorbed water during the freeze/thaw test, thus delaying the onset of carbonation in wet concrete. Another reason for reduced carbonation depth may be the reverse transport of $Ca(OH)_2$ during the sample in wet conditions.

Table 1 Mix constituent proportions for different strength classes

MIX CODE	STRENGTH CLASS[1)]	MIX CONSTITUENT PROPORTIONS, kg/m³							
		Water	Cement				Aggregates		
			PC	PFA	GGBS	LS/MK/SF[3)]	Fine	10 mm	20 mm
Concrete Mixes for 'Normal' Exposure Test									
BRE/35		160	230	95	-	0	715	385	770
DUN/35	C35	145	210	90	-	5	540	425	955
SIM/35		155	220	95	-	35	595	420	840
BRE/45		160	275	120	-	0	650	385	770
DUN/45	C45	145	255	110	-	0	470	435	960
SIM/45		155	265	115	-	35	470	440	880
Concrete Mixes for 'Highway' Exposure Test (Series 1)									
BRE/35		160	225	-	150	0	685	360	720
DUN/35	C35-air[2)]	145	205	-	140	0	495	355	965
SIM/35		150	170	-	115	40	600	400	805
BRE/40		160	190	-	125	0	695	410	820
DUN/40	C40	145	180	-	120	0	585	405	970
SIM/40		150	165	-	115	20	665	435	865
BRE/45		160	215	-	140	0	640	415	830
DUN/45	C45	145	205	-	135	0	510	435	970
SIM/45		150	180	-	120	35	630	430	855
Concrete Mixes for 'Highway' Exposure Test (Series 2)									
BRE/40		160	190	-	125	0	695	410	820
TSF/40	C40	160	160	-	125	30	695	410	820
TMK/40		160	150	-	140	50	705	400	800
BRE/45		160	215	-	140	0	640	415	830
TSF/45	C45	160	180	-	145	35	695	395	790
TMK/45		160	170	-	150	55	670	400	800

1) Two different strength classes were designed for the research purpose. The minimum required strength classes from BS8500 and BS206-1 are slightly different.

2) 0.22% air entrainer was used in the mixes.

3) This column refers to MK content in TMK mixes, SF in TSF mixes and LS in all other mixes.

Although carbonation development was delayed with the periodic freeze/thaw exposure, the periodic carbonation exposure, in verse, accelerated the Freeze/Thaw damage process. Figure 3 shows the comparison results based on equivalent exposures to the individual exposure tests. Figure 3a shows the carbonation depth relationship for 12 weeks accelerated exposure and equivalent exposure duration in the cycling test. The carbonation depth ratio of multi exposure to individual carbonation exposure was about 0.775. The scaling test results after 28 day equivalent exposure duration in the cycling test, in comparison with the scaling test results after the same exposure duration obtained in individual freeze/thaw tests is shown in Figure 3b. It can be seen that concrete suffered more damage in the cycling test than in the individual freeze/thaw tests. The average scaling ratio of multi exposure to equivalent period of individual freeze/thaw exposure was about 1.42.

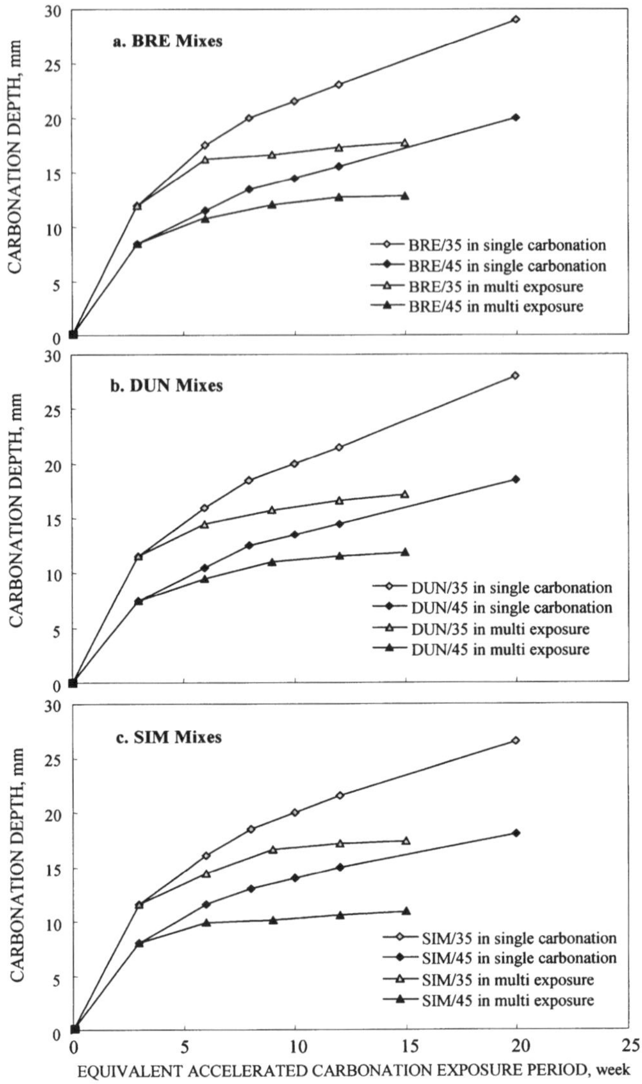

Figure 2 Carbonation development at equivalent exposure duration

Figure 4a shows weight changes of the concrete samples under cycling exposure tests. Because the samples absorbed water during the freeze/thaw process and lost water during the carbonation process, their weight changes cyclically with the exposure period. To judge the weight loss during the test, the samples should be compared at equal water content. Here the benchmark used was the room dry state (20°C, 55% RH), and the dotted lines shown on the Figure represent weight loss tendency of the samples during the cycling exposure. It can be seen that during a limited 5 cycles, nearly no weight loss for C45 concrete samples was observed, while that for C35 concrete, weight loss was about 2%.

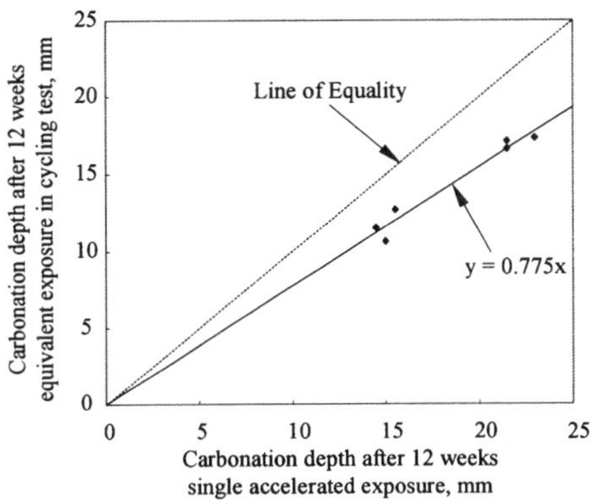

a. Carbonation depth

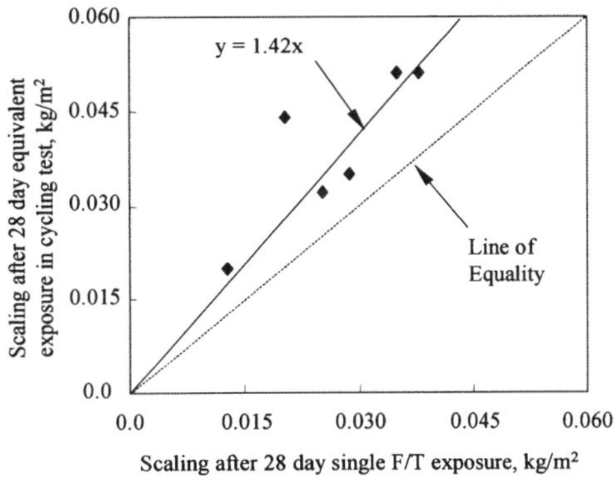

b. F/T scalling

Figure 3 Comparison of results in multi exposure test with results in equivalent individual exposure

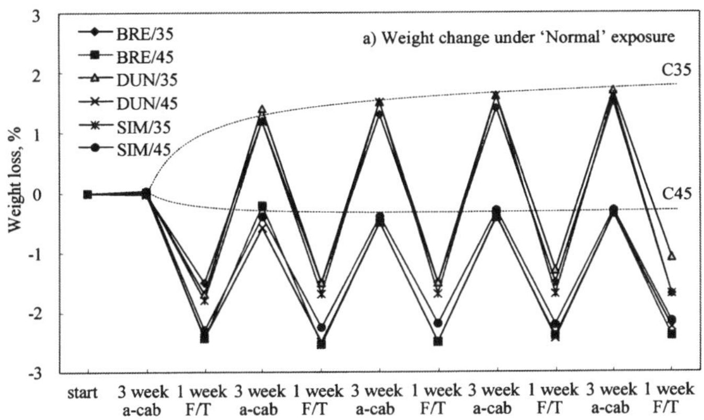

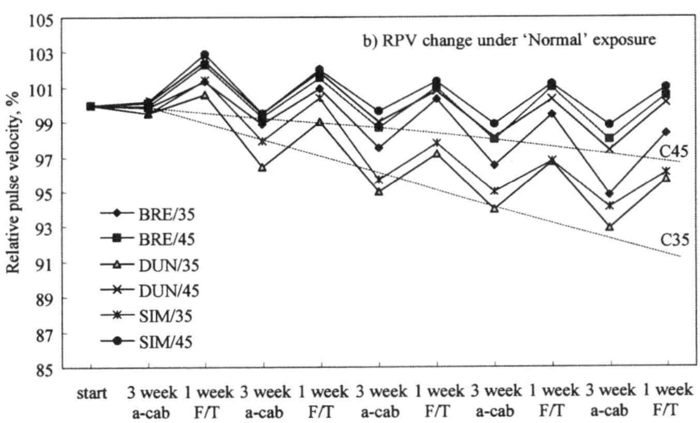

Figure 4 Weight and relative velocity changes during 'Normal' exposure cycle

Figure 4b shows relative pulse velocity (RPV) changes of the concrete samples under cycling exposure tests. Similar cyclical changes of (RPV) were seen during the cycling tests because the velocity of the ultrasound transmission in the concrete samples also changes with the water content of the samples. Ultrasound transmits faster in water saturated concrete than that in dry concrete. Again the benchmark used here is the room dry state, and the dotted lines shown on the Figure represent RPV change tendency of the samples during the cycling exposure. It can be seen that during the 5 cycles, RPV in all samples decreased. Greater reductions occurred in C35 samples than in C45 samples.

In summary, the designed 'Normal' multi-aggressive cycle exposure leads to reduced carbonation and increased freeze/thaw damage, when compared with their respective individual exposure tests. When the damage from Freeze/thaw is judged by weight or relative velocity changes, the effect of water content of the samples should be taken into account. The results suggest that the carbonation rate may be overestimated by the individual carbonation test and freeze/thaw induced damage underestimated for concrete exposed to a multi-aggressive environment.

Highway Exposure

Properties of carbonation depth and chloride profiles were tested after each single exposure phase of the 'Highway' exposure cycle. Comparing tests under individual exposure were also carried out.

Figure 5 shows the results of carbonation depth at equivalent exposure duration under cycling test in comparison with that in the individual carbonation test. Again, it can be seen that the carbonation depth was much lower than that in the individual test. In comparison with the results from 'Normal' exposure (Figure 3a), the carbonation depth ratios of multi exposure to individual one are even lower.

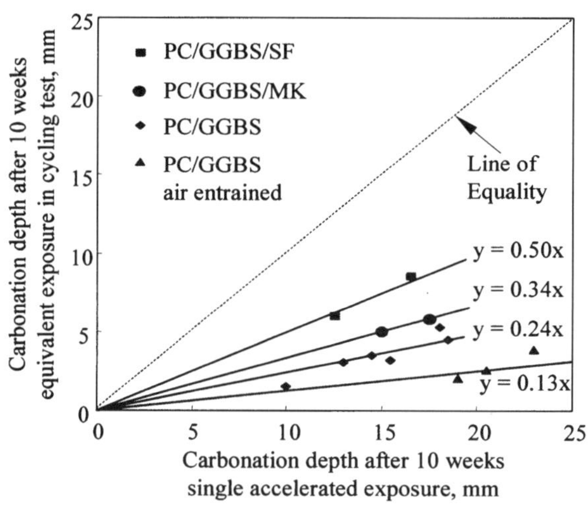

Figure 5 Comparison of carbonation results in 'Highway' multi exposure test with results in equivalent single exposure

In addition, it can be seen from Figure 5 that the carbonation depth ratios of multi to single exposure also vary with the cement combinations as well as the addition of air entrainer. The order of relative ranking of the ratio from lowest to highest was PC/GGBS with air entrainer (0.13), PC/GGBS binary (0.24), PC/GGBS/MK ternary (0.34) and PC/GGBS/SF ternary (0.50). Reasons for this phenomenon may include more wet duration during this 'Highway' cycling which means more time for the reverse transport of $Ca(OH)_2$, the total $Ca(OH)_2$ content in the original non-carbonised cement paste and, probably, entrained air caused less damage in the Freeze/thaw process.

Figure 6 shows the chloride profiles of 5 cycles 'Highway' exposure and equivalent 10 weeks single 3M chloride exposure for C40 concretes. It can also be seen an accelerated chloride ingress in the multi aggressive exposure in comparison with single exposure. For a given depth, chloride concentration is generally higher in multi exposure samples than that in single exposure samples. This is more significant between the depths from 8.5 to 18.5 mm into the sample during the 5 cycles' exposure. Three reasons could account for this phenomenon:

i) The other processes in the multi exposures, i.e. carbonation, freeze/thaw, rainfall and drying, do caused some damages in the microstructure of the concrete that accelerated the ingress of the chloride ions.

ii) The chloride ingress may continue go on inside the concrete samples in the other processes of the multi exposure when the sample is in wet condition.

iii) Carbonation can release bound chlorides and therefore increase the level of chlorides in the concrete. Although the carbonation depth was less than 4 mm for the majority of the samples under 5 cycles' 'Highway' exposure, it was reported that an increased concentration of chlorides in pore water even occurred 15 mm in advance of the carbonation front [20].

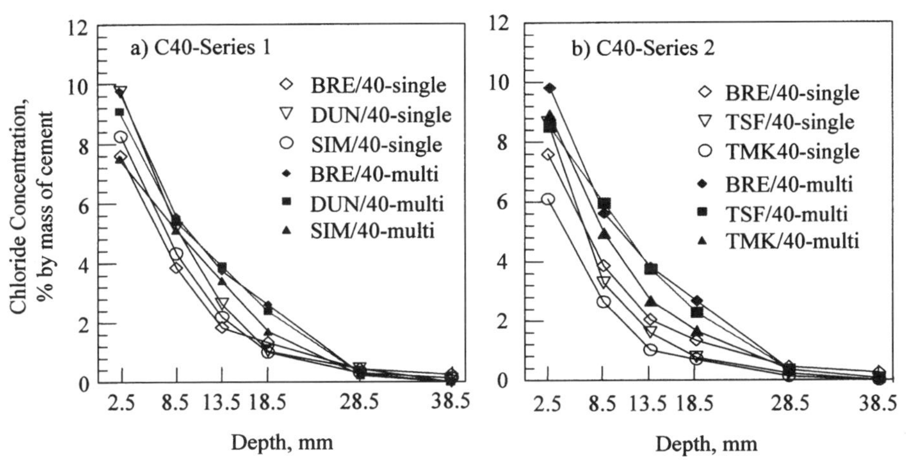

Figure 6 Chloride profiles of concrete in 'Highway' exposure and in equivalent single exposure

Figure 7 shows the chloride concentration at the depth of 13.5 mm inside the sample in 'Highway' exposure in comparison with that in equivalent single exposure. Although the data looks scattering, all points are above the equality line and the mean ratio is 1.68. Correspond to the carbonation results, it can be seen that the PC/GGBS/SF ternary has a higher ratio and air-entrained concrete a little bitter lower.

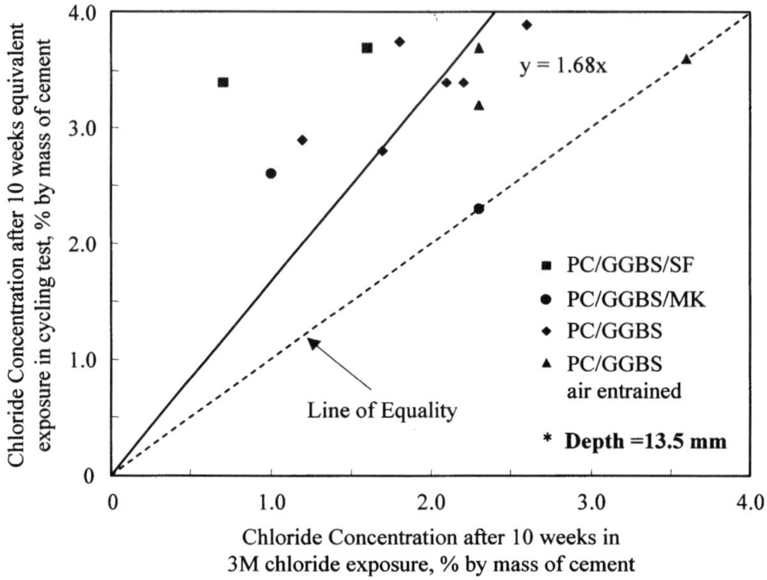

Figure 7 Comparison of chloride concentration in 'Highway' exposure test with results in equivalent single exposure

CONCLUSIONS

From this experimental investigation of concrete performance under multi-aggressive cycle exposure conditions in comparison that under individual durability tests, it can be found that there was much difference in concrete performance from different exposures.

From 'Normal' exposure, which was designed for the environment of carbonation and freeze/thaw combined attack, it was found that the carbonation rate was reduced, however, freeze/thaw damage increased, when compared with their respective individual exposure tests. In the designed 'Highway' exposure, the same reduced carbonation was observed, however, the chloride concentration was higher in comparison with single chloride exposure tests.

The reduction in carbonation can be measured by the carbonation depth ratio from multi to single exposures. Different ratios were observed from the different multi exposure regime as well as different cement combinations. Extended carbonation exposure, as in 'Normal' cycle, and cement combinations having higher carbonation rate, such as PC/PFA and PC/GGBS/SF, appeared less carbonation reduction in the multi exposure. However, entrained air may be beneficial to reduce the carbonation rate in the multi exposure when the exposure cycles include the Freeze/thaw process.

As for increased chloride ingress in the 'Highway' exposure, the mechanism may be quite complex since the observed chloride concentration ratios of multi to single exposures for a given depth in the specimen were scattering. On going chloride transport during the non-chloride exposure intervals and bound chloride release due to carbonation appeared to be the main reasons to account for this. Corresponding to the carbonation results, the worst increased chloride ingress appeared in PC/GGBS/SF samples.

All the above conclusions suggested that the data obtained from single durability test may be unrealistic and misleading due to the fact that nearly all concrete structures are exposed in the multi aggressive environments. The carbonation rate may be overestimated by the individual carbonation test. On the other hand, chloride and freeze/thaw induced damage may be underestimated. Further study is required for how to standardise the multi exposure processes and validate them against the real natural exposures which are still in progress with this research project. More realistic models will be expected for predict the deteriorate process of concrete exposed in multi aggressive environments when the validate data is available.

ACKNOWLEDGEMENTS

The authors would like to acknowledge the support provided by the UK government Department of Environment, Transport and the Regions, Appleby Group, Babtie Group Ltd, Castle Cement Ltd, IMERYS, MAC Spa, MBT Feb, Mott MacDonald, National Power, National Ash, PowerGen, Quarry Products Association, RMC Readymix Limited, Rugby Cement, ScotAsh Ltd, UKQAA and WS Atkins Consultants Ltd. The technical advice given by Professor R. K. Dhir of University of Dundee and Professor T. A. Harrison of Quarry Products Association is also greatly acknowledged. Thanks are also given to Dr. V Scorey for her excellent test work for 'Highway' exposures.

REFERENCES

1. SARJA A., Durability design of concrete structures – committee report 130-CSL, Materials and Structures, Vol. 33, 2000, pp 14-20
2. GEIKER M., Durability design of concrete structures: Minimising total life cycle costs – considerations and examples, Concrete Durability and Repair Technology, Edited by R K Dhir and M J McCarthy, Thomas Telford, 1999, pp 319-333
3. SARJA A., VESIKARI E., Durability design of concrete structures, RILEM Report 14, E & FN Spon, Chapman &Hall, 1996, 165pp
4. BRITISH STANDARDS INSTITUTION, BS EN 206-1:2000 Concrete. Specification, performance, production and conformity, London, 2001, 74pp
5. BRITISH STANDARDS INSTITUTION, BS 8500 Concrete – complementary British Standard to BS EN 206-1, B/517/1 Committee Draft, Draft 5, London, 2002

6. BRITISH STANDARDS INSTITUTION, BS 5328-1:1997 Concrete Part I. Guide to specifying concrete, London, 1997, 27pp

7. SUN W., MU R., LUO X., MIAO C., Effect of chloride salt, freeze-thaw cycling and externally applied load on the performance of the concrete, Cement and Concrete Research, Vol. 32, No. 12, pp 1859-1864

8. SUN W., ZHANG Y.M., YAN H.D., MU R., Damage and damage resistance with different strength grade under double damage factors, Cement and Concrete Compositions, Vol. 21, No. 5, 1999, pp 439-442

9. JONES M. R., DHIR R. K., MAGEE B J, Concrete containing ternary blended binders: Resistance to chloride ingress and carbonation, Cement and Concrete Research, 1997, Vol. 27, No. 6, pp 825-831

10. MARSH B. K., TEYCHENNE D. C., FRANKLIN R. E., ERNTROY H. C., Design of Normal Concrete Mixes, 2nd Ed., amended by Marsh B K, Building Research Establishment, UK, 1997

11. JONES M. R., ZHENG L., NEWLANDS M. D., A comparison of particle packing models for proportioning concrete constituents for minimum voids ratio, Materials and Structure, 2002, Vol. 35, pp301-309

12. JONES M. R., ZHENG L., NEWLANDS M. D., Estimation of the filler content required to minimise voids ratio in concrete, Magazine of Concrete Research, 2002, Accepted for publication

13. DEWAR J. D., Computer Modelling of Concrete Mixtures, E & FN Spon, London, 1999

14. JONES M. R., MCCARTHY M. J., DHIR R. K., Chloride resistant concrete. Proceedings of Concrete 2000 Conference, Economic and Durable Construction through Excellence, Dundee, 1993, E & FN SPON, London, 1993, Vol. II, pp. 1429-1444.

15. OH B.H., CHA S.W., JANG B.S., JANG S.Y., Development of high-performance concrete having high resistance to chloride penetration, Nuclear Engineering and Design, 2002, Vol. 212, No. 1-3, pp 221-231

16. DING J.T., LI Z.J., Effects of metakaolin and silica fume on properties of concrete, ACI Materials Journal, 2002, Vol. 99, No. 4, pp 393-398

17. BODDY A., HOOTON R.D., GRUBER K.A., Long-term testing of the chloride-penetration resistance of concrete containing high-reactivity metakaolin, Cement and Concrete Research, 2001, Vol. 31, No. 5, pp 759-765

18. JONES M. R., DHIR R. K., MAGEE B. J., Concrete containing ternary blended binders: Resistance to chloride ingress and carbonation, Cement and Concrete Research, 1997, Vol. 27, No. 6, pp 825 831

19. JONES M.R., MACPHEE D.E., CHUDEK J.A., HUNTER G., LANNEGRAND R., TALERO R., SCRIMGEOUR S.N., Studies using ^{27}Al MAS NMR of AFm and AFt phases and the formation of Friedel's salt, Cement and Concrete Research, 2003, Vol. 33, No. 2, pp 177-182

20. HO D.W.S., LEWIS R.K., The specification of concrete for reinforcement protection performance criteria and compliance by strength, Cement and Concrete Research, 1988, Vol. 18, No. 4, pp 584-594

TOUGHENING OF HIGH PERFORMANCE CEMENT SYSTEMS: INFLUENCE OF MATRIX – AGGREGATE BINDING

T F Stebbings

H W Chandler

D E Macphee

University of Aberdeen

United Kingdom

ABSTRACT. It has been shown recently that the bond between aggregate and matrix in cement mortars can influence composite toughness by controlling the extent of aggregate bridging taking place. Aggregate bridging is widely known to be an effective crack wake mechanism that increases the toughness of brittle materials. This paper describes the effectiveness of this mechanism in high strength cements which are known to exhibit increased brittleness due to the higher levels of strain energy stored prior to failure. Pore Reduced Cement (PRC) and silica fume blended cement are used as model high strength matrices. The increased density of conventional high strength pastes that results from the lower water cement ratio and better particle packing does not immediately lend itself to this toughening mechanism because of the increased physical bond developed between matrix and aggregate. However, the present study has shown that the measured toughness can also be used as a means of assessing the relative strength of the matrix- angular aggregate interaction.

Keywords: High strength concrete, Pore reduced cements, Toughness, Interfacial bond, Unbonded inclusions, Crack-bridging.

T F Stebbings is a graduate civil engineer currently undertaking a PhD project looking at the toughening of brittle matrices at Aberdeen University. The main focus of the project has been on toughening of high performance cements.

H W Chandler is the Head of the School of Engineering and Physical Sciences at Aberdeen University and has an established reputation in the areas of fracture in ceramics and in the forming of ceramics.

D E Macphee is a senior lecturer in the Department of Chemistry at Aberdeen University. His research interests centre around cement chemistry and its application to composite durability and mechanical performance.

INTRODUCTION

The link between the microstructure of a material and its mechanical properties has long since been established in materials science. Early expressions of this link can be found in Feret's Law [1], which relates the compressive strength of a material to its porosity, and in Griffith's law [2] of brittle failure, which links the critical size of microstructural flaws, i.e. zones in which there are discontinuities in the matrix, to flexural strength. Although these laws tend to hold for composite materials, the post-failure performance of composites is somewhat more complex than for single-phase materials and one must approach the problem with consideration being given to fracture mechanics.

This paper does not attempt to rigorously develop fracture mechanics theory but instead views fracture in a mechanistic way, principally to provide a means of explaining the enhanced post-failure mechanical performance of cementitious materials modified by incorporation of particulate inclusions. We have been able, in recent years, to demonstrate that the particulate inclusion approach can enhance the toughness of a range of brittle matrices such as glasses, thermosetting plastics, refractory ceramics and cement-based systems [3-6] and this is confirmed by Kovalev *et al* [7] and Lawn [8]. This paper describes the application of this approach to high performance cement systems.

The slightly ambiguous term 'high performance' can be applied to a range of cement-based materials. In many cases, the performance is measured in terms of strength or durability. In some cases however, the post-failure performance may be important and one can immediately see the benefits of a cement material which would be resistant to explosive spalling in a fire, crumbling on impact, or to spalling of cover concrete as a consequence of re-bar corrosion or other deleterious expansive processes.

The transition from 'normal' strength concrete to high strength concrete increases the risk of these processes occurring because the materials processing required to increase strength must, by virtue of the link between microstructure and strength, reduce the porosity. A direct consequence of this link is that stronger materials tend towards increased brittleness. It is with this trend in mind that we have addressed the toughening of high strength cement-based materials.

The materials selected for study are chosen as models for high strength concrete. The first is a mortar consisting of a blend of silica fume, OPC and crushed glass, formulated and cast to produce a low porosity, high strength matrix. The other is pore reduced cement (PRC). PRCs are prepared in the same way as conventional cement pastes, i.e. at normal water/cement ratio (w/c), but their water contents are subsequently reduced by water removal under high pressure in what is essentially a filter press (see Macphee [9]).

In this way lower porosity is achieved partly by reducing the amount of voids containing water and air but also by reducing the amount of water left to react with the cement. Additionally, in the process of pressing, the cement clinker is broken exposing fresh surfaces that can react with any residual pore fluid during the curing period. Crushed particulate glass inclusions are again used in the formulation. Conventional (normally processed) OPC mortars are included in the study to provide a reference.

PARTICULATE TOUGHENING

Relatively recent studies [6] have provided new insight into a toughening mechanism which operates in cement mortars and other composite materials. The principal mechanism relies on the presence of *unbonded* angular inclusions dispersed throughout the matrix. This approach increases the frequency of interaction between propagating cracks and the inclusions, which function as crack-bridges, consequently providing an increase in toughness with crack growth (see Figure 1). The function of the crack-bridge is to anchor matrix on either side of the propagating crack in such a way as to inhibit the separation of crack faces. The energy required to free the bridging grain from the matrix dissipates the energy of the crack thus slowing or stopping the advancing crack tip. The inclusion must be irregularly shaped to provide pinning points for the matrix to catch on to. It is important that the inclusion is unbonded as this ensures that the crack interacts with what appears to be an empty hole and passes un-deviated through the region of the inclusion. One practical implementation uses angular glass inclusions because their low reactivity in cementitious systems ensures that there is a weak bond between inclusions and the matrix that proves helpful to the subsequent debonding of the aggregate.

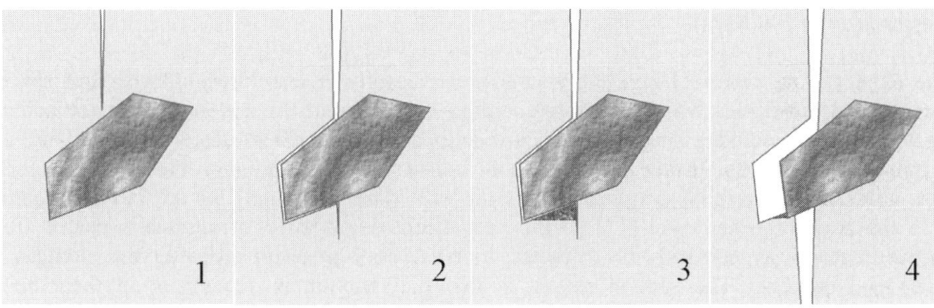

1. Propagating crack approaches a debonded inclusion. Debonded inclusion is perceived to be a pore.

2. Propagating crack passes inclusion. No deviation from original path because the inclusion is debonded and in a 'pore'.

3. Crack faces are pinned by angular inclusion. Angularity of inclusion creates pinning points in the cement matrix which must be broken if the crack is to proceed further.

4. Inclusion is pulled from the matrix, breaking it. Energy dissipated by crack in this process is measured and used as an indicator of the increased toughness.

Figure 1 Grain bridging using angular debonded inclusions.

MATRIX / INCLUSION BOND

The bond between a matrix and an aggregate/inclusion may be physical or chemical. Both bond types can exist separately or at the same time. For the above toughening mechanism to be active, any bond forming between cement matrix and glass inclusions must be weak. A previous publication reviews these effects [10] which can be briefly summarised here. A physical bond is considered as one in which no chemical bond is established across the interface and includes:

(i) Frictional interlocking *via* matrix penetration into surface texture of the inclusion.

(ii) Differential shrinkage/expansion effects leading to compressive loads across the interface.

(iii) Interfacial tension effects arising from moisture gradients.

Combinations of (i) and (ii) can give potentially strong physical interactions across the interface.

It can be anticipated that the PRC manufacturing process would be likely to encourage bonding of the type described under (i) above. Our previous paper highlighted the significant surface texture of the glass inclusions, with fissures having dimensions readily capable of admitting cement grains, and consequently, a mechanism for paste intrusion under pressing pressure (up to 200 MPa).

The existence of chemical bonding in the present study is less likely. Our earlier review summarised conditions which promote chemical reactivity of the inclusion with the cement system and this included such deleterious reactions as alkali-silica effects (ASR). There are two main reasons why surface degradation as a result of ASR is not a problem with the soda-lime-silica inclusions that are used in this project. The glass particles are large compared with the zone of penetration [11] so the very slight degradation of surface expected from inclusion reactivity can only be expected to be advantageous in reducing the strength of interfacial bonding. The second reason is that the reaction is dependent on there being residual pore fluid after hydration. In both types of matrices being used in this study, and in high strength concretes, the available water content is low.

EXPERIMENTAL METHOD

Pastes were prepared using cement, crushed soda-lime-silica glass and water. Where glass was surface coated, melted candle wax was applied, the wax being allowed to harden on the individual glass grains by tumbling the glass in an air flow. A w/c of 0.35 was used in the initial mixing of the paste for each specimen. In silica fume blended cements, the OPC/SF ratio was 90/10 and a superplasticiser was added (Cormix SP1; 2% by weight of binder). The glass used in the experiments was recycled bus shelter glass (donated by Adshel), crushed to 0.5 - 1 mm. A binder/inclusion ratio of 70/30 by weight was used. The components were mixed in a domestic food blender and the batches were cast into steel moulds to form notched plates (90 x 45 x 15 mm) and vibrated on a casting table in order to remove almost all the air bubbles.

After 3.5 hours, the PRC pastes were uncovered and demoulded. They were then placed in a die and loaded uniaxially (Figure 2) using a hydraulic press. The expressed pore fluid was drained through a hole in the base plate assisted by a carrier stream of compressed air. In this series of experiments the samples were subjected to a maximum pressure of 200MPa.

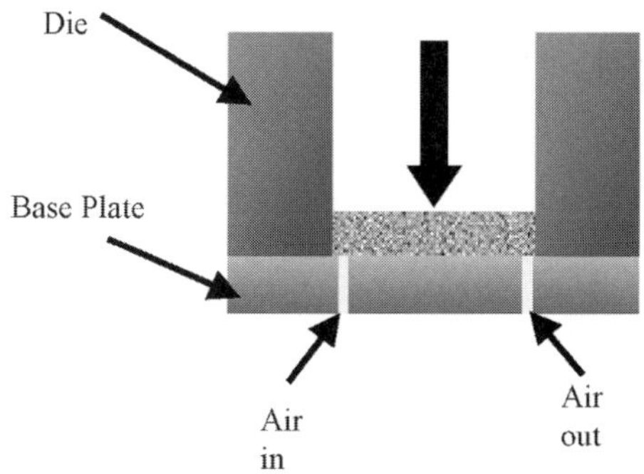

Figure 2 A schematic showing uniaxial loading of samples in a die press, acting as a filter press.

After the paste cylinders were pressed and removed from the die they were kept at constant humidity (≈ 100% RH) in a sealed bucket for 7 days. OPC/SF pastes were blended and cast as for the PRC pastes but after demoulding (at 24 hours from mixing) these were transferred to the curing buckets for a further 27 days.

Double Cantilever Beam Test

Fracture toughness tests were carried out on the notched plates using a test developed for brittle materials by Chandler et al [12] (Figure 3). The crack was extended by the application of a load using an Instron 1185 universal testing machine.

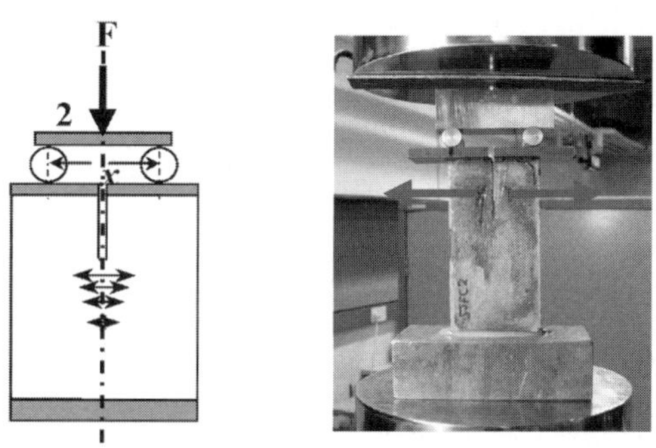

Figure 3 The double cantilever beam test configuration.

The test was carried out with a cross-head displacement of 0.2 mm/minute. The crack extension was measured visually. Toughness was obtained from the following expression:

$$K_{ap} = \frac{F}{w}\left(\frac{x}{w} - \frac{1}{4}\right)\sqrt{2}\sqrt{\frac{12}{1-\nu^2}}\frac{1}{B}$$

where: w is crack extension,
B is plate thickness,
ν is Poisson's ratio
F is the applied load
x has the meaning indicated in Figure 3.

RESULTS AND DISCUSSION

The results of the toughness testing of conventionally cast mortars are shown below in Figure 4. The positive slope of the graphs indicate that the mortars are increasing in toughness as the crack is forced to grow under the applied load. The different positions of the graphs reflect the intrinsic toughness of the matrix, indicated by the K_{1c} value (intercept with vertical axis). That the toughness increases with crack growth suggests the incorporation of angular inclusions has a general toughening effect, the extent of which is indicated by the gradient and this is consistent with earlier studies [4].

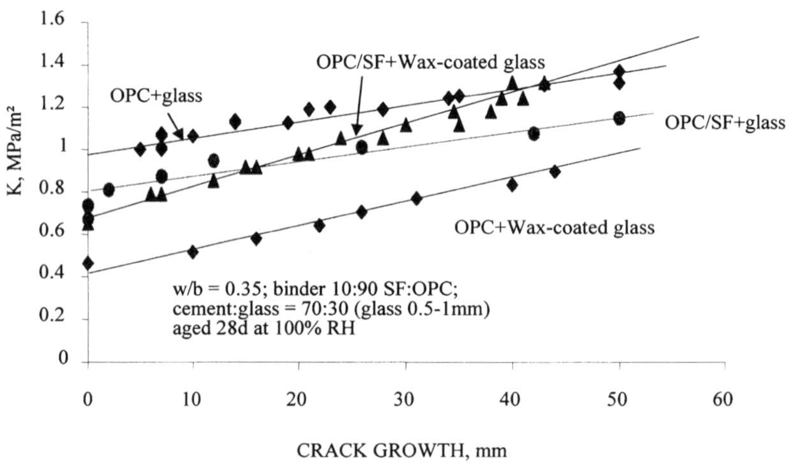

Figure 4 R-curve data for conventionally cast mortars containing wax-coated or untreated angular glass inclusions.

Figure 5 contains the corresponding data for the PRC products. Although the K_{1c} data are different for mortars with treated and untreated inclusions, the slopes are similar suggesting that the mechanism for optimisation of crack-inclusion interactions is equally active in both mortar types.

While it is important to note the different vertical scales in Figures 4 and 5, the significant increase in slopes observed for the PRC specimens suggest that the toughening mechanism is more active in these than in the conventionally cast mortars, despite the potential for increased physical binding between matrix and inclusion implied above.

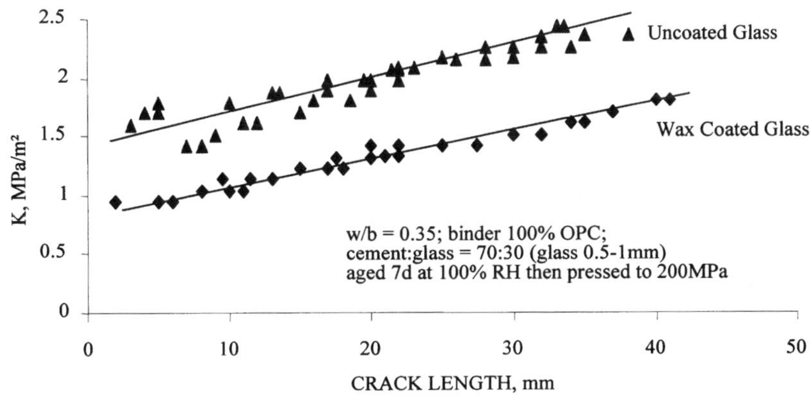

Figure 5 R curve data for PRC mortars containing wax-coated and untreated angular glass inclusions.

A convenient way to compare the interfacial bonding characteristics of the different systems studied is therefore to compare the slopes of the respective R-curves. Table 1 shows both K_{1c} and slopes for all systems studied using glass inclusions.

Table 1 Quantitative comparison of R-curve data for all systems containing glass inclusions

SYSTEM	TREATMENT	K_{IC}	SLOPE
PRC	Wax coated	1.418	0.027
PRC	Untreated	0.864	0.022
OPC/SF	Wax coated	0.686	0.014
OPC	Wax coated	0.430	0.010
OPC	Untreated	0.960	0.008
OPC/SF	Untreated	0.855	0.006

It is proposed that the slope of the R-curve is a measure of the frequency of crack-inclusion interactions which in turn reflects the strength of the interfacial bond. The weaker the bond, the more an *inclusion-filled hole* is represented and the more a 'flat-fracture path' in the region of the inclusion can be anticipated. This is better visualised with reference to Figure 6 which supports the flat fracture model for a porous material; the hemispherical holes are visible on the fracture surface.

Figure 6 Visualisation of a 'flat fracture' in foamed confectionary. The hemispherical holes indicate that the crack has selected the path of least resistance

As Figure 1 indicates, flat fracture (promoted by weak interfacial bonding) in the region of the inclusion is critical to optimisation of crack-inclusion interaction frequency and rising R-curve behaviour (increasing K_{app} with crack length).

This analysis offers insight into microstructural characteristics, specifically those related to hydration products adjacent to the inclusion grains. It is very difficult, particularly in high strength, low water content cement systems, to predict how the water will be ultimately distributed. Water availability is central to the hydration process and so restricted access through the sealing up of interconnected porosity will lead to water-deficient and incompletely hydrated zones.

This is a feature of PRC in which a significant proportion of the cement remains unreacted. This is because the pressing process causes pore structure collapse and any further reaction must involve residual water which is consumed by cement grains freshly exposed by the pressing process. In the present context, the availability of water for hydration is therefore unlikely to have been sufficient to provide strong binding at PRC-inclusion interfaces, with or without the wax coating, so a weak bond can be anticipated. This is consistent with the steeper gradients of the R-curve for both PRC specimens.

The hydration of OPC/SF blends progresses slightly differently in that it is only hydration processes which govern microstructural evolution. However, the water demand of SF is high and desiccation effects can be expected which may promote shrinkage of the products. With dimensionally stable glass inclusions, such shrinkage of the matrix onto the inclusions would promote physical bonding through compressive loading.

It might be anticipated that the strength of this binding would be influenced by how readily hydrated gel product could penetrate surface imperfections and this would be influenced by whether or not the glass was wax-treated. The R-curve data suggest that the wax coating does limit the degree of bonding with the untreated glass-matrix bond being the strongest of those studied. This would be consistent with a higher anticipated shrinkage in the OPC/SF blends than in OPC. Unfortunately, dimensional stabilities were not monitored during the study.

CONCLUSIONS

It is proposed that fracture toughness measurements may provide a means of assessing the strength of binding between cement matrix and aggregates. The relationship between interfacial bonding and toughness is based on the behaviour of propagating cracks in a composite containing angular inclusions. The strength of binding has been found to influence the crack path such that weak binding enhances the frequency of interaction between the propagating crack and the crack-bridging inclusions enhancing the fracture toughness.

Making the link between a fairly standard mechanical test and a microstructural property, largely inaccessable by direct methods, could have many advantages in that several durability issues which lead to deterioration of matrix-aggregate bond could be readily monitored. The present study has shown that high strength does not always correlate with brittleness but the exception identified in the present work represents a non-conventional processing route to high strength cement mortars. The expected trend for conventionally cast specimens was confirmed by comparing mortars made with OPC/SF with conventional OPC-based mortars, the use of wax coatings on the aggregate in some specimens allowing the correlation between interfacial binding and toughness to be more fully assessed.

REFERENCES

1. FERET, R., 'Etudes sur la constitution intime des mortiers hydrauliques', Bull. Soc. Encour. Ind. natn., Paris II, 1897, 160

2. GRIFFITH, A.A., 'The phenomena of rupture and flow in solids', Phil. Trans. Roy. Soc. (London), Vol. A221, 1920, 163

3. CHANDLER, H.W. MACPHEE, D.E., ATKINSON, I., HENDERSON R.J. AND MERCHANT, I.J., 'Enhancing the mechanical behaviour of cement-based materials', J. Eur.Ceram. Soc., Vol 20, (8), 2000, 1129.

4. CHANDLER, H.W., MACPHEE, D.E., MERCHANT, I.J. AND HENDERSON, R.J., 'Enhanced Crack-bridging by Unbonded Inclusions in a Brittle Matrix', J. Eur.Ceram. Soc., Vol 22, 2002, 129

5. MERCHANT, I.J., MACPHEE, D.E., CHANDLER H.W. AND HENDERSON, R.J., 'Toughening Cement-based Materials through the control of Interfacial Bonding', Cem. Concr. Res. , Vol 31, 2001, 1873

6. MERCHANT, I.J., CHANDLER, H.W., HENDERSON, R.J., MACPHEE, D.E. AND STEBBINGS, T.F., 'New Insights into Toughening Particulate Brittle Composites', Mat. Res. Soc. Symp. Proc., Vol 702, 2002.

7. KOVALEV, S., MIYAJIMA, T., YAMAUCHI, Y. AND SAKAI, M., 'Numerical Evaluation of Toughening by Crack-face Grain Bridging in Self-Reinforced Ceramics' J. Am. Ceram. Soc., Vol 83, (4), 2000, 817.

8. LAWN, B., 'Fracture of Brittle Solids', 2nd Ed. (Cambridge University Press, 1990), 216.

9. MACPHEE, D.E., 'PRC - Pore Reduced Cement: High Density Cement Pastes by Pore Fluid Extraction', Advances in Cements Research, Vol 3, (12), 1990, 135.

10. STEBBINGS, T., CHANDLER, H.W., HENDERSON, R.J. AND MACPHEE, D.E., ' Enhancing the toughness of high performance cement systems', Innovations and Developments in Concrete Materials Construction, 2002, 269.

11. BENTUR, A, DIAMOND, S, Fracture of glass fibre reinforced cement, Cement and Concrete Research, Vol. 14, No.1, 1984, pp 31 – 42

12. CHANDLER, H.W., HENDERSON, R.J., AL-ZUBAIDY, M.N., SARIBIYIK, M. AND MUHAIDI, A. 'A Fracture Test for Brittle Materials', Journal of the European Ceramic Society, Vol 17, 1997, 759 - 763

REPAIRING SLURRIES BASED ON A COMBINATION OF CEMENT-FLY ASH AND BITUMEN EMULSION

N Oikonomou

Aristotle University of Thessaloniki

Greece

ABSTRACT. Different kinds of slurries with cement and fly ash in combination with a non-ionic bitumen emulsion were prepared and tested. Properties of the fresh and hardened slurries, such as setting time, flow, compressive and flexural strength, water absorption, and abrasion resistance are presented in this paper. The results show that it is possible to produce suitable slurries in order to repair pavement surfaces or manufacture non-structural blocks.

Keywords: Slurry, Repairing, Pavement surfaces, Cement, Fly ash, Bitumen emulsion.

N Oikonomou is an Associate Professor in the Laboratory of Building Materials in the Department of Civil Engineering at the Aristotle University of Thessaloniki, Greece. His research focuses mainly on binder technology, pavement repair and maintenance, as well as recycled concrete aggregates.

INTRODUCTION

Slurry seal [1] is a mixture of dense graded aggregate, bitumen emulsion, water and mineral filters. It is used around the world as an economical treatment for sealing and extending the service life of asphalt or concrete pavements. Slurry seal requires the addition of filter additive (more often cement) in order to control the breaking time of the emulsion. A cationic bitumen emulsion is normally used for the application of slurry seal.

In previous work [2] we presented results of the use of fly ash, as a substitute for cement in microsurfacing, being a high quality slurry seal. The results of such a substitution showed that fly ash could be used and, in some cases, its usage would be advantageous. The present paper describes the use of fly ash-which is a by-product of coal-fired electric power generation – in slurry mixtures, which are based on a non-ionic bitumen emulsion in order to have sufficient time for their application for repairing pavement surfaces.

EXPERIMENTAL STUDY

Materials

Aggregates: The aggregates used were river sand (0-4mm) and crushed limestone coarse aggregate (4-10mm). Their characteristic properties met the requirements of the Greek Standards of Concrete Technology [3]. The gradation of the aggregate mixture was found within the specified limits as determined by ASTM [1] for type 3 mixture. The actual gradation of the aggregate mixture was as shown in Table 1.

Table 1 Aggregate mixture gradation

SEIVE SIZE, mm	PERCENTAGE PASSING, %
12.5	100
5.0	81
2.36	50
0.18	35
0.60	27
0.30	18
0.15	14
0.075	7

Cement: Ordinary Portland cement (OPC) type II 32.5 N from TITAN, a cement factory in Thessaloniki, was used. Its characteristics and composition are shown in Table 2.

Fly Ash: The Fly Ash (FA) used was from a coal-fired electric power generation plant situated in Ptolemaida, Northern Greece. Its characteristics and composition are also shown in Table 2.

Table 2 Characteristics and composition of OPC and FA.

COMPOUND	OPC	FA
	Percent, % by mass	
SiO_2	22.9	41.5
Al_2O_3	6.7	10.1
Fe_2O_3	2.4	3.4
CaO	57.2	37.8
MgO	2.5	3.5
K_2O	1.1	0.8
Na_2O	0.3	0.2
SO_3	3.0	2.5
CO_3	3.5	3.2
CaO (free)	1.3	11.4
Fineness, cm^2/g	3,300	4,740

Bitumen Emulsion: The bitumen emulsion used was a non-ionic bitumen emulsion (type SS-1) and it was supplied by the company BITOUMINA S.A. [4]. Its characteristics, which met the requirements of ASTM D977 [5], are shown in Table 3.

Table 3 Characteristic properties of bitumen emulsion

TESTS	LIMITS	RESULTS
Tests on Emulsion		
Viscosity, Saybolt Furol at 25°C, s	20-100	46
Storage stability, 24h, %	1	Pass
Cement mixing test, %	2	Pass
Residue by distillation, %	57	61
pH	-	6.8
Tests on Residue from Evaporation Tests		
Penetration, 25°C, 100g, 5sec, 0.1mm	100-200	134
Ductility, 25°C, 5cm/min, cm	40 (min)	Pass
Solubility in trichloroethylene, %	97.5 (min)	Pass

Plasticizer: A commercial plasticizer according to ASTM C 494, type A [6] was used as water-reducer.

Mix Design: The mix design was performed according to ASTM specifications [1]. The laboratory design, concerning representative mixes with OPC and PFA, is shown in Table 4. All mixes had the same consistence (approximately 15±1 cm) according to EN 1015-3 [7]. The river sand was from Axios river and the crushed aggregates were of limestone origin.

Table 4 Composition and test results of different mixes

COMPONENTS\CODE NO	C1	C2	C3	C4	C5	F1	F2	F3	F4	F5
Cement II 32.5N	1	1	1	1	1	0.5	0.5	0.5	0.5	0.5
FA						0.5	0.5	0.5	0.5	0.5
Non-ionic bit. emulsion (60%)		1	1	2	2		1	1	2	2
River sand (0-4mm)	1	1	1	1	1	1	1	1	1	1
Crushed sand (0-4mm)	1	1	1	1	1	1	1	1	1	1
Crushed coarse aggr. (4-10mm)	4	4	4	4	4	4	4	4	4	4
Water	0.80	0.50	0.35	0.40	0.30	0.95	0.60	0.42	0.46	0.37
Total water	0.80	1.10	0.95	1.60	1.50	0.95	1.20	1.02	1.66	1.57
Plasticizer (% on cement & FA)			1		1			1		1
Testing results										
Flow, cm	16	15	14	15	14	15	15	14	16	14
Water absorbability, 28d, %	7.3	5.2	4.8	4.5	4.1	6.8	4.9	4.4	4.2	3.6
Compressive strength, Mpa	29.5	13.4	17.7	10.8	13.7	19.7	12.2	14.3	5.4	10.5
Flexural strength, MPa	5.6	4.1	4.8	3.1	4.0	5.1	3.7	4.2	2.3	3.0
Wet track abrasion test, g/m2	fail	38.7	32.6	31.4	29.3	fail	45.6	41.7	38.6	35.3

Testing

The ten different mixes were tested in order to determine the water absorption (at the age of 28d and after 24hrs immersed in water), the compressive and flexural strengths (28d) according to EN 1015-11[8] and the weight loss by wet track abrasion test according to ASTM[1]. The results of testing are shown in Table 4. Setting time of the binders was determined by Vicat needles [9] and the results of this testing are shown in Table 5.

Table 5 Influence of bitumen emulsion on setting time

COMPONENTS	MIXES FOR DETERMINATION OF SETTING TIME					
Cement II 32.5N	1	1	1	0.5	0.5	0.5
FA				0.5	0.5	0.5
Non-ionic bitumen emulsion (60%)		1	2		1	2
Initial setting time, min	75	67	58	92	57	53
Final setting time, min	153	134	116	184	122	103

RESULTS AND DISCUSSION

Setting Time

The results obtained for the setting time showed that the addition of bitumen emulsion caused an accelaration of both initial and final setting time. This change was possibly due to the reaction of cement particles with bitumen particles and so there was an acceleration of cement hydration. This acceleration was higher for mixes containing FA maybe because of the presence of high amount of free CaO.

Absorption

Absorption of the hardened slurries with bitumen emulsion was lower than the ones with no emulsion. This change was expected because of the presence of the bitumen particles, which tightened the structure of hardened slurries. The addition of FA gave no important difference.

Compressive and Flexural Strength

Compressive and flexural strength of hardened slurries with bitumen emulsion were much lower than these of the hardened ones with no emulsion because of the presence of the bitumen. The more the bitumen content the lower the mechanical strength. The addition of FA again gave no important difference.

Wet Track Abrasion Test

Wet track abrasion test, after one hour soaking, was carried out according to ASTM [1]. Testing of specimens of mixes C1 and F1 failed because of too low consistency. Mixes with bitumen emulsion gave satisfactory results with very low figures of loss in weight (wear value). Mixes with higher amount of bitumen gave better abrasion resistance. The addition of FA also gave very low figures of loss in weight. The addition of plasticizer seemed to be positive concerning the characteristics (absorbability, mechanical strengths, abrasion resistance)of hardened slurries.

CONCLUSIONS

Different kinds of slurries containing cement, fly ash and non-ionic bitumen emulsion were prepared and tested. Although the strength of hardened slurries decreased, the satisfactory results from abrasion and absorption tests showed that such kinds of slurries can be used as repairing slurries for pavement surfaces (industrial floorings, parking areas, etc) and manufacturing of non-structural blocks. FA can be used without any problems in these slurries and so this by-product, the disposal of which causes a great environmental problem, can be an alternative material to cement reducing in a parallel manner the overall cost.

REFERENCES

1. ASTM D 3910. Design, Testing and Construction of Slurry Seal. Annual Book of ASTM Standards, Section 4, Construction, Vol. 04.03, 1993, pp 479-486.

2. NIKOLAIDES,A., OIKONOMOU,N., TSOHOS,G., The use of fly ash as a substitute of cement in micro surfacing. International Conference of Modern Technologies in Highway Engineering, Poznan, 2001.

3. GREEK STANDARD OF CONCRETE TECHNOLOGY. Specifications for aggregates, 1997 p 4.

4. BITOUMINA S.A. Thessaloniki, Greece.

5. ASTM D 977. Standard Specification for Emulsified Asphalt. Annual Book of ASTM Standards, Section 4, Construction, Vol.04.03, 1993, pp 170-1.

6. ASTM C 494. Standard Specification for Chemical Admixtures for Concrete. Annual Book of ASTM Standards, Section 4, Construction, Vol.04.02, 1996, pp 250-258.

7. EN 1015-3. Methods of test for mortar for masonry. Part 3: Determination of consistence of fresh mortar (by flow table), 1999.

8. EN 1015-11. Methods of test for mortar for masonry. Part 11: Determination of flexural and compressive strength of hardened mortar, 1999.

9. ASTM C 191.Standard Test Method for Time of Setting of Hydraulic Cement by Vicat Needles. Annual Book of ASTM Standards, Section 4, Construction, Vol.04.01, 1996, pp 160-162.

HIGH STRENGTH CONCRETE DURABILITY: FREEZING AND THAWING RESISTANCE AND SULFATE ATTACK

C Magureanu

A Popa

B Heghes

University of Cluj-Napoca

Romania

ABSTRACT. Freeze-thaw cycles and chemical attack pose a serious problem for concrete structures in cold and severe environments. This investigation deals with the exposure of high-strength concrete to freezing and thawing and sulfate attack in controlled environment. Experimental tests consisted of measuring the compression strength, tensile strength. Test results revealed that high-strength concrete was still durable after 50 cycles of freezing and thawing, and simulation of corrosive conditions in chemical solutions 2%$(NH_4)_2SO_4$.

Keywords: Compressive strength, Flexural strength, Splitting strength, Freeze-thaw durability.

Professor Cornelia Magureanu, is Director of the Reinforced and Pre-stressed Concrete Laboratory at Technical University of Cluj-Napoca, Romania. She specializes in high strength concrete, lightweight concrete, durability, reinforced and pre-stressed concrete and ferrocement as well.

Professor Augustin Popa, is Director at National Society of Geotechnical and Foundation for Transilvania, Romania, he is Head of Department of Geotechnical and Foundation at Technical University of Cluj-Napoca, Romania. He specializes in durability, rehabilitation, and soil mechanics and geotechnical engineering.

Bogdan Heghes, is assistent at Technical University of Cluj-Napoca, Faculty of Civil Engineering, Romania. He specilizes in high strength concrete and durability of concrete.

INTRODUCTION

Previous research has been carried out on the durability of high-strength concrete with regard to the chemical aspects of concrete durability, resistance to freezing and thawing, and corrosion of the reinforcing steel.

Research investigations conducted on the effect of freezing and thawing on properties of high-strength concrete have focused on compressive strength and tensile strength. However, a little information is available on the freeze-thaw effect and sulfate solution on tension properties of high-strength concrete. It has become increasingly evident that tension properties of concrete are major factor contributing to nonlinear behavior of reinforced concrete structures.

Experimental results [1] reveal that with a water/binder ratio of 0.3 and relevant microsilica contents (20-30% on binder basis) the concrete is freeze-thaw susceptible and air-entraining agents are required. Only with water/binder ratio lower than 0.25 does it seem that, due to a reduced porosity and the concrete may be frost and chemical attack resistant.

EXPERIMENTAL PROGRAMME

The programme aimed at investigating the mechanical properties of high strength/high performance concrete (HSC/HPC): compressive and tensile strength (splitting and flexural), modulus of elasticity in sulfate environmental. The different methods of curing after concreting are as follows:

- curing at 28 days under water, than in our standard storage, at 20°C±2°C and 60±5% relative humidity (in Table 2, curing method A);
- simulation of corrosive conditions: 28 days under water, than in chemical solutions: 1%$(NH_4)_2SO_4$ – curing method B; 2%$(NH_4)_2SO_4$ – curing method C;
- simulated freezing and thawing conditions: permanent in water at 20°C±2°C until at 400 days – curing method D; after 50 cycles freezing and thawing – curing method E.

Corrosive Conditions – Sulfate Attack

Altogether 12 tests were performed for each strength and time of testing.

The compressive and tensile strength were measured on specimens hardened at 28 days in curing method A, and after 90, 120 and 270 days at curing method A and C.

The modulus of elasticity are hardened at 28 days in curing method A and after 28 and 56 days in corrosive conditions B (when t_0=28 days).

Freezing and Thawing Conditions

Altogether 6 tests were performed for each strength at 28 days in curing method A, and at 400 days for curing method D and E.

HSC/HPC Mix Used

For the tests an HSC/HPC with strength class C60, was used for modulus of elasticity – curing method A and B – Mix a (Table 1) and C80 self compacting concrete was used for compressive tensile strength and freezing and thawing tests, - Mix b(Table 1). River sand and crushed gravel were used as aggregate.

Table 1 Concrete mix design

		MIX	
		a	b
Cement CEM I 52,5R	kg/m^3	520	520
Silica Fume (SF)	kg/m^3	52	52
River Sand 0..3mm	kg/m^3	530	530
Crushed gravel 3..7mm	kg/m^3	530	530
Crushed gravel 7..15mm	kg/m^3	706	706
Water	kg/m^3	150	150
Superplasticizer RAVENIT	l/m^3	11.44	12.30
W/(C+SF)		0.26	0.26

TEST RESULTS

Sulfate Attack

During the casting, the fresh concrete properties were evaluated using the slump test (60mm for Mix a) and by determining the fresh concrete density (2485 kg/m^3).

Modulus of elasticity

The modulus of elasticity tests according to our standard, were carried out after 28 days moist-cured t_0 and after 28, 56 days in curing method.

All concretes met the 28 days compressive strength – criterion of HSC/HPC of the guidelines of ENV206. In Table 2, the prismatic compressive strength „f_{cm}" modulus of elasticity „E_c" in curing method A and B at t_0=28 days and t_0=28, 56 days are given. The decrease of the modulus of elasticity after 56 days in solution 1%$(NH_4)_2SO_4$ is about 9%.

Table 2 Hardened concrete properties in curing method A and B

MATERIAL PROPERTIES	t_0=28 days CURING A	28 DAYS		56 DAYS	
		Curing A	Curing B	Curing A	Curing B
f_{cm} (N/mm^2)	66	73.6	72.7	73.7	63.7
E_c (N/mm^2)	46931	48077	47730	48000	43498
f^B_{cm}/f^A_{cm}	-	0.99		0.86	
E_c^B / E_c^A	-	0.99		0.91	

Compressive and tensile strength

Table 3 presents the evolution of compressive and tensile strength (splitting $f_{ct,sp}$ and flexural strength $f_{ct,fl}$) in curing method A and C.

Table 3 Hardened concrete properties in curing method A and C

MATERIAL PROPERTIES	t_0=28 days	90 DAYS		120 DAYS		270 DAYS	
	A	A	C	A	C	A	C
f_{cm} (N/mm^2)	80.3	88.5	72.0	92.4	83.16	103	91
$f_{ct,sp}$ (N/mm^2)	8.98	9.0	7.95	9.10	8.02	10.94	9.65
$f_{ct,fl}$ (N/mm^2)	8.54	-	-	10.71	9.32	10.99	9.712
f_{cm}^{C} / f_{cm}^{A}	-	0.813		0.90		0.88	
$f_{ct,sp}^{C} / f_{ct,sp}^{A}$	-	0.883		0.881		0.88	
$f_{ct,fl}^{C} / f_{ct,sp}^{A}$	-	-		0.870		0.88	

The increase in compressive strength between 28 days (t_0) and 90, 120 and 270 days in curing method A can be considerable (10%; 15%; 28%). In curing method C, the strengths (compressive and tensile) decreases comparative with curing method A, about 10-12%.

Freezing-thawing resistance

In Table 4 are presented the compressive and tensile strength (splitting and flexure). After 50 cycles of freezing and thawing at 400 days after casting, the concrete strengths are same or a little higher in comparative to the values at 28 days.

Table 4 Concrete properties after 50 cycles to freezing-thawing

MATERIALS PROPERTIES	TIME OF TESTING (days)	CURING METHOD A	D	E
f_{cm}(N/mm^2)	28	81.00	-	-
	400	83.00	85.00	80.00
$f_{ct,sp}$ (N/mm^2)	28	7.26	-	-
	400	7.78	8.28	9.35
$f_{ct,fl}$ (N/mm^2)	28	8.40	-	-
	400	9.30	10.03	11.00
f_{cm}^{D} / f_{cm}^{A}		1.024		
f_{cm}^{E} / f_{cm}^{A}		0.964		
$f_{ct,sp}^{D} / f_{ct,sp}^{A}$		1.064		
$f_{ct,sp}^{E} / f_{ct,sp}^{A}$		1.201		
$f_{ct,fl}^{D} / f_{ct,fl}^{A}$		1.078		
$f_{ct,fl}^{E} / f_{ct,fl}^{A}$		1.182		

CONCLUSIONS

The present paper focused attention on high-strength concrete through the use of additives such as silica fume (SF), superplasticizing admixtures. It has been demonstrated that significant improvements of strength and durability can be achived with such admixtures in concrete compositions. The research is to be continued.

AKNOWLEDGEMENTS

The research work has been supported by a Grant from the World Bank, the support of which is greatefully acknowledged.

REFERENCES

1. KRUMBACH, R., SEYFARTH, K., ERFURT, W., FRIEDMANN, K., High Strength Concrete – Durability Investigations by Using the CDF – Test – First Results. Leipzig Annual Civil Engineering Report, No. 4, 1999, pp. 57-71.

2. HAMA. Y., SENBU, O., TOMOSAWA, F., Effect of Curing Condition before Freezing and Thawing Test to Frost Resistance of High Performance Concrete HS/HP Concrete, vol. 2, Leipzig, June 2002, pp. 1017-1025.

3. BAROGHEL BOUNY, V., VILLAIN, G., ROUSSEL, P., Determination de la permeabilite aux gaz des betons. Influence du preconditionnement et calcul de la permeabilite intrinseque. Journees „durabilite" 8 et 9 mars 2002, Bordeaux – CETE du sud-ouest. LCPC, pp. 37-47.

4. SCHOKKER, A., BREEN, J.E., KREGER, M.E., High Performance Grouts for Durable Post-Tensioned Bridges. PCI/FHWA/FIB – International Symposium on HPC – Sept. 25-27, 2000, Orlando, Florida, USA, pp. 162-163.

5. AITCIN, P.C., High Performance Concrete E&SPON – 1998.

6. MAGUREANU, C., Reinforced Steel-Concrete Bond of High Strength Concrete. FIB, 2002, Osaka Congress, 13-19 Oct. 2002, Japan.

7. MAGUREANU, C., Mechanical Properties of HSC. 5-th International Symp. on Utilization of HS/HP Concrete 20-24 June, 1999, Sandefjord, Norway, pp. 1227-1237.

ROLE OF SUPER FINE FLY ASH ON HIGH PERFORMANCE CONCRETE

R Malathy

Kongu Engineering College

India

ABSTRACT. High performance concrete (HPC) is fastly getting acceptability for a wide range of applications in the construction of concrete structures. HPC is the concrete which is engineered to satisfy the criteria proposed to overcome limitations of conventional concrete or normal strength concrete. The parameters influencing the performance of concrete are strength, durability, permeability, shrinkage etc. This paper particularly focuses on the strength improvement of concrete using fly ash. Fly ash is a residue obtained by burning ground coal to produce electric power. The fly ash from coal burning power plant is used in concrete primarily because of its pozzolanic and cementitious properties. The principal reasons for using fly ash is economy and to improve the properties of concrete. The super fine fly ash, obtained from the sixth hopper of the thermal power plant and having a specific surface of 2000-4000sq.m/kg was used. This superfine fly ash was partially replaced for cement at 10-30% and the compressive strength, tensile strength, permeability, acid resistance and chloride resistance were checked compared with the normal concrete. From the experimental results, it was found that there was no improvement in strength of concrete with fly ash after 7 days. But at 28 and 56 days the strength improvement is significant when compare to normal concrete. Also the permeability and weight loss due to acid and chloride were also reduced for the concrete with superfine fly ash. Hence, to improve the strength and durability of concrete, super fine fly ash play a vital role like silica fume with no cost.

Keywords: High performance concrete, Super fine fly ash, Strength, Permeability, Acid and chloride resistance, Alkali Silica reaction.

R Malathy is a Senior Lecturer in Civil Engineering at Kongu Engineering College, India, with over 17 years of professional experience. She has also published over 20 technical papers.

INTRODUCTION

High Performance Concrete (HPC) is cement concrete, in which each ingredient performs effectively to contribute to fresh as well as hardened concrete properties. HPC can be designed to meet special performance requirements with regard to workability, strength and durability. To produce HPC, plasticizers are used for conventional concrete. But HPC requires high volume binding materials, which lead to shrinkage and large evolution of heat of hydration and this is also uneconomical. Hence cement is partially replaced by mineral admixtures, such as fly ash, ground granulated blast furnace slag (GGBS) and silica fume. These mineral admixtures not only improve the properties of fresh concrete but also enhance the durability characteristics. Mineral admixtures, also called cement replacement materials (CRM), act as pozzolanic materials as well as fine fillers, thereby the microstructure of the hardened cement matrix becomes denser and stronger. The concrete containing CRM improves acid resistance [1], early or later age strength, impermeability and reduces the heat of hydration and alkali-aggregate reaction [2]. However, the effect of CRM on the strength of concrete varies according to their properties, age and quality of replacement. The efficiency factor for getting strength at different ages when fly ash and slag are partially replaced are already investigated [3].

BACKGROUND

With the increase in temperature up to 250°C, the behaviour of fly ash concrete is similar to that of non fly ash concrete [4]. Durability of concrete is also increased when fly ash is added to concrete. As per ASTM C1012, fly ash has sulfate resistance and reduces corrosion of reinforcing steel exposed to chlorides in a wet environment [5]. A percentage of fly ash below 30% of the cement mass seems to be the best in resisting acids to prevent corrosion [6]. Concrete with a high proportion of fly ash to cement has been used successfully in many places [7]. Where thermal cracking due to excessive heat of hydration should be avoided such as dam construction, large foundation and particularly nuclear reactor building[8]. Concrete containing 15-30% fly ash carbonated to a slightly greater extent than OPC concrete of the same strength grade, but difference were generally small [9]. The performance of the fly ash based HPC mixes are good. The slight reduction in the workability of the fly ash based mixes shows that fly ash collection and processing technology in India need to be improved to get acceptable quality fly ash [10].

RESEARCH SIGNIFICANCE

To produce High Performance Concrete, there are many types of mineral admixtures are used to improve its strength and performance. While adding mineral admixture silicafume plays an important role to increase strength in aggressive condition. Hence today SF is the better choice for getting High Performance Concrete to withstand worst exposure conditions. But this silica fume is found to be very costly. Hence there is an attempt made to replace superfine fly ash instead of silicafume. This super fine fly ash is best alternative to silicafume in high early strength concrete. Its fineness around 1500 m^2/kg against 400 m^2/kg in fly ash. Also, its durability should be checked. Hence, in this paper the replacement of silicafume and super fine fly ash were considered and their strength and durability were compared.

EXPERIMENTAL PROCEDURE AND MATERIALS

Cement

Ordinary Portland cement namely 43 grade cement conforming toIS : 8112 – 1989 and similar to ASTM type III (C150 – 95) was used for the present experimental investigation. Its specific gravity was found to be 3.15.

Fine aggregate

Natural river sand with fraction passing through 4.75mm sieve and retained on 600 micron sieve the presence of inorganic and silt particles in the adopted sand fraction. The fineness modulus of sand was 2.81 with a specific gravity of 2.60.

Water

Potable tap water available in the laboratory with pH value of 7.0 ± 1 was used for mixing concrete and curing the specimens as well.

Silica Fume

Silica fume is a by - product resulting from the reduction of high purity quartz with coal in electric are furnaces in the manufacture of silicon and ferrosilicon alloys. The fume has a high content of amorphous silicon dioxide from the gases escaping from the furnace. Its physical properties and chemical properties are shown in table 1 & 2 respectively.

Super fine fly ash

Super fine fly ash with low calcium was obtained from last hopper of the Mettur Thermal power plant. It is a very good mineral admixtures, having very good pazzolanic properties. It reacts with free lime produced during the hydration of cement to form additional cementitious products. Its physical properties and chemical properties are shown in table 1 & 2 respectively.

Table 1 Physical Properties

PROPERTY	SILICA FUME	SUPER FINE FLY ASH
Colour	Light Grey	Light Grey
Specific Gravity	2.5	2.65
Bulk Loose Density	250 Kg/m^3	400kg/ m^3
Fineness	20,000 m^2/kg	2000 m^2/kg
Particle size	0.1 μ m	10μ m

Super plasticizer

The super plasticizer helps in maintaining workability even at low water cement ratios and also helps in improving strength. In this investigation, complast SP 337 (Complies with IS 9103- 1999, BS, ASTM Standards) was used.

Table 2 Chemical Properties

PROPERTY	SILICA FUME(%)	FLY ASH (%)
SiO_2	80-86	50-60
Al_2O	0.5-3	0.3
Fe_2O_3	0.2-0.8	0.6
MgO	0.5-1.5	0.03
CaO	0.1-0.5	1.00
Na_2O	0.2-0.7	0.10
K_2O	0.4-1	0.03
C	0.5-1.4	-
S	0.1-0.4	-
LOI	0.4-2.5	0.50

Mix Proportions

In order to study the influence of super fine fly ash compared with silica fume, their content 0%,10% & 15% were adopted. 500 kg/m^3 of cement, 650 kg/m^3 of sand was taken and w/b ratio varied from 0.3 to 0.35 at a constant rate of 0.05.

TESTING OF SPECIMENS

Compression test

Standard metallic cube moulds (150 mm) were used for the preparation of concrete specimens for compressive strength. A table vibrator was used for compaction of the filled concrete cubes. The specimens were demoulded after 24 hours an subsequently immersed in water till the time of testing. Three specimens were used for the determination of average compressive strength.

Split tensile strength test.

The cylinder specimen is of size 150 mm × 300mm, if the target nominal size of the aggregate does not exceed 20 mm.This is an indirect method of finding tensile strength of concrete.The split tensile strength has been estimated after 28 days by the following relationship

$$f_t = 2P / \pi DL$$

where

f_t = Split tensile strength of concrete in MPa.

P = Load at failure in Newtons ,

D = Diameter of cylinder = 150 mm.

L = Length of cylinder = 300 mm.

Permeability Test

Standard metallic cubes of 150 mm size were cast and cumed for 28 days. After curing period, it was taken out and allowed to dry for 2 days and paintedthe four faces of cube to prevent penetration of water from sides. Then the top surface is effectively sealed to achieve water tightness. Compressor was started and the pressure was applied at $0.50/mm^2$ to the water column and the quantity of water collected , time of observation were recorded. The co-efficient of permeability may be calculated as

$$K=QL/ATH$$

Where K=Co-efficient of permeability in cm/sec
Q= Quantity of water in mm
A = Effective area of specimen in cm^2
T = Time in sec. Over which Qis measured
H = Pressure head in cm
L = Length of specimen in cm.

Acid Resistance Test

Cube specimens of edge size 150mm were used. After 14 days curing, the specimens are taken out and allowed to dry for two days. The weights of specimens were noted. Then they are immersed in 1% H_2SO_4 solution for 2 months and taken out and surfaces cleaned and again weighted. The loss in weight was calculated in percentage.

Salt water Resistance Test

This is the same as the previous test. The specimen were immersed in 3% sodium chloride sodium chloride solution and the percentage difference in weight was noted.

Alkali-Silica Reactions

The alkali-silica reactions were observed on mortar bars kept under 90 percent humidity without direct contact with water. The temperature was maintained at 38 °±2°C throughout the test. The expansion of mortar bars due to alkali-silica reactions was measured.

TEST RESULTS

The test results are given in Tables 3-9 and Figures 1-7.

Table 3 Compressive Strength of concrete with w/b ratio 0.3

CURING TIME	COMPRESSIVE STRENGTH in N/mm² (0.3 water binder ratio)			
	Normal Concrete	10% fly ash	20% fly ash	30% fly ash
7days	44.4	43.7	38.52	36.9
28days	53.18	55.2	56.1	54.11
56days	59.2	64.3	63.3	60.1

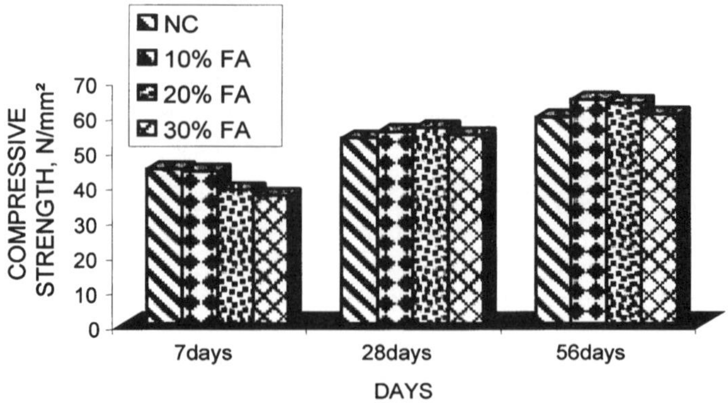

Figure 1 Age of Concrete Vs Compressive Strength of Concrete for different % of fly ash

Table 4 Compressive Strength of concrete with water binder ratio 0.3

CURING TIME	COMPRESSIVE STRENGTH, N/mm^2 (0.3 water binder ratio)				
	Normal Concrete	10% Silica fume	10% fly ash	15% Silica fume	15% fly ash
3days	33	46.2	32.1	44.4	30.5
7days	40	54.55	38.2	53.6	36.4
28days	50.36	59.25	56.5	55.11	53.2
56days	55.44	61.33	61.22	57.77	59.26
90days	57.18	62.62	62.52	58.36	59.62

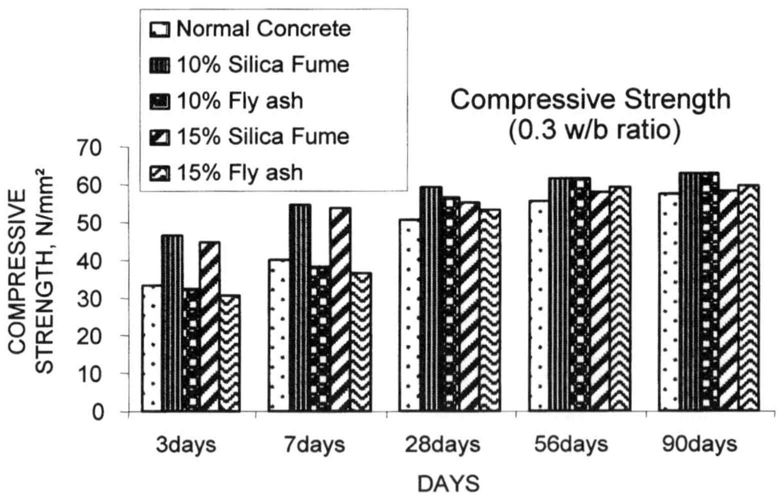

Figure 2 Age of Concrete Vs Compressive Strength of Concrete for different % of fly ash and silica fume

Table 5 Tensile Strength

	TENSILE STRENGTH in N/mm^2 (28 days)				
	Normal Concrete	10% Silica fume	10% fly ash	15% Silica fume	15% fly ash
0.30 w/b	3.798	3.916	3.87	3.82	3.617
0.35 w/b	3.49	3.585	3.49	3.5	3.485
0.40 w/b	3.42	3.49	3.467	3.48	3.42

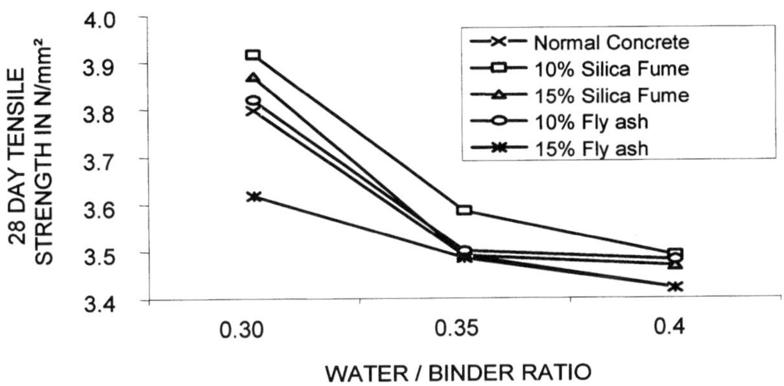

Figure 3 Water/binder ratio Vs Compressive Strength of Concrete for different % of fly ash and silica fume

Table 6 Permeability Test

MIX	PERMEABILITY COEFFICIENT x 10^{-7} cm/sec	
	Silica fume	Fly ash
0% Admix	8.1	8.1
10% Admix	7.2	7.1
15% Admix	6.5	6.8

Table 7 Acid Resistance Test

MIX	% WEIGHT LOSS
NC	2.17
10% SF	1.2
15% SF	1.1
10% FA	1.25
15% FA	1.05

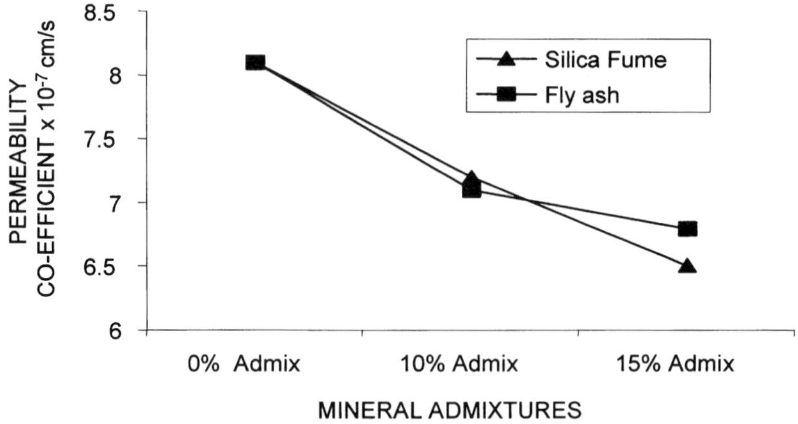

Figure 4 Permeability Co- efficient for different % of mineral admixtures.

Table 8 Salt Resistance Test

MIX	% WEIGHT GAIN DUE TO SALT PENETRATION
NC	1.612
10% SF	0.81
15% SF	0.92
10% FA	1.1
15% FA	1

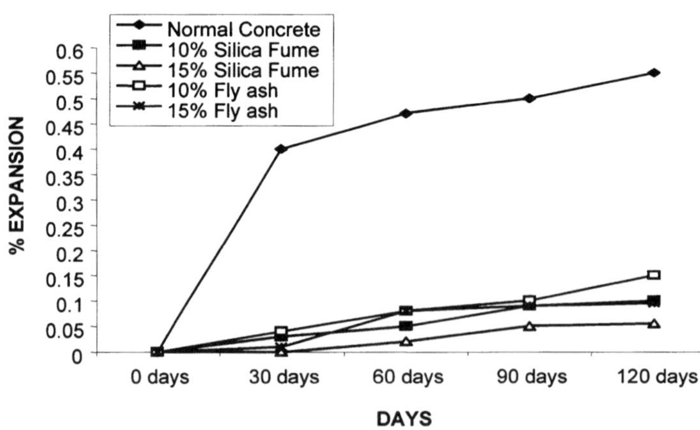

Figure 5 Age of Concrete Vs % expansion of mortar bars for different % of fly ash and silica fume

DISCUSSION OF TEST RESULTS

The compressive strength of concrete with 20% replacement of super fine fly ash was the maximum of 56 MPa at 28 days and 63.3 MPa at 56 days shown in Table 3 and Figure 1. The improvement of strength at early ages of superfine fly ash concrete was less than that of normal concrete but after 28 days of curing , the compressive and tensile strength of HPC with super fine fly ash increases to the same as that of SF(refer Table 4 and 5 and Figure 2 and 3). The permeability reduction of both SF and super fine fly ash is almost same but with 15% super fine fly ash, reduction of permeability of concrete is less, compared to SF shown in Table 6 and Figure 4. The performance against salt attack and Acid attack of both SF and super fine fly ash was almost same from Table 7 and 8. The expansion due to alkali silica reaction of concrete also reduced by adding superfine fly ash which almost equal to the performance of silica fume from Table 9 and Figure 5.

CONCLUSION

- The addition of SF and Super fine fly ash does contribute considerably to improve the strength properties in both compression and tension.
- For both SF and Super fine fly ash, the replacement to 10% cement gave better results than 15% replacement.
- The addition of Super fine fly ash improve the strength of concrete as that of SF.
- The permeability resistance of SF was good and for super fine fly ash is also almost same.
- The resistance against acid attack, salt attack and alkali silica reaction of SF and Superfine fly ash are good and the performance was almost same.
- The expansion of concrete prisms due to alkali silica reaction of super fine fly ash was very less and almost equal to silica fume.
- The optimum replacement of SF and Super fine fly ash are 10% to achieve high strength and durability of concrete.

REFERENCES

1. SWAMI R.N., High Performance and durability through design : Act SP-159 1996: 209-30

2. SHAH SP, AHMAD SH., High Performance Concrete: Properties and applications : McGraw Hill 1994.

3. B.H.BHARATHKUMAR., et.al Mix Proportioning of HPC : Cement & Concrete. Composites 23 (2001)71-80.

4. M.POTHA RAJU AND A.JANKI RAO., Effect of temperature on residual compressive strength of fly ash concrete : The Indian Concrete Journal, May 2001.

5. ROY H.KECK AND ENGENE H.RIGGS., Specifying fly ash for durable concrete : Concrete Internation.

6. N.DE BELIE, V.DE COSTER AND D.VAN NIEUWENBURG., Magazine of Concrete Research: 1997,49,No.181, Dec., 337-344.

7. DON PROCTER., How much Fly ash ?- Concrete Construction : Dec 2001.

8. CENGIZ DURAN ATIS., Heat evolution of high-volume fly ash concrete: Cement and Concrete Research 32(2002) 751-756.

9. M.D.A THOMAS, J.D.MATTHEWES., Corbonation of fly ash Concrete: Magazine of Concrete research, 1992, 44, No.160, September, 217-228.

10. S.GOPALAKRISHNAN., et.al Effect of partial replacement of cement with fly ash on the strength and durability of HPC : The Indian Concrete Journal May 2001.

NUMERICAL SIMULATION OF SURFACE PROPERTIES

E A B Koenders

K van Breugel

Delft University of Technology

Netherlands

ABSTRACT. Material properties of building and construction elements are requested to last for an increasingly longer lifetime. Long-term quality of cementitious building materials and material properties finds its roots in the development of the microstructure in a structure's early lifetime. The microstructure's quality plays an essential role in current and future maintenance plans as well. Degradation of material properties in the longer term is, to a large extent, the result of a change of the microstructure's quality. Aggressive substances might have succeeded in passing through the surface cover layer of structural elements and in affecting the microstructures integrity. The capillary pore structure that enables aggressive substances to intrude into the cement paste microstructure is dominated by the permeability of the concrete cover. It facilitates the potential movement of substances through the hardening paste. In the paper, numerical simulations will be presented, providing results about the relationship between the interface permeability of structural elements and the evolution of the micro-structural properties. Simulations of cement pastes with different water/cement ratios, clearly demonstrate a movement of capillary water through the concretes cover layer. The simulation models are part of a multi-scale approach to predict the short and long-term performance of concrete structures.

Keywords: Modelling, ITZ, Permeability, Numerical simulation, Moisture transport

Dr ir E A B Koenders is a researcher at the concrete structures group of Delft University of Technology in The Netherlands. His main topics are: Numerical modeling moisture transport processes and volume changes of cement paste systems and young concrete.

Professor dr ir K van Breugel is employed at the Delft University of Technology, Faculty of Civil Engineering and Geosciences in The Netherlands, since 1980. In 1999 he became a full professor at the Department of Concrete Structures. His main topics are Numerical simulation of the behavior of concrete at early ages, durability and the design of concrete structures for en environmental protection. He participates in several task groups of FIB, RILEM and CEN.

INTRODUCTION

When it comes to numerical simulation of micro-structural formation of cementitious materials one faces a complex multi-dimensional problem. Due to this complexity, in the past, the focus has been on the individual parameters that play a role in formation of microstructure and its associating development of material properties. This approach of material science has let to single parameter research where interactions of parameters were omitted. Integration of fundamental processes is therefore the direction to go while seeking for new and fundamental solutions for concrete science.

Performing research, for gaining solutions for macro-scale structural problems, normally finds its basis at a more detailed investigation level. In general, as regards research on composite materials, three different levels of investigation are distinguished, viz. micro, meso and macro level (see Figure 1). The most fundamental level in this respect is the micro-level, where processes are investigated and modeled on a micro-structural level. However, today, the nano-level of modeling is becoming more in the foreground and is taking over the position of the micro-scale level as being the most fundamental modeling level. The meso-level on the other hand, is representing a laboratory scale level of investigation and makes distinction between aggregates, bulk material and the transition zone situated between the bulk paste and the aggregates. Macro-level investigations, finally, is the level at which structural elements are investigated and modeled. It is the engineering level at which structural problems are described.

As regards the applicability and reliability of numerical simulation models for the prediction of materials properties of composite systems at early ages, the multi-scale modeling approach is inevitable. Bridging the gap between the different length scales creates the opportunity to relate macro-scale observations of material behavior towards the cause that lies behind these observations, at a more detailed level of investigation. This process starts with a coupling of the various numerical simulation models, while bridging the gap between the different detail levels of modeling.

In this paper, a numerical simulation model, called *ribbon*, is presented that bridges the gap between the micro and meso level of modeling, while emphasizing the potential of numerical simulation models as regards the prediction of surface properties of composite materials. When it comes to the formation of microstructure and the development of its material properties, the model is based on HYMOSTRUC modeling software [1]. The results presented show the potential of the ribbon model on predicting a structures surface permeability and its effects on the development of the physical and material parameters. The paper ends with an outlook and conclusions on the applicability of multi-scale models in general and micro-level models in particular, while focusing on the durability of the material quality.

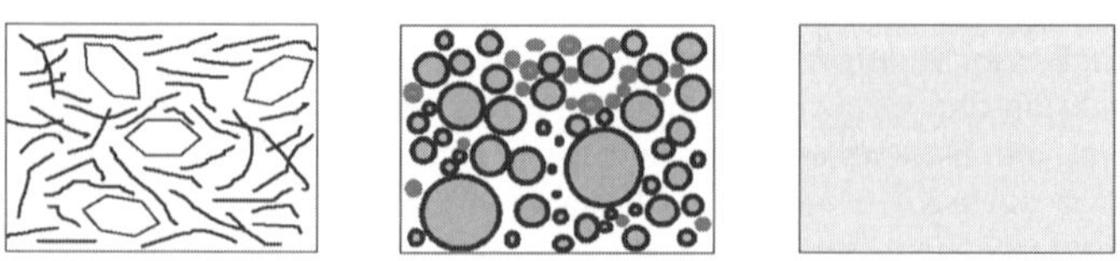

Figure 1 Definition of (a) micro-, (b) meso- and (c) macro-level for cement and concrete [2].

MULTI-SCALE MODELLING

In the recent past, a large number of numerical simulation models have been developed for all three distinct modelling levels. These models do actually not compete or overlap with each other, but, on the contrary, they have the potential to be complementary. Output calculated from a certain modelling level provides input information for a particular simulation model that prevails on a lower modelling level. Detailed information (e.g. micro-level) can then be put at the disposal of the meso-modelling level in order to calculate the material behaviour on this lower level of modelling, while performing these calculations with material information that finds its roots on a more detailed modelling level.

The advantage of simulating material behaviour on a less detailed modelling level is the fact that the geometrical scale of the elements in view is larger. Larger-scale problems can be simulated possessing larger physical dimensions. This supports the idea that cooperation between the numerical simulation models with respect to material data transfer is a challenge for future model improvements.

Modelling material behaviour of cementitious material on a micro scale level can be performed by application of the so-called *ribbon* model, which is based on the 3D HYMOSTRUC model [3,4]. The model finds its basis in a generated random microstructure and includes several model features of the basic HYMOSTRUC model [1]. The ribbon model represents a predefined volume of cement paste located either between two opposing aggregate particles (B) or located adjacent to the outer surface (A) of a full-scale structural element (see Figure 2). These typical conditions can be analysed by applying the right set of boundary and starting conditions to the micro-structural model. These conditions include the stereological boundary conditions for stacking of the cement particles in the predefined volume. Other conditions are the thermal boundary conditions during hydration and the boundary conditions that determine the climate at which the structural element is exposed to during its lifetime. This also includes the hygral conditions, determining the relative humidity of the structural element during and after hydration.

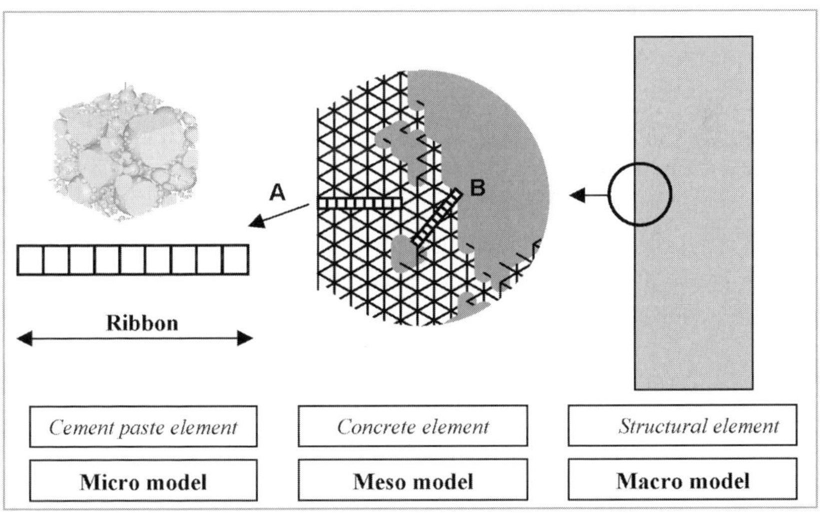

Figure 2 Schematic representation of the relationship between the micro- meso- and macro modeling level.

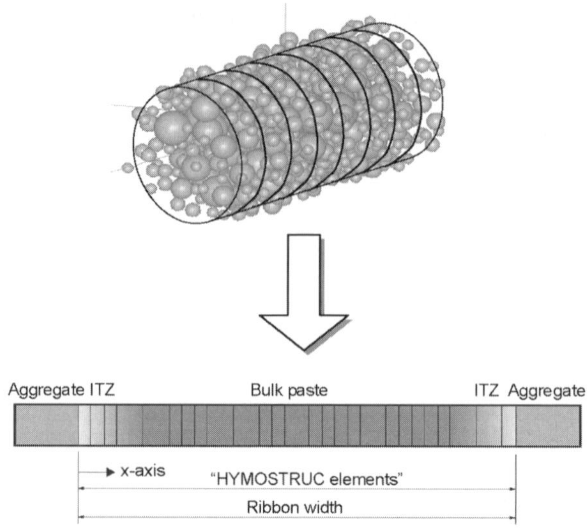

Figure 3 Ribbon model. Top: an impression of a generated random particle structure Bottom: a schematic representation of the numerical simulation schematisation.

The ribbon model can be applied for analyzing the material properties of a cementitious system situated between two aggregate particles. From the particle size distribution of the spherical shaped cement particles, stacked at random positions in the confined tubular shaped envelope, the distribution of the water-cement ratio can be calculated. Due to the physical boundary conditions at the aggregate surfaces, cement particles are not allowed to protrude these boundary planes, resulting in a disturbance of the packing of the cement particles in these particular regions. These packing disturbances are called the "wall-effect" and are responsible for the existence of the interfacial transition zones, where the material properties differ substantially from the properties of the areas situated in the central bulk paste of a cement paste system (see Figure 4).

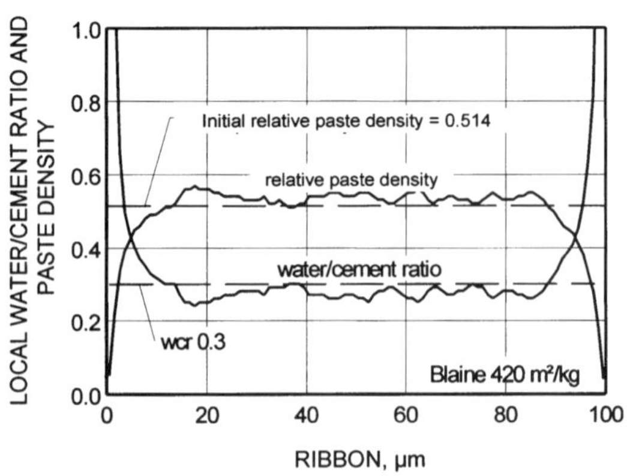

Figure 4 Distribution of paste density and water-cement ratio for ribbon paste sample.

MODEL DESCRIPTION

Each element within the ribbon model, in-fact, follows a blue print of the basic HYMOSTRUC model [1], where adding a marginal note, that, instead of the statistical particle positions model, the ribbon model's particle structure is based on random particle positions. The particles are stacked in a predefined envelope volume, following a certain distribution. Once the cement particles start to hydrate, a microstructural system is formed possessing a continuously densifying capillary pore structure. Within this pore system, distinction can be made between capillary water bounded physically to the pore walls and so-called free capillary water, which remains available in the pores for ongoing hydration. Both, the physically bounded water (adsorbed) and the free water contribute to the state of the capillary water, which, together with the development of the empty pore space are modelled within the HYMOSTRUC model explicitly. It forms the basis for modelling volume changes driven by drying, during hardening.

The movement of capillary water throughout the hardening cement paste system is explicitly modelled within the HYMOSTRUC model [1,4,5]. As hydration proceeds, water is chemically and physically bounded to the cement while forming hydration products that contribute to the development of a load bearing microstructure. As a result, capillary water is consumed and taken out of the system, causing unbalances in the reduced pore pressures. These unbalances are forcing capillary water to move from the water-rich area's (ITZ) towards the water lacking area's in the central parts of the cement paste matrix. This process leads to a redistribution of the capillary water in the paste, while influencing the material's properties development. It introduces a so-called drying shrinkage in the hardening cementitious system.

The movement of the capillary water can be modelled by adopting the capillary water as the aqueous medium that may move [4], depending on the development of reduced pressures in the pores, that are emptied due to the withdrawal of water by hydration. The movement of capillary water V_w can be described by:

$$\frac{1}{V}\frac{\partial Vw}{\partial t} = \frac{1}{\gamma_w}\frac{\partial}{\partial x}\left(k_w \frac{\partial p_w}{\partial x}\right) \tag{1}$$

where p_w is the actual pressure in the capillary water, γ_w the specific water density and k_w the water permeability. The actual pressure in the capillary water p_w is related to the gas pressure p_g in the emptied pores by:

$$p_g = p_w + \frac{2\sigma}{r} \tag{2}$$

where σ is the actual tension in the surface layer and r the radius of the largest capillary pore filled with water. A change of the volume of the capillary water, due to withdrawal of water by hydration, inherently balances with an identical change of the gas volume in the emptied pores. This process is shown schematically in Figure 5. The figure shows a capillary pore, partly filled (emptied) with water available for further hydration of the cement particles. The removal of the capillary water by hydration is captured by vapour, following a proportional lower vapour pressure.

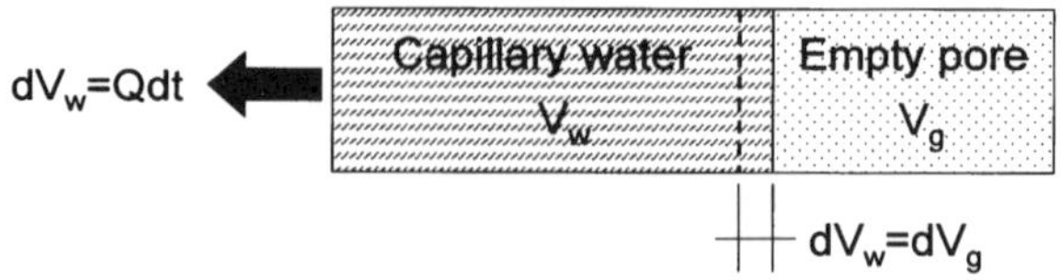

Figure 5 Schematic representation of a change of the volume of capillary water and gas due to withdrawal of the capillary water by hydration (dV_w).

The process of changing capillary water volume in relation with the change of volume of the empty pores has to be modelled explicitly. Within an incremental time step of constant porosity and assuming the gas in the emptied pores to behave in accordance width the ideal gas law, the relation between the change of the gas volume and the accompanying change of the gas pressure at constant temperature holds:

$$dV_g = \frac{V_g}{p_g} dp_g \tag{3}$$

While combining the equations (1), (2) and (3) the incremental change of the pore pressure can be calculated, driven by the consumption of capillary water due to hydration. Capillary water will be accumulated in regions in the paste, where the pores are relatively narrow and where substantial suction pressures can develop.

TOWARDS SURFACE PROPERTIES

The consumption of capillary water due to hydration, and the accompanying emptying of the capillary pores, is influencing the availability of capillary water at certain locations in the cementitious system. This will, therefore, also affect the degree of hydration, being a measure for the total amount of cement that has actually reacted with water. In
Figure 6, it can be observed how the redistribution of the capillary water affects the evolution of the degree of hydration of the ribbon paste. The results show that the movement of capillary water causes a small reduction of the degree of hydration at the ITZ regions whereas the degree of hydration in the central bulk paste is increasing roughly up to about 20%. The attraction of additional water leads to an improvement of the material properties at the inner paste at the expense of the quality at the ITZ-regions.

As regards the boundary conditions of the moisture transport model, it is assumed that the aggregate's at the outer ends of the ribbon paste are impermeable, imposing the flux towards these aggregates is equal to zero. Due to this assumption, the volume of capillary water in the ribbon system is conserved and no additional water can enter or escape from the system via these outer boundaries. By applying these impermeable boundary conditions to the system, a ribbon is simulated where the paste is confined by two impermeable boundaries (see Figure 2, B). If the surface properties of an open system would be simulated, the outer boundaries should simulate an open condition which associates with the conditions that belong to the surrounding air pressure (see Figure 2, A). This may, therefore, affect the pressure in the emptying pores as well as the actual relative humidity in the cement paste system, since the relative humidity outside the outer surface of a concrete element has to be in balance with the relative humidity inside the element. This part of the model is still under development and results can not be presented yet.

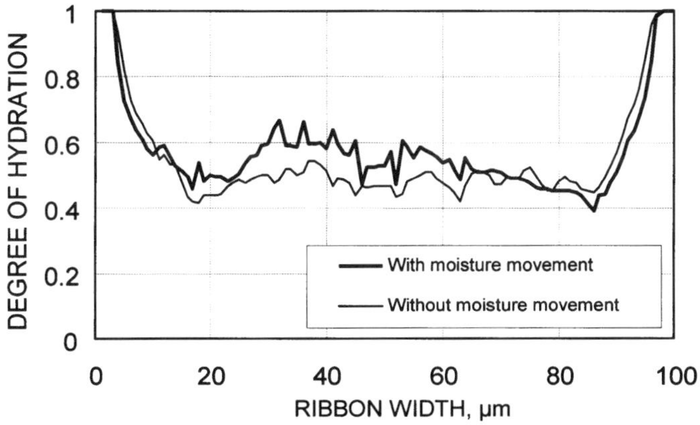

Figure 6 Degree of hydration with and without allowance for movement of capillary water through the hardening ribbon paste.

The ability of movement of the capillary water through the ribbon system is driven by pressure gradients on the one hand and by the openness of the pore system on the other. It is determined by the systems permeability, which is continuously changing as hydration proceeds (see Figure 7, left). The permeability is a measure for the ability of the capillary water to travel throughout the ribbon system. This can also be observed from a side view of the initial situation of the three-dimensional particle structure (see also top part of Figure 3). The observed pores (irregular white spots) narrows as hydration proceeds (Figure 7, right).

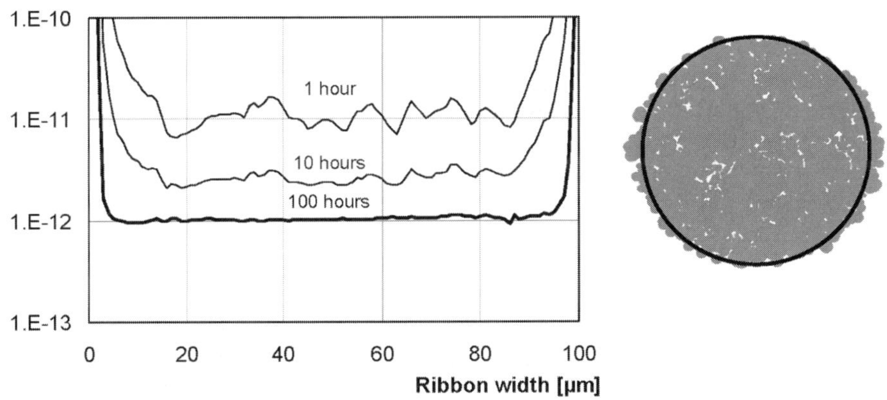

Figure 7 Left: Calculated permeability for the ribbon paste [4]. Right: Cross-section of the initial ribbon paste, where "through pores" can be observed.

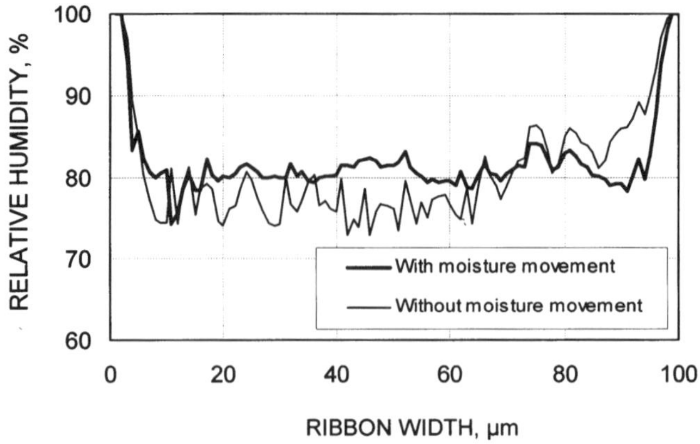

Figure 8 Relative humidity with and without allowance for the capillary water to move through the hardening ribbon paste, while hardening.

The permeability as calculated with the HYMOSTRUC model is presented in Figure 7 is in good agreement with permeability measurements [4,8,9]. The movement of capillary water throughout the ribbon system also affects the relative humidity in the emptied pores. As hydration proceeds, capillary water is consumed to form hydration products, and is taken out of the system. Pores become emptied while experiencing a reduction of the vapor pressure in these emptied pores. This reduced vapor pressure forms the basis for the relative humidity and can be calculated according to the Kelvin equation [1,4,6,7].

The influence of the hydration-driven capillary water movements throughout the hardening ribbon paste is provided in Figure 8. The results show that the redistribution of capillary water enlarges the relative humidity in the central bulk paste and reduces the relative humidity at the ITZ regions. It indicates a reduction of internal drying in the central bulk paste at the expense of the drying at the ITZ regions, close to the aggregates surfaces. This additional drying inherently leads to an increase of the likelihood of the occurrence of micro cracking at the interface between the aggregate particles and the adjacent cement paste system.

OUTLOOK AND CONCLUSIONS

When is comes to modeling interface phenomenon of composite systems, moisture movements of capillary water through the system may not be neglected. The moisture movements determine whether there are potential locations in structural concrete elements that may dry out, while simultaneously increasing the risk of micro cracking in general and reducing the concrete's quality in particular. It emphasizes the need and value of numerical simulation models for research of fundamental topics of composite materials, while aiming for an improvement of its durability.

The multi-scale modeling approach as referred to in this paper is a way to relate damage, as may be observed from a surface of a (macro) structural element, to a mechanism that find its origin at a micro-scale level. This challenges the opportunity for researchers to get feedback from the engineering practice on the one hand, and creates more added values for the research carried out by R&D institutes, on the other hand.

REFERENCES

1. BREUGEL, K. van., Simulation of Hydration and Formation of Structure in Hardening Cement-Based Materials, PhD thesis, Delft University of Technology, Delft, The Netherlands, 1991, pp. 295.

2. WITTMANN, F.H., Structure of Concrete with respect to Crack Formation, Fracture Mechanics of Concrete, Elsevier, London/New York, 1983, 43.

3. YE, G., HU, J., BREUGEL, K. Van, STROEVEN, P., Characterization of the development of microstructure and porosity of cement-based material by numerical simulation and ESEM image analysis, Journal of Materials and Structures, 2003.

4. KOENDERS, E.A.B., Simulation of volume changes in hardening cement-based materials, PhD thesis, Delft University of Technology, Delft, The Netherlands, 1997, pp. 171.

5. KOENDERS, E.A.B. and BREUGEL, K. van, Numerical Simulation of ITZ Quality, Proceedings of the conference on Structural Faults + Repair 2003, London, 2003.

6. SETZER, M., J., Einfluss des Wassergehalts auf die Eigenschaften des erhärteten Betons, Deutscher Ausschuss Für Stahlbeton DAfSt ,Heft 280, 1978, pp. 43-79.

7. MEIJERS, S.J.H., Computational Modelling of Chloride Ingress in Concrete, PhD thesis, Delft University of Technology, Delft, The Netherlands, 2003, pp. 171.

8. NEVILLE, A.M., Properties of Concrete, Pitman, 1978.

9. POWERS, T.C., COPELAND,L.E.,HAYERS,J.C. and MANN, H.M., Permeability of Portland Cement Paste, Research and Development Laboratories of the Portland Cement Association, Bulletin 53, Chicago, 1955.

THEME FIVE:

SPECIAL AND NON-PORTLAND CEMENTS

Keynote Paper

SCIENTIFIC AND SOCIETAL ISSUES INVOLVED IN DEVELOPING SUSTAINABLE CEMENTS

E Gartner
Lafarge Central Research Laboratory
France

ABSTRACT: This paper discusses the practicality of replacing Portland cements with alternative hydraulic cements that could result in lower total CO_2 emissions per unit volume of concrete of equivalent performance. Currently, the cement industry is responding rapidly to the perceived societal need for reduced CO_2 emissions by increasing the production of blended Portland cements, using supplementary cementitious materials that are principally derived from industrial by-products such as blast-furnace slags and coal-combustion fly ashes. However, the supplies of such by-products of suitable quality are limited, so more radical solutions to the CO_2 emissions problem may ultimately be needed. In this paper, we show that the most promising alternative systems currently appear to be those based at least in part on calcium sulfates, the availability of which is increasing due to the widespread implementation of sulfur dioxide emission controls. These include calcium sulfoaluminate-based cements, especially those that can be made from relatively common raw materials. They also include cementing systems that make better use of the potential synergies between calcium sulfate, calcium silicate and calcium aluminate hydrates. However, a great deal more R&D, mainly oriented towards establishing the usage properties and durability of concretes made from such cements, will be needed to provide the data necessary to modernise our construction codes and standards if we are to effectively use these potentially more "CO_2 efficient" technologies on a large enough scale to have a significant global impact.

Keywords: Cement, Energy, Carbon dioxide, Environment, Sustainable development

Ellis Gartner is a Principal Scientist at the Lafarge Central Research Laboratory, near Lyon, France. In 1974 he obtained a PhD in Physical Chemistry from the University of Cambridge, but his interest in cement science began during his subsequent period as a Higher Scientific Officer for the UK Department of the Environment's Building Research Establishment in Garston, followed by several more years at the Portland Cement Association in Skokie, Illinois, USA. He later joined the Central Research Laboratory of W. R. Grace & Co. in Columbia, Maryland, where his work led to a series of new cement and concrete additives. He is a Fellow of the Institute of Materials, Minerals and Mining (UK) and the American Ceramic Society; and also a Member of the American Chemical Society and the Materials Research Society (USA).

INTRODUCTION

The Portland Cement Manufacturing Industry is under close scrutiny these days because of the large volumes of CO_2 emitted in the manufacture of Portland cement clinker. However, CO_2 emissions alone are only a minor part of the larger issue of the role of the construction industry in the sustainable development of our civilisation. According to the World Business Council on Sustainable Development, the three "pillars" of sustainable development are Economic Growth, Ecological Balance and Social Progress, and one cannot have one without the others, as is symbolised by the image below:

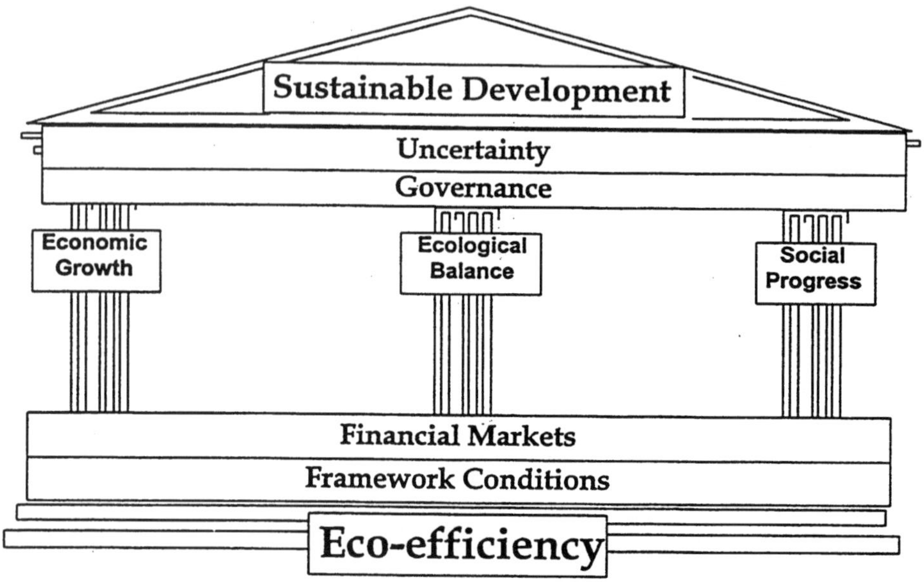

Thus, in looking for ways to reduce the global CO_2 emissions related to cement manufacture, we must not consider the cement manufacturing process alone, but also where and how the raw material resources will be found, and where and how the product will be used, with what overall impact on the global community. Lafarge takes the issue of sustainable development very seriously, as is evidenced by our recent publication of a corporate report on Economic, Societal and Environmental Performance [1] and our public commitment, endorsed by the World Wildlife Fund, to reduce our total global CO_2 emissions by 15%, and our mean specific CO_2 emissions per tonne of cement by 20%, over the period 1990-2010.

Concrete, based mainly on Portland cement clinker, is the most widely-used material on Earth. Current estimates of world cement manufacture are of the order of 1.7×10^9 tonnes/yr., enough to produce over 6 cubic km of concrete per year, or about one cubic metre per person. Of course, not all of this cement is pure Portland cement; much of it already contains secondary cementitious materials and fillers. This makes our challenge all the greater. Is it possible to go beyond the current approach of the cement industry, that of maximising the use of supplementary cementitious materials such as granulated slag, PFA and natural pozzolans, to an approach based on alternative cementitious systems that are inherently less CO_2-intensive? This fundamental question is the basis for the rest of this paper.

AN HISTORICAL PERSPECTIVE

It should be remembered that a similar pressure for change arose in the mid 1970's, when the OPEC oil embargo forced the cement industry to reduce its dependence on oil, and, more generally, improve its overall energy-efficiency [2]. This crisis gave rise to a period of intense R&D activity, much of it process-oriented, that led to the rapid refinement and commercialisation of what are still the current "best" Portland cement-manufacturing technologies, that is, high-capacity multi-stage preheater/precalciner kilns coupled with vertical raw mills (to recuperate waste heat for drying the raw feed) high-efficiency closed-circuit finish mills, and improved clinker coolers. There was also a rapid change away from oil and towards coals, cokes and a wide variety of waste fuels as the primary combustible in the kiln. The shock provided by the OPEC oil embargo provided the short-term economic incentive necessary to force the industry to adopt new process technologies; and these new technologies turned out to have long term benefits for the overall efficiency of the industry.

CO_2 Emissions vs. Energy Consumption

The comparison of the current situation with the 1970's Oil Shock is important, because most anthropogenic CO_2 emissions result directly from the combustion of fossil fuels to produce usable forms of energy. Thus, almost any approach to decreasing fossil fuel consumption should have a similarly beneficial effect in reducing CO_2 emissions per unit of product. Clearly, a modern cement plant produces much less CO_2 per ton of product than an average plant predating 1974, and the strong incentive to minimise the relatively high fraction of cement manufacturing costs associated with energy has in most cases also served to minimise average specific CO_2 emissions. We are therefore starting from a relatively low baseline level for the "Best Available Technology" (BAT) in terms of Portland cement manufacturing, and it seems very unlikely that new manufacturing-process research can offer any very significant further improvements in BAT with regard to CO_2 emissions. Nevertheless, many cement plants around the world still operate at well below the BAT level, which still leaves considerable room for improvement provided that the capital is available for investment.

It has been suggested in some quarters that it might be possible to produce portland cements with lower associated CO_2 emissions by processes that do not directly make use of fossil fuels, e.g. electric furnaces, or even solar or nuclear-powered furnaces. Suffice it to say that a many such ideas were seriously investigated in the 1970's but with no apparent practical success. Electric furnaces are rarely the most energy-efficient approach for the bulk thermal processing of materials, because most electricity is itself made from fossil fuels or other thermal sources at an energy-efficiency of generally less than 40%. Nuclear reactors are clearly not a solution that would please most environmentalists, and, as far as I am aware, no one has seriously considered a direct nuclear approach (other than one using nuclear-generated electricity) because it looks so unattractive for the same reasons that nuclear electric power generation is no longer being considered as a serious option by most societies. Finally, although it should be possible, theoretically, to make Portland cement clinker in a specially-designed solar furnace, the practical difficulties would be enormous. A huge solar collection and focussing device would be required to run a normal size cement plant, and it could only operate in strong direct sunlight, so it would lose all of the advantages of continuous operation; intermittent operation clearly leads to severe energy and operational inefficiencies.

However, in addition to the relatively straightforward issue of energy-efficiency, we are also faced with the problem of raw materials and the "potential CO_2" contained within them. Portland cements are very rich in calcium, and the principal concentrated source of calcium in the Earth's crust is limestone, which is composed essentially of calcium carbonate, ($CaCO_3$). To make Portland cement clinker, in which the principal active component is tri-calcium silicate, (Ca_3SiO_5, or "C_3S" in the simplified oxide notation used by mineral chemists) limestone is heated with a source of silica in a kiln at temperatures of well over 1350°C. The simplest form of the basic reaction is shown in eq. (1), *(mass proportions given in italics)*:

$3CaCO_3$	+	SiO_2	$\Rightarrow$	Ca_3SiO_5	+	$3CO_2$	(1)
300		*60*	=	*228*		*132*	
calcium carbonate	+	silica	$\Rightarrow$	C_3S (or alite)	+	carbon dioxide gas	

Two important facts are to be observed here. Firstly, for every ton of C_3S produced from limestone in this way, 579 kg of carbon dioxide gas is liberated as a chemical by-product, irrespective of the process used and its fuel efficiency. Secondly, the enthalpy change, ΔH, of this reaction is about +416 kJ/mol at 20°C, which means that the theoretical minimum amount of thermal energy required to produce one ton of C_3S from limestone and quartz is 1.83 GJ. (In practice, one cannot hope to come very close to this minimum, as explained later.) In addition, the reaction will not proceed at all at temperatures lower than about 1250°C, (even with a catalyst) and it is usually necessary to heat the materials to well over 1350°C in order to get the reaction to go at a high enough rate to be commercially useful.

One might pose the question as to whether a source of calcium other than limestone could be used, but it turns out that there are no other sufficiently widespread and sufficiently concentrated sources of calcium available except perhaps for calcium sulfates (gypsum and anhydrite) The manufacture of portland cement clinker from calcium sulfate has historically been practised in certain locations, but its major complication is that the principal by-product is sulfuric acid, as shown by the following (intentionally simplified) reaction equation:

$3CaSO_4$	+	SiO_2	+	$3H_2O$	$\Rightarrow$	Ca_3SiO_5	+	$3H_2SO_4$	(2)
408		*60*		*54*		*228*		*294*	
anhydrite	+	silica	+	water	$\Rightarrow$	C_3S (or alite)	+	sulfuric acid	

This process was practised industrially primarily as a means of producing sulfuric acid, with the by-product being Portland cement clinker; but it fell out of favour relative to other less complicated and less expensive methods of sulfuric acid manufacture. As with all processes that attempt to obtain two useful products at the same time, relative economic fluctuations in the raw materials supply and in the market for both products made it difficult to operate profitably for long periods of time. In addition, the overall thermal efficiency and environmental impact of the cement-sulfuric acid process was considerably worse than that of the normal cement manufacturing process. The total global demand for Portland cement is roughly an order-of-magnitude greater than that for sulfuric acid, meaning that it would never be practical to produce more than a small percentage of cement in this way. Moreover, good deposits of calcium sulfate are less abundant than good deposits of limestone, making this an important issue in locating a factory. Nevertheless, the process could be looked at again for suitable locations, if the market conditions were also right.

It therefore appears almost impossible to make the majority of Portland cements from sources other than limestone; and limestone-based cement manufacture inherently gives rise to very high CO_2 emissions even if the minimum theoretical energy consumption can be obtained. Therefore, we decided that it would be worth looking at alternative cementing systems to see whether or not any of them might in future be capable of offering many of the advantages of Portland cement but with significantly lower inherent CO_2 emissions. It should be added that the principal economic advantage of Portland cement is its low delivered cost in most important locations, due principally to the ready availability of the necessary raw materials. One can make cements with better performance for specific applications, but none of them can currently be produced at a competitive price for "ordinary" construction applications.

RAW MATERIALS CO_2 VS. FUEL-DERIVED CO_2

It is clear from the previous discussion that we have to separate the CO_2 emissions from any manufacturing process into two categories: those simply derived from carbon compounds in the raw materials that are converted to CO_2 during the manufacturing process, which we refer to as "raw materials" CO_2, (RM$\bar{C}$), and those produced by the combustion of fossil fuel necessary to drive the manufacturing process, which we refer to as "fuel-derived" CO_2 (FD$\bar{C}$). The former is much easier to predict than the latter, since one usually knows very precisely the composition of the raw materials and the products. Thus, in the previous example, manufacture of C_3S from limestone and silica produces 0.58 g/g of RM$\bar{C}$, whereas manufacture of the identical product from gypsum and silica produces essentially no RM$\bar{C}$. On the other hand, calculation of the FD$\bar{C}$ emissions per unit of product requires a detailed knowledge of the processing equipment to be used and the way it is operated. It also depends on the specific fuel chosen. Thus, it is hard to obtain reliable data on FD$\bar{C}$ emissions except for industrial processes that are already operating and for which good statistical data can be obtained. As a result, this paper does not attempt to make any very accurate estimates of FD$\bar{C}$ emissions for most of the alternative cementing systems discussed.

Table 1: Simplified ideal heat balance for the manufacture of Portland cement clinker [3]

POSITION IN KILN SYSTEM	TEMPERATURE RANGE	PROCESS	HEAT REQUIRED, kJ/g CLINKER
Preheater	20-900°C	Heating raw feed to 900°C	+1.53
	About 450°C	Dehydration of clays	+0.17
	20-900°C	Cooling CO_2 and H_2O	-0.59
Calciner	About 900°C	Dissociate of calcite	+1.99
	About 900°C	Reaction of dehydrated clays	-0.04
Rotary Kiln	900-1400°C	Heating feed from 900 to 1400°C	+0.52
	900-1400°C	Formation of clinker phases	-0.31
Clinker cooler	400-20°C	Cooling of clinker to 20°C	-1.51
		Net hear required	+1.76

Notes:

a) Assumes dry limestone and clay as kiln feed

b) Industrially, it is not generally considered practical to recover heat from kiln exit gases below about 120°C, due to the problems engendered by condensation of water.

In order to explain why it is so difficult to calculate the likely FDC̄ emissions of a manufacturing process from fundamental physico-chemical data, it is sufficient to use portland cement manufacture, which we understand very well, as an example. In practice, the thermal energy requirements for clinkering vary widely from plant-to-plant depending on the type of kiln system used. The most efficient modern coal-fired preheater (dry-process) plants use as little as 3.06 GJ of fuel energy per ton of clinker (GJ/t), whereas some older wet-process plants can consume more than twice that amount. The wet process uses a lot of heat to evaporate water, but there are better alternatives even where the raw materials are naturally wet. The main justification for continued use of wet kilns is simply the high replacement cost.

Table 1 shows a calculation of the theoretical thermal energy requirement for the manufacture of Portland cement clinker in the absence of heat losses. The total heat requirement is 1.76 kJ/g (or GJ/t) if we start with all raw materials at 20°C, end up with all products at 20°C and there are no heat losses to the surroundings. However, this is impossible in industrial reality. Not only is there the problem of insulation of the kiln shell, (it is not possible to make or use a perfectly insulated kiln because suitable materials do not exist) but there is also the problem of combustion thermal efficiency. In other words, any fuel that we use has to be burned (in air) to produce a certain volume of hot combustion gases (ideally consisting only of N_2, CO_2 and of H_2O). These hot gases then transfer some of their heat to the kiln charge by radiation and convection, but, clearly, this transfer can only occur when there is a significant difference in temperature between the two. As can be seen readily from Table 1, about 2.2 GJ/t of heat is needed at temperatures above about 900°C for calcination and clinker formation, whereas only about 1.1 GJ/t is needed below 900°C for preheating dry raw materials. It turns out that the calcination temperature acts a kind of bottleneck in the process, because no matter how much heat we have available below that temperature, it cannot be used to drive either the calcination or the clinkering reactions.

Ideally, therefore, we need a fuel that burns to give gases that are very much hotter than 900°C, so that a significant proportion of the heat can be used for clinkering and calcining. One ton of good quality bituminous coal has a gross heating value of 32.6 GJ and produces 2.94 tons of CO_2 when burned stoichiometrically in air. In addition, the combustion gases include 8.26 tons of N_2 from the combustion air and 0.50 tons of water vapour from hydrogen in the coal. If the coal and air are initially at ambient temperatures (about 20°C), 19.5 GJ/(tonne of fuel) of this heat is available above 900°C, i.e. about 60% of the gross heating value. However, this percentage is further degraded by the fact that typically about 10% excess air is used to assure complete combustion, and by the fact that heat transfer from the gases to the solids requires a significant temperature difference. Thus, to provide 2.2GJ/(tonne of clinker) above 900°C we would need to burn coal of heating value equal to about 3.7 GJ.

This situation is made even worse by any other heat losses (e.g. kiln shell losses) above 900°C; but we also have one additional source of heat that makes a very useful contribution, and that is the clinker itself. By preheating the combustion air in the clinker cooler, we can also recover a significant amount of heat above 900°C. The combination of all of these factors leads to the estimation that the practical thermal efficiency limit for a perfectly-tuned coal-fired kiln operating with minimal excess air is about 3 GJ/t, (which gives about 0.28 t/t of fuel-derived CO_2) relative to the theoretical 1.76 GJ/t, (i.e. about 0.16 t/t fuel-derived CO_2). If more energy-rich and carbon-poor fuels, such as methane, could be used, that would help a little. In theory, the energy consumption and thus fuel-derived CO_2 emissions could also be further reduced by:

a) Improving kiln shell insulation: However, this would be limited to saving at most about 0.3GJ/t even if perfect insulating materials could be developed for this application, (and this is not a practical proposition for many reasons).

b) Reducing clinkering times and temperatures, e.g. by means of catalysts: This would indirectly help to reduce kiln shell losses above 900°C, but in practice the energy savings will be only a small fraction of the maximum possible savings (0.3GJ/t) estimated for eliminating completely those kiln shell losses. Thus, they are likely to be insignificant.

c) Reducing the volume of exhaust gases: Typically, the kiln exhaust gases are well over 60% nitrogen, and this carries a lot of heat away with it into the atmosphere, especially since it is not industrially practical to cool these gases to below about 120°C in order to avoid condensation of liquid water. If the fuel could be burned in pure oxygen, a much higher percentage of its heat would be available over the bottleneck temperature of 900°C, and much lower total volumes of exhaust gases would be created. Thus, overall thermal efficiency could, in theory, be greatly improved. However, the practical difficulties are enormous; the materials required for the kiln burner and clinker cooler would have to be capable of resisting the extremely high temperatures developed and the risk of fire in metallic components (since most metals burn fairly readily in pure oxygen). In addition, air fractionation also consumes significant amounts of electrical energy, which would counterbalance some of the energy savings in the kiln system. Nevertheless, partial oxygen enrichment of the combustion air has been and may still be considered as one possible approach to further decreasing cement kiln fuel requirements and thus help further reduce the fuel-derived CO_2 emissions.

d) Reducing the lime saturation factor of the kiln feed: This essentially means making clinkers with lower alite and higher belite contents. Clearly, the higher the belite content, the less limestone will be required. However, on going from a typical modern OPC with about 65% alite to a belite-rich OPC without much alite, the overall reduction in limestone consumption is no more than 8% of the total. Even allowing for the ensuing reductions in burning temperature, the likely maximum CO_2 emissions savings are only of the order of 10%, and this must be balanced against the fact that low-alite OPC clinkers give very slow strength development that would be considered unsatisfactory by most customers.

Looking at the above list, it becomes clear that it is either very difficult or impractical to make major gains in the fuel-efficiency of the cement manufacturing process without changing the clinker chemistry radically to one that is much less limestone-intensive. Essentially, this is because both the RMČ and the FDČ are roughly proportional to the limestone content of the kiln feed. This realisation leads us to look more closely into the question of raw materials

NATURAL RESOURCES SUITABLE FOR MAKING HYDRAULIC CEMENTS

Given the large volumes of Portland cement required world-wide, and its very low unit cost, any alternative would also have to be made from raw materials that are extremely abundant and widespread, since transportation costs would otherwise dominate. If one looks at the relative abundance of minerals on the surface of the Earth, one finds that limestone is one of the most abundant and widely-distributed. It is also a relatively pure and concentrated source of calcium, which itself is one of the most abundant metals in the Earth's crust. But what other minerals might conceivably be available in many locations and in such large quantities? A study made using public US Geological Survey data suggests that, on this basis, only the following materials are reasonable alternative candidate raw materials for hydraulic cements:

1. Clays (e.g. kaolinite): a source of Al and Si
2. Calcium Sulfates (gypsum or anhydrite): a source of S and Ca
3. Iron oxides (iron ores): a source of Fe
4. Silica (quartz, chert, etc.): a source of Si
5. Sodium salts (sodium carbonate; sodium chloride): a source of Na
6. Coal ash (including desulfurization products): a source of Si, Al, and S.

In reality, other sources of the same elements may be available, but any particular source may be much less widespread. For example, bauxite is readily available in some locations but not very widespread, so we have not included it here. The same is true for magnesite, and for phosphate ores. Thus, alternative cements that are very rich in Al, Mg or P seem less likely to be viable on a global basis, although they might be useful for local or specialised applications.

Based on the elements and minerals discussed above, we have calculated the RM$\bar{C}$ values for the known cementing phases that might be produced. These data are given in Table 2. RM$\bar{C}$ values are listed in two different units; the first, more commonly-used form, is on a mass basis, i.e. the mass of raw materials CO_2 liberated per unit mass of the resultant cementing phase. The second unit is relative to the unit volume of the cementing phase, i.e. in grams of CO_2 per ml (= kg/l or tons/m^3) of the pure, anhydrous cementing phase. The interest of this second unit is that it may relate better to the rheology of the resulting fresh concrete, since increased paste volume is generally associated with a more fluid concrete.

It is interesting to note the wide spread in RM$\bar{C}$ values between these different compounds. In particular, there is a factor of about 2.7 (on a mass basis) between alite and calcium sulfoaluminate (3.2 on a volume basis). However, the difference between alite and belite is only a factor of 1.13 on a mass basis, and even less (1.06) by volume. This is another reason why the idea of making belite-rich portland cements seems to be of rather limited interest.

Table 2: Raw Materials CO_2 emitted in the manufacture of "pure" cement compounds

CEMENT COMPOUND	RAW MATERIALS USED	RM CO_2 in g/g	RM CO_2 in g/ml
M (magnesia, periclase)	magnesite	1.092	3.91
C (calcia, quicklime)	limestone	0.785	2.63
C_3S (alite)	limestone + silica	0.578	1.80
β-C_2S (belite)	limestone + silica	0.511	1.70
C_3A (tricalcium aluminate)	limestone + alumina	0.489	1.50
C_4AF (calcium aluminoferrite)	as above + iron oxide	0.362	1.29
NS (sodium metasilicate)	soda + silica	0.361	-
CA (monocalcium aluminate)	limestone + alumina	0.279	0.83
$C_4A_3\bar{S}$ (calcium sulfoaluminate)	as above + anhydrite	0.216	0.56

[Note that some useful cementing phases (e.g. plaster, most pozzolans, and certain calcium phosphates) have essentially zero RMČ values, and are thus not included in this table.]

However, simplistic comparisons such as this can only get us so far, because it is unrealistic to imagine making pure cementing phases, and it is also unreasonable to ignore the benefits that accrue from certain favourable combinations. A good example is that of pozzolans. Since the CO_2 emissions associated with the mining of natural pozzolans are very small, the advantage of pozzolanic cements is clear, even though pozzolans must be activated, e.g. either by lime or by Portland cement clinkers with reasonably high alite contents. Similarly, since calcium sulphates also have negligible associated CO_2 emissions, cementing systems that make use of significant proportions of calcium sulphate have very significant advantages.

Based on the above list of cement phases and potentially viable raw materials, we have concluded that the most promising low-CO_2 alternatives to Portland cement are probably to be chosen from the following major known classes of cementing system:

A) Pozzolan-based cements
B) Calcium sulfoaluminate-based cements
C) Calcium sulphate-based cements

These three classes are analysed in more detail below. Note that we have deliberately omitted blast-furnace slags from this list, mainly because their production is slowly declining; most good sources of such slags are already either being used or are earmarked for use in the manufacture of blended cements; and the technology for this is already very well established.

POZZOLAN-BASED CEMENTS

We define "pozzolans" as reactive aluminosilicates (A_xS_y) which can be either glassy (such as natural volcanic glasses, or man-made glasses such as fly ash or silica fume), or semi-crystalline (such as metakaolin). The best known and best understood class of pozzolan-based cements is that based on lime activation, which includes activation by Portland cement. This is based on the following generalised reaction:

$$A_xS_y + mCH + nH \Rightarrow \text{C-S-H} + \text{C-A-H} \quad (3)$$
$$\text{pozzolan + lime + water} \Rightarrow \text{calcium silicate hydrates + calcium aluminate hydrates}$$

The precise composition of the C-S-H and C-A-H formed varies greatly with the reaction conditions and the presence of other components, (e.g. sulphates, which preferentially enter the C-A-H phases). Mixed C-A-S-H phases can also form, especially in the absence of sulphates.

Reaction (3) is classically referred to as the "pozzolanic reaction", and it consumes lime, pozzolans and water to make new hydrated phases that occupy more volume that the original solid phases. The lime itself can also be obtained from an indirect source, which is usually the hydration of the alite *(and to a lesser extent the belite)* in Portland cement:

$$C_3S\ \mathit{(C_2S)} + 5.3\ \mathit{(4.3)}\ H_2O \Rightarrow C_{1.7}SH_4 + 1.3\ \mathit{(0.3)}\ CH \quad (4)$$

Traditional pozzolanic cements, as developed by the Romans, used lime as the activator; but this reaction is too slow to be of use for most modern concrete applications. The advantage of using alite as the activator is that its hydration is sufficiently rapid to give reasonable early concrete strengths, while the slower pozzolanic reaction can then be used to "mop up" all of the free lime produced. It should be noted that the C-S-H formula used in Eq. (4) is only valid for the system in which an excess of CH is present. In pozzolanic cements, the ultimate C-S-H may well have a lower C/S ratio and this allows even more C-S-H to be formed (i.e. the high C/S C-S-H can contribute some of its own lime to the pozzolanic reaction.)

Considering these reactions in conjunction with the RMC̄ values in Table 2, and making the most conservative assumption that the C-S-H has the highest possible C/S ratio, it can be seen that the overall CO_2-efficiency of portland cements is greatly improved by using pozzolans. For example, the hydration of alite plus silica fume should give the following stoichiometry:

$$C_3S + 0.76\,S + 7.04\,H_2O \Rightarrow 1.76\,C_{1.7}SH_4 \qquad (5)$$

In this case, we produce 1.76 times as much C-S-H per unit of alite reacted. Since we assume silica fume to have zero RMČ, the effective RMČ for C-S-H production from this pozzolanic cement is therefore reduced by 43% relative to the value for pure alite. Essentially, this argument depends on the fact that most people accept C-S-H to be the major binding phase in conventional concrete, and that CH (portlandite) is a less effective and less desirable hydrate. Thus, every mole of CH persisting in concrete can be considered to represent one mole of CO_2 emitted unnecessarily, and it should ideally be replaced by either C-S-H or C-A-H. This argument has its limitations, however, because some excess CH can be useful in certain cases, e.g. to ensure that the pH remains high enough to avoid steel corrosion in reinforced concrete.

In addition to lime activation, certain pozzolans can also be activated by reaction with other bases. A particularly interesting case is the reaction between alumina-rich pozzolans (such as metakaolin) and alkali silicate solutions *[RS_z (aq.)]* :

$$AS_y + RS_z\,(aq.) + nH = RAS_{(y+z)}H_n \qquad (6)$$

This reaction has been referred to by some as "geopolymerization," but the term has no special scientific significance. A recent scientific study has clearly shown the alkali-silicate/metakaolin reaction to be a kind of condensation reaction in which the dissolved alkali silicate cross-links with the reactive aluminosilicate to form a kind of "low temperature alkali aluminosilicate glass" with chemistry close to that of a zeolite (in which each aluminium atom is in a tetrahedral environment and requires an alkali cation to balance its charge) but with an amorphous (glass-like) structure [4]. The products of this reaction are apparently very microporous but have a rather low chemically-bound water content, so it is not really appropriate to refer to it as a hydration reaction. The reaction does have some cementing value, because the alkali silicate is initially present as a solute, and thus the overall solid volume increases significantly during the reaction. However, the kinetics are very different to those typical of cement hydration, and it is more difficult to obtain long working times, which makes it hard to imagine such systems replacing Portland cement in most ordinary applications. Under the assumed name of "geopolymers," a number of potential applications are currently under investigation world-wide, some making use of fly ash as the main raw material; but few good large-scale applications have yet been found and an enormous amount of further research needs to be done to understand these systems to the level where than can be considered truly predictable and reliable. Regarding overall CO_2-efficiency, the activator itself has significant RMC̄ (see typical value in Table 1) and FDC̄ values (alkali silicates are made in furnaces similar to those used to make ordinary "bottle" glass, and the FDČ should therefore be similar to that of ordinary glass.)

The pozzolan may have very low associated CO_2-emissions if it is a natural pozzolan or a by-product, like fly ash. However, metakaolin and related products are artificially heat-treated clays with non-negligible FD$\bar{C}$ values.

A third possibility for the activation of pozzolans is by reaction with magnesium hydroxide (or oxide) following the same type of chemistry as given in Eq. (3) but substituting Mg for Ca. It has been claimed recently that such chemistry is feasible for the manufacture of concrete products [5]. However, it seems unlikely to be of widespread value as a low-CO_2 cementing system due to (a) the very low solubility of magnesium hydroxide, which will render its reaction with pozzolans very slow; (b) the very high RM$\bar{C}$ and FD$\bar{C}$ values associated with the manufacture of MgO from magnesite, and (c) the relative global scarcity of good magnesite deposits. Claims that MgO-based cements carbonate more rapidly than portland cements have yet to be verified publicly, but both lime-based and magnesia-based hydrates will fully carbonate in air (equivalent to the reversing reaction (1) if exposed for a long enough time, giving calcium or magnesium carbonate plus silica and alumina gels, and absorbing all of the RM$\bar{C}$ emitted during their manufacture. The main limitation of this reaction is that it is diffusion-controlled, so that, in large concrete pours, it may take many thousands or even millions of years. It is already well-known that small-to-medium-sized Portland-cement-based concrete products can be hardened by accelerated carbonation under factory conditions, but this is not a practical approach for concrete use on most construction sites, and such carbonated concretes are also not compatible with steel reinforcement.

CALCIUM SULFOALUMINATE-BASED CEMENTS

As shown in Table 1, calcium sulfoaluminate, ($C_4A_3\bar{S}$,) has very low RM$\check{C}$ relative to the other calcium-based cementing phases. This phase hydrates readily to give a combination of C-A-H, C-A-\$-H and AH phases, depending on the availability of sulphate and lime. In terms of hydrate volume creation per unit of "clinker," its reaction with portlandite and gypsum to give ettringite is one of the most effective hydration reactions:

$$\begin{array}{lllllll} C_4A_3\check{S} & + 6CH & + 8C\check{S}H_2 & + 74H_2O & \Rightarrow & 3C_6A\check{S}_3H_{32} & \\ 610.3 & + 444.6 & + 1377.5 & + 1333.2 & \Rightarrow & 3765.9 & (g) \\ 234.1 & + 198.3 & + 596.3 & + 1333.2 & \Rightarrow & 2121.6 & (ml) \end{array} \quad (7)$$

Volume of hydrates per mass of initial dry powder = **0.87 ml/g**
Solid volume ratio (hydrates/dry powder) = **2.06**
Volume of hydrates per mass of $C_4A_3\check{S}$ = **3.48 ml/g**
Volume of hydrates per mass of ($C_4A_3\check{S}$ + CH) = **2.01 ml/g**

Compare this to the complete hydration of a mixture of alite and reactive silica:

$$\begin{array}{llllll} C_3S & + 0.76\,S & + 7.04\,H_2O & \Rightarrow & 1.76\,C_{1.7}SH_4 & \\ 228.3 & + 46.0 & + 127.2 & \Rightarrow & 401.5 & (g) \\ 73.1 & + 20.9 & + 127.2 & \Rightarrow & 200.8 & (ml) \end{array} \quad (8)$$

Volume of hydrates per mass of initial dry powder = **0.73 ml/g**
Solid volume ratio (hydrates/dry powder) = **2.14**
Volume of hydrates per mass of C_3S = **0.88 ml/g**

Both of the above reactions are similarly effective in the creation of hydrate volume from total solid reactants, but reaction (7) is much more efficient in terms of its use of "clinker" phases, because it also combines a lot of gypsum that does not have to be heat-treated in a kiln. This makes it a much more efficient cementing system in terms of total CO_2 emissions.

However, reaction (7) is not the most typical embodiment of calcium sulfoaluminate cement technology, as there are many practical problems with such ettringitic cementing systems, especially the problem of controlling the expansion associated with the reaction. If one wishes to make use of ettringite as a major cementing phase, there is an older technology that still has great interest: that of "supersulphated" slag cements. Such cements, based on ground granulated blast-furnace slag (GGBFS) with fairly large amounts (10-20%) of gypsum or anhydrite, plus a very small amount of lime or Portland cement as a catalyst, can give volume- stable concretes in which ettringite and C-S-H are the main hydrate phases. However, their strength development is slow compared to that of Portland cements, and the concretes are more easily carbonated, sometimes leading to strength loss and corrosion of reinforcing steel. As a result, they have fallen out of favour with concrete users and most of the available GGBFS is instead currently used in blends with Portland cement. Nevertheless, given that GGBFS is an industrial by-product and can thus be considered to have zero associated CO_2 emissions, supersulphated cements still appear to be a promising area for additional research.

Although calcium-sulfoaluminate-based cements are widely used in a number of specialised applications where high early strengths and self-stressing or shrinkage compensation are required, their more general application to concrete is limited to China, where a wide range of $C_4A_3\bar{S}$-based cements have been developed and normalised under the name of the "Third Cement Series" (TCS) [6]. In the Chinese literature, it is stated that these cements, which are typically based on clinkers containing $C_4A_3\bar{S}$, belite and ferrite as their major phases, can be used in a wide variety of applications depending on their different phase compositions and on the amount of gypsum or anhydrite interground to make the final cement. However, as far as we are aware, the TCS technology currently practised in China is based mainly on clinkers with fairly high $C_4A_3\bar{S}$ contents (60-70%), aimed at the pre-stressed concrete sector in which the rapid strength development at moderate curing temperatures, plus self-stressing, are economic advantages. Such cements must be manufactured using bauxite as a major raw material, and this makes them relatively expensive compared to Portland cements. Thus, it is not yet clear whether or not such cements are really capable of offering very significant global CO_2 savings, but further research in this area does seem warranted.

CALCIUM SULFATE-BASED CEMENTS

The hydration of calcium sulphate hemihydrate ("plaster") to give gypsum is probably the best-known and most ancient cementing reaction, and has the advantage of very rapid but controllable setting and hardening rates, good global raw materials availability and very low process energy requirements, although it creates less hydrate volume than reactions (7) or (8).

$$\begin{array}{ccccccl} C\check{S}H_{0.5} & + & 1.5H_2O & \Rightarrow & C\check{S}H_2 & & (9) \\ 145.2 & + & 27.0 & \Rightarrow & 172.2 & (g) & \\ 53.1 & + & 27.0 & \Rightarrow & 74.5 & (ml)\ ; & \end{array}$$

Volume of hydrates per mass of initial dry powder = **0.51 ml/g**
Solid volume ratio (hydrates/dry powder) = **1.40**

The RM$\overline{C}$ is zero and the FD$\overline{C}$ is very low even compared to calcium sulphoaluminate cements. The main disadvantage is the softness and high solubility of gypsum, which results in a low durability of gypsum-based mortars in outdoor exposure. Nevertheless, very dense (low-porosity) gypsum (alabaster) can be fairly durable and really quite strong. So, there may well be some circumstances in which gypsum-based cementing systems could be of interest for replacing Portland cements, especially when the gypsum produced by hydration of plaster or anhydrite can be combined with moderate amounts of relatively inexpensive additives such as calcium aluminates or silicates to help further stabilise it towards softening by water. In some respects, supersulfated slag cements can be considered to be good examples of such systems. It has also been shown that addition of as little as 25% of a suitable pozzolanic cement to gypsum plasters can strengthen and stabilise the material under completely wet conditions [7].

Of course, gypsum-based cements are not well-suited to use with ordinary steel reinforcement since they offer very little corrosion protection. However, they are very compatible with glass fibre reinforcement, which could allow for the production of low-cost, low-energy gypsum-based composites. Thus, such systems appear to be another promising research avenue in the long term quest to reduce the global CO_2 emissions associated with construction.

CONCLUDING REMARKS

Based on the analysis that we have made in this paper, the most promising "low-CO_2" alternative cementing systems currently appear to be those that effectively stabilise hydrated calcium sulphates. This is mainly because natural calcium sulphate sources are relatively abundant, contain no CO_2, and require very little processing energy for their activation. In addition, large volumes of by-product calcium sulphates are now being produced world-wide by gas desulphurisation processes, mostly associated with coal- and oil-fired electric power plants; and such sources are likely to continue to increase very rapidly as more and more societies adopt stringent sulphur dioxide emission controls. It is interesting to note that most SO_2 control technologies are actually net CO_2 emitters because they make either direct or indirect use of limestone as the neutraliser, the overall reaction being:

$$2SO_2 + 2CaCO_3 + O_2 \Rightarrow 2CaSO_4 + 2CO_2 \quad (10)$$

Thus, for each molecule of SO_2 captured, one molecule of CO_2 is emitted. Nevertheless, the environmental benefits of this process would be very evident to anyone living near the plant in question; SO_2 is a toxic gas that causes severe respiratory problems at quite low doses, whereas CO_2 is completely non-toxic, and, indeed, is essential for all life processes.

We have seen that there are a number of alternative cementing systems that could result in overall reductions in CO_2 emissions per unit volume of concrete relative to pure Portland cements, and even, to some extent, relative to Portland cements that have been fully reacted with pozzolanic materials to convert all free lime into calcium silicate and aluminates hydrates. However, we have not yet attempted to make any practical performance comparisons of such alternative cementing systems. This is partly due to lack of space in this article, but is also, and more importantly, due to the fact that we have very limited performance data for the systems that appear to be of most interest for CO_2 emissions reduction per unit volume of concrete. Clearly, if any alternative system is ultimately to have a real impact on global CO_2 emissions related to the construction industry, it will have to have performance and durability characteristics at least as good as the current generation of Portland-based cements, and probably even better since it is likely to be, at least initially, more expensive to the consumer. To establish the performance and durability of alternative cements and concretes to the level required for the introduction of the appropriate new Standards and Construction Codes is a very expensive undertaking, since the number of tests involved is likely to be very large. It will require the full participation and co-operation of industry, government, the scientific community, and members of the general public. It is only by such a concerted effort that our society can hope to bring about the long-term changes necessary to make our built environment truly sustainable.

REFERENCES

1. "Building a Sustainable World," a First Report on Lafarge's Economic, Societal and Environmental Performance, Lafarge, 61 rue des Belles Feuilles, Paris, France, 2001

2. "Energy Conservation Potential in the Cement Industry," Federal Energy Administration, Washington DC, USA, 1975. (Available as NTIS PB 245159)

3. See p126 in "The Chemistry of Cement and Concrete, (3rd Edn,)" by F. M. Lea, Arnold, London, (1970)

4. RAHIER H., MELE B.V., BIESEMANS M., WASTIELS J., and WU X., "Low-Temperature Synthesised Aluminosilicate Glasses" J. Mater. Sci., **31**, 71-79, (1996)

5. PEARCE F., "Green Foundations," New Scientist, 13 July, 2002, pp 39-40

6. ZHANG L., SU M., and WANG Y., "Development and use of Sulfo- and Ferro-Aluminate Cements in China," Adv. Cem. Res. Vol. 11, No15 (1999)

7. BENTUR A., KOVLER K. and GOLDMAN A., "Gypsum of Improved Performance using blends with Portland Cement and Silica Fume," Adv. Cem. Res. **6** [23], 109-116 (1994)

PROPERTIES OF HIGH PERFORMANCE MATERIALS BASED ON CALCIUM SULFOALUMINATE CEMENT

J Ambroise

J Péra

INSA Lyon

France

ABSTRACT. Calcium sulfoaluminate cements have essentially been developed in China in the 1970's. Designed by the CBMA (China Building Materials Academy), they were intended to the manufacturing of self-stress concrete pipes due to their swelling properties. Then, they were applied in other fields: quick setting concrete, fibre-reinforced composites, concrete with high sulfate resistance. This paper presents three innovative utilizations of calcium sulfoaluminate cement : (i) development of concrete with high early strength, (ii) design of self-levelling topping mortar, (iii) glass-fibre reinforced cement (GFRC) composites. A new type of high-performance concrete was designed to reach the following requirements: (a) high workability: slump higher than 200 mm for at least 50 minutes, (b) high-early strength: 40 MPa, 6 hours after its preparation, and higher than 55 MPa after 24 hours. The topping mortar was designed to present a time of workability higher than 30 minutes, a set within 75 minutes and a low drying shrinkage (< 250μm/m). GFRC composites can be demolded four hours after casting and present high ductility and durability after ageing in different weathering conditions.

Keywords: Calcium sulfoaluminate, Rapid-hardening cement, High-early strength, Self-levelling, Curling, Glass-fibre reinforced cement composite, Repair, Workability.

Jean Ambroise is Assistant-Professor at the National Institute of Applied Sciences, Lyon, France, where he carries out research on the design of innovative building materials.

Jean Péra is Professor of Civil Engineering at the National Institute of Applied Sciences, Lyon, France. He leads the research team on building materials.

INTRODUCTION

Calcium sulfoaluminate cements have essentially been developed in China in the 1970's. Designed by the CBMA (China Building Materials Academy), they were intended to the manufacturing of self-stress concrete pipes due to their swelling properties. Sulfoaluminate cements contain the phases belite (C_2S), yeelimite, or tetracalcium trialuminate sulfate ($C_4A_3\bar{S}$), and gypsum ($C\bar{S}H_2$) as their main constituents. They also contain other phases like C_4AF, $C_{12}A_7$, C_3A, and C_6AF_2 [1-3].

When CSA cement hydrates, ettringite ($C_6A\bar{S}H_{32}$) is formed according to the following reactions [4]:

$$C_4A_3\bar{S} + 2C\bar{S}H_2 + 36H\ \bar{S}\ 3C_6A\bar{S}_3H_{32} + 2AH_3,$$ in absence of calcium hydroxide, (1)

$$C_4A_3\bar{S} + 8C\bar{S}H_2 + 6CH + 74H\bar{S} \rightarrow 3C_6A\bar{S}_3H_{32},$$ in presence of calcium hydroxide, (2).

The microstructure of ettringite is strongly dependent on the presence of lime [5]. Ettringite produced by the reaction in Eq. 2 is expansive and this property is exploited in special applications such as shrinkage-resistant and self-stressing cements [6]. Ettringite formed in the absence of lime by the reaction in Equation 1 is non-expansive and generates high early strength in cementitious systems [7].

This last property was exploited to develop a concrete with high early strength (40 MPa after 6 hours of age) and workability maintained for 60 minutes.

The property of limited shrinkage due to the production of expansive ettringite was used to develop self-levelling screed without curling and self-levelling repair mortar.

Glass fibre reinforced cement (GFRC) is an interesting construction material of considerable potential. But its development has been more moderate than expected due to the high alkalinity of Portland cement matrices. Borosilicate E-glass is chemically destroyed in Portland cement and even alkali-resistant glass fibre is subjected to physical attack by calcium hydroxide crystals. Pozzolanic additions like metakaolin or microsilica are used to decrease the calcium hydroxide content and therefore guarantee the durability of GFRC composites [8-9]. Less alkaline non-Portland cement matrices have been developed for reinforcement by glass fibres: high-alumina cement, supersulfated cement and calcium sulfoaluminate cement [10-11]. This paper presents the preliminary results obtained on white sulfoaluminate cement reinforced by AR glass fibres.

CONCRETE WITH HIGH EARLY-AGE STRENGTH

The following materials were used in the study:

- calcium sulfoaluminate cement (CSA) produced by Carrières du Boulonnais, containing 80% calcium sulfoaluminate clinker and 20% phosphogypsum. The composition of calcium sulfoaluminate clinker, assessed by X-ray diffraction and chemical analysis is given in Table 1.
- ordinary Portland cement (OPC), class CEM I 52.5, according to the European Standard EN 197-1. Its Bogue composition is given in Table 1.

- limestone powder (98% $CaCO_3$),
- liquid-based polycarboxylate with a dry matter content of 30%,
- lithium carbonate (Li_2CO_3), which acts as an accelerator of ordinary Portland cement and calcium aluminate cement [12]. It was a pure PROLABO powdered product.
- silico-calcareous river sand (0/5 mm) and coarse aggregate (5/14 mm).

Table 1 Composition of CSA and OPC clinkers (w_t %)

CLINKER	$C_4A_3\bar{S}$	C_3S	C_2S	C_3A	C_4AF	PEROVSKITE
CSA	53.5		21.2		16.3	9.0
OPC		61.6	21.5	7.6	8.4	

In concrete, the cementitious material (CM) was a mixture of 80% CSA and 20%OPC. The introduction of OPC in the blend prevents the loss of strength at later ages [13]. Three CM contents were investigated: 400, 500, and 600 kg/m^3. Powdered limestone was introduced in the mixture to enhance workability. Its content was 7.5% of the CM content. The W/CM was 0.37 and the superplasticizer dosage was 2.5%. The Li_2CO_3 dosage was 0.05%. Table 2 shows the composition and workability of the different mixtures.

Table 2 Mixture proportions of concrete

CONTENT, kg/m^3	MIXTURE		
	1	2	3
CM	400	500	600
Powdered limestone	30	37.5	45
Sand (0/5 mm)	940	840	740
Gravel (5/14 mm)	940	840	740
Water (W/CM = 0.37)	150	185	220
Superplasticizer	10	12.5	15
Li_2CO_3	0.020	0.025	0.030
Time of workability (minutes)	50	100	65

Good workability was maintained for at least 50 minutes. As illustrated in Figure 1, the 6-hr strength, measured on cylinders (ϕ = 110 mm, h = 220mm), was higher than 40 MPa, except for mixture 2. This is certainly due to an extended time of workability (100 minutes instead of 50 or 65 minutes). After 7 days of hydration, high-strength concrete was obtained regardless of the cementitious material content. As shown in Figure 1, the long-term strength development is not hindered and the strength increase between 1 day and 60 days is higher than 30%.

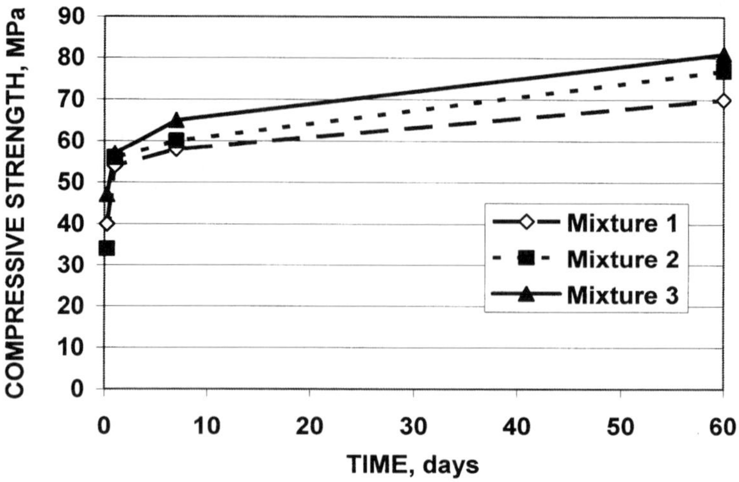

Figure 1 Strength development of concrete

SELF-LEVELLING TOPPING MORTAR

Another sulfoaluminate clinker was used in this study, presenting the following composition:

- $C_4A_3\overline{S}$ = 66%,
- C_2S = 17%,
- Perovskite = 9.9%
- $C_{12}A_7$ = 7.1%.

The topping mortar was designed to fulfil the following requirements

- self-levelling,
- time of workability: 30 minutes,
- final setting within 75 minutes,
- no curling and limited shrinkage,
- thickness: 2 to 30mm,
- 24-h strength: > 35 MPa (measured on 40x40x160mm prisms).

Five different mixtures were studied and are shown in Table 3. Different proportions of phosphogypsum were introduced in the mixtures to study the influence of the gypsum content on the engineering properties of the topping mortar. The phosphogypsum/sulfoaluminate clinker varied from 10/90 to 30/70. In this study, other parameters were kept constant. The water to total cementitious material ratio was 0.42.

The five mixtures presented the required time of workability (30 minutes) and were self-levelling, as pointed out in Figure 2.

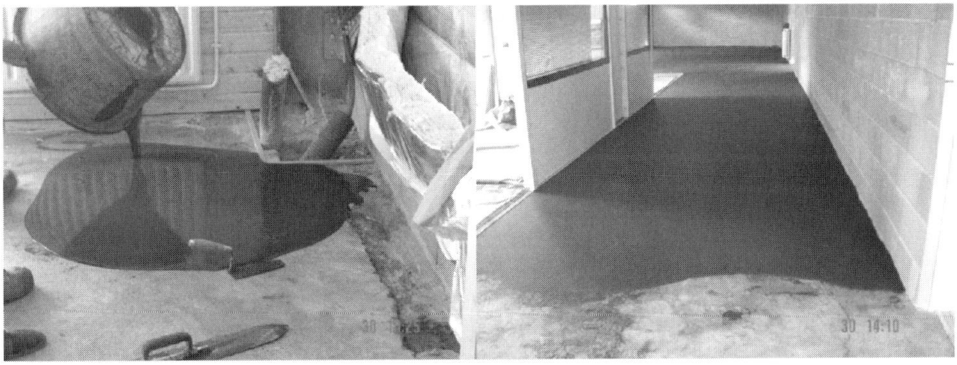

Figure 2 Flowing properties of topping mortar

Table 3 Mixture proportions of topping mortars

CONTENT, kg/m^3	MIXTURE				
	1	2	3	4	5
Sulfoaluminate clinker	810	765	720	675	630
Phosphogypsum	90	135	180	225	270
OPC	32	32	32	32	32
Sand (0/2mm)	900	900	900	900	900
Admixtures	65	65	65	65	65
Water	390	390	390	390	390
Specific gravity	2.01	2.01	1.99	1.97	1.96

As shown in Figure 3, the gypsum content has a slight effect on the compressive strength of mortar. The five mixtures presented a 24-h strength very close to 35 MPa. After 28 days of age, a higher amount of gypsum led to higher strength.

The influence of gypsum content is more effective on length variations of specimens, as shown in Figure 4. Those length changes were measured on mortars cured at 20°C, and 50% RH. Mixtures 1 to 3 started shrinking after one day of curing and reached unacceptable values at 28 days of age (> 250 μm/m). The best compromise between a good strength and a limited shrinkage was obtained by mixture 4, which contained 75% sulfoaluminate clinker and 25% gypsum.

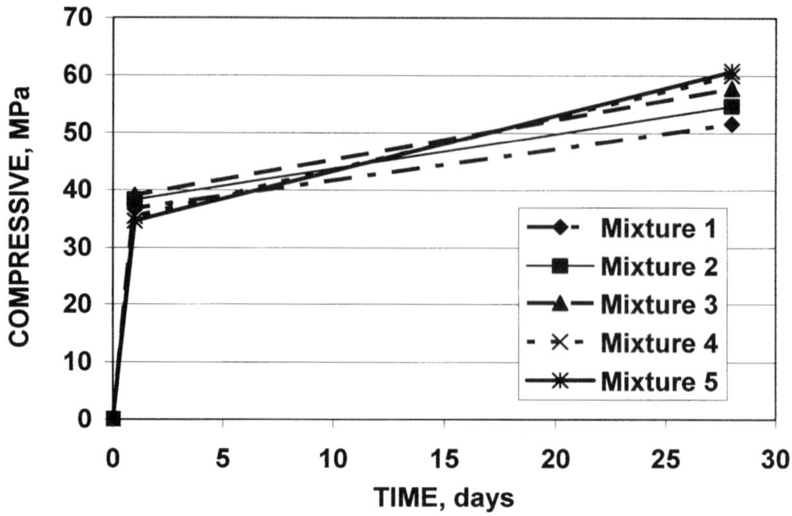

Figure 3 Compressive strength of topping mortar

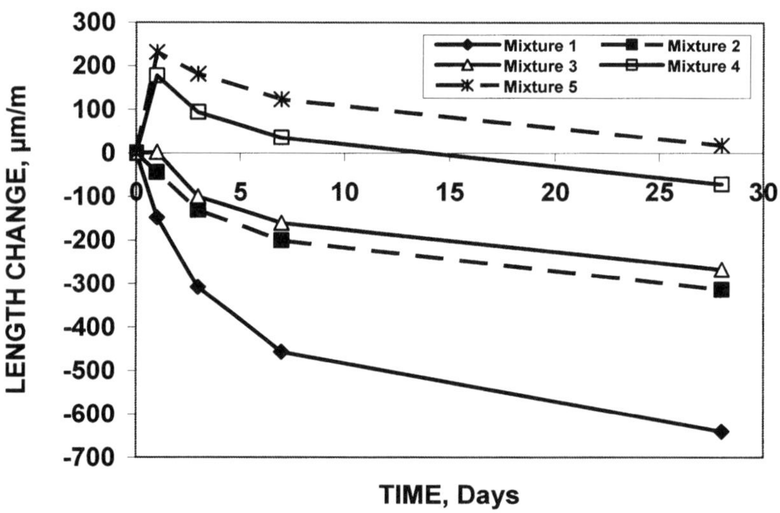

Figure 4. Length changes of mortars cured at 20°C, and 50% RH

DEVELOPMENT OF GLASS FIBRE REINFORCED COMPOSITES USING WHITE SULFOALUMINATE CEMENT

As the hydration of calcium sulfoaluminate cement does not yield any calcium hydroxide, it should be interesting to use such cement in the production of glass fibre reinforced composites. Moreover, in such matrix efflorescences due to the carbonation of calcium hydroxide cannot appear.

A white calcium sulfoaluminate clinker was prepared in a rotary kiln, from a mixture of residual aluminium hydroxide and phosphogypsum. The phase composition of this clinker was as follows:

- belite (C_2S): 11 %,
- yeelimite ($C_4A_3\overline{S}$): 84 %,
- ferrite (C_4AF): 4 %,
- perovskite (C_3FT_2): 1 %.

Calcium sulfoaluminate cement was obtained by intergrinding 70 % of that clinker and 30 % phosphogypsum.

The mixture proportions of the matrix are shown in Table 4

Table 4 Mixture proportions of the matrix.

Cement/Sand	1.00
Water/Cement	0.37
Superplasticizer (% cement)	1.00
Accelerator (% cement)	0.05
Retarder (% cement)	0.27
Water-retaining agent (% cement)	0.06

The sand was a very fine siliceous sand (< 500 μm). All chemical admixtures were powdered products:

- superplasticizer: polycarboxylate,
- accelerator: lithium chloride,
- retarder: sodium tetraborate,
- water-retaining agent: polyethylene-glycol.

Short fibres (l = 11 mm) and mesh derived from the same AR-glass (Vetrotex - Cem-Fil). The properties of fibres are presented in Table 5.

Table 5 Properties of short fibres

Filament diameter	14 μm
Strand tensile strength	1700 MPa
Elastic modulus	72 Gpa
Density	2.68g/cm^3
Strand	38 tex (102 filaments)

Cem-Fil 120/1 had a nominal weight of 120 g/m^2 and was produced using 320 tex and 640 tex Cem-Fil direct rovings. The mesh density was 8 x 8 strands per 10 cm. The short fibre content was 2.83 % of total weight (cement + sand). The mesh was placed at 5 mm from the extreme tensile fiber of the composite. Thin plates (350 mm x 350 mm x 14 mm) were cast and cured at ambient temperature. They were demolded 90 minutes after casting. Specimens of 70 mm in width were cut in each plate and subjected to 3-point loading on a span of 200 mm at a cross-head speed of 1 mm/min. The load deflection was recorded during the test and allowed the calculation of the following parameters:

- limit of proportionality (LOP): stress corresponding to the first crack of the matrix,
- modulus of rupture (MOR): maximum stress recorded,
- several toughness indices.

These indices were those described by the ASTM C 1018 85 [14]: I_5, I_{10}, I_{20} and I_{30}. The residual strength factors $R_{20,10} = 10\ (I_{20} - I_{10})$ and $R_{10\text{-}5} = 20(I_{10} - I_5)$ were also calculated. Tests were performed at different ages: 3 hours, 24 hours, 7 and 28 days. Specimens tested after 3 hours of age were sealed in plastic bags at 20°C until the date of tests.

From the load-deflection curves recorded, equivalent elastic bending stress versus deflection curves were derived. Typical stress-deflection curves are shown in Figures 5. All composites developed ductile behaviour as shown in Figure 6: all indices were higher than the required values and remained constant with time.

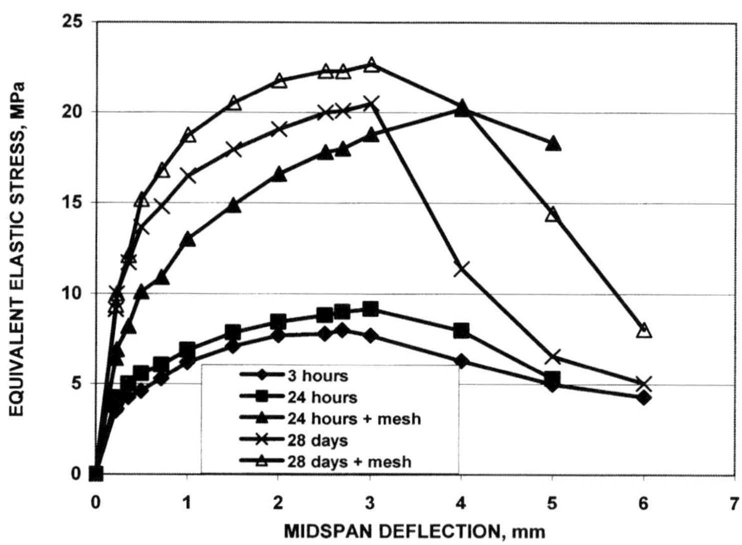

Figure 5 Flexural behaviour of GFRC composites at different ages.

As shown in Figures 5, the presence of mesh increased the performances of composites, and especially at 24 hours of age. The MOR values obtained at different ages are summarized in Table 6. The main remarkable result was the performance obtained after 3 hours of hydration: MOR = 8 MPa, ductile behaviour. This is very important for the precast industry: such performance allows quick demolding and rapid rotation of molds.

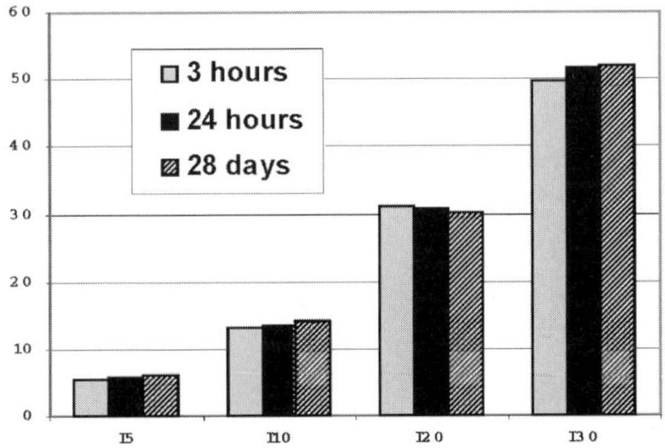

Figure 6. Indices of toughness at different ages for composites reinforced by short fibres

Table 6 MOR values recorded at different ages.

AGE	MOR (MPa)
3 hours	8.0
24 hours	9.3
24 hours with mesh	20.2
28 days	20.5
28 days with mesh	22.8

At 28 days of age, the performances of the composite are similar to those obtained for a white OPC composite, reinforced by 3 % fibres and mesh, containing metakaolin and a polymer. To get such a result, the composite had to be cured at 40°C for 3 days, before being stored at 20°C in sealed plastic bags.

CONCLUSIONS

From this limited research work, it can be concluded that a wide range of innovative building materials can be developed, using sulfoaluminate cement: high early strength concrete, self-levelling topping mortar, and high performance glass-fibre reinforced composites. Those materials are based on three main properties exhibited by calcium sulfoaluminate cement: rapid formation of ettringite, expansive property of ettringite, and low alkalinity of the pore solution of the cementitious matrix. Other materials are under development: self-levelling screed with limited curling, fire-resistant materials, materials including large amounts of industrial by-products.

REFERENCES

1. ZHANG, L., GLASSER, F. P., New concretes based on calcium sulfoaluminate cement, Proc. Int. Conf. Modern Concrete Materials: Binders, Additions and Admixtures, Ed; R. K. Dhir and T. D. Dyer, Thomas Telford, 1999, pp. 261-274.

2. CHATTERJEE, A. K., Special cements, Structure and Performance of Cements, Ed. J. Bensted and P. Barnes, Spon Press, 2002, pp. 226-231.

3. SU, M., WANG, Y., ZHANG, L., LI, D., Preliminary study on the durability of sulfo/ferro-aluminate cements, Proc. 10th Int. Congress on the Chemistry of Cement, Gothenburg, Sweden, June 2-6, 1997, Ed. H. Justnes, Amarkai AB and Congrex, Vol. IV, p.4iv029, 12 p.

4. ODLER, I., Cements containing calcium sulfoaluminate, Special Inorganic Cements, Ed. A. Bentur, S. Mindess, E & FN Spon, 2000, pp. 69-87.

5. MEHTA, P. K., Mechanism of expansion associated with ettringite formation, Cement 0and Concrete Research, Vol. 3, 1973, pp.1-6.

6. SU, M., KURDOWSKI, W., SORRENTINO, F., Development in non-Portland cements, Proc. 9th Int. Congress on the Chemistry of Cement, New Delhi, India, 1992, Vol. 1, pp; 317-354.

7. BERETKA, J., SHERMAN, N., MARROCCOLI, M., POMPO, A., VALENTI, G. L., Effect of composition on the hydration properties of rapid-hardening sulfoaluminate cements, Proc. 10th Int. Congress on the Chemistry of Cement, Gothenburg, Sweden, June 2-6, 1997, Ed. H. Justnes, Amarkai AB and Congrex, Vol. II, p.2ii029, 8p.

8. AMBROISE, J., DEJEAN, J., PERA, J.,Metakaolin blended cements: an efficient way to improve GRC durability and ductility, Proc. 6th biennal Congress of the GRCA, Edinburgh, 1987, pp. 19-27.

9. BENTUR, A., DIAMOND, S., Effects of direct incorporation of microsilica into GFRC composites on retention of mechanical properties after ageing, Proc. Symposium on Durability of Glass-Fibre Reinforced Concrete, Chicago, 1985, p. 337-347.

10. MAJUMDAR, A. J., LAWS, V., Glass fibre reinforced cement, Oxford, BSP Professional Books, 1991, pp. 130-142.

11. SHEN, R., Research and development of some new fibre reinforced cement composites in China, Proc. RILEM Symposium on Developments in Fibre Reinforced Cement and Concrete, Sheffield, 1986, Vol. 2, pp. 32-40.

12. DAMIDOT, D., RETTEL, A., CAPMAS, A., Action of admixtures on Fondu cement, Journal of Advances in Cement Research, 1996, Vol. 8, pp. 111-119; 1997,, Vol. 9, pp; 127-134.

13. JANOTKA, I., KRAJCI, L., RAY, A., MOJUMDAR, S. C., The hydration phase and pore structure formation in the blends of sulfoaluminate-belite cement with portland cement, Cement and Concrete Research, 2003, Vol. 33, N° 4, pp. 489-497.

14. JOHNSON, C. D, Methods of evaluating the performance of fibre-reinforced concrete, Proc. Symposium on Fiber-Reinforced Cementitious Materials, Materials Research Society, Boston, 1990, Vol. 211, pp. 15-24.

STUDY ON SOLIDIFICATION OF INTERMEDIATE LEVEL RADIOACTIVE WASTE IN LARGE POURED ALKALI-ACTIVATED SLAG CEMENT

J Zhou   S Yan

Nanjing University of Technology

China

ABSTRACT. Non α low and intermediate level radioactive waste was solidified in large volume poured alkali-activated slag cement with 80% zeolite and 20% slag in the whole of the waste-solidified material. The solidified form with its load of 13.5% nitrate and water to solid ratio of 0.34 not only possesses excellent liquidity but has 10.3 MPa of compressive strength. In deionized water, the leaching rates of Cs^+, Sr^{2+} in 28 days reached 3.1×10^{-4} and 2.2×10^{-5} $g\cdot cm^{-2}\cdot d^{-1}$ at 90℃ or 5.6×10^{-4} and 3.0×10^{-5} $g\cdot cm^{-2}\cdot d^{-1}$ at 150℃ (MCC-1P), and the leaching percentages were 0.11%, 3.5% and 0.2% or 6.2% and 0.3% respectively. They were not significantly changed in salt solution. It shows that this kind of solidified form has strong ability to retain Cs^+, Sr^{2+} and its other properties also makes it suitable for large volume poured cement.

Keywords: Alkali-activated slag cement, Large volume poured cement, Cement solidification, Non α low, Intermediate level radioactive waste

Zhou Jidong graduated from the University of Science and Technology of China in 1985, now is pursuing a doctorate degree of materials science of cement and concrete with the major in cementitious solidification of wastes.

Yan Sheng graduated from Nanjing Institute of Chemical Technology of China in 1975. He is a professor at Nanjing University of Technology, and as a director at the Institute of Chemical and Building Material Design & Research. He is mainly engaged in the research of advanced materials and new technology. At one time or another, he presided and was involved in many key scientific research issues. He has published many monographs such as the Quality Analysis and Control in Cement Production, Study on Alkali Activated Slag Cement Immobilization of High Radioactive Waste, and Special Low-calcium Fly Ash Cement, and Cement Plant Technology Design.

INTRODUCTION

With the rapid growth of nuclear power industry in China, the accumulated radioactive waste stored in underground single-shell tanks recently potentially poses severe environmental threats In this regard, it is demanded to carefully propose the final disposal of Chinese High Level Liquid Waste (HLLW). After the procedures of TRU elements recovered by TRPO process, ^{90}Sr deleted by extraction, ^{137}Cs removed by absorption, HLLW transforms to Non α Low And Intermediate Level Radioactive Waste (LILW) which can be solidified by cement and be disposed in shallow land [1].

Although cement had been used largely in disposal of LILW recently, research went on in immobilization of HLLW. However, its inherent porous property makes the solidified form retaining Cs^+ more difficult in water [2,3,4]. Alkali-activated slag cement (AASC), a kind of high strength cementitious material, with high content of zeolite added, could be applied as large volume poured cementitious material based on non α low and intermediate level radioactive waste.

MATERIALS AND MOLDING, CURING METHODS

Raw Materials

Blast furnace slag (BFS) was from Nanjing No. 9424 Iron Plant. Its chemical composition is showed as follows.

Table 1 Chemical composition of slag

Oxide	SiO_2	Al_2O_3	CaO	MgO	Fe_2O_3	TiO_2	MnO_2	K_2O+Na_2O
Quality Percentage (%)	34.80	10.45	41.06	10.34	1.50	0.32	0.22	1.31

Its hydraulic modulus K=(wt.%CaO + wt.%MgO + wt.%Al_2O_3)/(wt.% SiO_2 + wt.% MnO_2 + wt.% TiO_2) = 1.75>1.2, and the alkalescence modulus M=(wt.%CaO + wt.%MgO)/(wt.%Al_2O_3 + wt.% SiO_2) = 1.14>1. Its specific surface area was about 700m^2/kg.

Checheng zeolite from North China's Hebei province, which mainly consists of clinoptiolite and mordenite, ground to 1% of 0.080mm sieve residue, was used in the experiment. Its chemical composition is showed in Table 2.

Table 2 Chemical composition of zeolite

Oxide	SiO_2	CaO	MgO	Fe_2O_3	Al_2O_3	K_2O	Na_2O	LOSS	TOTAL
Quality percentage (%)	67.64	3.55	0.36	1.61	13.10	2.55	0.93	10.52	100.26

Activated Agent And Water Reducer

Chemical reagent $Na_2SiO_3 \cdot 9H_2O$(for chemical purification) or water glass with chemical reagent NaOH was used as the alkaline activator, with a dosage of 3 wt % (by Na_2O against slag and zeolite). The water reducer used was a condensation product of β - naphthalenic sulfuric acid and formaldehyde, with a dosage of 0.5 wt %.

Radioactive Waste Liquid

A Non α LILW came from HLLW treated by TRPO, ^{90}Sr extraction and ^{137}Cs absorption processes, whose composition is listed in Table 3.

Table 3 Initial main composition of LILW

COMPONENT	CONCENTRATION (g/L)	COMPONENT	CONCENTRATION
Na	8.53	Sr	0.25 (g/L)
Al	1.73	Cs	0.12 (g/L)
Fe	1.71	HNO_3	$0.5\ mol \cdot L^{-1}$
Cr	0.21	TRU	$9.4\times10^{4}\ Bq \cdot L^{-1}$
Ni	0.86	^{137}Cs	$1.1\times10^{9}Bq \cdot L^{-1}$
K	0.24	^{90}Sr	$4.8\times10^{7}Bq \cdot L^{-1}$
Ba	0.0073	^{99}Tc	$4.3\times10^{6}Bq \cdot L^{-1}$
Mn	0.014		

^{137}Cs and ^{90}Sr were substituted by ^{133}Cs and ^{88}Sr in order to confect simulated LILW in the experiment. The nitrate in LILW was removed by formic acid, a kind of denitratic reducing agent, and LILW was concentrated by heat treatment simultaneously. After neutralized and alkali-activated by alkali, the transformed waste could be mixed with cementitious materials and be solidified directly.

Molding And Curing

Both grinded BFS and zeolite mixed with each other at the required weight ratios were admixed with water, alkali-activating-agent and water reducer respectively, then added to the alkali-activated concentrated waste form. Water to solid ratio was adjusted according to the experimental demand or to the flowability required by large volume poured cement workabilities[5], and waste load was fixed at 13.5% (by the quantity of nitrate in original waste liquid against solid form), the paste was mixed in the blender and filled into molds of 2cm×2cm×2cm cubes or 5cm×5cm cylinders with jolts of 30 times. After set in the standard curing room for nearly 48 hours, the specimens were demolded and cured in saturated humid air at 25℃ for 28 days as packed in plastic bag.

RESULTS AND DISCUSSIONS

Leachabilities

Nuclide leaching properties played an important role on evaluating the properties of the waste solidified forms in terms of safety of immobilization. The leach test was in accordance with the MCC-1 Static Leach Test of Materials Characterization Center of the United States of

America, and the concentrations of Cs^+, Sr^{2+} in leachates were analyzed by Atomic Absorption Spectroscopy equipped with a graphite oven. Table 4 shows the leaching results of various solidified forms with different contents of zeolite abided MCC-1P at 90°C or 150°C in 28 days' period in deionized water or Salt solution. In this experiment, water to solid ratios of all specimens was adjusted to pastes' liquidity index determined by the workability of the large volume poured cement (190mm).

Table 4 Leaching rates of Cs^+ and Sr^{2+} in solidified waste forms in 28 days and their relevant leaching percentage (Cs^+/Sr^{2+})

SPECIMEN	25°C		90°C		150°C	
	Leaching rate ($g\cdot cm^{-2}\cdot d^{-1}$)	Leaching percentage (%)	Leaching rate ($g\cdot cm^{-2}\cdot d^{-1}$)	Leaching percentage (%)	Leaching rate ($g\cdot cm^{-2}\cdot d^{-1}$)	Leaching percentage (%)
Portland cement (in deionized water)	7.6×10^{-4} / 2.7×10^{-4}	8.4/2.1	6.8×10^{-3} / 2.6×10^{-3}	76.8/23.6	7.8×10^{-3} / 2.9×10^{-3}	84.3/26.4
AASC (in deionized water)	3.0×10^{-4} / 2.2×10^{-4}	3.3/1.7	3.7×10^{-3} / 6.4×10^{-4}	41.8/5.8	5.3×10^{-3} / 1.5×10^{-3}	59.8/13.6
AASC with30% zeolite (in deionized water)	1.1×10^{-3} / 2.5×10^{-5}	1.2/0.2	6×10^{-3} / 5.0×10^{-5}	18.1/0.5	2.4×10^{-3} / 1.2×10^{-4}	27.1/1.1
AASCwith 80% zeolite (in deionized water)	4.5×10^{-5} / 1.4×10^{-5}	0.5/0.11	3.1×10^{-4} / 2.2×10^{-5}	3.5/0.2	5.6×10^{-4} / 3.0×10^{-5}	6.3/0.3
AASCwith 80% zeolite (in salt solution)	4.5×10^{-5} / 1.4×10^{-5}	0.5/0.11	2.5×10^{-4} / 3.7×10^{-5}	2.8/0.3	4.9×10^{-4} / 2.1×10^{-5}	5.5/0.2

It can be seen from Table 4 that AASC is well capable to retain Cs^+ and Sr^{2+} compared to Portland cement from 25°C to 150°C attributed to its low porosity, low solubility of hydration products, and with zeolite added to the solidified form, the leaching rates of $Cs^+$and Sr^{2+} especially Cs^+ decreased dramatically demonstrating that the zeolite has priority to immobilize Cs^+ and Sr^{2+} in solidified waste forms. When the zeolite content reached 80wt.% in the cementitious material, $Cs^+$and Sr^{2+}leaching rates of the solidified form after 28 day were only 4.5×10^{-5} and 1.4×10^{-5} $g\cdot cm^{-2}\cdot d^{-1}$ at 25℃, 3.1×10^{-4} and 2.2×10^{-5} $g\cdot cm^{-2}\cdot d^{-1}$ at 90℃, 5.6×10^{-4} and 3.0×10^{-5} $g\cdot cm^{-2}\cdot d^{-1}$ at 150℃,and 0.5% and its leaching percentages being 0.11%, 3.5% and 0.2% or 6.2% and 0.3% in deionized water respectively, and they were not significantly changed in salt solution. All this proves that zeolite has strong ability to suppress leaching.

Microstructure

Two types of solidified waste forms containing 80wt.% zeolite were prepared for analyzing their surface morphology, type 1 solid form (T1) molded at a water to solid ratio as high as 0.34, adjusted to the liquidity index of 190mm. In contrast, type 2 solid form (T2) molded at a pressure of 30kPa, whose water to solid ratio was less than 0.09 leading to most of the slag in this type of solidified forms not being hydrated. Leaching test results show that there is a great distinction between these two types of solid forms, the Cs^+ leaching rate of T1 is 3.1×10^{-4} $g\cdot cm^{-2}\cdot d^{-1}$ at 90℃, but the T2's is 3.2×10^{-3} $g\cdot cm^{-2}\cdot d^{-1}$ which is one order of magnitude higher. This means the hydrated paste also plays an important role in retaining Cs^+.

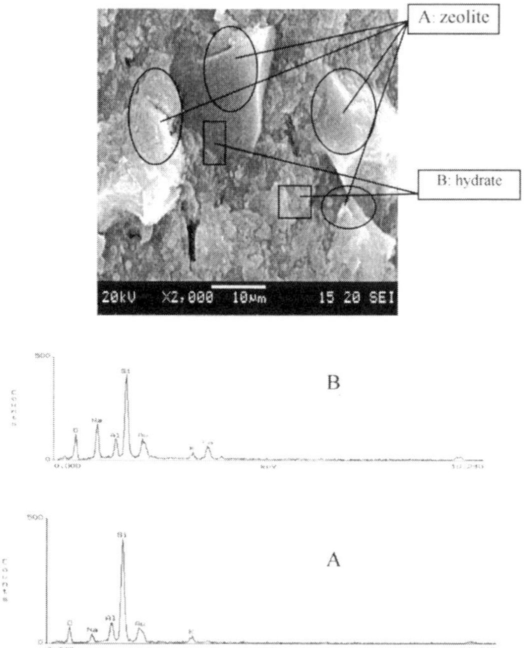

Figure 1 SEM photograph and EDS of type 1 solid form

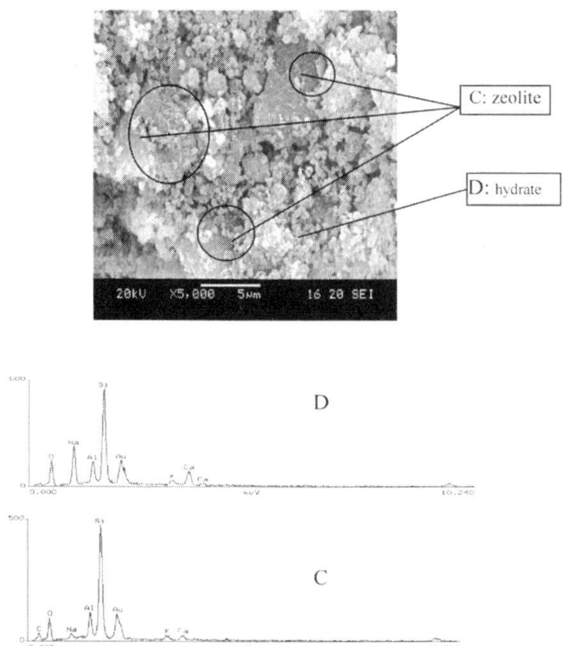

Figure 2 SEM photograph and EDS of type 2 solid form

Figure 1 and Figure 2 shows the surface morphology by scanning electron microscopy. Figure 2 shows the microstructure of a T2 solid form, without connection between slag and zeolite due to the low hydration degree. In contrary, Figure 1 highlights the high of hydration giving the zeolite linked with the hydrated slag that makes T1 solid form so close that the zeolite surface is wrapped with the hydration product of the slag which is able to block the Cs^+ leaching channels in zeolite leading the Cs^+ leaching to the rate of T1 much lower than that of T2.

Other Requirements for the Properties of Large Volume Poured Cement

T1 was chosen as specimen (5cm×5cm cylinder) to be examined for its properties such as compressive strength, anti-striking ability, bleeding, freeze-thaw resistance, setting time, and flow index, time of lost of flowability according to Chinese National regulations of GB14569.1-93, GB1346-77 and GB2419-86 respectively, which are similar to the relevant ASTM regulations. Its compressive strengths as a function of its curing time are shown as follows.

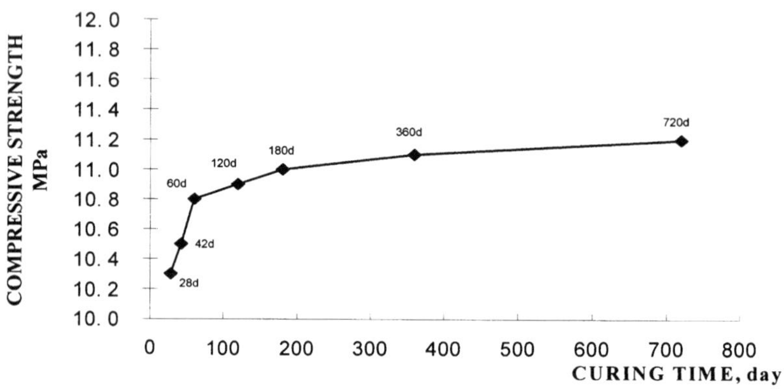

Figure 3 Compressive strength change of T1

It can be drawn from Figure 3 that the compressive strength increased with the hydration time elapsed which could increase sharply within the first 60 days. The compressive strength at 28d is 10.3 MPa, with the 60d's being 10.8 MPa, and also enhanced further in demonstrating that there is no case of shrinkage of strength for this type of waste solidified form. So the conclusion can be given in that T1 poses character of long stability. Table 5 lists the main properties from the test results of the T1 solid form.

Results show that as far as the sample which contented with 80wt.% zeolite in solidified form as well as with the 0.5wt.% water reducer could meet the requirement of large volume poured cement.

Table 5 Main properties of the T1 solid form

ITEMS	PROPERTIES
Load Of Waste (%)	13.5
28d Sr Leaching Rate ($g \cdot cm^{-2} \cdot d^{-1}$)	2.2×10^{-5}
28d Cs Leaching Rate ($g \cdot cm^{-2} \cdot d^{-1}$)	3.1×10^{-4}
28d Compressing Strengthen (MPa)	10.3
Anti-Striking	No breakage
Dissociating Liquid	Without
Anti-Soaking	Good
Anti-Freezing And Melting (MPa)	>90%(9.6 MPa)
Liquidity Index (mm)	190mm
Time Of Lost Flowability (min)	180
Time Of Early Coagulation (hr)	32
Time Of Final Coagulation (hr)	44

CONCLUSIONS

(1) Containing 80wt.% zeolite, the solidified form could be loaded with 13.5wt.% of radioactive waste (by nitrate in the original liquid) possessing 10.3 MPa compressive strength, without bleeding on its surface, strong anti-striking ability and freeze-thaw resistance.

(2) Its molding technology and formula meet the requirement of large volume poured cement in cement paste liquidity index, time of lost flowability, early and final setting time which was tested to be 190mm, 180min, 32hr and 44hr respectively.

(3) In deionized water, $Cs^+$and leaching rates of the solidified form at 28 day were only 4.5×10^{-5} and 1.4×10^{-5} $g \cdot cm^{-2} \cdot d^{-1}$ at 25°C, 3.1×10^{-4} and 2.2×10^{-5} $g \cdot cm^{-2} \cdot d^{-1}$ at 90°C, 5.6×10^{-4} and 3.0×10^{-5} $g \cdot cm^{-2} \cdot d^{-1}$ at 150°C, and its leaching percentages being 0.5% and 0.11%, 3.5% and 0.2% or 6.2% and 0.3% respectively, and they did not change significantly in salt solution. All evidences prove this kind of solidified form has strong anti-leaching ability.

REFERENCES

1. SONG C, JIN G, WANG J, ZHU Y., Treatment of Chinese HLLW by Partitioning, Tsinghua Science and Technology, Vol. 1, 1996, pp 81-85

2. WU X, YAN S, SHEN X,TANG M., Alkali-activated slag cement based radioactive waste forms. Cement Concrete Research, 1991; pp 16 - 20

3. YAN S, SHEN X, WU X, TANG M, YANG L., The Properties Of Alkali-Activated Slag Cement Based High-Level Waste Forms. 3rd Beijing Int. Symp. Cem. Concr., Vol 2, Beijing, China,1993, pp 1054-1059

4. SHEN X, YAN S, WU XN, TANG M, YANG L., Immobilization of Simulated High-Level Waste into Waste Form. Cem. Concr. Res, 1994, pp 133-138

5. HO W., An Evaluation of Waste Disposal by Shale Fracturing . ORNL-TM-5209, Oak Ridge , Tennessee, 1976

RAPID HARDENING COMPOUND CEMENTITIOUS MATERIAL

Z Wang S Cui

Beijing University of Tcchnology

L Zhang

China Building Materials Academy

China

ABSTRACT. The compound system of ordinary Portland cement and sulpho-aluminate cement shows some special properties which are imperative for rush repair work. The setting time of the compound system is adjustable from 10 minutes to hours by the combined use of accelerator and retarder simultaneously, and the time between initial and final setting is very short. The strength develops rapidly after setting, compressive strength at 30 minutes can reach as high as 8~10 MPa. The final strength at 28 days is over 50 MPa, and continue develops afterwards. The influences of temperature, water solid ratio and other chemical admixtures on the properties the compound cementitious material was also examined and discussed. The strength and free expansion rate of concrete made of the cementitious material were measured. The mortar made from the cementitious material showed good corrosion resistance to erosive solution and good frost resistance. By means of XRD analysis and porosity measurement, the mechanism of high early strength was discussed. The compound cementitious material has been used in several rush repair engineering.

Keywords: High early strength, Sulfo-aluminate, Compound cement, Repairs, Ettringite

Z Wang is a Professor in the College of Material Science and Engineering at the Beijing University of Technology, China.

S Cui is a research worker in the College of Material Science and Engineering at the Beijing University of Technology, China.

L Zhang is a research worker at the China Building Materials Academy.

INTRODUCTION

In many cases the repairing and consolidation of old concrete structures must be carried out while not affecting their routine service. Road and airport runway, bridge, railway tunnel and railway substrate are usually required to be repaired while in service, so the repair work must be finished in a short interval and the strength gain of the repair materials must be quick enough to withstand the load. Presently there are about 2 thousand kilometers of railway tunnels in China. However, because of cracks, corrosion and deformation of concrete lining etc., a lot of tunnels have exhibited percolating phenomena, which result in dropping of lining, moving of roadbed, sinking of railway etc. The railway substrate in the 14 km long Dayao Mountain Tunnel in the Jingguang line, the longest tunnel in Asia, was severely damaged and became hollow. The speed of the train was strictly limited to 60 km/h when passing through the tunnel. But the repair of the railway substrate faces a lot of limitations. The Jingguang line is the busiest railway line of China, over 90 matches of trains running on this line everyday, and the longest internal (skylight time) between the consecutive two trains is only 70 minutes. So the grout operation must be finished in 30 minutes and the strength of the grout material must reach as high as 8MPa in the following 30minutes, and the strength at 28 day is required to be not below 50MPa. Furthermore, the grout material should possess some anti-washout ability and good groutability as well as be environmentally friendly, for the construction site is half sealed. At first, an organic grouting material of polyurethane from Denmark was used to reinforce the tunnel, but this material was finally rejected for its high cost and environmental concerns. So, on the basis of a systematic survey, a cementitious grout material with high 1 hour strength and good self-leveling ability was proposed and developed to meet the needs of urgent repairing works in railway substrate engineering and for rapid consolidation works.

EXPERIMENTAL DETAILS

Portland cement (PC) from Jidong Cement Co. Ltd., and Sulfo-aluminate cement (CAS) made by Tangshan Polar Bear Special Cement Co., were used as basic components, whose chemical compositions are listed in Table 1. Citric acid, boric acid and lithium salt, industrial grade, were used as admixtures to adjust the setting time and strength development rate. Test methods were referred to the Chinese National Standard GB8076-1997 Concrete Admixtures, GB/T 8077-2000 Test Method for Uniformity of Concrete Admixtures, and industrial specification JC476-2001 Concrete Expansive Agent.

Table1 Chemical compositions of PC and CSA clinkers

Type	LOI	SiO_2	Al_2O_3	Fe_2O_3	CaO	MgO	SO_3	TiO_3	total
PC	1.80	21.48	5.20	3.26	64.04	1.47	0.30		97.55
CAS	1.41	8.65	31.80	3.08	41.69	3.76	7.32	1.39	99.10

RESULTS AND DISCUSSIONS

Formulations of the compound cementitious material

The compound cementitious material basically consisted of PC, CSA cement, an expansive agent and other supplementary elements. Its typical formulations are shown in Table 2. The proportion of different components affects the setting time and strengths.

Table 2 Compositions of compound cementitious material

CSA	PC	UEA	Admixtures	SETTING TIME, min		COMPRESSIVE STRENGTH, MPa	
				Initial	Final	30min	60min
81	15	4	3.3	5	6	17.7	18.6
66	30	4	2.3	9	11	18.6	20.5
51	45	4	2.8	11	14	16.3	19.1

Fluidity, setting time and workability of cementitious pastes

The flow and fluidity maintaining ability with the elapse of time of the paste determine its groutability. Work time means the maximum allowable time from the point of mixing with water to the beginning of initial setting of the grout paste. At room temperature the flow and its time dependent property change with the water to solid ratio (w/s). The setting time of cementitous paste prolongs with the increase of w/s. But the time gap between initial and final setting is hardly affected by the variation of w/s and temperature, as shown in Table 3. Tests were carried out under the condition where environmental temperature was 24° C and the mixing water temperature was 19° C, Results indicate that the compound cement paste has very good fluidity and filling capacity.

Table 3 Properties of cementitious pastes at different water solid ratio

ITEMS w/s	FLOW, mm				OPERATING TIME, min	SETTING TIME, min	
	3min	5min	8min	10min		initial	final
0.32	160	160	200	60	9	10	12
0.35	200	200	240	200	10	12	14
0.37	205	205	270	205	11	13	15
0.40	235	235	300	265	11	13	15

30–minute-strength and strength development

After mixing with water, the cement paste was poured into a 4×4×16cm mould in 5 minutes, and cured to specified age. For the 30 minute strength test, specimens were demoulded in 25-28 minutes after adding water. Strength develops very quickly after setting, and the 30 minute compressive strength can reach as high as 8~10MPa, in the case of keeping the cement paste in flowable state in the first 12 minutes, see Fig. 1 and Fig. 2. The compressive strength at 1 day is in the range of 35~45MPa, and strength developing steadily afterwards. 3 days strength of specimens (w/c<0.45) exceeds 45MPa and strength at 28 days exceeds 50MPa.

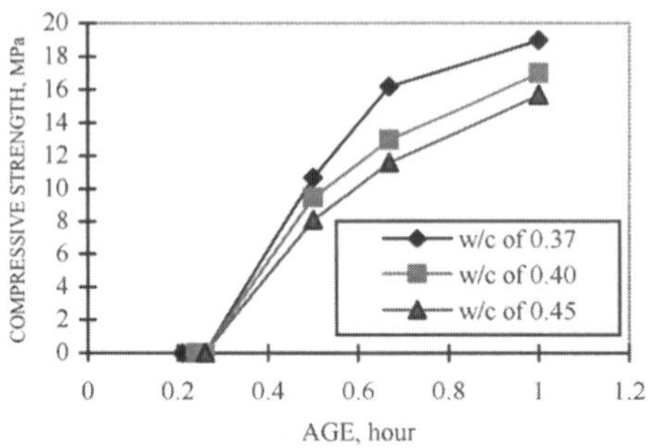

Figure 1 Compressive strength development at early ages

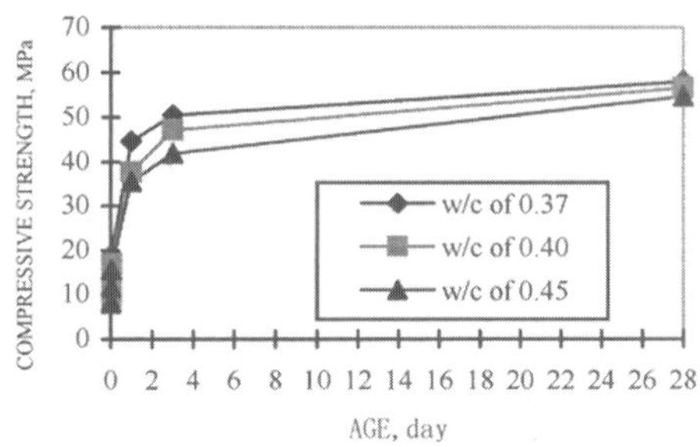

Figure 2 Relation between compressive strength and age under different w/s ratio

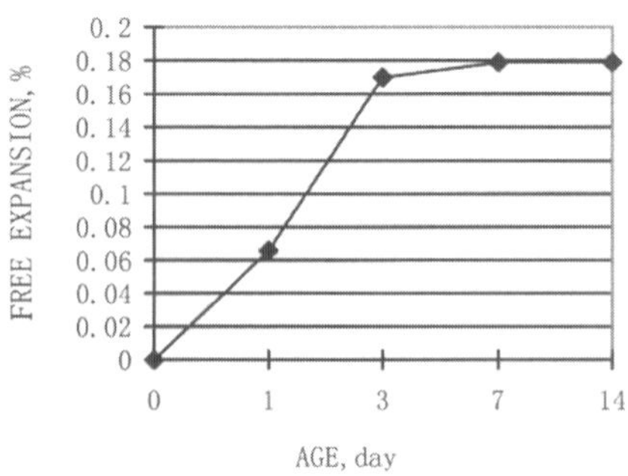

Figure 3 Free expansion rate vs age

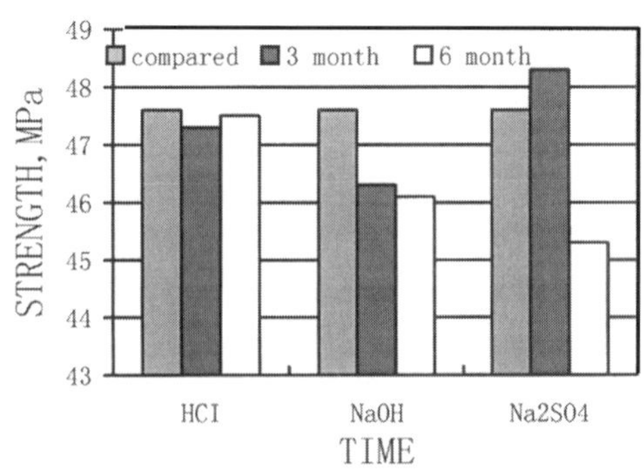

Figure 4 Strength variation after subjected to corrosive media

Free expansion of hardened grout paste

Ordinary cement paste as grout material tends to shrink during the hardening process, thus leads to cracks (including secondary cracks), which enable the passage of water and other corrosive media. To avoid the happening of shrinkage, UEA expansive agent was employed in the cement-based grout material, which would form a certain amount of ettringite during the hydration process and compensates the contraction tendency. The free expansion rate of hardened grout paste made at w/c 0.4 was measured at different ages. The Prismatic specimens (25×25×275mm) were demoulded at 1 hour, and their initial lengths were measured. Test results are shown in Fig. 3. Most of the expansion occurred in the first 3 days, and the expansion process become stable in 7 days.

Table 4 Curing conditions and rate of expansion of concrete

RESTRICTION STATE	CURING CONDITION	RATE OF EXPANSION ($\times 10^{-4}$)		
		3d	7d	14d
Free	Water	0.4	1.24	1.53
Free	Fog	0.25	1.08	1.46
Limiting	Water	0.12	0.90	1.29

The expansion rate of concrete made with compound cementitious binder is shown in Table 4. Mix proportioning of concrete was as follows: 1: 2.5 : 2.5 : 0.45. The result of the test has ascertained the micro-expansion ability, which is important for concrete to improve capacity of anti-cracking and compactness.

Corrosion resistance

After being cured in standard conditions for 28 days, the hardened compound cement pastes were immersed in 5% NaOH solution, saturated Na_2SO_4 solution, and 5% HCI solution for 3 months and 6 months, then their appearance and strength loss were examined. The results are shown in Fig. 4. The specimens subjected to HCI solutions showed apparent erosion in the surface, no change was observed for the other two specimens. Strength losses of all the specimens were not obvious that showed that the hardened grout pastes were corrosion resistant.

Mechanism of ultra-high early strength

The results of x-ray diffraction test are shown in Figures 5-8. We can see that ettringite formed at 30min, at the same time, there was a lot of non-hydrating gypsum, calcium sulfoaluminate, a little tricalcium silicate and some calcium hydroxide formed during the early hydration stage. After 5h, the quantity of ettringite still increases, but calcium sulfoaluminate starts to decrease. At the third day, calcium sulfoaluminate continues lessening and the amount of gypsum also decreases slightly, but there is a little secondary ettringite. Meanwhile, there are other hydrated substances. However, the main products are ettringite, secondary ettringite and residual non-hydrating gypsum and Alite etc.

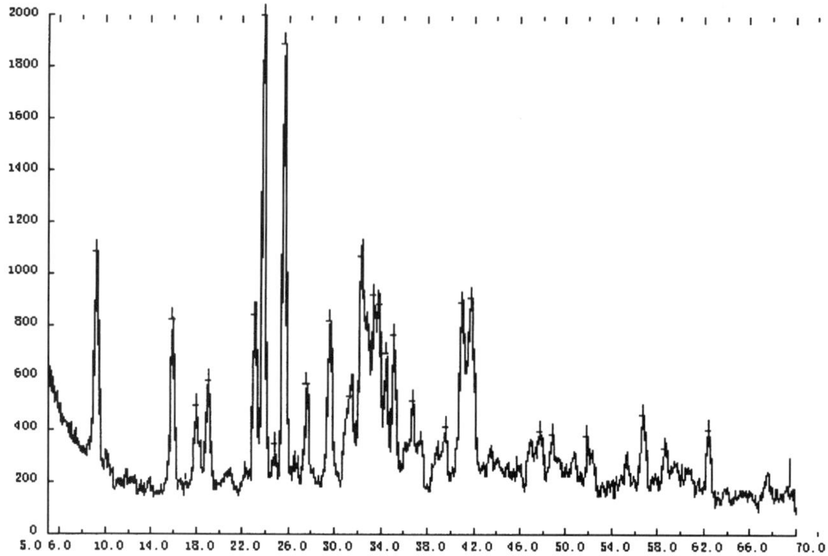

Figure 5 X-ray diffraction pattern of compound cement paste at 30 minute

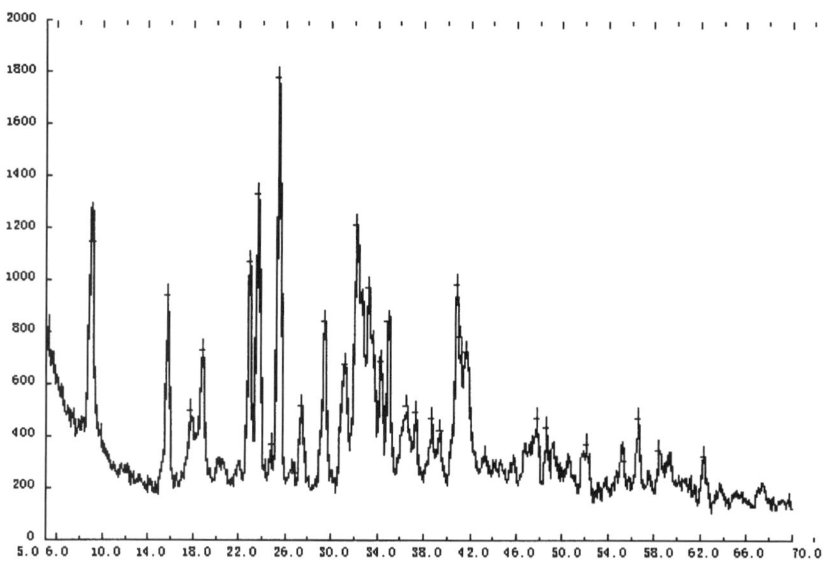

Figure 6 X-ray diffraction pattern of compound cement paste at 60 minute

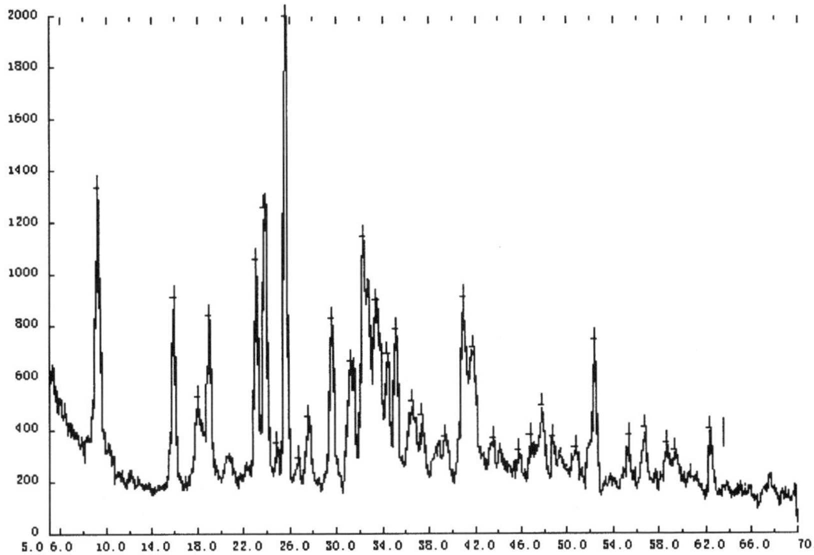

Figure 7 X-ray diffraction pattern of compound cement paste at 1 day

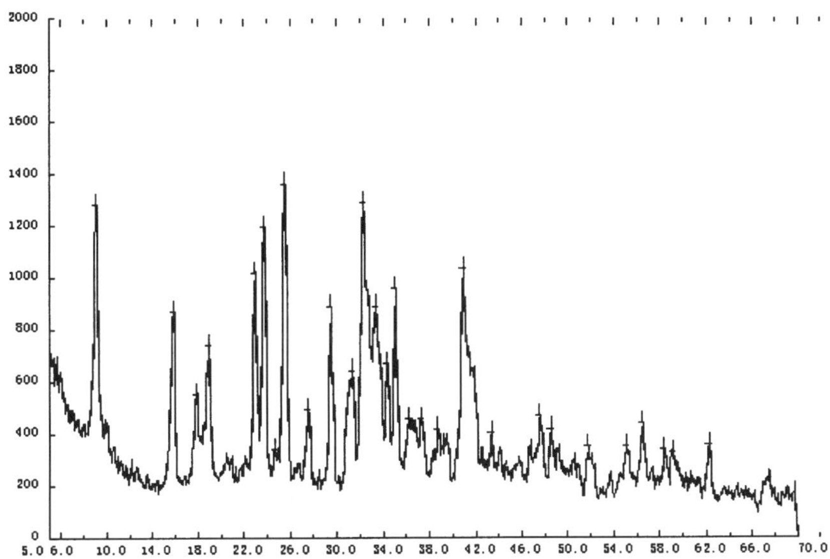

Figure 8 X-ray diffraction pattern of compound cement paste at 3 day

From the pore size distribution, we know that compound cement paste has developed a dense structure after 2h, which has a total mercury-intruded volume of 0.2854ml/g. The dimension of capillary pores, apparent diameter of pores and even pore diameter are respectively 815Å, 69Å and 254Å. Most pores range between 80 and 3000Å. Three pore peaks are observed in the range of 1000 to 2000Å, 400 to 500Å and 100 to 200Å. After 1day, total mercury-intruded pore volumes decreased to 0.2368ml/g, and the dimensions of pores (578Å) and porosity (33.7%) got smaller, but apparent diameters (103Å) and even pore-diameter (266Å) values became slightly larger. The number of pores with diameters of 400 to 500Å increases; total mercury-intruded pore volumes decrease to 0.22ml/g at 7 day, dimensions of pore, apparent diameter, even pore-diameter and porosity decrease to 407Å, 88Å, 208Å and 30.7% respectively. Pores in the range of 1000Å to 2000Å in diameter drop remarkably, while micro pores with the diameter of 400Å to 500Å and 100Å to 200Å increase notably. Pore size mainly ranges between 50 and 1500Å.

CONCLUDING REMARKS

1. The compound cementitious material can meet the demands of good fluidity in 10~12 minutes, short time gap between initial and final setting, and 9~11minutes placing time.

2. The 30 minute and 1 hour compressive strength of hardened cement pastes can reach as high as 8~10MPa and 15~19MPa, respectively, and the long term strength developed steadily. Strength at 28 day was over 50 MPa.

3. The hardened grout paste showed good micro expansive property. Most of the free expansion happened in the first 3 days and expansion became steady afterwards.

4. Adequate amount of ettringite and gels had formed at 30 minute after adding mixing water, which are the source of high one hour strength.

5. The compound cementitous material is suitable for rush repair work as waterproof grout, and for urgent reinforcing work. Satisfactory results have been obtained from practical repair engineering.

p

REFERENCES

1. ZHANG, L. J., WANG, Z. W., Development and uses of GRM ultra high early strength self-leveling cement-based grouting material (Appraisal documents), December, 2000, Beijing (in Chinese), p6

2. WANG, Y.M., SU, M.Z., Sulphoaluminate Cement, Publication of Beijing University of Technology, 1st edition, Beijing , December 1999, pp149

THEME SIX:

EXPLOITING SCIENCE

Keynote Paper

THE PRODUCTION AND USE OF CEMENT IN A SUSTAINABLE CONTEXT

W van Loo
CEMBUREAU
Belgium

ABSTRACT. Cement is an indispensable constituent for all sectors of the building industry, from residential building to large infrastructural projects. It is therefore produced and used in large quantities, about 175 million tonnes in the EU and 1.75 billion tonnes worldwide. Consequently huge amounts of materials and energy are needed. CEMBUREAU – The European Cement Association – plays a co-ordination role in many of those initiatives. This paper will briefly deal with the following issues (i) substitution of fossil fuels with alternative fuels, (ii) European standardisation and use of alternative materials (iii) development of Best Available Techniques for the abatement of emissions (iv) Reduction of greenhouse gas emissions (v) fire safety of concrete structures (vi) life cycle assessment of concrete

Keywords: Cement, Sustainability, Alternative fuels, Emissions, Fire safety

Willem van Loo is a Technical Director with CEMBUREAU, the representative organisation of the cement industry in Europe.

INTRODUCTION

Cement is an indispensable constituent for all sectors of the building industry, from residential building to large infrastructural projects. It is therefore produced and used in large quantities, about 175 million tonnes in the EU and 1.75 billion tonnes worldwide. Consequently huge amounts of materials and energy are needed. - The cement industry has taken many initiatives to contribute positively to sustainable development. CEMBUREAU – The European Cement Association – plays a co-ordination role in many of those initiatives. This paper will briefly deal with the following issues:

- Substitution of fossil fuels with alternative fuels
- European standardisation and use of alternative materials
- Development of Best Available Techniques for the abatement of emissions
- Reduction of greenhouse gas emissions
- Fire safety of concrete structures
- Life Cycle Assessment of concrete

THE USE OF ALTERNATIVE FUELS IN THE CEMENT INDUSTRY

The ambition that drives the cement industry to promote the use of alternative fuels and raw materials has two facets. First, there is of course the saving of costs by substituting fossil fuels with alternative fuels. Waste is usually cheaper than primary fossil fuels; in certain cases, alternative fuels may even be a negative cost item. Such cost varies of course with each type of waste and each set of local conditions. The saving that can thus be obtained is important to maintain the sustained development of the cement industry in Europe where energy prices are among the highest in the world. But, this economic benefit is not the end of the story. Far from it. In the context of Corporate Social Responsibility, it is the commitment of the cement industry to save natural resources, both minerals and fuels by substituting these with alternative raw materials and fuels, being mostly by-products and wastes from other processes. In the EU the degree of substitution has reached the level of 12% on average, but varies widely from country to country, Table 1.

Table 1 Substitution Rate Fuels

EU	12%
Austria	29%
Belgium	30%
Denmark	4%
Finland	3%
France	27%
Germany	30%
Ireland	0%
Netherlands	72%
Poland	1%
Portugal	1%
Switzerland	31%
United Kingdom	6%

In order to further promote the use of such alternative fuels and materials within the cement industry, CEMBUREAU has defined an Action Plan for the Use of Waste describing a long-term programme with a broad range of actions. Among the 24 actions of the Plan, three clusters of activities have been defined as high priority actions to be dealt with. Firstly, "Communication" is of major importance when relating with various stakeholders in the public domain about the use of waste, regarding new initiatives, operational aspects and complaints. Secondly, the various aspects related to the cement making process will be considered, i.e. the preparation of detailed statistics (Table 2) and other data, quantitative and qualitative, with respect to all relevant aspects of existing waste streams and technical limitations linked to the use of waste in the cement process. These statistics and data are aimed for an internal use within the cement industry. And thirdly, reference documents, for external use, will be prepared related to the process emissions (stack) and the product (leaching of concrete) comparing the situation with and without the use of alternative raw materials and fuels. (Figure 1).

Table 2 CEMBUREAU statistics – Energy, alternative fuels 2002

	EUROPEAN UNION	CEMBUREAU
Total Energy	14,780 Tcal	16,557 Tcal
	% Alternative Used	
Used Tyres	20.6	19.9
Used Oil	23.2	24.1
Plastics	45.	4.8
RDF	0.0	0.4
Sewage Sludge	0.8	1.2
Solvents	10.0	9.8
Other Hazrdous Materials	10.4	10.1
INdustial Wastes	10.3	9.2
Others incl Animal Meals	20.2	20.5

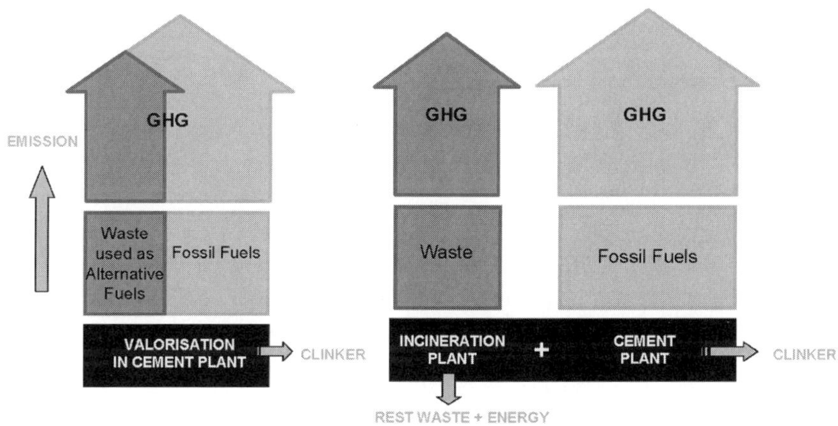

Figure 1 Energy recovery

EUROPEAN STANDARDISATION AND THE USE OF ALTERNATIVE MATERIALS IN CEMENT

By the end of the year 2002, about 96% of all cement produced in the 15 EU Member States was CE-marked. This can be considered a major success for European Standardisation after the Cement Standards EN 197-1 and EN 197-2 were issued as the first harmonised standards under the Construction Products Directive. National standards were withdrawn by 1 April 2002. This standard describes the composition and the requirements of the 27 different cement types in Europe (Table 3). This means that nowadays already 20% of clinker, the major cement constituent, has been replaced with alternative materials like blastfurnace slag, fly ash and limestone, Table 4.

Table 3 The 27 products in the family of common cements, EN 197-1

Main types	Notation of the 27 products (types of common cement)		Composition (percentage by mass [a])										
			Main constituents										Minor additional constituents
			Clinker	Blast-furnace slag	Silica fume	Pozzolana		Fly ash		Burnt shale	Limestone		
						natural	natural calcined	siliceous	calca-reous				
			K	S	D [b]	P	Q	V	W	T	L	LL	
CEM I	Portland cement	CEM I	95-100	–	–	–	–	–	–	–	–	–	0-5
CEM II	Portland-slag cement	CEM II/A-S	80-94	6-20	–	–	–	–	–	–	–	–	0-5
		CEM II/B-S	65-79	21-35	–	–	–	–	–	–	–	–	0-5
	Portland-silica fume cement	CEM II/A-D	90-94	–	6-10	–	–	–	–	–	–	–	0-5
	Portland-pozzolana cement	CEM II/A-P	80-94	–	–	6-20	–	–	–	–	–	–	0-5
		CEM II/B-P	65-79	–	–	21-35	–	–	–	–	–	–	0-5
		CEM II/A-Q	80-94	–	–	–	6-20	–	–	–	–	–	0-5
		CEM II/B-Q	65-79	–	–	–	21-35	–	–	–	–	–	0-5
	Portland-fly ash cement	CEM II/A-V	80-94	–	–	–	–	6-20	–	–	–	–	0-5
		CEM II/B-V	65-79	–	–	–	–	21-35	–	–	–	–	0-5
		CEM II/A-W	80-94	–	–	–	–	–	6-20	–	–	–	0-5
		CEM II/B-W	65-79	–	–	–	–	–	21-35	–	–	–	0-5
	Portland-burnt shale cement	CEM II/A-T	80-94	–	–	–	–	–	–	6-20	–	–	0-5
		CEM II/B-T	65-79	–	–	–	–	–	–	21-35	–	–	0-5
	Portland-limestone cement	CEM II/A-L	80-94	–	–	–	–	–	–	–	6-20	–	0-5
		CEM II/B-L	65-79	–	–	–	–	–	–	–	21-35	–	0-5
		CEM II/A-LL	80-94	–	–	–	–	–	–	–	–	6-20	0-5
		CEM II/B-LL	65-79	–	–	–	–	–	–	–	–	21-35	0-5
	Portland-composite cement [c]	CEM II/A-M	80-94	<---------- 6-20 ---------->									0-5
		CEM II/B-M	65-79	<---------- 21-35 ---------->									0-5
CEM III	Blastfurnace cement	CEM III/A	35-64	36-65	–	–	–	–	–	–	–	–	0-5
		CEM III/B	20-34	66-80	–	–	–	–	–	–	–	–	0-5
		CEM III/C	5-19	81-95	–	–	–	–	–	–	–	–	0-5
CEM IV	Pozzolanic cement [c]	CEM IV/A	65-89	–	<------ 11-35 ------>				–	–	–	0-5
		CEM IV/B	45-64	–	<------ 36-55 ------>				–	–	–	0-5
CEM V	Composite cement [c]	CEM V/A	40-64	18-30	–	<---- 18-30 ---->		–	–	–	–	0-5
		CEM V/B	20-38	31-50	–	<---- 31-50 ---->		–	–	–	–	0-5

a The values in the table refer to the sum of the main and minor additional constituents.
b The proportion of silica fume is limited to 10 %.
c In Portland-composite cements CEM II/A-M and CEM II/B-M, in pozzolanic cements CEM IV/A and CEM IV/B and in composite cements CEM V/A and CEM V/B the main constituents other than clinker shall be declared by designation of the cement (for example see clause 8).

Table 4 Use of cementitious materials

CEMENT TYPE (EN197-1)	CLINKER CONTENT (%)	MARKET SHARE (2000) Tonnes/yr (x1000)	%
CEM I	95-100	80118	34.2
CEM II	65-95	121.815	51.9
CEM III	5-64	12.885	5.5
CEM IV	45-89	11740	5.0
CEM V	20-64	7.970	3.4
TOTAL		234.528	100

This major step having been taken, the industry will complete the process by standardising special cements (low heat, sulphate resisting, low early strength, calcium aluminate and masonry cement).

But new challenges lie ahead. The initial ideas concerning the second generation of standards have been recently developed in CEN/TC51 (*"Comité Européen de Normalisation"*). It is the firm view of CEMBUREAU Members that the relevant environmental, health and safety aspects of the Cement Standard should be incorporated by the time the revision of EN 197-1 in 2005 is completed. The design of a test method for soluble chromium VI and the leaching of trace elements together with the format of an Environmental Product Declaration (EPD) have already been set. CEMBUREAU is leading on this matter.

Other major standardisation issues are related to the implementation of the (voluntary) Concrete Standard (EN 206), the regulations concerning concrete in contact with drinking water and the revision of the Fly Ash Standard EN 450-1. A crucial element of this revision is the effect of co-combustion in power plants and the related quality specifications of fly ash, as a constituent in cement or in concrete mixes.

BEST AVAILABLE TECHNIQUES

Under the so-called IPPC Directive (Integrated Pollution Prevention and Control) a number of industry sectors are required to develop Best Available Techniques for their operation. For the cement industry, the BREF – BAT Reference Document was finalised in 2001 and describes BAT levels and abatement techniques for energy consumption and the emission of dust, NOx and SO_2: **(i) Alternative Fuels, (ii) Emission of Metals, VOC, Dioxins/ Furans.**

Whilst dust abatement has been widely applied for more than 50 years and SO_2 abatement is a plant specific issue, the abatement of NOx is nowadays the main issue for the cement industry. Many cement plants have adopted general primary measures, such as process control optimisation, use of modern gravimetric solid fuel feed systems, optimised cooler connections and the use of power management systems. These measures are usually taken to improve clinker quality and lower production costs but they also reduce the use of energy and air emissions.

The best available techniques for reducing NOx emissions are a combination of general primary measures to control NOx, staged combustion and selective non-catalytic reduction (SNCR). The BAT emission level associated with the use of these techniques is 500-800 mg/m^3. There is also a view that selective catalytic reduction is a BAT with an associated level of 100-200 mg/m^3. However, this SCR technique has not proven to be a technically and economically viable solution for the cement industry.

BATs for reducing SO_2 emissions are a combination of general primary measures and absorbent addition for initial emission levels *not higher* than about 1200 mg/m^3 and a wet or dry scrubber for initial emission levels *higher* than 1200 mg/m^3. The associated BAT level is 200-400 mg/m^3. SO_2 emissions from cement plants are primarily determined by the content of the volatile sulphur in the raw materials. Kilns that use raw materials with little or no volatile sulphur have SO_2 emission levels well below this level without using abatement techniques.

BATs for reducing dust emissions are a combination of general primary measures and efficient removal of particulate matter from point sources by application of electrostatic precipitators and/or fabric filters. The associated BAT level is 20-30 mg/m^3.

BATs for reducing waste are to recycle collected particulate matter to the process wherever practicable. When the collected dusts are not recyclable, the use of these dusts in other commercial products, when possible, is considered a BAT.

The current BAT Reference Document (BREF) will be revised, in particular regarding NOx abatement, the use of alternative fuels and emission of metals, VOC (Volatile Organic Compound) and dioxins/furans. This revision is expected to be finalised in 2005.

REDUCTION OF GREENHOUSE GAS EMISSIONS

In Europe, the cement industry contributes to approximately 3% of greenhouse gas (GHG) emissions, in particular carbon dioxide (CO_2). In the EU, the cement industry together with other industries covered by the proposed Directive on emissions trading represent 46% of all CO_2 emissions. These industries are challenged to reduce quite considerably their emissions, Figure 2. The European Commission is developing a trading system for CO_2 in order to facilitate and optimise the efforts necessary to reach the targets of the different Member States under the Burden Sharing Agreement within the framework of the Kyoto Protocol. CEMBUREAU has been actively involved in the preparation of the proposed Directive on emissions trading, on its own and in an alliance with 5 other energy-intensive industries (lime, glass, pulp and paper, non-ferrous metals and steel industries). The following aspects have been emphasised:

- Reward for early action, meaning that efforts already undertaken during the nineties are properly taken on board;
- An allocation of CO_2-allowances free of charge, i.e. no auctioning;
- A provision to allow the further development of the use of alternative fuels;
- Intra-group and intra-company transfer.

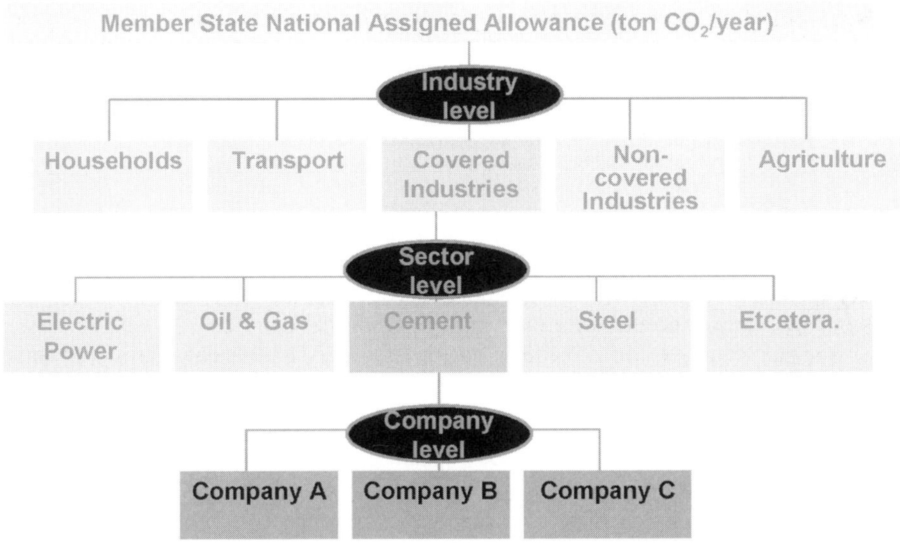

Figure 2 Allowance allocation from country to company

Overall these objectives have been achieved, with the exception that, in the second commitment period, a maximum of 10% of auctioning is left open to Member States. In order to achieve a level playing field, CEMBUREAU has started to develop common guidelines for the allocation of allowances and, in close collaboration with the World Business Council for Sustainable Development (WBCSD) and CEN ("Comité Européen de Normalisation"), protocols for the monitoring of GHG emissions. Regarding the use of alternative fuels, the CEMBUREAU strategy is to obtain 100% allowances free of charge for the corresponding CO_2 and to have no reduction target for this part of the (inevitable) CO_2-emission. In the meantime, Member States have started to develop their national plans for the allocation of allowances, which will have to be finalised and published by 31 March 2004. It goes without saying that, in order to have the trading system in place by January 2005, a lot of questions will have to be answered. CO_2 reduction will, therefore, remain a top priority for the cement industry over the coming years.

FIRE SAFETY WITH CONCRETE

With the participation of BIBM ("Bureau International du Béton Manufacturé") and ERMCO (European Ready Mixed Concrete Association), CEMBUREAU is carrying out a comprehensive Action Plan. Highlights are as follows:

EU proposal of Directive on minimum safety requirements for tunnels

The EU Commission proposes to harmonise the organisation of safety in road tunnels at national level. The objective is to guarantee by means of a set of harmonised minimum safety standards a high level of protection for users of road tunnels longer than 500 meters located in the Trans-European Transport Network.

The Proposal encompasses the main technical and operational safety-related aspects. Nevertheless, the technical specifications for infrastructure do not refer to road pavement. It goes without saying that in this respect the merits of concrete pavement are unquestionable.

Partner in Innovation Project

CEMBUREAU is partner in a 3-year follow-up research project funded by the UK government with BCA (British Cement Association) as a major contributor and led by BRE (Building Research Establishment). The project started in October 2002. The main tasks are to evaluate the results in detail of the fire testing on Cardington seven-storey concrete frame structure co-funded by CEMBUREAU and to investigate the parameters influencing the spalling of reinforced concrete.

Fire Safety Engineering

The Benefeu study commissioned by DG Enterprise and completed earlier in 2002 evaluates the benefits of future initiatives, at European level, on fire safety engineering aimed at the safeguarding of persons' life and health and, to a lesser extent, the protection of property and goods. The EU Commission intends to support performance based fire codes at national level in relation to construction works and products instead of existing national prescriptive rules and regulations. The trend towards performance based regulation is less favourable to concrete, as high resistance to fire becomes a less significant factor.

ECO-CONCRETE

In 2002, a Joint Project by BIBM, ERMCO, UEPG (European Aggregates Association officially left the Joint Project Group in June 2002), EUROFER (European Confederation of Iron and Steel Industries), EISA (European Independent Steelworks Association) (EISA was taken over by EUROFER on 1 August 2002) and CEMBUREAU on LCI/LCA of Concrete completed **EcoConcrete**, the new and tailor-made LCA (Life-Cycle Analysis) software tool providing fully-fledged life cycle assessments of 10 selected concrete applications (functional units), according to three different LCA methodologies Figure 3

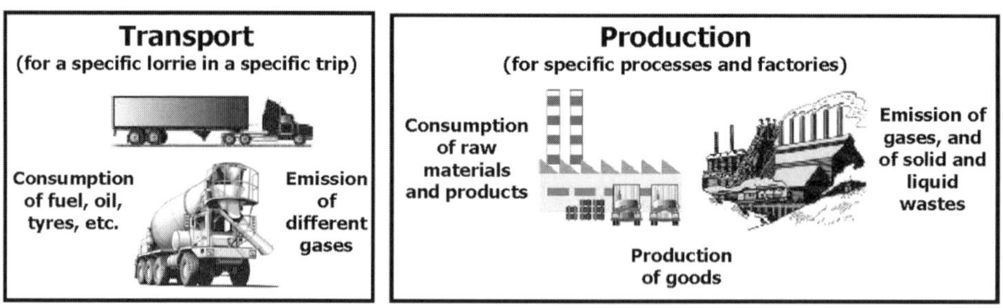

Figure 3 Ecoconcrete LCA software tool - LCA stands for Life Cycle Assessment and is a method that assesses quantitatively the environmental

EcoConcrete is a user-friendly development of Microsoft® Excel to be used by both non-specialists and specialists for the purpose of promoting Life Cycle thinking within the cement/concrete industry and related products industries (Figure 4).

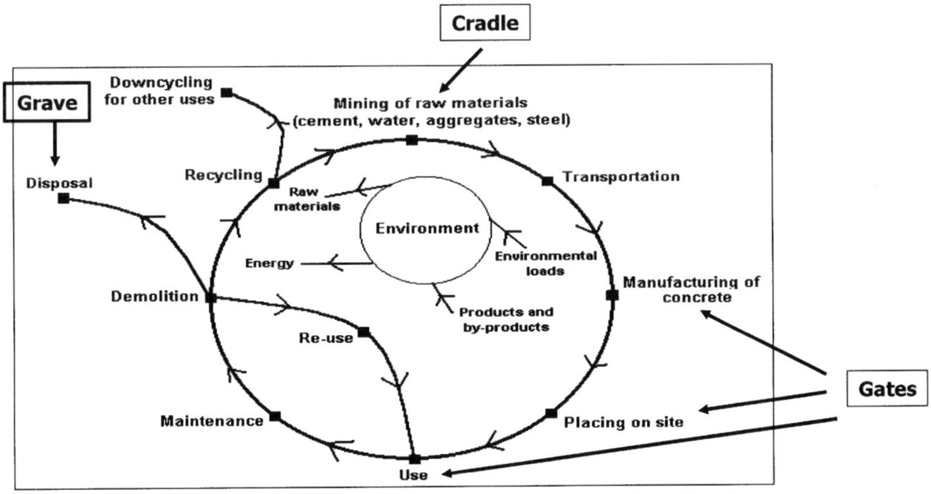

Figure 4 Key aspects in an LCA: Life Cycle. The LCA of a product must take into account the environmental impact produced in all its life (from cradle to grave).

EcoConcrete is being disseminated only within the cement/concrete and related products industries through a licensing system including a training programme (Multiplier System). Each licensee is allowed to use the tool only after having been educated during a one-day seminar called "EcoConcrete Workshop" and after having trained similarly six other potential users.

The external communications strategy (towards competitors) is based on the "shield function" i.e. providing guidelines/recommendations on how to use the information in the LCI database as a shield against denigration from competitors.

CONCLUSIONS

- One of the conclusions from a Pan-European Perception Study of Cement and Concrete by CEMBUREAU was that public audience is not very familiar with the concept of "sustainable development" and tends to interpret this concept primarily in terms of environmental challenge, putting industry on the defensive.

- The manufacture and use of cement in a Sustainable Context means more than a responsible environmental attitude by the industry. Corporate Social Responsibility rests on three pillars: a social, an ecological and an economical one. In other words: People, Planet and Profit.

- Valorisation of waste in the cement industry is beneficial and safe. This is a reality, which should be an element of any sound environmental waste management policy. The challenge is to get the message across to all the EU Member States to turn this into a reality and a priority.

- The increased utilisation of cementitious materials in cement contributes significantly to the reduction of the use of natural resources and thereby of the emission of greenhouse gases, in particular CO2.

- Clear and transparent communication on the environmental performance of the cement industry and its products will have a positive contribution to the perception of a valuable and indispensable building material: **concrete.**

CONTRIBUTION OF CEMENT PRODUCTION AND USE OF SUSTAINABLE DEVELOPMENT

G Thielen

G Locher

B Hauer

FIZ

Germany

ABSTRACT. Sustainable development has an ever growing impact on all parts of society. It covers environmental, economic and social aspects. The contributions of the cement industry to sustainable development are widespread and include raw material supply, clinker production and the use of blended cements. Far-sighted quarry management and the use of industrial by-products save resources and minimise the impact on the environment. While the thermal energy efficiency is already very high, greenhouse gas emissions can further be reduced by promoting the application of alternative fuels. Production and use of blended cements give a double chance: the reduction of resource and energy consumption during the production of cement and the increase in the choice between different cement types with respect to their performance characteristics well fitting to the intended use. Therefore both, the improvement of efficiency during cement production and the supply of variety of different cement types - composed of standardised constituents - contribute to sustainable developments. This contribution will give an overview of the corresponding state of the art on the basis of practical cases and examples.

Keywords: Sustainability, Eco-efficiency, Cement production, Nature conservation, Alternative raw materials, Energy efficiency, Alternative fuels, Blended cements

G Thielen is a civil engineer and received his engineering education and doctor's degree at the TU München. After a post-doc year at the UC-Berkeley he served several years as structural and R & D engineer in the building industry and 3 years as Tech. Director with CEB. In 1987, he joined the FIZ as Managing Director. Since 1996, he is Honorary Professor at the RWTH in Aachen.

G Locher is a process engineer and received his education at the RWTH in Aachen, the ETH in Zürich where he obtained his doctor's degree and at the MIT (Cambridge, USA). After four years of R&D in the industry he joined the FIZ where he is in charge of joint environmental issues and modelling. The focus of his work is on European environmental law, modelling of the clinker burning process and cement hydration.

B Hauer is a physicist and received his education at the University of Bonn, the Research Centre Jülich and the Saarland University where he obtained his doctor's degree. He joined the FIZ in 1996, where he is in charge of sustainable construction. The focus of his work is on life cycle assessment and on environmental labels and declarations.

INTRODUCTION

Sustainable development which "meets the needs of the present without compromising the needs of future" [1] cannot be discussed without taking into account building and construction. Buildings and infrastructure are of decisive importance for man's survival and welfare. Housing and schools, offices and factories, theatres and sport centres provide protection, comfort, business and pleasure [2]. Furthermore infrastructure as roads and railways, drinking water supply and energy transmittance belong to the core activities of the construction sector. On the other hand, buildings and the built infrastructure require high economic efforts and cause heavy impacts on the environment. The importance of the construction sector for sustainable development is paralleled by the importance of cement. Cement bound materials as concrete play probably the most important role within the building materials today. As for all other building materials the eco-efficiency of cement, i. e. the ratio of its beneficial contribution to construction to its economic cost and environmental impacts, needs continuous optimisation.

The German cement industry and its research institute have been dealing with these issues for many years. As an example, the reduction of emissions during cement clinker production [3, 4, 5] and the research on high performance cement bound materials [6, 7, 8, 9] shall be mentioned. In addition the possibilities of life cycle assessment have been explored and intensive research on the overall environmental performance of cement and concrete has been performed [10, 11, 12]. The current state of contributions of cement and concrete to sustainability and possible further steps towards this goal have been summarised in extended reports [13, 14]. Within this contribution, recent developments with regard to the main aspects of cement production and use shall be highlighted.

In order to achieve an increased eco-efficiency of products one has to increase their benefits and to reduce the pertaining environmental impacts and costs. Since cement is an intermediate product, its ecological impacts, its economic expenditures and its social benefits can only be assessed on the basis of end products (e.g. a building or a road), which serve as functional unit for sustainability considerations. Within this context the cement industry has identified two main areas where a contribution to an improved eco-efficiency can be made. One area is related to the reduction of environmental impacts during the cement production, where, on a local level the land use for quarry operation is often in the focus and where on a global perspective, the use of energy and the emissions of carbon dioxide as a greenhouse gas seem to be the most important aspects.

These topics will be dealt with below. The reduction of environmental impacts must be achieved without impeding the potentials to provide and to improve high quality products as a prerequisite to economic and environmentally sound buildings and infrastructure which finally is the main goal of sustainable construction. Therefore the other area is related to the influence of cement on the durability and service life of buildings. The cement industry is requested to offer cements on the building market which show maximum performance and quality aspects with a minimum of economic expenditure and ecological impact. As described below blended cements offer a solution for this optimisation. The cement industry is aware of its responsibility for both approaches: the enhancement of eco-efficiency of the cement production in cement plants and of the cement use in the construction of buildings.

PRODUCTION OF PORTLAND CEMENT CLINKER

Quarrying and Nature Conservation

The life cycle of cement starts in the quarries of the cement industry. According to the interrelations mentioned above, the availability of such quarries is necessary for the present and future cement production. This is important to be mentioned since in discussions, quarries are often reduced to their environmental impact disregarding their high economic importance. Nowadays, in Germany approx. 40 million tons per year of limestone are needed for the cement clinker production. Besides the depletion of the natural limestone resource, the related land use is of particular concern. For Germany, e.g., a total land use of 0.652 km^2 for cement quarries has been estimated for 1997 [15]. Even though this is less than 0.0002 % of the total area of Germany, these quarries impose locally a serious change in the use of land. However, the lifetime of quarries is limited and, already before starting the operation, the use of the area after extraction has to be determined. According to a survey of the German cement industry on the land use before and after quarrying in all German cement plants [16], more than 50 % of the quarries which have been exploited are dedicated to nature conservation (Fig. 1). Taking into consideration, that the land use before quarrying mainly was agricultural, the environmental value of the land has increased after quarrying. Furthermore, also active quarries can be beneficial for many animals and plants. In some cases they even protect some species which are threatened by extinction. Recommendations for quarry management issued in Germany focus on measures to enhance the value of quarries for nature conservation even during quarry operation [17]. They cover both, the use and the restoration phase of quarries and address, e.g., important aspects of steep faces, wetlands and intermediate restorations. Further, as these recommendations support the application of recognised measures for nature conservation, they are fully in line with recommendations for action issued by the World Business Council for Sustainable Development (WBCSD), which are published within the "Project cement" [18].

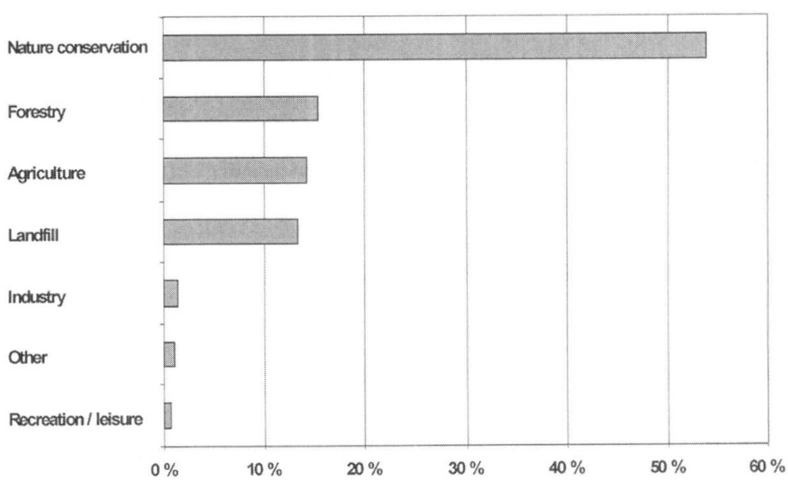

Figure 1 Land use after extraction in quarries of the German cement industry [16]

Alternative Raw Materials

The utilisation of alternative raw materials allows for a reduction of natural raw materials and consequently for a decreased land use related to quarrying. Consequently the use of alternative raw materials has also been recommended by the WBSCD [18]. Alternative raw materials are fractions of waste or industrial by-products, which contain calcium oxide, silicon dioxide, iron oxide and/or aluminium oxide as main constituents. Such materials are, e.g., lime sludge from drinking water and sewage treatment, used foundry sand or fly ash. Such materials can be combined with natural raw materials in particular if the raw material situation prevailing at a cement work necessitates corrective materials in order to meet the preconditions of the raw material composition. More than 1.3 mil tons of such alternative raw materials have been utilised in German cement plants in 2001 (Table 1) [19]. In addition to the saving of resources, their application helps to valorise materials, which in other processes cannot be utilised in a comparably useful manner.

Table 1 Secondary raw material components for the Portland cement clinker production in Germany in 2001 [19]

GROUP	RAW MATERIAL	INPUT QUANTITY [1.000 t/a]
Ca	Lime sludge from drinking water and sewage treatment, hydrated lime, foam concrete granulates, calcium fluoride	396
Si	Used foundry sand	111
Al	Residues from reprocessing salt slag, aluminium hydroxide	43
Si/Al	Residues from coal pre-treatment	15
Fe	Roasted pyrite, contaminated ore, iron oxide/fly ash blends, dusts from steel plants, mill scale	198
Si/Al/Ca	fly ash, paper residuals, ashes from incineration processes, mineral residuals (e.g. soil contaminated by oil)	569
Total		1.332

Energy Efficiency of the Burning Process

As already indicated above, secondary raw materials contribute significantly to the saving of primary materials and the ecologically sound utilisation of waste materials. Against the background of greenhouse gas discussions, however, energy savings are even more in the focus of public perception. During cement production, fuel energy is mainly used for Portland cement clinker burning. As this process is one of the most energy intensive industrial processes, high energy costs have forced the cement industry for decades to increase the energy efficiency. As a consequence, the specific fuel energy consumption of the German cement industry has been reduced by more than 50 % during the last 50 years. Figure 2 shows the trend in Germany from 1950 to 2001. From 1987, also the new federal states are included. Today the energy efficiency of the total production process amounts to more than 70 %. According to the diagram, the burning process is highly optimised and a further appreciable reduction cannot be expected from process engineering measures.

An evaluation of German cement works shows that even a theoretical replacement of all existing installations with modern installations on a "greenfield site" basis would lead to a fuel energy reduction of only 9%. This high efficiency has been the result of a continuous optimisation process following the investment cycles of a capital intensive industry [3, 20, 21]

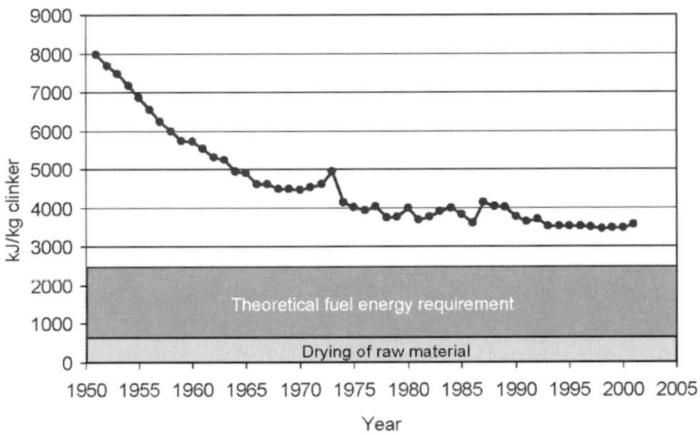

Figure 2 Specific fuel energy consumption
(until 1986 West Germany, thereafter entire Germany)

Alternative Fuels

Eco-efficieny can further be increased by the utilisation of wastes as secondary fuels. They replace fossil fuels like hard coal or lignite in the Portland cement clinker burning process. Accordingly, the consumption of primary energy resources can be reduced. In the German cement industry, the share of secondary fuels increased from about 7% in 1990 to approximately 30 % in 2001.

Energy-related CO_2 emissions and their reduction are in focus of the discussions about the utilisation of secondary fuels. Prepared from waste, their use in the Portland cement clinker production contribute significantly to a reduction of greenhouse gas emissions, since wastes would otherwise have to be land-filled or incinerated, often under worse energy-efficiency conditions. Several life cycle assessment studies investigated the effect of diverse options of waste utilisation on total CO_2 emissions, e.g. for plastic waste and for used tyres [22, 23]. The studies analyse total life cycles from waste recovery to waste use or disposal comparing the waste as secondary fuel for the Portland cement clinker production with alternative uses. Emissions of other greenhouse gases like methane, which are emitted in pre-stages, e.g., during coal mining, have also been taken into account. If plastic waste replaces hard coal in the clinker burning process, it can be shown that, depending on the composition of the waste, the total emission of greenhouse gases is smaller by almost up to 100 % CO_2 equivalents relative to the actual CO_2 emissions of the cement works compared to the land-fill of the waste. If the plastic waste replaces lignite in the clinker burning process, the reduction rate increases to more than 105 %. Finally, if the incineration of waste in a refuse incineration plant is considered as the reference scenario, far more than 100 % savings can be achieved. Such high reduction rates can be achieved, in particular, when CH_4 or NO_2 emissions are avoided [3]. These results support the existing political agreements in several European countries, e.g. in Germany, allowing to regard secondary fuels as CO_2 - neutral in a global sense.

BLENDED CEMENTS

Environmental Benefits of Blended Cements

In addition to the use of alternative raw materials and fuels in the cement clinker production the production and application of blended cements provide effective possibilities to improve the eco-efficiency of cement and cementious products. The European cement standard EN 197-1 allows the use of several main constituents which may substitute Portland cement clinker. Granulated blast furnace slag, siliceous and calcareous fly ash, unburned limestone, burnt shale, puzzolanic materials and silica fume are recognised as main constituents also in other parts of the world and facilitate the production of a huge variety of blended cements.

While Portland cement CEM I only contains Portland cement clinker as a main constituent, CEM II cements contain 6% to 35 %, CEM IV/B- and CEM V-cements more than 35 % of those other materials as main constituents. CEM III-cements contain 36 to 95 % of blast furnace slag. On the German market almost 60 % of the total cement production are used for in situ concrete with more than 80 % as Ready Mixed Concrete. More than 80 % of this cement belongs to the 32.5 strength class. It is this portion of the market which offers the most possibilities for blended cements. During the last years, from 1998 to 2002 the production of CEM I 32.5 cements decreased from 35.5 % of the total production in Germany to 23.6 %, whereas the corresponding volume of CEM II 32.5 cements doubled from 12.4 to 25.0 %, with a constant volume of about 11 to 12 % for CEM III 32.5 cements (Figure 3). About 30 % of the total German cement production is used in prefabrication plants, the majority (more than about 80 %) of these cements belong to the 42.5 and 52.5 strength class. Within these strength classes the portion of CEM II cements amounts to about 15 % with only a small increase within the last 5 years.

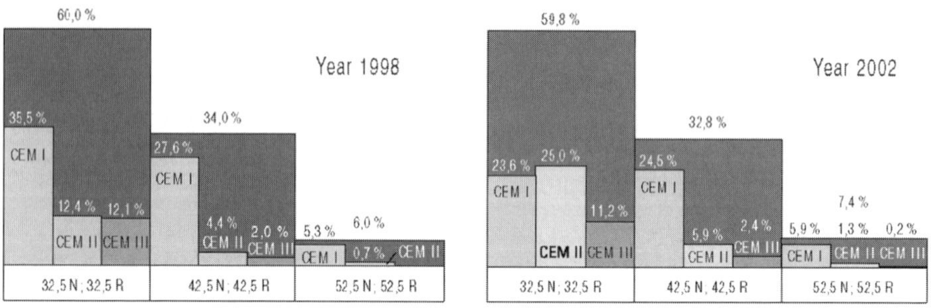

Figure 3 Share of cement types and strength classes in the domestic delivery of cements manufactured in Germany in 1998 (left) and 2002 (right)

It is of decisive importance for the cement market that the cement meets all performance characteristics required in the concrete it is used for. Therefore the increase of the volume of CEM II 32.5 cements mentioned above has only been possible under the condition that the strength, workability and durability performance aspects of the concretes with CEM II cements remain comparable or are superior to those when using CEM I cements. Early age strength development and frost resistance have been of crucial importance when replacing

CEM I 32.5 with CEM II 32.5 cements. In addition – and this is particular for the German cement market - CEM II 32.5 cements must allow fly ash as concrete addition and partial cement replacement without impeding the early strength and frost performance of concrete.

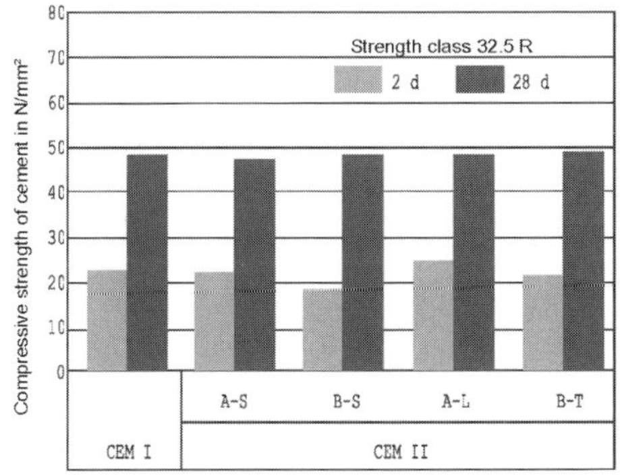

Figure 4 Average values of cement strengths of the strength class 32.5 R

Figure 4 shows the mean values of the early (2 d) and standard strength (28 d) for CEM I 32.5 R and for the mostly used CEM II 32.5 R cements produced in Germany and proves their comparable performance.

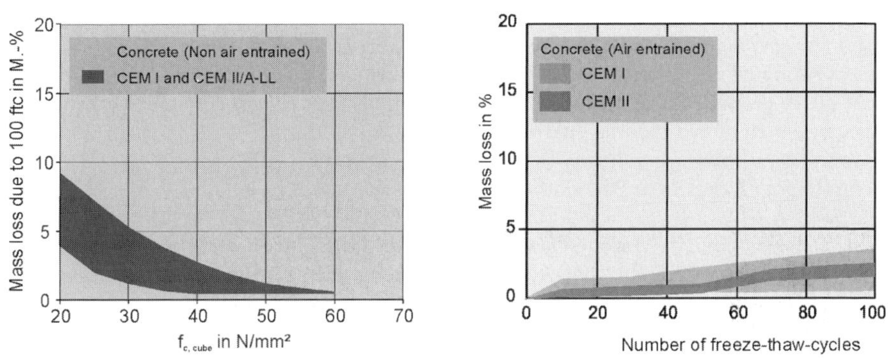

Figure 5 Freeze-thaw resistance of non air entrained concrete after 100 freeze thaw cycles (ftc) in dependence on the concrete strength (left) and freeze-thaw resistance with de-icing salt of air entrained concrete (right), both tested by the cube method [24]

Figure 5 show results of frost resistance tests, carried out on non air entrained and on air entrained concrete specimens with CEM I 32.5 and CEM II 32.5 cements. Whereas air-entrained concrete shows basically the same frost resistance independent on the type of cement, non air entrained concretes differ in their frost resistance with the type of cement. However CEM II-cements can show frost performance comparable to CEM I-cements if the granulometry and reactivity of the main constituents are correctly chosen [25]. These results explain why Ready Mixed Concrete plants in Germany have accepted to replace CEM I 32.5 cement by CEM II 32.5 cements as standard cement for the bulk of their concrete production.

By this switch from CEM I to CEM II cements the portion of Portland cement clinker in German cements has dropped from 83 % in 1998 to 78 % in 2001. Although the use of non-clinker main constituents requires also some additional energy, the total energy consumption remains smaller. As an example, the use of blast furnace slag generates additional energy consumption due to drying, transport and higher specific energy for grinding. However, the total energy consumption per ton of cement in Germany is reduced by 4 % with the decrease in Portland cement clinker. Figure 6 shows the decrease in energy demand per ton of cement for CEM II and CEM III cements compared with CEM I cements [21].

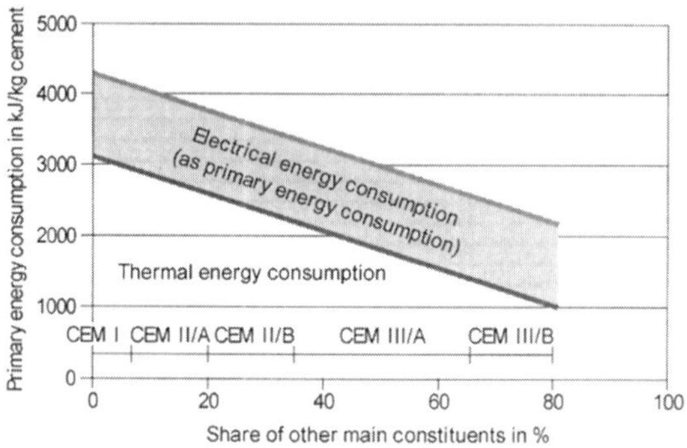

Figure 6 Primary energy consumption of cements containing several main constituents

CO_2 emissions are even more reduced due to the fact that the amount of CO_2 resulting from decarbonation of the limestone during the Portland cement clinker production amounts to about 60 % of the total, hence the effect of Portland cement clinker replacement on CO_2-emisson is even higher than on energy demand.

Technological Benefits of Blended Cements

Portland cement and blended cements offer a wide variety of performance characteristics which allow the concrete producer to choose the cement type which fits best for the intended use. For in situ concrete – ready mixed or site mixed – blended cements belonging to the 32,5 strength class fit almost any performance requirements for the construction of private and commercial buildings. In situ concrete for civil works as bridges, tunnels etc is mainly made with cements belonging to higher strength classes and consequently with a higher portion of CEM I as compared to CEM II or CEM III cements. However, CEM II/A and CEM III/A cements become more and more competitive also in these fields of use. Concrete plants for the prefabrication of structural elements or for concrete paving products predominantly use cement of the 42.5 and 52.5 strength class. Blended cements of these strength classes may improve the form stability of green concrete used for paving products without impeding their final strength and durability. In addition blended cements are used as special cements with low heat of hydration to reduce early age cracking or as special cements with a low effective alkali-content against alkali-silica-reaction or with a high resistance against sulphate attack. The increased possibilities of the most appropriate choice of cement for the intended use are important for the enhancement of eco-efficiency.

Summing up the increased number of cements offers not only ecological but also remarkable technological advantages since the cement can be better chosen according to the demands of the intended use, improving the performance during the entire life cycle of concrete.

CONCLUSIONS

The cement industry contributes by two main strategies to the enhancement of the eco-efficiency of cement: the reduction of environmental impacts and costs related to the production of cement and the supply of a wider variety of cements with optimal performance characteristics for their intend use in building, civil works or products. The examples described above show some of the possibilities related to both approaches. The conflict between the land use for quarries and nature conservation can be diminished to a large extend by a far-sighted management of quarrying and restoration. The use of alternative raw materials supports this goal and helps to solve problems related to waste management in other industrial sectors. The utilisation of alternative fuels allows a further decrease of the consumption of fossil fuels, which is already largely supported by a high efficiency of the clinker burning process. In addition these measures contribute very effectively to a reduction of greenhouse gas emissions as CO_2. The increased use of blended cement contributes also to the eco-efficiency in the concrete sector. Especially the large variety of cements provides an excellent basis for the optimal performance of building components and structures in their intended use during their total service life. New developments in cement production and in the application of blended cements will lead to further improvements of eco-efficiency.

REFERENCES

1. THE WORLD COMMISSION ON ENVIRONMENT AND DEVELOPMENT, "Our common future", Oxford University Press, Oxford, 1987, p 400

2. THE EUROPEAN COMMISSION, Sustainable construction final report, http://www.europa.eu.int/comm/enterprise/construction/suscon/sustcon.htm.<rev.14/06/2002>

3. HOENIG, V., SCHNEIDER, M., CO_2 reduction in the cement industry. Proceedings of the 5th International VDZ Congress 2002, Verein Deutscher Zementwerke (ed.), Düsseldorf, in progress

4. HOENIG, V., HOPPE, H., BODENDIEK, N., Options and limitations in NO_x abatement in the cement industry, ZKG INTERNATIONAL, Vol. 54, 2000, no.4, pp 165-171, no.7, pp 356-361

5. SCHÄFER; S., HOENIG, V., Effect of kiln operation on the behaviour of mercury in the clinker burning process. Proceedings of the 5th International VDZ Congress 2002, Verein Deutscher Zementwerke (ed.), Düsseldorf, in progress

6. GRUBE, H., RICKERT, J., Self compacting concrete – another stage in the development of the 5-component system of concrete, in: Concrete Technology Reports 1998-2000, Verlag Bau+Technik, Düsseldorf, 2001, pp 39-47

7. KORDTS, S., GRUBE, H., Controlling the workability properties of self compacting concrete used as ready-mixed concrete [German], Beton, vol. 52, 2002, pp 217-223

8. THIELEN, G., ALONSO, M: T:, Strength Development and Shrinkage of High Strength Concrete [German], in: 6. Münchner Massivbau Seminar 11.-12.4.2002, Springer, Düsseldorf, 2002, pp 159-183

9. ALONSO, M. T., High Strength Concrete [German], Beton, vol. 52, 2002, pp 563-567 and pp 621-624

10. WISCHERS G., KUHLMANN K., Eco-Balance of Cement and Concrete, ZKG INTERNATIONAL, vol. 44, no.11, 1991, pp 454-553

11. BUNDESVERBAND BAUSTOFFE - STEINE UND ERDEN (BBS) (ed.), „LCA of Building Materials: Impact Assessment and Interpretation in the Building Materials Industry", BBS, Frankfurt/M., 2000, p 57

12. SCHNEIDER M., HAUER B., Criterion for the assessment of buildings [German], Beton, vol. 51, 2001, pp 22-24

13. SOZIALPOLITISCHE ARBEITSGEMEINSCHAFT DER DEUTSCHEN ZEMENTINDUSTRIE, ET AL (ed.), Sustainability and the cement industry [German], Verlag Bau+Technik, Düsseldorf, 2002, p 52

14. REINHARDT, H.-W., ET AL, Sustainable Construction with Concrete - Status report [German], Deutscher Ausschuss für Stahlbeton (ed.), Beuth, Berlin, 2001, p 94

15. BUNDESANSTALT FÜR GEOWISSENSCHAFTEN UND ROHSTOFFE (BGR) (ed.), Flächenbedarf für den Abbau von oberflächennahen Rohstoffen (Steine und Erden, Braunkohle und Torf) im Jahr 1997. Commodity Top News no. 9, BGR, Hannover, 2000

16. BUNDESVERBAND DER DEUTSCHEN ZEMENTINDUSTRIE; VEREIN DEUTSCHER ZEMENTWERKE (ed.), Nature conservation and the cement industry, part 1: analysis of a survey [German], Verlag Bau+Technik, Düsseldorf, 2001, p 23

17. BUNDESVERBAND DER DEUTSCHEN ZEMENTINDUSTRIE; VEREIN DEUTSCHER ZEMENTWERKE (ed), Nature conservation and the cement industry, part 3: management recommendations [German], Verl. Bau+Technik, Düsseldorf, 2001, p 26

18. MISRA, K. K., Toward a sustainable development. Substudy 11: Management of Land Use, Landscape, and Biodiversity, World Business Council for Sustainable Development, Geneva, 2002

19. VEREIN DEUTSCHER ZEMENTWERKE (VDZ) (ed.), Environmental data of the German Cement Industry, VDZ, Düsseldorf, 2002, p 51

20. HOENIG, V., SCHNEIDER, M., German Cement Industry's Voluntary Efforts on the Issue of Climate Change – A Success Story, in: CANMET/ACI International Symposium on Sustainable Development of Cement and Concrete: 3 San Francisco 16.-19.09.2001, American Concrete Institute, Farmington Hills, MI, 2001, pp 15-32

21. VEREIN DEUTSCHER ZEMENTWERKE (VDZ) (ed.), activity report 1999-2001, VDZ, Düsseldorf, 2001, p 178

22. HEYDE, M., KREMER, M., Utilisation of plastics waste form sales packaging in the cement industry – ecological analyses by the LCA principle [German], Fraunhofer Institute of Foods Technology and Packaging, Freising, 1997, p 75

23. GIEGRICH, J., FEHRENBACH, H., ORLIK, W., SCHWARZ, M., Ökologische Bilanzen in der Abfallwirtschaft, Umweltbundesamt (ed.), Berlin, 1999, p 208

24. DEUTSCHER AUSSCHUSS FÜR STAHLBETON (ed.), Prüfung von Beton – Empfehlungen und Hinweise als Ergänzung zu DIN 1048, Beuth, Berlin, 1991

25. LUDWIG, H.-M., General report 1; Influence of process technology on the manufacturing of market-oriented cements Proceedings of the 5th International VDZ Congress 2002, Verein Deutscher Zementwerke (ed.), Düsseldorf, in progress

THE PROBLEM OF HIGHLY EFFECTIVE AIR PURIFICATION FROM DUST

V A Batluk **V K Batluk**
Lviv National University Lvivska Polytechnika
Y Y Shelyukh
Lviv State Institute of Fire Safety
Ukraine

ABSTRACT. The article deals with the problem of providing high performance apparatuses for cleaning air from dust in various branches of industry in order to reduce hazardous emissions to the level conforming to sanitary-hygienic norms. The article describes new trends in the development of dust catching apparatuses based on the use of centrifugal-inertial forces, permitting to improve significantly the effectiveness of dust catching.

Keywords: Emissions, Polluting the atmosphere, Neutralization, Cyclone operations

V A Batluk is a Professor at the Lviv University, Ukraine.

V K Batluk is a Professor at the Lviv University, Ukraine.

Dr Yu Ye Shelyukh is the Chairman of Service Organisation at the Lviv State Institute of Fire Safety in Ukraine.

INTRODUCTION

Along with various natural phenomena (volcanic eruptions, forest fires, soil weathering and erosion, etc.) causing pollution of the atmosphere human activities connected with the use of natural resources and the development of industry, agriculture, construction and transport play an ever growing role in resolution of this problem. Due to the insufficient level of knowledge, inadequate technology or absence of comprehensive prognostic analysis these activities are accompanied by unwanted processes – emission into the atmosphere of the whole range of harmful by-products. Polluting the atmosphere these by-products cause tremendous, often unjustified harm not only to the environment, but to humans as well.

The problem of prevention of atmosphere contamination has long ago crossed the borders of separate nations and even entire continents, and has acquired the status of the international problem and become common for practically all countries of the world. This has been determined by a number of reasons: solidarity of interests in the development of measures and means aiming at reduction of the levels or complete elimination of emission of hazardous admixtures into the atmosphere, by the importance of collaboration and exchange of information on the development of methods and means of prevention of harmful consequences of atmosphere contamination and losses the create and, finally, by the fact, that the hazardous materials are carried over great distances regardless of the state borders.

Once it becomes evident, that the improvement of technological processes, establishment of sanitary-hygienic zones, increased height of smokestacks and other measures can not reduce the level of pollution of atmospheric air to the maximum permissible concentration level, emissions must be purified to such a level, that in the end they do not exceed the norm of maximum permissible concentration.

ANALYSIS OF LATEST RESEARCH WORKS

At present cleaning of contaminated air and gases is the main method of protection of air basin from pollution, applied in all cases when the use of active cleaning methods is still impossible or economically unfeasible. The task of industrial gas cleaning is neutralization of the air basin from organized gas emissions and stationary pollution sources.

Cleaning of emissions is simplified considerably if emitted gases, for example, stack gases, move along the gas ducts. In many cases, however, for example, in operating mills, pickling installations, electroplating tanks, in dressing and cleaning castings, in plastic, glass-reinforced plastic, fabric-based laminate materials processing, etc. it is necessary to use special measures to avoid a direct emission of these materials into the atmosphere. In other cases the admixtures not only become dangerous for the operators and adverse for the product quality, but become hard to catch.

For this reason catching hazardous materials directly from the source of emission is the guarantee of not only improved hygienic conditions of labor at work places, but of high effectiveness of dust catching installations, which, as a rule, consist of the following elements:

- Collecting device, which may consist of one or several dust collectors;
- Grid of pipe-lines;
- Fan, drawing off dust-laden or gas-contaminated air via pipe-lines to the dust-cleaning installation.

The systems of cleaning and decontamination of gas emissions can be conventionally divided into 2 large groups:

- Group I – installations for cleaning from toxic gas admixtures (chemical cleaning);
- Group II - installations for cleaning gas emissions from sprays (dust, smoke, droplets or fog).

STATEMENT OF TASK

The greatest achievements in centrifugal catching of solid particles from a gas flow have been noted in the field of hardware design (engineering) and not in the field of scientific projects, which is explained, on the one hand, by accumulation of long experience of operation of industrial installations and, on the other hand, by exceptional complexity of description of separate phenomena and characteristics of heterogeneous systems: rigid body – gas in centrifugal force field. For this reason the theory of cyclone operation is still underdeveloped and does not provide an opportunity to calculate cyclones of various designs. For example, the problem of the most suitable shape of a cyclone till now is resolved only empirically or experimentally.

Different requirements are presented to the level of final dust content in gas flows depending on physico-chemical properties of dust, technological parameters of dust-gas flows, especially with respect to fine toxic dust. For this reason it is possible to find a range of decisions when solving the problem of design of centrifugal force apparatuses for separation of solid particles from gas flows. The analysis we have carried out showed a great number of various designs of centrifugal force apparatuses, whose design most often has not been supported by theoretical and experimental investigations of the structure of the formed flows and the process of separation the particles from them.

We shall note, that the main effect on efficiency of separation of dust solid particles in centrifugal force apparatuses is produced by improvements made on the apparatus body, whose purpose is, predominantly, to increase the centrifugal factor of separation or reduce the harmful effect of increased centrifugal factor of separation, or reduce the harmful effect of radial removal of dust.

A large number of various constructions of dust catchers for the same purpose, which do not have any clear specifications, stand in the way of selecting an appropriate system of cleaning in the process of design of dust catching devices, and absence of standardized dust cleaning equipment is an obstacle on the way to resolve the problem of organization of manufacture of such equipment.

To simplify the process of choosing the type of dust catchers for each specific case an attempt has been made to standardize, arrange in an order all known constructions of dust catchers, placing them by way of comparative testing in a small specific order by their efficiency, hydrodynamic resistance and specific quantity of metal per structure with equal power, air consumption in order to specify design of the apparatus with the best operating characteristics, which may become a prototype for design of the new type apparatuses.

PRESENTATION OF PRINCIPAL DATA OF RESEARCH

For this purpose we have used the "Uniform techniques of comparative testing of dust catchers", which include the item of preparation of experimental dust, determination of its parameters, means of contamination with dust of the air supplied to the dust catcher, degree of diagglomeration of dust in case of artificial contamination of the air with dust, established as mandatory for conducting of bench tests of various types of dust catchers: air filters, dry-type cyclones, wet-type dust catchers, fabric filters, individual duct catching apparatuses.

Principal parameters dry-type dust catchers with similar power and air consumption, efficiency and pressure $1 \cdot 10^3$ Pa are shown in Table 1. Efficiency is shown in the form of relation of dust removal from the given cyclone to dust removal from cyclone CN-11 (standard). Specific quantity of metal per structure is the value of the cyclone surface per 1000 m^3/h of the cleaned air. To determine the place of domestic dust catchers we have provided specifications of the best foreign apparatuses in Table 2.

Table 1 Indices of dry dust catchers

CYCLONE TYPE	DEGREE OF DUST REMOVAL, %	DIAMETER, mm	HEIGHT, mm	SPECIFIC QUANTITY OF METAL PER STRUCTURE, m^2	SURFACE AREA, m^2
LIOP	1.23	1.17	0.8	2.07	1.05
SIOP	1.05	1.62	0.7	1.82	0.92
VNDIOP	1.05	1.25	0.91	2.22	1.05
T-4/630	0.96	1.14	1.57	3.64	1.85
CN-11	1.00	1.00	1.00	1.96	1.00
CN-15	1.05	0.90	0.93	1.67	0.85
CN-15У	1.26	0.92	0.70	1.29	0.66
CN-24	1.35	0.75	0.72	1.09	0.55
Matryoshka	1.62	1.43	0.92	3.67	1.86
UC-38	0.78	1.60	1.25	3.86	1.73
4B CSh	0.82	1.10	0.53	2.47	1.26

Modern gas-cleaning units for cleaning exhaust industrial gases to the degree, which meets requirements of sanitary-hygienic norms, and installations, intended for cleaning of technological process, as a rule, can not operate with high efficiency, because most of gas-cleaning units are rather complicated installations, consisting of a number of gas-cleaning apparatuses of various types, installed in several stages.

Table 2 Specifications of foreign apparatuses

TYPE OF DUST CATCHER	PRODUCTIVITY, m^3	HYDRODYNAMI C RESISTANCE, Pa	EFFICIENCY, %
D1-DIV Poland	800-100000	1350	92.2
TGL Germany	800-5760	1200	90
SGA Czech. Rep.	800-100000	1650	90.8
SHA England	800-100000	1730	91.2
TOSHIBA Japan	800-100000	850	94.1

Using our theory of the process of separation of heterogeneous systems [1,2] by way of destruction of turbulent vortexes in dust catchers we can determine the minimal size of the particle moving under the influence of centrifugal force to the body wall upon its separation from the vortex and the maximal size of the vortex itself, thus determining dimensions of the apparatus, specifically the distance from the external to the internal wall of the body. Fine dust particles are under the influence of turbulent efficiency, which does not permit them to separate from the flow and they continue to move with the flow to the internal wall. The main task at this stage is freeing the particle from the influence of vortex, which is possible only in case of destruction of the latter. This can be achieved provided, that the internal wall of the apparatus is made in the form of shutter-type separator instead of one-piece wall.

Taking into consideration mathematical model of the process of separation in centrifugal-inertial force dust catchers we have proposed a new group of apparatuses, whose operating principle is based on the combination of centrifugal, inertial forces and weight. The know-how of all groups is presence in the apparatus body of the second cleaning stage – shutter-type separator, in which dust-and-gas flow is subjected to additional cleaning passing through the shutters.

Cleaning of air from dust in the apparatuses we have designed takes place as follows (Figure 1). Dust-laden air comes in tangentially predominantly to the upper part of the apparatus, where at first under the influence of centrifugal forces the flow is separated layer by layer: larger (by size and weight) particles of spray are deflected to the internal wall of the apparatus body and continue their circular motion along the wall, while fine particles move in circles along the internal single-piece outlet pipe for cleaned air. One half of revolution is sufficient for this process, which has been proved experimentally.

At this section 1 of the apparatus cleaning of the air from dust occurs in the way similar to cyclone operation. At section 2 of the apparatus additionally to this process secondary cleaning of the air from dust begins along the shutter separator placed coaxially to the apparatus body. Gas flow flows past the separator shutters and turns at the angle over 90° but less than 180° and goes out to clean air outlet pipe through the openings between them. Fine dust particles which have not been separated from the flow by centrifugal forces also approach the separator, but owing to their inertia they lag behind gas and hit the separator shutters and, depending on their location on the shutters they are either repelled to the wall of the body and caught up by the dust flow moving there, or again are dragged by gas going to the separator for additional cleaning, hit the shutters again and so on, till they finally get into dust flow moving along the walls of the apparatus body, which will transport them to dust outlet pipe. The number of collisions depends on the parameters of dust (dispersion composition, physico-chemical and morphometric indices), gas flow (velocities: when entering the apparatus, of movement in the apparatus body, of passage through the shutter separator) and design features of the shutters, determining the angle of attack (angle between the direction of movement of gas flow and area of each shutter) and clear opening area.

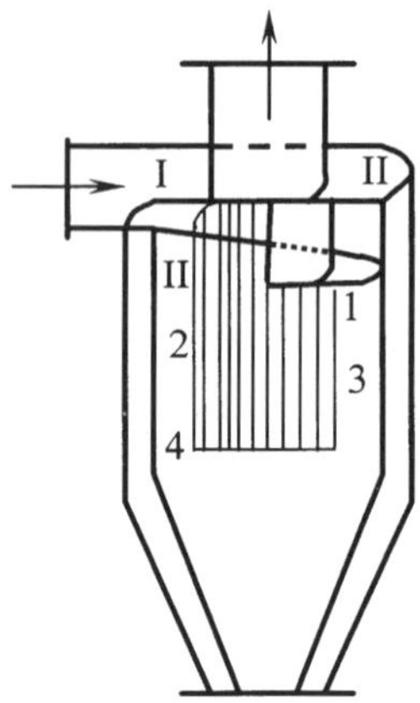

Figure 1 Dust catcher

During many years we have been improving design of these groups of apparatuses, upgrading their efficiency and structural elements in principle. We have most dramatically and effectively have improved construction of the separator's shutters. Construction features of dust catchers and their elements are protected by author's certificates and patents [3, 4]. All experimental investigations were carried out using the standard experimental stand of the State University «Lviv Polytechnic» in accordance with the «Uniform manual» for comparative specifications of dust catchers using standard quartz dust with clearly specified median diameter.

In the result of experiments it has been determined:

1) Effectiveness of dust catching increases with the increase of: total consumption of the air in the stand; median diameter of dust particles entering the dust catcher; hydrodynamic resistance;

2) fraction effectiveness of dust catchers increases with the increase of median diameter of dust particles;
3) hydrodynamic resistance of the dust catcher increases with the increase of consumption of the air in the stand.

Summing up the above one can insist, that effectiveness of dust catching will be increasing with:

- higher velocity of the dust particle before the separator;
- greater weight of the dust particle;
- greater density of the dust particle;
- smaller angle between the direction of movement of the flow and area of the shutter;
- lower velocity of movement of cleaned air flow to the separator opening;
- greater clear opening area of the shutter separator;
- greater flexibility of dust particle;
- greater stability of the velocity of dust-air mixture along the separator, ensuring constant velocity of air passage through the separator.

Table 3 Introduction of apparatuses into production

TYPE OF APPARATUS	ENTERPRISES WHERE THEY HAVE BEEN INTRODUCED
Cylindrical-conic	■ Woodworking, Tychy, Poland; ■ Production Association «Lvivprombud», Lviv, II stage; ■ Building Materials Plant 2; ■ Sandblaster, foundry of Production Association «Avtonavantazhuvach», Lviv; ■ Cement industry; ■ Ashphalt-concrete manufacture; ■ Flax treatment, I stage ■ Metalwork, Production Association «Electron», Lviv, ■ «Mechanical Plant», Lviv; ■ Production Association «Conveyer» ■ Production Association «Chemical Plant», minium manufacture.
Cylindrical-conic with hopper	■ Ceramic tile, Production Association «Ceramic Plant», Lviv ■ All ash catchers: Lviv Railways; Tukhlya, Lviv region; ■ Reinforced concrete constructions Plant, Stryy, Lviv region; ■ Reinforced concrete constructions Plant, Lviv; ■ Flax treatment, II stage; ■ Metalwork, «Mechanical Plant», Lviv; ■ Clay powder production, Production Association «Chemical Plant»

CONCLUSIONS AND PERSPECTIVES OF FURTHER SCIENTIFIC STUDIES

Comparative investigations have proved, that effectiveness of operation of the proposed dust catcher surpasses effectiveness of operation of cyclone CN-11 (standard) by 15-17%, hydrodynamic resistance is reduced by 150-250 Pa, and dimensions by 1.5 times, which permits to make a statement of considerable reduction of power and metal consumption.

Further works in this field should be conducted towards improvement of the design of shutter separator as the second stage of cleaning, towards optimization of the dimensions of dust catcher body and improvement of the processes of separation for further increase of effectiveness of dust catching, development of new energy saving technologies and protection of the environment.

In order to develop high performance equipment for fine dust catching it is necessary to continue research works in the field of development of acoustic, magnetic and magnetic-acoustic dust catching equipment.

REFERENCES

1. BATLUK V.A., AZARSKYY K.I. Mathematical software for selection of optimal equipment for cleaning air from dust with the aid of computers // Ukrainian journal of medical engineering and technology. - Kiev. - 2000. -#2. pp. 92-94.

2. BATLUK V.A. Mathematical models of the processes of separation of heterogeneous systems in dust cleaning // International scientific practical conference «New machines for manufacture of building materials and constructions, modern building technologies». Poltava. - 2000. - pp. 87-91.

3. BATLUK V.A., BATLUK V.K., SHELYUKH YU.YE. Ukr. Patent No 54119 A (2003).

4. BATLUK V.A., SHELYUKH YU.YE. Ukr. Patent No 53864 A (2003).

STUDY OF THE HYDRATION PRODUCTS OF MIXTURES OF CEMENT AND ASPHALT EMULSION USED IN RECYCLING WORKS OF FLEXIBLE PAVEMENTS

V Kasselouri-Rigopoulou S Kolias
M Louverdi A Karahalios
National Technical University of Athens
Greece

ABSTRACT. The method of recycling of flexible pavements with cement or with bituminous emulsion and cement is finding increase use in cases of rehabilitation of old flexible pavements needing improvement not only of their functional but also of their structural characteristics. These recycling methods are recognized for their contribution to the sustainable development. In this work the results of a study of the hydration products of mixtures of cement and of bituminous emulsion and cement used in recycling works are presented. Mixtures of cement and bituminous emulsion in various percentages as well mixtures of a) aggregates with cement and b) aggregates and combinations of bituminous emulsion and cement are analysed in order to study the hydration products and the cohesion characteristics between i) cement and aggregates ii) cement and bitumen and iii) cement and aggregates in presence of bitumen. For this purpose X-ray Diffraction, Optical Microscopy and Scanning Electronic Microscopy analysis are used. It is found that the bituminous emulsion is not prohibiting the hydration cement but is delaying it. It seems also that the cohesion characteristics between hydration products and bitumen, and/or aggregates is not impaired by the presence of bitumen.

Keywords: Cement hydration, Bituminous emulsion, Recycling of flexible pavements.

V Kasselouri–Rigopoulou is a Professor in the Department of Chemical Engineering, NTUA. Her specialist interests are in the areas of Inorganic Chemistry and High Temperature Chemistry and Technology.

S Kolias is an Associate Professor in the Civil Engineering Department of NTUA. His specialist area is Pavements and Pavement Materials.

M Louverdi is a Post Graduate student in the Department of Chemical Engineering, NTUA.

A Karahalios is a Post Graduate student in the Department of Civil Engineering, NTUA.

INTRODUCTION

Cold in-situ recycling of distressed flexible pavements with cement, or with emulsion and cement, or with foamed asphalt and cement, are methods which are finding increasing use and are recognized for their contribution to the sustainable development.

In particular for the method of recycling with bituminous emulsion, the addition of cement has been found that it produces, in most cases, mixes with enhanced mechanical characteristics in comparison with the mixes without cement. While there is an abuntance in data regarding the mechanical characteristics of the recycled mixes with emulsion and cement [1, 2, 3, 4], there is very few information [5] regarding the mechanism of cement reaction in the mix and its hydration products.

In this work an effort is made to investigate this subject using X-ray Diffraction, Optical Microscopy and Scanning Electronic Microscopy.

EXPERIMENTAL

Two series of mixtures were prepared:

a) Mixtures containing cement, bituminous emulsion and aggregates
b) Mixtures containing cement and emulsion only.

The w/c ratio was kept to 0.4.

The materials used were :

- Limestone aggregates
- Cationic bituminous emulsion KE-5 (50% water)
- Portland cement CEM II A 32.5N

The composition of samples prepared is presented in Table 1.

Table 1. Composition of mixtures

SAMPLE CODE	% CEMENT	% EMULSION	% AGGREGATE	% WATER
DOC0	4.0 (3.62)	0	90.58	5.8
DOC1	4.0 (3.62)	2.0	90.58	3.8
DOC2	2.0 (1.81)	2.0	92.40	3.8
DOC3	50	50	-	-
DOC4	58	29	-	13
DOK5	60	-	-	40

All samples cured at temperature 22±2°C and RH>95% for 7, 14, 28 and 40 days. X-ray Diffraction, Optical Microscopy and Scanning Electron Microscopy used for the investigation of the behaviour of the mixtures during their hydration. XRD analysis of samples containing only cement and emulsion gave representative results due to the presence of appropriate amounts of both materials resulting to higher peaks than those in the samples with aggregates. For this purpose in this paper only the XRD patterns concerning pure cement-emulsion mixtures are presented in this report.

RESULTS AND DISCUSSION

X-ray Diffraction (XRD) study

In Figure 1 XRD patterns of pure cement paste at ages of 14 and 28 days are presented, where a normal hydration is observed.

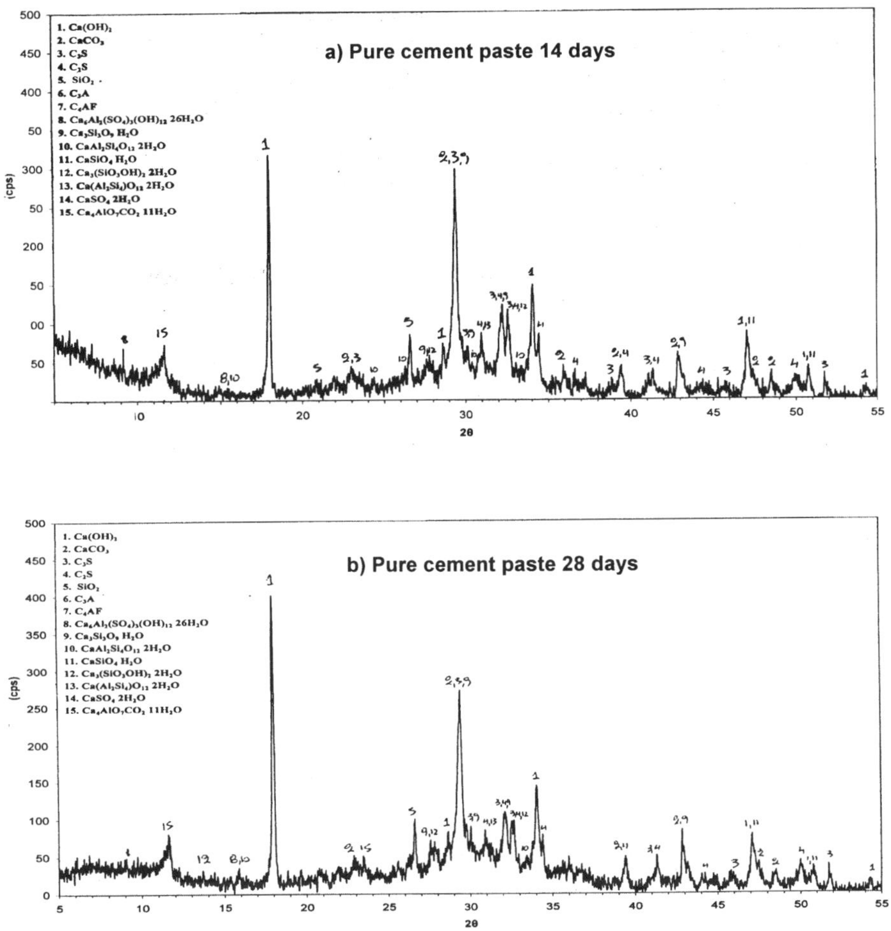

Figure 1 XRD patterns of pure cement pastes a) 14 days and b) 28 days of hydration.

In Figure 2, XRD patterns of cement-emulsion mixtures cured for 7 and 28 days concerning the mixtures with cement/emulsion=1:1 and 2:1, are presented,

At 7 days (DOK112 (1:1), DOK113 (2:1)) it seems that the emulsion influences the hydration, indicating a slight retarding. Peaks concerning calcium aluminate hydrates (No 15 in the list of compounds in Fig.1) and calcium silicate hydrates (No 9, 12 in Fig.1) are more intense in sample containing less emulsion. The same is observed at the age of 28 days. DOK123 (2:1) appears better hydration behavior, than DOK 122 (1:1).

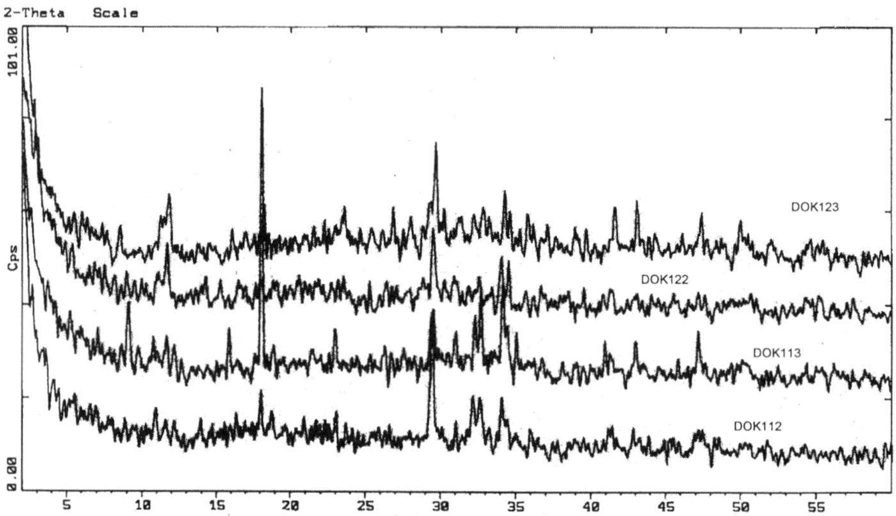

Figure 2 XRD patterns of hydrated cement-emulsion mixtures. DOK122(1:1):28 days, DOK123(2:1):28 days, DOK 112(1:1):7 days, DOK 113(2:1):7 days

In Figure 3, XRD patterns of samples with cement/emulsion ratio 1:1 (DOK122), 2:1 (DOK123) and 3:1 (DOK124) after 28 days of hydration, are presented. No significant differences between DOK123 and DOK124 are observed, while in DOK122 a retarding effect is observed. What that means is that after 28 days of hydration, mixtures containing less than 50% emulsion follow a normal rate of hydration, while mixtures with 50% emulsion appear a hydration retarding. The presence of ettringite (not earlier than 28 days (No 8 in the list of Figure 1) confirms this retarding.

Optical Microscopy (OM) study

The study was performed in samples containing cement, emulsion and aggregates. Figures 4 and 5 present OM photos concerning DOK0 (4% cement, 0% emulsion) after 40 days of hydration. It is shown that there is a good cohesion between cement paste and aggregates, without large pores and cracks. Figure 6 shows OM photo concerning DOK1 (4% cement and 2% emulsion). A good cohesion between cement paste and aggregates can be seen. A thin film also of bitumen around the aggregates is observed. In Figure 7 representing an OM photo of DOK2 (2% cement and 2% emulsion) the above observations are also confirmed. It seems that the presence of bitumen doesn't inhibit the cement hydration.

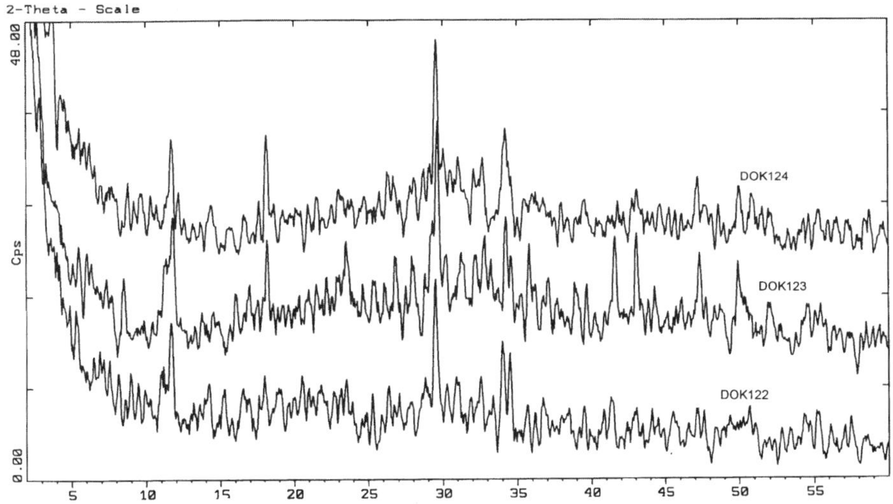

Figure 3 XRD patterns of hydrated cement-emulsion mixtures.
DOK122(1:1): 28 days, DOK123(2:1) : 28 days, DOK 124(1:3): 28 days

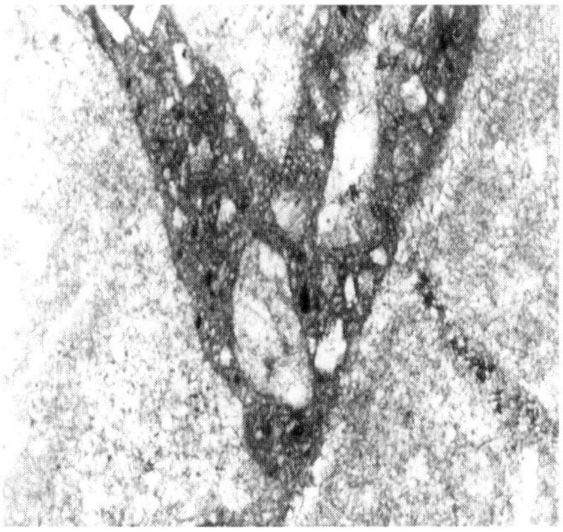

Figure 4 and 5 OM photo of DOK0 after 40 days of hydration.

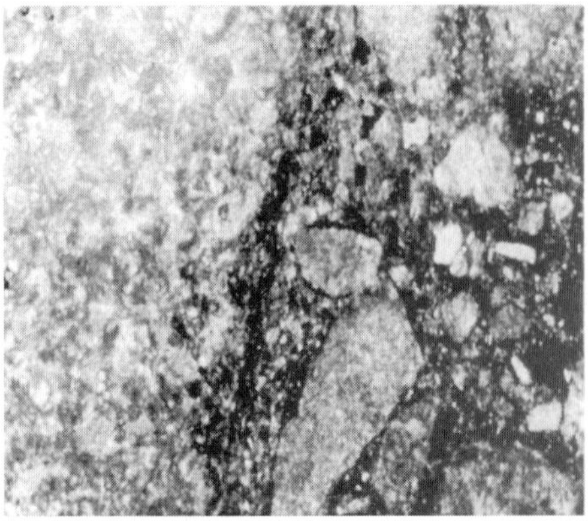

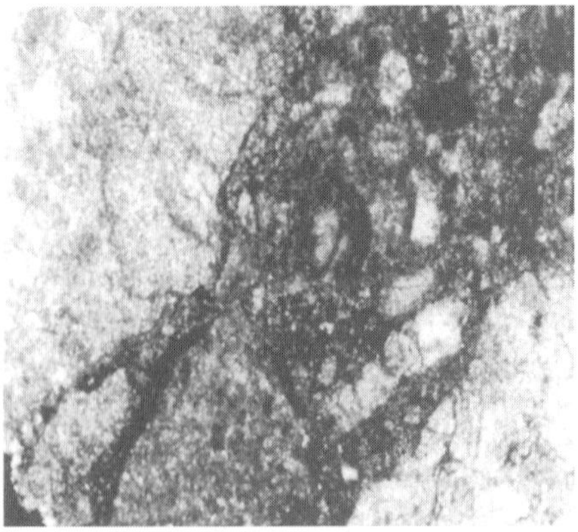

Figure 6 and 7 OM photo of DOK1 and DOK2 after 40 days of hydration, respectively

Scanning Electronic Microscopy (SEM) study

The mixture consisted of 50% cement and 50% bituminous emulsion has been studied at the age of 14 and 28 days.

Figures 8, 9, 10 and 11 concern SEM photos at the age of 14 days. There are observed areas with cement grains covered by amorphous bitumen (8, 9). Between the covered grains uncovered ones have been hydrated (10, 11). Figure 12 shows hydrated cement mixture at the age of 28 days. It seems that the hydration of cement is in progress. It is also clear that the influence of bitumen is weaker.

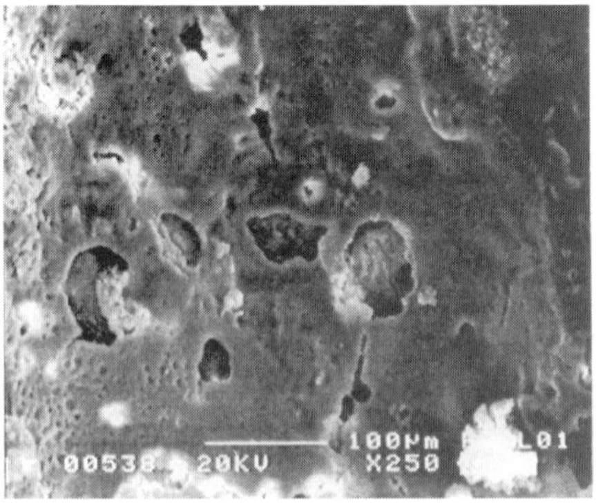

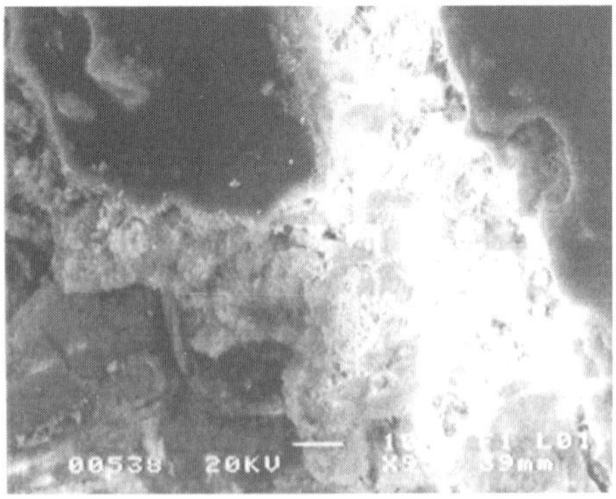

Figure 8 and 9 SEM photo of DOK3 after 14 days of hydration

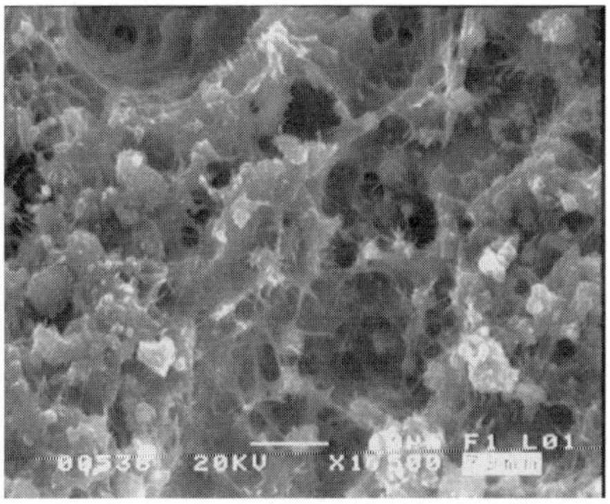

Figure 10 and 11 SEM photo of DOK3 after 14 days of hydration

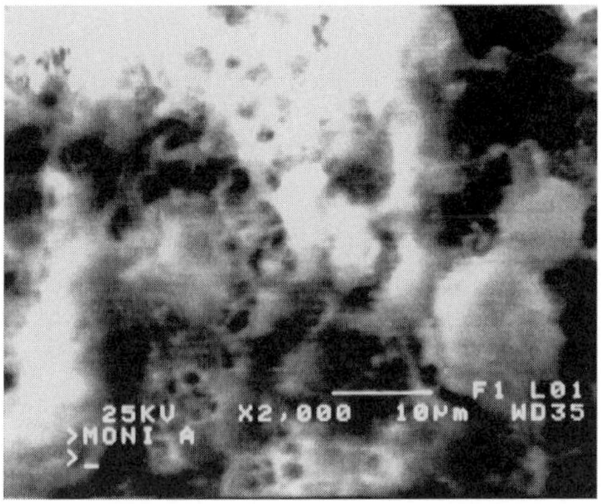

Figure 12 SEM photo of DOK3 after 28 days of hydration

CONCLUSIONS

The hydration of cement-bituminous emulsion mixtures is not inhibited by the contained bitumen, but it appears a slight delay up to 28 days. The cement/emulsion ratio influences this delay, but for emulsion quantities used in practice this delay maybe considered as not significant. Cohesion characteristics between hydration products and bitumen and/or aggregates are not impaired by the presence of bitumen.

REFERENCES

1. JEAN – FRANCOIS CORTE, "Cold–in–place recycling with emulsion or foamed bitumen", PIARC Seminar on Road Pavement Recycling, Warsaw, 10-11/Oct/2002.

2. KONRAD ROSSBERG, "In-situ pavement recycling using cement and bitumen combined as binder – strength parameters and long-term performance", 1st Int. Symposium on subgrade stabilization and in situ pavement recycling using cement, Salamanga, 1-4/Oct/2001.

3. SIMONE, A BONINI, C. SANGIORGI, "The influence of Portland cement on cold-mix asphalt recycling", 1st Int. Symposium on subgrade stabilization and in situ pavement recycling using cement, Salamanga, 1-4/Oct/2001.

4. STEYN, WJVDM, "Conceptual performance model for deep in situ recycled pavements with cement and bitumen additives", 1st Int. Symposium on subgrade stabilization and in situ pavement recycling using cement, Salamanga, 1-4/Oct/2001.

5. Y. KANNO, T. KUROKAWA, T. MARUSHIMA, "Application of cement and asphalt emulsion to in situ recycling base course method", 1st Int. Symposium on subgrade stabilization and in situ pavement recycling using cement, Salamanga, 1-4/Oct/2001.

6. F. GIULIANI, "X-Ray diffraction method for studying cement-modified bitumen-emulsion mixtures in asphalt pavement cold recycling", 1st Int. Symposium on subgrade stabilization and in situ pavement recycling using cement, Salamanga, 1-4/Oct/2001.

CLOSING PAPER

MAJOR CHANGES IN CEMENTS: CURRENT STATUS AND THE EFFECTS ON A SUSTAINABLE FUTURE

D M Roy

B E Scheetz **P J Tikalsky**

Pennsylvania State University

United States of America

ABSTRACT. This paper discusses recent studies carried out to investigate the broad variation in characteristics of currently manufactured Portland cements, ASTM Types I and II, and their effects on concrete durability. Twenty-one diverse cements and clinkers were statistically selected from 114 commercially available North American cements, chosen to span the characteristics that were identified as the most likely to significantly impact the long-term durability of the concretes in which they were used. After detailed characterization and evaluation, laboratory studies were made of primarily the four properties: sulfate resistance, alkali-silica reaction, rapid chloride permeability, and delayed ettringite formation. Other reference properties such as heat of hydration and strength are of course included. Relationships between some 72 measured cement characteristics and various properties were evaluated. Such information provides a solid starting point for understanding the behavior of Portland cements augmented with the addition of supplementary cementitious materials such as fly ash (Class C and F), silica fume, ground granulated blast furnace slag, and metakaolin and other pozzolans. The use of the latter materials has at least two goals; (i) utilization of by-product materials from other industries in the manufacture, (ii) allowing greater tolerances in variation of Portland cement composition. This provides considerable savings to the environment and savings in the use of raw materials in cement production, which would otherwise be rejected.

Keywords: Cement characteristics, Sulfate resistance, Alkali-silica reaction, DEF, Concrete durability, Raw materials, Supplementary cementitious materials

D Roy is Professor Emerita of Materials Science, Materials Research Institute, Materials Research Laboratory, The Pennsylvania State University. Her recent research focuses on cement characterization, alkali-activated cements and concrete durability.

B Scheetz is Professor of Materials, Materials Research Laboratory, and Civil and Nuclear Engineering, The Pennsylvania State University. His present research emphasizes environmental remediation and materials characterization.

P Tikalsky is Associate Professor of Civil Engineering. His recent research focuses on durability of concrete structures and evaluation of performance of structures.

INTRODUCTION

Scientists have attempted to understand the mechanisms behind the reactions governing the manufacture and usage of Portland cement for 150 years [1-5]. The knowledge in this field is at a point where specific chemical compounds have been targeted as affecting certain concrete properties. Yet, in many cases, the exact mechanism under which the specific chemical reacts, as well as the extent to which other chemical species contribute, is not completely understood. Gonnerman and Lerch [6] conducted the first major study on the interrelation between cement and concrete properties for cements produced in the United States. They studied the changes in characteristics of Portland cement as exhibited by laboratory tests over the period 1904-1950. The change in chemical and physical properties of Portland cement over two centuries was reviewed by Corish and Jackson [7]. However, the most comprehensive and thorough study on the interrelation between cement and concrete properties of the cements produced in the United States was done by Blaine and co-researchers at the National Bureau of Standards (NBS) from data collected in 1953, the first major report occurring in 1968 [8]. This comprehensive study on the vast number of North American Portland cements paved the way for changes in the American Society for Testing and Materials (ASTM) C150 specification.

The changes in cement manufacturing and physical characteristics have prompted the need for a new look into the interrelations between cement and concrete properties. ASTM conducted a survey of Portland cements manufactured in the United States in 1994 (Gebhardt [9]). The primary purpose of this survey, submitted to ASTM Committee C-1, was to determine the characteristics of the Portland cement marketed in North America to meet the requirements of ASTM C150. Similar studies were performed in the U.K. with the Concrete Society Working Party [10] and Osborne and Hardcastle [11]. According to the latter, the major changes in Portland cement composition over the years can be summarized as follows:

1. C_3S contents have risen steadily, whereas corresponding C_2S levels have fallen. The proportions of the other main components, C_3A and C_4AF, have not changed much.
2. Modern cements have become finer. The change in particle size distribution of the cement particles is not well documented over time.
3. The alkali level of cements has increased due to changes in cement manufacture from wet processes to more energy-efficient dry processes, and the recycling of cement kiln dust into the raw feed.
4. Heats of hydration of cements have increased – more specifically, the heat of hydration at earlier ages of curing.
5. Increases in the C_3S/C_2S ratio have resulted in increases in compressive strength at early ages.

Currently, although many concrete bridge decks, pavements, highway structures and other concrete structures have been built with concrete mixtures incorporating Portland cements, meeting these requirements, they have exhibited varying degrees of premature deterioration caused by factors such as internal and external sulfate attack, alkali aggregate reaction, freezing and thawing deterioration, reinforcing steel corrosion, and more recently, delayed ettringite formation. There has been considerable debate about the contribution of Portland cement to this deterioration, particularly in view of the adoption of cost-effective, energy-efficient, and environmentally beneficial cement production methods that seem to influence cement characteristics.

Although a great deal of research has been performed to address the various aspects of concrete durability, this research has not provided clear conclusions concerning the combinations of Portland cement characteristics that will enhance concrete durability. Such research was needed to assess the effects of Portland cement characteristics on concrete durability; identify combinations of chemical, physical and mineralogical requirements to ensure durability; and develop recommendations to help improve Portland cement specifications.

Described in the following pages is an experimental study that was undertaken in order to address this problem, which has been summarized in a larger report [12]. In the current presentation, the focus is upon the relationship of cement characteristics to the three specific properties as representative of concrete durability: sulfate resistance, alkali-silica reaction and delayed ettringite formation.

EXPERIMENTAL STUDIES

Cements and Their Characterisations

Our recent study was undertaken to evaluate the effect of current cement characteristics on concrete durability. A survey and sampling of North American cements was conducted with the assistance of the Portland Cement Association. A statistical analysis model based on the design of a weighted central response surface was developed using these data. For this study, twenty-one cements were selected from the 114 ASTM Type I and Type II cements manufactured in the United States. Selection was made on a statistical basis to provide a systematic coverage of the range in variability that bounded the extremes of manufactured cements for the following parameters:

- Fineness
- C_3A
- Na_2Oeq
- SO_3

Extreme low, high and mean values of four characteristics were chosen from this analysis. Data from the North American survey of cement manufacturer's data revealed typical high, low, average and extreme values. These values are displayed in Table 1 (as target values).

Table 1 Target values for selected cement characteristics

RESULT	PROPERTY			
	SO_3 %	Equivalent Alkalis %	C_3A %	Fineness Blaine
Extremely High	4.2	1.1	15	4600
High	3.6	0.85	12	4200
Average	3	0.6	9	3800
Low	2.4	0.35	6	3400
Extremely Low	1.8	0.1	3	3000

Table 2 gives the analytical data for the twenty-one selected cements, analyzed in accordance with ASTM C 114 procedures.

Table 2 Chemical Characteristics of Portland Cements in the study

Code	SiO_2	Al_2O_3	Fe_2O_3	CaO	MgO	SO_3	LOI	Na_2O	K_2O	TiO_2	P_2O_5	ZnO	Mn_2O_3	B_2O_3	BaO	Cr_2O_3	NiO	SrO	Cl
1	20.91	3.97	2.99	62.97	3.27	2.83	1.46	0.10	0.45	0.21	0.03	0.105	0.134	0.030	0.046	0.012	0.011	0.147	< 0.001
2	21.45	4.01	4.02	64.65	1.39	2.38	0.59	0.20	0.49	0.25	0.16	0.018	0.067	0.015	0.018	0.028	0.016	0.090	0.001
3	20.26	4.45	3.20	63.81	1.55	4.13	1.52	0.16	0.46	0.23	0.04	0.022	0.420	0.020	0.022	0.155	0.010	0.054	< 0.001
4	19.79	5.46	1.61	64.49	1.16	3.77	1.64	0.15	0.92	0.28	0.11	0.014	0.052	0.021	0.020	0.016	0.011	0.106	0.005
5	20.56	5.02	1.94	63.77	3.77	2.73	1.23	0.09	0.56	0.34	0.19	0.024	0.095	0.016	0.025	0.018	0.014	0.328	< 0.001
6	20.45	5.61	2.90	63.94	1.88	3.00	1.02	0.18	0.12	0.55	0.07	0.024	0.078	0.062	0.045	0.032	0.020	0.108	0.001
7	19.81	6.03	2.54	64.46	0.88	3.04	1.37	0.24	0.39	0.63	0.21	0.032	0.061	0.015	0.041	0.018	0.009	0.300	< 0.001
8	20.08	5.47	2.90	63.77	1.15	3.13	2.11	0.12	0.79	0.29	0.12	0.017	0.156	0.012	0.068	0.011	0.007	0.178	0.008
9	20.49	5.14	2.54	63.10	1.74	3.84	1.41	0.18	1.00	0.30	0.11	0.018	0.151	0.014	0.021	0.010	0.007	0.121	0.002
10	21.12	4.42	2.48	63.89	1.99	2.81	1.44	0.47	0.67	0.24	0.16	0.012	0.048	0.010	0.034	0.012	0.017	0.198	0.001
11	19.61	5.25	2.60	62.43	2.11	4.35	1.71	0.20	1.09	0.25	0.08	0.040	0.090	0.017	0.016	0.025	0.010	0.108	0.003
12	20.54	4.56	3.18	61.01	5.01	3.48	1.12	0.25	0.82	0.26	0.05	0.011	0.101	0.018	0.033	0.011	0.007	0.076	< 0.001
13	21.33	4.04	2.88	63.85	1.25	2.68	1.52	0.13	0.36	0.18	0.20	0.017	0.240	0.019	0.044	0.010	0.009	0.230	0.002
14	21.08	4.70	3.12	63.96	1.22	2.68	1.46	0.10	0.61	0.20	0.09	0.013	0.069	0.009	0.026	0.020	0.007	0.050	0.003
15	20.20	4.58	2.59	62.65	2.09	3.26	1.74	0.20	0.93	0.23	0.18	0.041	0.092	0.022	0.016	0.025	0.008	0.113	0.006
16	20.77	4.55	1.52	63.71	2.48	3.05	1.46	0.06	0.02	0.36	0.15	0.040	0.156	0.022	0.021	0.029	0.009	0.040	0.002
17	21.25	4.29	2.44	63.37	2.04	3.19	0.84	0.39	0.71	0.27	0.18	0.017	0.116	0.017	0.024	0.029	0.011	0.250	0.002
18	19.72	5.27	1.97	63.14	3.27	3.43	0.73	0.32	1.15	0.22	0.23	< 0.005	0.063	0.028	0.020	0.010	0.006	0.320	0.011
19	20.29	5.54	2.90	64.97	1.04	2.73	0.70	0.11	0.83	0.28	0.17	0.006	0.143	0.021	0.048	0.010	< 0.005	0.144	< 0.002
20	20.68	4.84	3.23	64.65	1.85	2.90	0.70	0.04	0.54	0.30	0.07	< 0.005	0.024	0.019	0.022	< 0.005	< 0.005	0.050	< 0.002
21	20.40	4.63	2.91	61.76	3.96	3.18	1.36	0.32	0.57	0.28	0.10	0.010	0.041	0.018	0.021	0.016	< 0.005	0.098	< 0.002

Using these values as guidelines, ASTM Type I and Type II cements were selected from a coded manufacturer inventory sheet. Five "average" cements were selected in order to form a basis for comparison. Eight cements were selected as "corner points" for the model to represent high and low value combinations for the characteristics of interest. Another eight cements were selected as "star points" representing the extreme high and extreme low values for each of the four characteristics.

Additional physical and chemical characteristics determined for the cements investigated are given in Table 3 [13] along with the Bogue mineralogical composition.

Other cement characterization studies made included microscopic characterization of each clinker, by reflected light incroscopy, Normarski differential interference microscopy, scanning electron microscopy, and computer controlled SEM point counting, reported elsewhere [14]. Also, quantitative phase analysis was determined by x-ray diffraction methods (GM Quant and Rietveld methods), also reported elsewhere [15].

Table 3 Physical and chemical characteristics of Portland cement in the study

PROPERTY													
Bogue Chemical Composition % By Mass					Bet Surface Area	Blaine Fineness	Wet Sieve Analysis ASTM C430	Normal Consistency ASTM C187	Isothermal Cal 24 Hrs	Heat Of Hydration Astm C186 (Kj/Kg)		
Code	C_3S	C_2S	C_3A	C_4AF	m^2/g	m^2/kg	% by mass	% by mass	kj/kg	7 day	14 day	28 day
1	58.4	15.9	5.5	9.1	1.83	4084	94.11	24.5	180	341	418	431
2	60.7	15.7	3.8	12.2	1.05	3622	93.66	24.5	65	289	293	418
3	59.5	13.2	6.4	9.7	1.73	4048	97.32	25.5	180	257	344	347
4	62.4	9.7	11.7	4.9	1.48	3282	89.95	24.7	172	355	367	380
5	59.1	14.4	10.0	5.9	1.44	3871	89.81	24.3	168	392	475	468
6	54.5	17.5	10.0	8.8	1.32	3831	94.36	25.0	160	209	455	456
7	59.0	12.2	11.7	7.7	1.80	3705	96.30	25.7	178	429	479	523
8	57.2	14.4	9.6	8.8	1.82	4009	91.28	25.0	198	334	385	451
9	52.0	19.5	9.3	7.7	1.54	3871	89.03	25.0	151	287	380	497
10	58.3	16.5	7.5	7.5	1.41	4345	95.97	25.0	213	403	478	497
11	53.7	15.7	9.5	7.9	1.42	4029	92.67	24.5	176	428	581	598
12	47.2	23.3	6.7	9.7	1.50	3950	96.88	25.7	150	222	226	234
13	58.9	16.7	5.8	8.8	1.54	3768	95.60	25.0	171	311	401	485
14	56.5	17.8	7.2	9.5	1.36	3600	97.13	25.1	169	348	362	410
15	57.8	14.3	7.8	7.9	1.47	4179	91.80	25.3	196	468	502	535
16	60.1	14.2	9.5	4.6	1.59	3850	94.71	24.9	217	357	362	***
17	55.1	19.4	7.2	7.4	1.17	3384	95.82	24.6	157	321	394	431
18	59.2	11.9	10.6	6.0	1.03	3663	95.21	26.8	165	301	456	426
19	61.1	12.0	9.8	8.8	1.07	3803	89.25	24.6	172	314	385	497
20	60.6	13.6	7.4	9.8	1.05	3930	95.67	25.2	176	322	401	435
21	52.0	19.2	7.3	8.9	1.37	4104	95.56	27.0	173	322	460	481

DURABILITY STUDIES

Sulfate Resistance

Sulfate attack causes deterioration in concrete structures exposed to moist, sulfate-rich environments. The resistance to sulfate attack has long been known to be connected to the characteristics of the cementitious material [2, 16-19], however the complex reactions in hydrating cement have made it difficult to isolate more than one variable at a time. The statistical relationship of multiple cement characteristics to sulfate attack is the subject of this research. The research study performed sulfate resistance testing on mortar specimens prepared from twenty-one cements with statistically broad chemical and physical characteristics, and developed multivariate model relationships between these characteristics and the results of sulfate resistance testing using Na_2SO_4 as the aggressive medium.

Mortar mixtures were prepared according to ASTM C 1012, "Standard Test Method for Length Change of Hydraulic Cement Mortars Exposed to a Sulfate Solution." For each cement, between six and eight mortar bars were immersed in a 50 g/L sodium sulfate (Na_2SO_4) solution. Comparator readings were taken at 1, 2, 3, 4, 8, 13, and 15 weeks, and at 4, 6, 9, and 12 months. Measurements were made using a length comparator according to ASTM C 157.

Mortar cubes (51 mm) were made using the same mortar. Half of the cubes were immersed in the sodium sulfate solution, while the others were cured in a lime bath. These cubes were evaluated for strength according to ASTM C 109. Additional information is given in Ref. [20].

Data compiled during this study were cataloged and analyzed using several computer software programs. Raw data were entered in a formatted database and subsequently analyzed for pertinent properties and trends. The raw data were then inserted into the database with the cement characterization data. A series of plots were made using one independent variable (cement characteristic) and one dependent variable (experimental result). Visual inspection of these plots, when combined with statistical analysis results, allowed for first order relationships between cement characteristics and expansion to be further understood.

Results

Expansion rates differed significantly among the cements in the study. The 12 month data were not useful in modeling statistical behavior because more than half of the cements had expanded to fracture. The distribution of 6-month expansions are shown in Figures 1 and 2 where it is seen that there is a strong (negative) correlation between calculated C_4AF content and 6-month expansion. Similar results exist in correlation with Fe_3O_3 content.

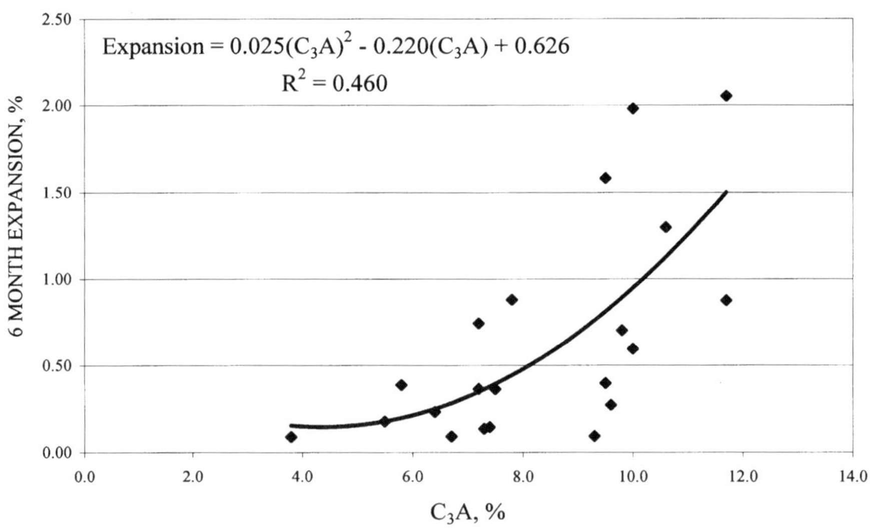

Figure 1 Correlation between C_3A and sulfate expansions

The research study looked to improve on the correlation using multivariate analysis using the 6-month expansion data. Multivariate analysis of potential predictive variables was performed after 6 months using a commercially available statistical software package. The 6-month expansions were evaluated for all first and second order effects [12]. The most relevant variables were combined to evaluate potential third order correlations. Using polynomial equations, Fe_2O_3, C_4AF, C_3A, total equivalent alkalies, K_2O, Na_2O, Al_2O_3, C_3S/C_2S, C_3A/C_4AF, and SrO were correlated with sulfate expansions. In nearly every case, equations containing Fe_2O_3 or C_4AF were the best predictive equations. The most accurate predictions came from the use of Fe_2O_3 and total alkalies. The R^2 value of 0.84 is very good for an empirical relationship and is substantially better than any predictive equation containing C_3A. The best predictive equation containing C_3A involved the relationship containing the C_3A/C_4AF ratio, with an R^2 value of 0.78, slightly improved over C_4AF alone [13].

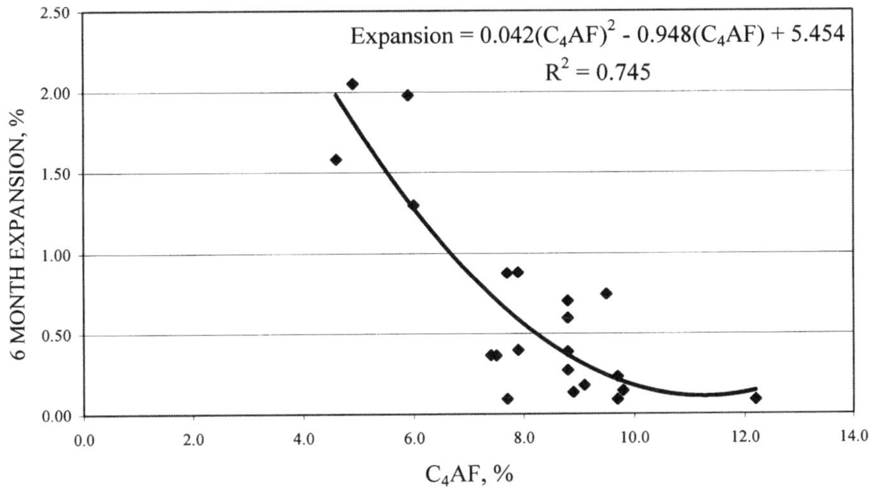

Figure 2 Correlation between C_4AF and sulfate expansions

In conclusion, although C_3A content has typically been assumed the characteristic responsible for the lack of sulfate resistance, this study has confirmed that it is not the strongest controlling factor for expansion in alkali sulfate environments. The very high negative correlations between both Fe_2O_3 and C_4AF and sulfate attack expansion reveal that the cements with high iron contents are the most beneficial in controlling sulfate attack. The iron-rich ettringite formed from these high-iron cements is not as expansive. The roles of alkalies and alumina are clearly part of the sulfate attack mechanism, but iron has a strong mitigating effect.

Alkali Silica Reaction

The objective of this part of the investigation was to statistically examine the effects of cement characteristics on alkali silica reactivity of modern concretes. To this end, test mortars were prepared as nearly as possible to being identical and allowing only the cement composition to change from one concrete mixture to the other [20].

Alkali silica reaction (ASR) occurs when soluble alkali ions present from the cement, admixtures, or other sources, such as sodium and potassium, react with certain types of silica structures present in the aggregate [21]. This reaction results in the formation of a silica gel that is capable of imbibing large quantities of water, causing internal stresses, which ultimately result in cracks. The cracks then fill with additional gel, increasing the stress on the microstructure, which forms a "weak link" in the matrix. Three conditions are necessary for ASR to occur: (1) a cementitious mixture sufficiently high in soluble alkalis, (2) the presence of a soluble form of silica, and (3) the presence of water. It has also been found that a relative humidity of 80 percent or greater referenced to 21°C to 24°C is required to support the ASR chemical reactions [22]. In addition, the chloride ion may contribute to the alkali-silica reaction if it is present in the pore solution in significant amounts [23, 24].

Materials and procedures

The materials used in this study include Portland cement, fine aggregate, and crushed Pyrex cullet. The Pyrex cullet used for the ASTM C 441 test was obtained from the original inventory used by the U.S. Bureau of Mines. The non-reactive aggregate is a locally derived siliceous sand that tested non-reactive and possesses an excellent field performance history. As a part of the testing, two test protocols to monitor the development of ASR were followed in this study, ASTM C 441 and C 227 [12]. Because the results from ASTM C 227 did not show significant expansion at periods up to 1 year, no further discussion of these results will be made. Mortar bars were prepared using a specified gradation of Pyrex glass, which optimizes the reaction resulting in rapid screening of the cements.

For the ASTM C 441 test, the bars were suspended above water, with an adequate fabric wicking system providing a stable humidity level throughout the sealed outer container to promote the maximum amount of expansion [25]. Three mortar bar samples were prepared for each cement, and a set of control bars was prepared with non-reactive natural sand aggregate as well for comparison. The storage containers were placed in a 38°C chamber for the duration of the experiment. Length measurements were made at the end of a two-week undisturbed time period in the storage container, and subsequently at 4, 6, 8 and 12 weeks. Additional details are given elsewhere [12, 20].

Results

The general trends of ASR expansion for all 21 cements are determined from the obtained data, Table 4. At 8 weeks, total equivalent alkali content and K_2O appear to be the only two significant contributors to alkali-silica expansion. The regression results reveal that all other cement characteristics have significantly lower r^2 values. The majority of the 21 cements showed no additional expansion between the 8-12 weeks of exposure periods, consistent with the use of the 14-day period as a test criterion for estimating expansion.

Table 4 ASR Mortar Bar Expansion at Various Exposures by ASTM C 441 Test

CEMENT N°	PROPERTY					ASR EXPANSION, %				
	PCA Number	SO_3	Eq. Na_2O	C_3A	Blaine Finenes s	Weeks 2	4	6	8	12
1	2414	2.83	0.39	5.50	4084	0.01	0.01	0.01	0.01	0.01
2	2312	2.38	0.52	3.80	3622	0.10	0.12	0.12	0.13	0.12
3	2403	4.13	0.46	6.40	4048	0.16	0.19	0.18	0.19	0.18
4	1056	3.77	0.76	11.70	3282	0.36	0.39	0.39	0.39	0.24
5	1043	2.73	0.46	10.00	3871	0.05	0.07	0.07	0.07	0.07
6	1016	3.00	0.26	10.00	3831	0.00	0.00	0.01	0.02	0.01
7	1013	3.04	0.50	11.70	3705	0.18	0.22	0.22	0.22	0.22
8	1024	3.13	0.64	9.60	4009	0.20	0.21	0.28	0.22	0.22
9	1039	3.84	0.84	9.30	3871	0.32	0.34	0.36	0.36	0.37
10	1015	2.81	0.91	7.50	4345	0.30	0.39	0.40	0.41	0.50
11	1028	4.35	0.92	9.50	4029	0.36	0.37	0.37	0.38	0.43
12	1051	3.48	0.79	6.70	3950	0.24	0.28	0.27	0.22	0.28
13	1104	2.68	0.37	5.80	3768	0.00	0.01	0.01	0.02	0.01
14	1060	2.68	0.50	7.20	3600	0.08	0.09	0.09	0.09	0.09
15	1027	3.26	0.81	7.80	4179	0.37	0.38	0.37	0.38	0.38
16	1059	3.05	0.07	9.50	3850	0.01	0.01	0.00	0.01	0.00
17	1017	3.19	0.86	7.20	3384	0.22	0.24	0.24	0.24	0.24
18	1007	3.43	1.00	10.60	3663	0.35	0.37	0.38	0.38	0.38
19	1020	2.73	0.66	9.80	3803	0.13	0.14	0.14	0.14	0.14
20	2322	2.90	0.39	7.40	3930	0.02	0.02	0.02	0.03	0.02
21	2334	3.18	0.70	7.30	4104	0.15	0.18	0.19	0.19	0.19

Single variant analysis of these cements suggests that the single most important chemical factor related to the maximum expansion is equivalent Na_2O (Figure 3). The cements that show the maximum expansion have high equivalent Na_2O (above 0.70). Exceptions to this are the cements 17 and 21. These cements have equivalent Na_2O above 0.70 % but their expansions are not large and lack of expansion could not be correlated to any other property except SiO_2. Both have relatively high SiO_2 contents [low C_3S/C_2S ratios] which may reduce the otherwise anticipated expansion. Figure 4 compares the expansion data as a function of the single variables, Na_2O and K_2O. According to this plot, K_2O shows a higher single-variable correlation to Na_2O than ASR. However, the highest correlation obtained is still for total equivalent Na_2O rather than the individual alkalis. This suggests that controlling the total alkali content is more important for controlling ASR. Low r^2 values for the combinations in which equivalent Na_2O and SO_3 are not one of the variables show that, C_3A, Blaine fineness and C_3S/C_2S ratio have relatively little influence on ASR expansion.

The second most important chemical factor correlating to the ASR expansion was believed to be the SO_3 content of Portland cement (Figure 3). All the cements that show high expansion have SO_3 content above 3.25 %. The only exception to this was the cement number 3. This cement has very high SO_3 (4.10 %), but low alkali. SiO2 data showed a slight negative correlation. There was no correlation observed between C_3A, Blaine and ASR expansions. The cements that show high ASR have reasonably high Blaine fineness (exception number 4). But high Blaine fineness by itself does not have any relation to ASR, if equivalent Na_2O was very low.

Cement number 8 can be cited as an example to prove the above observation. These observations suggest that there may be a combination of factors responsible for ASR in concrete.

Response surface modeling

When these data were subjected to the full multi-variant non-linear modeling, only the equivalent alkali and the silica content of the clinker exhibit any significant correlation. A response surface model was fit to these data [20], which yielded the following relationship:

$$\% \text{ Expansion} = 1.05934 + 0.47883\ Na_2O_{eq} - 0.05634\ SiO_2$$

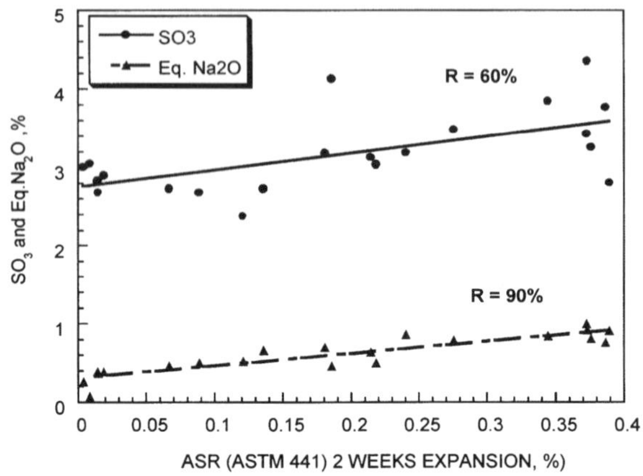

Figure 3 ASR Expansion at 2 Weeks (ASTM C 441 Test) versus Na_2O_{eq} and SO_3 Content of Portland Cements.

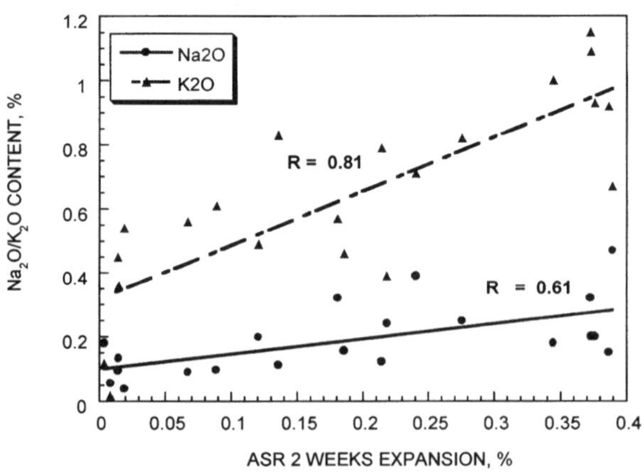

Figure 4 ASR Expansion at 2 Weeks (ASTM C 441 Test) versus Na_2O and K_2O Figure Content of Portland Cements

The study further determined that total equivalent alkali, Na_2O_{eg}, however, more strongly correlates to expansion confirming the contribution from sodium to the overall expansion. When viewed in context with all of the 75 variables examined in this study, the silica content of the cement in combination with Na_2O_{eg} demonstrated the most contribution to expansion. Cements with lower C_3S/C_2S ratios (higher SiO_2 contents) produce C-S-H compounds with lower C/S, which do not contribute as greatly to ASR as has been noted [20, 26].

Delayed Ettringite Formation

Among the mechanisms of degradation of concrete from the presence of excess sulfate identified in recent years has been the phenomenon of delayed ettringite formation (DEF). This phenomenon is a degradation process that sometimes is identified in concrete that has been exposed to curing at temperatures above 70 C [27]. However, suspect cases have also been identified at lower temperatures. The process, however, is complex and is still not thoroughly understood, even with intense research interest in the past decade that focused on understanding the process. According to some interpretations, any ettringite formed prior to curing at above 70 C is destroyed during the initial heat treatment, and the damage in the concrete structure is associated with the formation or reformation of ettringite during the subsequent service at ambient temperatures. The current study has attempted to extend the potential area of concern to more normally achieved field conditions [28].

For this study both air-entrained and non-air entrained mortar specimens were prepared according to ASTM C 305 as detailed in Table 5. To minimize variables, all aggregates used were non-expansive. A modification to the Stark method [29] was used to evaluate the expansion produced by SEF/DEF. The mortar bar samples were treated in three different exposure conditions as seen in Table 6. After initial curing at room temperature in saturated lime water, 1) air entrained mortar bar samples cured at 60°C (A60), 2) air entrained mortar bar samples cured at 80°C (A80), 3) and non-air entrained mortar bar samples cured at 60°C (NA60). Each drying period was followed by room temperature saturation and then freezing and thawing. The time for the total 3 cycles of curing/exposure was 242 days.

The data obtained for the total dimensional change after 3 cycles of the different exposure conditions show that only very limited net expansions were generally observed; and then in only a few of the 21 cements. The most expansive is shown in Figure 5, while a very non-expansive one is shown in Figure 6. The total data set is given in Ref. [12].

Table 5 Proportions of Mixtures Prepared for DEF Studies

COMPONENTS	AIR ENTRAINED	NON-AIR ENTRAINED
Sand	1650g	1650g
Cement	600g	600g
Water	319g	319g
Air entraining admixture	0.55mL	-

Table 6 Modified Stark SEF/DEF test

CYCLE	DAYS	ACTION
Initial	1-28	curing at RT in saturated limewater
	29-43	drying at 60C or 80C
1	43-99	curing at RT in saturated limewater
	100	24 hour freeze-thaw cycling
	101-114	drying at 60C or 80C
2	115-170	curing at RT in saturated limewater
	171	24 hour freeze-thaw cycling
	172-185	drying at 60C or 80C
3	186-241	curing at RT in saturated limewater
	242	24 hour freeze-thaw cycling

The majority of the cements, 17 in all, exhibited a net shrinkage, as a result of the thermal cycle in the testing sequence, that was larger than any expansion caused by ettringite formation. Statistical modeling with all of the chemical and mineralogical variables examined in this program resulted in a positive single variable correlation between expansion and sulfate of 70% for the air entrained samples cured at 80°C of the form:

$$Exp = 1.27 \times 10^{-2} - 5.66 \times 10^{-2} \times SO_3 + 1.19 \times 10^{-2} \times (SO_3)^2$$

and 67% for the non-air entrained samples cured at 60°C, with the form:

$$Exp = -7.9 \times 10^{-2} + 3.24 \times 10^{-3} \times SO_3 + 4.74 \times 10^{-3} \times (SO_3)^2.$$

The best correlation was observed between DEF expansion and the ration of SO_3/Al_2O_3 for all three sets of test specimens even though Al_2O_3 alone does not correlate with the observed expansions. This relationship took the form:

$$Exp = 1.9 \times 10^{-1} + 0.363346 \times SO_3/Al_2O_3 - 0.153024 \times (SO_3/Al_2O_3)^2$$

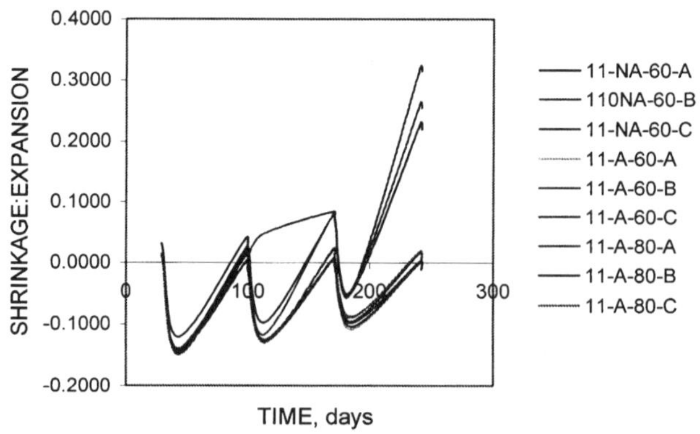

Figure 5 Normalized expansion data for Cement 11 for 3 cycles of DEF testing.

Scanning electron microscopic [SEM] characterization coupled with energy dispersive x-ray [EDX] analyses of the non-air entrained specimens identified massive ettringite formation in extrinsic porosity formed in the specimens [12, 28]. The chemistry of the pore filling solids was determined by EDX to be consistent with ettringite. In these studies, only massive formation was observed in open porosity (e.g. cement #11).

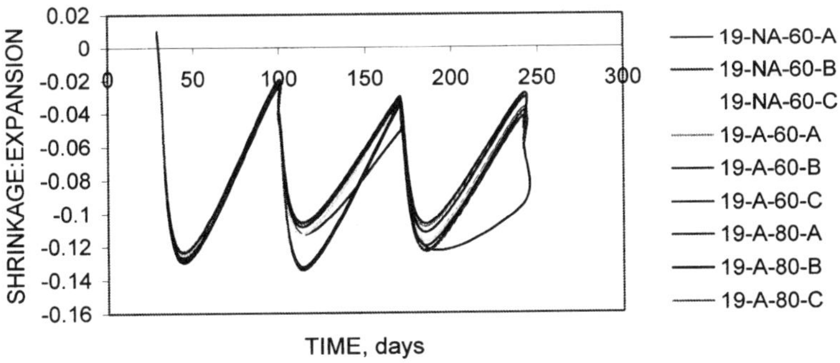

Figure 6 Normalized expansion of cement number 19 (non-expansive) after 3 cycles in the DEF testing.

DISCUSSION, CONCLUSIONS AND FUTURE IMPLICATIONS

Detailed characterization studies were made of 21 cements and clinkers statistically selected from some 114 Type I and Type II cements manufactured in the United States. Analyses of the relationships between the 71 measured cement characteristics and concrete durability properties was performed, including multi-variant and response surface modeling. The following oxide compositions have shown an increase in the current cement samples as compared to the cements used in the earlier National Bureau of Standards (NBS) study [8]: 1) SO_3 content is 40 percent higher, 2) K_2O content is 45 percent higher; 3) equivalent Na_2O content has increased by 8 percent, 4) current cements are finer, 5) setting times are faster, and 6) C_3S/C_2S ratio is higher.

Historically, sulfate attack has been attributed to a high C_3A content and a high SO_3 content, but the present study has identified a somewhat stronger negative correlation between C_4AF content of Portland cement and sulfate attack expansion in concrete. The correlation to sulfate attack was even stronger when Fe_2O_3 and alkalis are used in a multivariate model or the ratios of C_3A/C_4AF are taken into account. The study has found that maintaining the C_3A/C_4AF ratio below 1 can control sulfate attack.

Although potassium oxide has exhibited a stronger correlation to ASR than sodium oxide, the influence was even stronger when the combination of alkalis (equivalent alkalis) was taken into account. The second major characteristic responsible for ASR expansion was the SO_3

content of Portland cement. The higher ASR expansion observed for cements that contain high SO_3 and high equivalent alkalis supports the view that the presence of alkalis in the form of sulfates may contribute mainly to ASR expansion. Also, C_3S/C_2S ratio combined with equivalent alkali showed a high correlation with ASR.

A modified Stark method [29] was used to evaluate the potential expansion produced by SEF/DEF. After room temperature cure, mortar bar samples were treated in three cycles of different exposure conditions, including dry heating/wet curing and freezing-thawing: 1) air entrained mortar bar samples exposed to 60 °C(A60) drying, 2) air entrained mortar bar samples exposed to 80 °C (A80) drying, and non-air entrained mortar bar samples exposed to 60 °C(NA60) drying for a total of 242 days (for the total 3 cycles). The total length changes were small, and most were a net negative. The exceptions were NA60 specimens with 5 cements which experienced a marginal positive expansion. The cements that showed positive expansion after NA60 treatment conditions were categorized by the following chemical variations: 1) SO_3 content above 3.5 percent, 2) CaO content below 63 percent, 3) SO_3/Al_2O_3 ratio above 0.7, and 4) SO_3/CaO ratio above 0.06. The present study therefore supports the finding that there does not appear to be a major DEF expansion problem if: a) the cement contains low SO_3 (less than 3.5 percent), has an SO_3/Al_2O_3 ratio less than 0.7 and SO_3/CaO ratio less than 0.06; b) the concrete is air-entrained; and c) the exposure temperature does not exceed 60 °C.

In sum, this study overall showed that the concrete degradation mechanisms investigated are indeed influenced mainly by the SO_3, C_3A, C_4AF, and equivalent alkali contents of the Portland cement. Other factors such as C_3S content (and SiO_2 content), fineness, and heat of hydration have secondary effects but these have proven more difficult to quantify; therefore, significant specific recommendations for these factors are not available as a result of the present study. Based on the study, the following optimum contents of SO_3, C_3A, C_4AF, equivalent alkali or combination of these variables may be recommended for durable Portland cement concrete:

- **ASR control and compressive strength development**: Na_2O equivalent content below 0.6% and K_20 content below 0.8%
- **Sulfate attack control:** C_3A/C_4AF ratio less than 1.0%. In chemical terms, Al_2O_3 should be less than 5.0%, Fe_2O_3 between 2.5 and 3.5%.
- **DEF/SEF expansion control:** the cement preferably contains low SO_3 (less than 3.5%), has a SO_3/Al_2O_3 ratio less than 0.7 and an SO_3/CaO ratio less than 0.06; and the concrete is air-entrained and the exposure temperature does not exceed 60 °C.

It is clear that a challenging future exists for use of supplemental cementitious materials, as they have generally been shown to increase the tolerance in variation of some of the above parameters of Portland cements. The use of the supplemental, cementitious materials either as admixtures or in blended binary or ternary cements has been shown to increase the performance of concretes [30, 31]. Similar detailed studies need to be made to develop guidelines for achieving durability with various cementitious materials combinations. The objective of such studies is to both develop superior performance and to provide considerable savings to the environment.

ACKNOWLEDGEMENTS

The authors acknowledge the participation of J. Rosenberger, P. Arjunan, R. Malek and T. Cavalline in the studies, which were supported by the National Cooperative Highway Research Program.

REFERENCES

1. MEHTA, P.K., Durability of Concrete - Fifty Years of Progress? Durability of Concrete, Second International Conference, Montreal, Canada. Vol. I, 1991, pp 1-3

2. TAYLOR, H.F.W., Cement Chemistry, Thomas Telford, 1997

3. POWERS-COUCHE, L., MILLER, F. M., A Tale of Two Kiln Burners. Proceedings of the 18th International Conference on Cement Microscopy, Houston, TX: International Cement Microscopy Association, 1996, pp 74-84

4. REGOURD, M., Crystal Chemistry of Portland Cement Phases, In: Structure and Performance of Cements, P. Barnes (ed), Essex, England: Applied Science Publishers Ltd., 1983, pp 109-136

5. SAMET, B, BONEN, D., SARKAR, S. L., Effect of Burning Conditions on the Clinker Microstructure and Paste Characteristics. Proceedings of the 15th International Conference on Cement Microscopy, Dallas, TX, 1993, pp 72-88

6. GONNERMAN, H. F., LERCH, W., Changes in Characteristics of Portland Cement as Exhibited by Laboratory Tests over the Period 1904-1950, Philadelphia: American Society for Testing Materialsl, 1952, p56

7. CORISH, A.T., JACKSON, P.J., Cement Properties Past and Present, Concrete, Vol. 16, No.7, (1982), pp 16-18

8. BLAINE, R. L. ARNI, H. T., (eds) Interrelations Between Cement and Concrete Properties, Part 3, Washington D.C.: National Bureau of Standards, Building Science Series 8, 1968

9. GEBHARDT, R. F., Survey of North American Portland Cements: 1994, Cement, Concrete and Aggregates CCAGDP 17(2), 1995, pp 145-189

10. CONCRETE SOCIETY WORKING PARTY REPORT, Changes in Properties of Ordinary Portland Cement and Their Effects on Concrete, Concrete Society Technical Report, No. 29, 1987, p 30

11. OSBORNE, G.J., HARDCASTLE, J.L., Changes in Properties of Portland Cement and Their Effects on Concrete Durability. Proceedings, 4th CANMET/ACI International Conference on the Durability of Concrete, Sydney, Australia, Vol. 2, (1997), pp 1411-1432

12. ROY, D.M., TIKALSKY, P.J., SCHEETZ, B. E., ROSENBERGER, J.L., Relationship of Portland Cement Characteristics to Concrete Durability, Final Report to National Cooperative Highway Research Program, Transportation Research Board, Washington, DC, 20418, 2001.

13. TIKALSKY, P. J. JR., ROY, D. M., SCHEETZ, B. E, KRIZE, T., Redefining Cement Characteristics for Sulfate Resistant Portland Cement, Cement Concrete Research, 32(8), 2002, pp 1239-1246

14. MALEK, R. I., ROY, D. M., SCHEETZ, B. E., TIKALSKY, P. J., Influence of Clinker Fabric on Properties of Concrete. Proceedings, 11th Congress on the Chemistry of Cement, Durban, South Africa, 11-16 May 2003, pp 1054-1062

15. ARJUNAN, P., ROY, D. M., SCHEETZ, B. E., TIKALSKY, P. J., BADGER, S, Quantitative Phase Analysis of Portland Cement. Proceedings, 11th Congress on the Chemistry of Cement, Durban, South Africa, Vol. 3, 1-16 May 2003, pp 1535-1545

16. Hewlett, P.C. (ed.), Lea's Chemistry of Cement and Concrete. 4th Edition, Arnold/ohnWiley and Sons, New York, NY, 1998.

17. ROY, D.M., AND G.M. IDORN, Concrete Microstructure. Strategic Highway Research Program, Report SHRP-C-340, National Research Council, Washington D.C., 1993

18. SKALNY, J., PIERCE, J., Sulfate Attack Issues: An Overview, Materials Science of Concrete: Sulfate Attack Mechanisms, Special Volume, J. Marchand, J. Skalny, Eds., American Ceramic Society, Westerville, OH, 1999, pp 49-64

19. BROWN, P.W., TAYLOR, H.F.W.,. The Role of Ettringite in External Sulfate Attack, Materials Science of Concrete: Sulfate Attack Mechanisms, Special Volume, J. Marchand and J. Skalny, Eds., American Ceramic Society, Westerville, OH, 1999, pp 73-98

20. ROY, D.M., TIKALSKY, P.J., SCHEETZ, B. E., ROSENBERGER, J.L., KRIZE T., AND ARJUNAN, P., Influence of Portland Cement Characteristics on Alkali Silica Reactivity, TRB, Washington, D.C, 2003

21. SWAMY, R.N., The Alkali – Silica Reaction in Concrete, Van Nostrand Reinhold, 1992

22. STARK, D., The Moisture Condition of Field Concrete Exhibiting Alkali-Silica Reactivity. Durability of Concrete, Second International Conference, Montreal, Canada, 1991, pp. 973-988

23. DIAMOND, S., PENKO, M., Alkali Silica Reaction Processes: The Conversion of Cement Alkalies to Alkali Hydroxide. G.M. Idorn International Symposium on the Durability of Concrete, 1992, pp 153-168

24. HELMUTH, R., Alkali-Silica Reactivity: An Overview of Research. Strategic Highway Research Program, Report SHRP-C-342, 1993

25. HOOTON, R.D., Effect of Containers on ASTM C 441 - Pyrex Mortar Bar Expansions. Proceedings of the 7th International Conference on Concrete Alkali-Aggregate Reactions, Ottawa, Canada, 1986, pp 351-357

26. CHATTERJI, S., Concrete Durability and CaO/SiO2 Mole Ratio of CSH. Cement and Concrete Research, Vol 25, No. 5., 1995, pp 929-932

27. YAN, R., LAWRENCE, C. D., LYNSDALE, C. J., SHARP, J. H., Delayed ettringite formation in heat cured Portland cement mortars, Cement Concrete Research, Vol 29, 2000, pp 17-25

28. ROY, D. M., SCHEETZ, B. E., TIKALSKY, P. J., CAVALLINE, T, Delayed Ettringite Formation and the Influence of Cement Characteristics. Proceedings, 11th Congress on the Chemistry of Cement, Durban, South Africa, 11-16 May 2003, pp 75-83

29. STARK, J., BOLLMANN, K., Ettringite Formation – A Durability Problem of Concrete Pavements. 10th ICCC, Gothernburg, Sweden, 4iv062, 1997

30. OZYILDIRUM, H.P.C., Bridge Decks in Virginia. Concrete International, Vol. 21, 1999, pp 259-60

31. ROY, D.M., High –Performance Concrete: Superior Microstructure for Long-Term Durability. International Workshop on High Performance Concrete, P. Zia, (ed), Bangkok, Thailand, ACI SP-159, 1996, pp 393-408

INDEX OF AUTHORS

SUBJECT INDEX

This index has been compiled from the keywords assigned to the papers, edited and extended as appropriate. The page references are to the first page of the relevant paper.